Variational Methods For Nonlocal Fractional Problems

This book provides researchers and graduate students with a thorough introduction to the variational analysis of nonlinear problems described by nonlocal operators. The authors give a systematic treatment of the basic mathematical theory and constructive methods for these classes of nonlinear equations, plus their application to various processes arising in the applied sciences. The equations are examined from several viewpoints, with the calculus of variations as the unifying theme. Part I begins the book with some basic facts about fractional Sobolev spaces. Part II is dedicated to the analysis of fractional elliptic problems involving subcritical nonlinearities, via classical variational methods and other novel approaches. Finally, Part III contains a selection of recent results on critical fractional equations.
A careful balance is struck between rigorous mathematics and physical applications, allowing readers to see how these diverse topics relate to other important areas, including topology, functional analysis, mathematical physics, and potential theory.

Encyclopedia of Mathematics and Its Applications

This series is devoted to significant topics or themes that have wide application in mathematics or mathematical science and for which a detailed development of the abstract theory is less important than a thorough and concrete exploration of the implications and applications.

Books in the **Encyclopedia of Mathematics and Its Applications** cover their subjects comprehensively. Less important results may be summarized as exercises at the ends of chapters. For technicalities, readers can be referred to the bibliography, which is expected to be comprehensive. As a result, volumes are encyclopedic references or manageable guides to major subjects.

ENCYCLOPEDIA OF MATHEMATICS AND ITS APPLICATIONS

All the titles listed below can be obtained from good booksellers or from Cambridge University Press. For a complete series listing visit

www.cambridge.org/mathematics.

114 J. Beck *Combinatorial Games*
115 L. Barreira and Y. Pesin *Nonuniform Hyperbolicity*
116 D. Z. Arov and H. Dym *J-Contractive Matrix Valued Functions and Related Topics*
117 R. Glowinski, J.-L. Lions and J. He *Exact and Approximate Controllability for Distributed Parameter Systems*
118 A. A. Borovkov and K. A. Borovkov *Asymptotic Analysis of Random Walks*
119 M. Deza and M. Dutour Sikirić *Geometry of Chemical Graphs*
120 T. Nishiura *Absolute Measurable Spaces*
121 M. Prest *Purity, Spectra and Localisation*
122 S. Khrushchev *Orthogonal Polynomials and Continued Fractions*
123 H. Nagamochi and T. Ibaraki *Algorithmic Aspects of Graph Connectivity*
124 F. W. King *Hilbert Transforms I*
125 F. W. King *Hilbert Transforms II*
126 O. Calin and D.-C. Chang *Sub-Riemannian Geometry*
127 M. Grabisch *et al. Aggregation Functions*
128 L. W. Beineke and R. J. Wilson (eds.) with J. L. Gross and T. W. Tucker *Topics in Topological Graph Theory*
129 J. Berstel, D. Perrin and C. Reutenauer *Codes and Automata*
130 T. G. Faticoni *Modules over Endomorphism Rings*
131 H. Morimoto *Stochastic Control and Mathematical Modeling*
132 G. Schmidt *Relational Mathematics*
133 P. Kornerup and D. W. Matula *Finite Precision Number Systems and Arithmetic*
134 Y. Crama and P. L. Hammer (eds.) *Boolean Models and Methods in Mathematics, Computer Science, and Engineering*
135 V. Berthé and M. Rigo (eds.) *Combinatorics, Automata and Number Theory*
136 A. Kristály, V. D. Rădulescu and C. Varga *Variational Principles in Mathematical Physics, Geometry, and Economics*
137 J. Berstel and C. Reutenauer *Noncommutative Rational Series with Applications*
138 B. Courcelle and J. Engelfriet *Graph Structure and Monadic Second-Order Logic*
139 M. Fiedler *Matrices and Graphs in Geometry*
140 N. Vakil *Real Analysis through Modern Infinitesimals*
141 R. B. Paris *Hadamard Expansions and Hyperasymptotic Evaluation*
142 Y. Crama and P. L. Hammer *Boolean Functions*
143 A. Arapostathis, V. S. Borkar and M. K. Ghosh *Ergodic Control of Diffusion Processes*
144 N. Caspard, B. Leclerc and B. Monjardet *Finite Ordered Sets*
145 D. Z. Arov and H. Dym *Bitangential Direct and Inverse Problems for Systems of Integral and Differential Equations*
146 G. Dassios *Ellipsoidal Harmonics*
147 L. W. Beineke and R. J. Wilson (eds.) with O. R. Oellermann *Topics in Structural Graph Theory*
148 L. Berlyand, A. G. Kolpakov and A. Novikov *Introduction to the Network Approximation Method for Materials Modeling*
149 M. Baake and U. Grimm *Aperiodic Order I: A Mathematical Invitation*
150 J. Borwein *et al. Lattice Sums Then and Now*
151 R. Schneider *Convex Bodies: The Brunn–Minkowski Theory (Second Edition)*
152 G. Da Prato and J. Zabczyk *Stochastic Equations in Infinite Dimensions (Second Edition)*
153 D. Hofmann, G. J. Seal and W. Tholen (eds.) *Monoidal Topology*
154 M. Cabrera García and Á. Rodríguez Palacios *Non-Associative Normed Algebras I: The Vidav–Palmer and Gelfand–Naimark Theorems*
155 C. F. Dunkl and Y. Xu *Orthogonal Polynomials of Several Variables (Second Edition)*
156 L. W. Beineke and R. J. Wilson (eds.) with B. Toft *Topics in Chromatic Graph Theory*
157 T. Mora *Solving Polynomial Equation Systems III: Algebraic Solving*
158 T. Mora *Solving Polynomial Equation Systems IV: Buchberger Theory and Beyond*
159 V. Berthé and M. Rigo (eds.) *Combinatorics, Words and Symbolic Dynamics*
160 B. Rubin *Introduction to Radon Transforms: With Elements of Fractional Calculus and Harmonic Analysis*
161 M. Ghergu and S. D. Taliaferro *Isolated Singularities in Partial Differential Inequalities*
162 G. Molica Bisci, V. Radulescu and R. Servadei *Variational Methods for Nonlocal Fractional Problems*
163 S. Wagon *The Banach–Tarski Paradox (Second Edition)*

Encyclopedia of Mathematics and Its Applications

Variational Methods for Nonlocal Fractional Problems

GIOVANNI MOLICA BISCI

Università 'Mediterranea' di Reggio Calabria, Italy

VICENTIU D. RADULESCU

Institute of Mathematics of the Romanian Academy

RAFFAELLA SERVADEI

Università degli Studi di Urbino 'Carlo Bo', Italy

Cambridge
University Press

CAMBRIDGE
UNIVERSITY PRESS

University Printing House, Cambridge CB2 8BS, United Kingdom

Cambridge University Press is part of the University of Cambridge.

It furthers the University's mission by disseminating knowledge in the pursuit of education, learning, and research at the highest international levels of excellence.

www.cambridge.org
Information on this title: www.cambridge.org/9781107111943

First published 2016

A catalogue record for this publication is available from the British Library.

ISBN 978-1-107-11194-3 Hardback

"Io credo, per l'acume ch'io soffersi
del vivo raggio, ch'i' sarei smarrito,
se li occhi miei da lui fossero aversi"

Dante, *La Divina Commedia*
Paradiso, Canto XXXIII (vv. 76–78)

Dedicated to Antonella
Bianca, Teodora
and
Antonietta, Enzo

Contents

Foreword

Although fractional derivatives for functions of one variable can be traced to the origin of calculus and were already introduced rigorously in the nineteenth century by Liouville and Riemann, the development of linear and nonlinear equations involving fractional derivatives of functions of one or several variables, and in particular fractional Laplacians, is a more recent phenomenon.

Rich mathematical concepts allow in general several approaches, and this is the case for the fractional Laplacian, which can be defined using Fourier analysis, functional calculus, singular integrals, or Lévy processes. Its inverse is closely related to the famous potentials introduced by Marcel Riesz in the late 1930s. In contrast to the Laplacian, which is a local operator, the fractional Laplacian is a paradigm of the vast family of nonlocal linear operators, and this has immediate consequences in the formulation of basic questions such as the Dirichlet problem.

If the Laplacian has been and still is a stimulating cornerstone of the theory of linear partial differential equations and linear operators, the twentieth century has seen an increasing and outstanding activity in the study of the Dirichlet or other boundary value problems for nonlinear perturbations of the Laplacian and for its quasi-linear or fully nonlinear extensions. All techniques of nonlinear functional analysis, such as iterative, topological, monotonicity, or variational methods, have been essentially tested on those problems and have received, at this occasion, important developments.

It is therefore a natural question to see which results "survive" when the Laplacian is replaced by the fractional Laplacian. It is also a fruitful question because the extension of classical results to new situations also sheds light on a better and deeper understanding of the classical results.

Giovanni Molica Bisci, Vicenţiu Rădulescu, and Raffaella Servadei, the authors of this monograph, have all three, jointly or separately, contributed in a significant way to the use of modern critical point theory to nonlinear perturbations of a fractional Laplacian. But their monograph goes much beyond a presentation of their own results in book form. It starts with a substantial, clear, and systematic presentation of the fractional Sobolev spaces, where the considered problems will live, including the corresponding spectral theory. Then follows, for subcritical and critical perturbations, respectively, a panorama of how the methods of critical point theory apply successfully, at the expense of overcoming new technical difficulties, to many of the important problems considered in the Laplacian case.

For subcritical problems, mountain pass and linking-type nontrivial solutions are obtained, as well as Ricceri's solutions for parametric problems, followed by equations at resonance and the obtention of multiple solutions using pseudoindex theory. Kirchhoff-type and Schrödinger equations are also considered. For critical problems, emphasis is put on extending to the new setting the important results of Brezis–Nirenberg and related ones, as well as the case of concave-convex nonlinearities. One should notice that many results are obtained in situations where the fractional Laplacian is replaced by a more general nonlocal operator.

With its bibliography of some 200 items collecting the references to the original Laplacian perturbation results as well as to the corresponding fractional Laplacian extensions, and with its useful index, this carefully and clearly written monograph is and will remain a fundamental reference for any mathematician interested in the variational approach to nonlinear perturbations of nonlocal linear operators and, in particular, of the fractional Laplacian. Because those problems have been motivated by important applications, the book also will be useful to scientists interested in the mathematical techniques and ideas, allowing them to treat rigorously new models involving such nonlocal operators.

Jean Mawhin, July 2015

Preface

A very interesting area of nonlinear analysis lies in the study of elliptic equations involving fractional operators. Recently, great attention has been focused on these problems, both for pure mathematical research and in view of concrete real-world applications. Indeed, this type of operator arises in a quite natural way in different contexts, such as the description of several physical phenomena.

The current literature on these abstract tools and on their applications is therefore very interesting and, up to now, quite large. Motivated by this increasing interest on these subjects, this book deals with some classes of fractional problems widely investigated by many mathematicians and scientists.

This monograph is divided into three parts and is based on results obtained by ourselves or through direct cooperation with other mathematicians. More precisely, the first part deals with some basic facts about fractional Sobolev spaces, and the second part is dedicated to an analysis of fractional elliptic problems involving subcritical nonlinearities via classical variational methods and other novel approaches. Finally, in the third part of the book we give a selection of recent results on critical fractional equations, studied in the recent literature, also in relation to the celebrated Brezis–Nirenberg problem. Of course, there are many other interesting applications and theoretical aspects of fractional nonlocal problems, but it is not our ambition to treat all these topics here.

This book is addressed to researchers and advanced graduate students specializing in the fields of fractional elliptic equations, nonlinear analysis, and functional analysis. We also emphasize that the bibliography does no escape the usual rule, being incomplete. Indeed, we have listed only papers that are closer to the topics discussed in this book. But we are afraid that even for these arguments, the references are far from being exhaustive. We apologize for possible omissions.

We emphasize that this book would never have appeared without the encouragements of some dear friends and colleagues. It is a pleasure to thank some of them, especially Rossella Bartolo, Xavier Cabré, Philippe Ciarlet, Bernard Dacorogna,

Alessio Fiscella, Jean Mawhin, Giuseppe Mingione, Giampiero Palatucci, Patrizia Pucci, Gaetana Restuccia, Biagio Ricceri, Dušan Repovš, Sandro Salsa, Simone Secchi, Francesco Tulone, Enrico Valdinoci, Gianmaria Verzini, and Binlin Zhang.

We also thank the colleagues and administrative staff of the Department Patrimonio, Architettura, Urbanistica of the *Mediterranea* University of Reggio Calabria for their help and the warm and creative atmosphere we always feel during our stay at that institution. A special mention goes to the Director Francesca Martorano for her extraordinary and constant friendship.

Finally, a heartily thanks goes to Massimiliano Ferrara and to Rector pro-tempore of the *Mediterranea* University of Reggio Calabria, Pasquale Catanoso. We warmly thank Roger Astley and Clare Dennison from Cambridge University Press for their constant and kindest help at all stages of the editing process.

Part I

Fractional Sobolev spaces

1
Fractional framework

Recently, great attention has been focused on the study of fractional and nonlocal operators of elliptic type, both for pure mathematical research and in view of concrete real-world applications. This type of operator arises in a quite natural way in many different contexts, such as, among others, the thin obstacle problem, optimization, finance, phase transitions, stratified materials, anomalous diffusion, crystal dislocation, soft thin films, semipermeable membranes, flame propagation, conservation laws, ultrarelativistic limits of quantum mechanics, quasi-geostrophic flows, multiple scattering, minimal surfaces, materials science, water waves, chemical reactions of liquids, population dynamics, geophysical fluid dynamics, and mathematical finance (American options). The fractional Laplacian also provides a simple model to describe certain jump Lévy processes in probability theory. In all these cases, the nonlocal effect is modeled by the singularity at infinity. For more details and applications, see [13, 35, 47, 52, 55, 75, 140, 216, 217, 218, 219] and the references therein.

From a physical point of view, nonlocal operators play a crucial rule in describing several phenomena. As a general reference in this topic, we cite the recent paper of Vázquez [217]. In that paper, the author describes two models of flow in porous media, including nonlocal (long-range) diffusion effects, providing a long list of references related to physical phenomena and nonlocal operators. The first model is based on Darcy's law, and the pressure is related to the density by an inverse fractional Laplacian operator. The second model is more in the spirit of fractional Laplacian flows but nonlinear: contrary to the usual porous medium flows, it has infinite speed of propagation.

Moreover, the fractional power of the Laplace operator has been studied in relation to the obstacle problem that appears in many contexts, such as in the study of anomalous diffusion, in the so-called quasi-geostrophic flow problem, and in pricing of American options governed by assets evolving according to jump processes (see, e.g., the papers [54, 188, 189]).

For the sake of completeness, we mention that fractional nonlocal problems have been considered recently under certain Neumann boundary conditions using different methods and approaches (see, among others, the papers [78, 79, 80, 160, 209]). All these different Neumann problems for nonlocal operators recover the classical Neumann problem as a limit case, and most of them have clear probabilistic interpretations as well. In this setting, in [85], the authors propose an intriguing approach to studying Neumann problems with a variational structure.

In this chapter we sketch the basic facts on fractional Sobolev spaces and fractional nonlocal operators. Our treatment is mostly self-contained, and we tacitly assume that the reader has some knowledge of the basic objects discussed here. More precisely, the main purpose of this section is to present some results on fractional Sobolev spaces and nonlocal operators in the form in which they will be exploited later on. Since this is an introductory chapter to convey the framework we work in, the rigorous proofs will be kept to a minimum. Some extra reading of the references may be necessary to truly learn the material.

Here we will consider a nonlocal fractional framework, providing models and theorems related to nonlocal phenomena. The results of this chapter are based on the papers [83, 147, 148, 198, 199, 200].

1.1 Fourier transform of tempered distributions

In this section we just recall briefly the notion of Fourier transform of a tempered distribution. First of all, we consider the Schwartz space $\mathscr{S}$ of rapidly decaying $C^\infty(\mathbb{R}^n)$ functions whose topology is generated by the seminorms $\{p_j\}_{j\in\mathbb{N}}$ defined as

$$p_j(\varphi) := \sup_{x\in\mathbb{R}^n} (1+|x|)^j \sum_{|\alpha|\leq j} |D^\alpha \varphi(x)|,$$

where $\varphi \in \mathscr{S}(\mathbb{R}^n)$. More precisely, $\mathscr{S}$ contains the smooth functions φ satisfying

$$\sup_{x\in\mathbb{R}^n} |x^\alpha D^\beta \varphi(x)| < +\infty,$$

for all multi-indices α and $\beta \in \mathbb{N}_0^n$.

The natural locally convex topology on $\mathscr{S}$ can be characterized by the following notion of convergence:

the sequence $\{\varphi_j\}_{j\in\mathbb{N}}$ *converges to* 0 *in* $\mathscr{S}$ *if and only if*

$$\lim_{j\to+\infty} x^\alpha D^\beta \varphi_j(x) = 0, \text{ for all } \alpha \text{ and } \beta \in \mathbb{N}_0^n.$$

We denote by

$$\mathscr{F}\varphi(\xi) := \frac{1}{(2\pi)^{n/2}} \int_{\mathbb{R}^n} e^{-i\xi\cdot x}\varphi(x)dx \tag{1.1}$$

the *Fourier transform* of a function $\varphi \in \mathscr{S}$. Note that, for every $\varphi \in \mathscr{S}$, one has that $\mathscr{F}\varphi \in \mathscr{S}$.

It may be readily verified that the Fourier transform (1.1) and the inverse Fourier transform, given by

$$\mathscr{F}^{-1}\varphi(x) := \frac{1}{(2\pi)^{n/2}} \int_{\mathbb{R}^n} e^{ix\cdot\xi\cdot}\varphi(\xi)\,d\xi, \tag{1.2}$$

are both continuous on $\mathscr{S}(\mathbb{R}^n)$ into $\mathscr{S}(\mathbb{R}^n)$. Moreover, since

$$\mathscr{F}^{-1}\mathscr{F}\varphi = \mathscr{F}\mathscr{F}^{-1}\varphi = \varphi,$$

each of them is, in fact, an isomorphism and a homeomorphism of $\mathscr{S}(\mathbb{R}^n)$ onto $\mathscr{S}(\mathbb{R}^n)$.

Now let $\mathscr{S}'$ be the topological dual of $\mathscr{S}$. As usual, a tempered distribution is an element of $\mathscr{S}'$. If $T \in \mathscr{S}'$, the Fourier transform of T can be defined as the tempered distribution given by

$$\langle \mathscr{F}T, \varphi\rangle := \langle T, \mathscr{F}\varphi\rangle,$$

for every $\varphi \in \mathscr{S}$, where $\langle\cdot,\cdot\rangle$ denotes the usual duality bracket between $\mathscr{S}$ and its dual $\mathscr{S}'$.

By using definition (1.1), one has

$$u \in L^2(\mathbb{R}^n) \quad \text{if and only if} \quad \mathscr{F}u \in L^2(\mathbb{R}^n) \tag{1.3}$$

and

$$\|u\|_{L^2(\mathbb{R}^n)} = \|\mathscr{F}u\|_{L^2(\mathbb{R}^n)}, \tag{1.4}$$

for every $u \in L^2(\mathbb{R}^n)$. Formula (1.4) is the so-called Parseval–Plancherel formula, which will be crucial in what follows for proving the equivalence between the fractional spaces $H^s(\mathbb{R}^n)$ and $\widehat{H}^s(\mathbb{R}^n)$ (see Corollary 1.15).

For a detailed introduction to the classical theory of distribution and Fourier transform, we refer to the monograph [187] and the recent book [69] for several applications to elliptic problems of linear and nonlinear functional analysis.

1.2 Fractional Sobolev spaces

Let Ω be a possibly nonsmooth, open set of the Euclidean space $\mathbb{R}^n$ and $p \in [1, +\infty)$. For any $s > 0$, we would define the fractional Sobolev space $W^{s,p}(\Omega)$. In the literature, fractional Sobolev-type spaces are also called *Aronszajn*, *Gagliardo*, or *Slobodeckij spaces*, by the names of the ones who introduced them, almost simultaneously (see [15, 110, 207]).

If $s \geq 1$ is a positive integer, we denote by $W^{s,p}(\Omega)$ the classical Sobolev space equipped with the standard norm

$$\|u\|_{W^{s,p}(\Omega)} := \sum_{0\leq|\alpha|\leq s} \|D^\alpha u\|_{L^p(\Omega)},$$

for every $u \in W^{s,p}(\Omega)$, where here and in what follows $\|\cdot\|_{L^p(\Omega)}$ denotes the usual norm in $L^p(\Omega)$, and D^α stands for the α-distributional derivative. This section is devoted to the definition of fractional Sobolev spaces; that is, here we are interested in the case where $s \notin \mathbb{N}$.

For a fixed $s \in (0,1)$, we recall that the Sobolev space $W^{s,p}(\Omega)$ is defined as follows:

$$W^{s,p}(\Omega) := \left\{ u \in L^p(\Omega) : \frac{|u(x)-u(y)|}{|x-y|^{n/p+s}} \in L^p(\Omega \times \Omega) \right\}.$$

It is endowed with the natural norm

$$\|u\|_{W^{s,p}(\Omega)} := \left(\int_\Omega |u(x)|^p \, dx + \int_{\Omega\times\Omega} \frac{|u(x)-u(y)|^p}{|x-y|^{n+sp}} \, dx \, dy \right)^{1/p}, \tag{1.5}$$

where the term

$$[u]_{W^{s,p}(\Omega)} := \left(\int_{\Omega\times\Omega} \frac{|u(x)-u(y)|^p}{|x-y|^{n+sp}} \, dx \, dy \right)^{1/p} \tag{1.6}$$

is the *Gagliardo seminorm* of u.

When $s > 1$ and $s \notin \mathbb{N}$, we can write $s = m + \sigma$, where $m \in \mathbb{N}$ and $\sigma \in (0,1)$. We can define $W^{s,p}(\Omega)$ as follows:

$$W^{s,p}(\Omega) := \left\{ u \in W^{m,p}(\Omega) : D^\alpha u \in W^{\sigma,p}(\Omega) \text{ for any } \alpha \text{ s.t. } |\alpha| = m \right\}.$$

In this case, $W^{s,p}(\Omega)$ is endowed with the norm

$$\|u\|_{W^{s,p}(\Omega)} := \left(\|u\|^p_{W^{m,p}(\Omega)} + \sum_{|\alpha|=m} \|D^\alpha u\|^p_{W^{\sigma,p}(\Omega)} \right)^{1/p},$$

for every $u \in W^{s,p}(\Omega)$. All in all, the space $W^{s,p}(\Omega)$ is well defined and is a Banach space for every $s > 0$.

As in the classical case (i.e., $s \in \mathbb{N}$), any function in the fractional Sobolev space $W^{s,p}(\mathbb{R}^n)$ can be approximated by a sequence of smooth functions with compact support. Indeed, for any $s > 0$,

$$\overline{C_0^\infty(\mathbb{R}^n)}^{\|\cdot\|_{W^{s,p}(\mathbb{R}^n)}} = W^{s,p}(\mathbb{R}^n);$$

that is, the space $C_0^\infty(\mathbb{R}^n)$ is dense in $W^{s,p}(\mathbb{R}^n)$.

In general, if $\Omega \subset \mathbb{R}^n$, the space $C_0^\infty(\Omega)$ is not dense in $W^{s,p}(\Omega)$. Hence, we denote by $W_0^{s,p}(\Omega)$ the closure of $C_0^\infty(\Omega)$ with respect to the norm $\|\cdot\|_{W^{s,p}(\Omega)}$; that is,

$$W_0^{s,p}(\Omega) := \overline{C_0^\infty(\Omega)}^{\|\cdot\|_{W^{s,p}(\Omega)}}.$$

With this definition, we can also construct $W^{s,p}(\Omega)$ when $s < 0$. Indeed, for $s < 0$ and $p \in (1, +\infty)$, we can define

$$W^{s,p}(\Omega) := \left(W_0^{-s,q}(\Omega) \right)';$$

that is, $W^{s,p}(\Omega)$ is the dual space of $W_0^{-s,q}(\Omega)$, where $1/p + 1/q = 1$.

1.2.1 Embedding properties

This subsection is devoted to the embeddings of fractional Sobolev spaces into Lebesgue spaces. We point out that Sobolev inequalities and continuous (compact) embeddings of the spaces $W^{s,p}$ into the classical Lebesgue spaces L^q are exhaustively treated in [83, sections 6 and 7] (see also [3]). Here we recall briefly some basic facts.

Proposition 1.1 Let $p \in [1,+\infty)$ and let Ω be an open set in $\mathbb{R}^n$. Then the following assertions hold true:

(a) If $0 < s \le s' < 1$, then the embedding

$$W^{s',p}(\Omega) \hookrightarrow W^{s,p}(\Omega)$$

is continuous. Hence, there exists a constant $C_1(n,s,p) \ge 1$ such that

$$\|u\|_{W^{s,p}(\Omega)} \le C_1(n,s,p)\|u\|_{W^{s',p}(\Omega)},$$

for any $u \in W^{s',p}(\Omega)$.

(b) If $0 < s < 1$ and Ω is of class $C^{0,1}$ (i.e., with Lipschitz boundary) and with bounded boundary $\partial\Omega$, then the embedding

$$W^{1,p}(\Omega) \hookrightarrow W^{s,p}(\Omega)$$

is continuous. Hence, there exists a constant $C_2(n,s,p) \ge 1$ such that

$$\|u\|_{W^{s,p}(\Omega)} \le C_2(n,s,p)\|u\|_{W^{1,p}(\Omega)},$$

for any $u \in W^{1,p}(\Omega)$.

(c) If $s' \ge s > 1$ and Ω is of class $C^{0,1}$, then the embedding

$$W^{s',p}(\Omega) \hookrightarrow W^{s,p}(\Omega)$$

is continuous.

Proof See propositions 2.1 and 2.2 and corollary 2.3 in [83]. □

Now we recall some basic properties about continuous (compact) embeddings of the fractional Sobolev spaces into Lebesgue spaces. In what follows, we need the following definition:

Definition 1.2 For any $s \in (0,1)$ and any $p \in [1,+\infty)$, an open set $\Omega \subset \mathbb{R}^n$ is an *extension domain* for $W^{s,p}$ if there exists a positive constant $C := C(n,p,s,\Omega)$ such that *for every function* $u \in W^{s,p}(\Omega)$, *there exists* $\mathscr{E}_u \in W^{s,p}(\mathbb{R}^n)$ *such that* $\mathscr{E}_u(x) = u(x)$ *for any* $x \in \Omega$ *and*

$$\|\mathscr{E}_u\|_{W^{s,p}(\mathbb{R}^n)} \le C\|u\|_{W^{s,p}(\Omega)}.$$

Note that any open set of class $C^{0,1}$ with bounded boundary is an extension domain for $W^{s,p}(\mathbb{R}^n)$; see [83, theorem 5.4] for a direct proof.

For the sake of completeness, we also recall an interesting result, proved in [83, lemma 5.1], about the construction of the extension $\mathscr{E}_u$ to the whole of $\mathbb{R}^n$ of a function u defined on an open set $\Omega \subset \mathbb{R}^n$.

Lemma 1.3 *Let Ω be an open set in $\mathbb{R}^n$, and let $u \in W^{s,p}(\Omega)$ with $s \in (0,1)$ and $p \in [1,+\infty)$. If there exists a compact subset $\mathscr{K} \subset \Omega$ such that $u \equiv 0$ in $\Omega \setminus \mathscr{K}$, then the extension function $\mathscr{E}_u$, defined as*

$$\mathscr{E}_u(x) := \begin{cases} u(x) & \text{if } x \in \Omega \\ 0 & \text{if } x \in \mathbb{R}^n \setminus \Omega, \end{cases}$$

belongs to $W^{s,p}(\mathbb{R}^n)$, and

$$\|\mathscr{E}_u\|_{W^{s,p}(\mathbb{R}^n)} \leq C\|u\|_{W^{s,p}(\Omega)},$$

where C is a suitable positive constant depending on $n, p, s, \mathscr{K}$, and Ω.

Now we are ready to discuss the embedding properties of $W^{s,p}$. For this purpose, we distinguish three different cases, that is, $sp < n$, $sp = n$, and $sp > n$. We refer to [83, sections 6–8] for a proof of these results.

Case 1: $sp < n$.

Theorem 1.4 *Let $s \in (0,1)$ and $p \in [1,+\infty)$ such that $sp < n$. Then there exists a positive constant $C := C(n,p,s)$ such that, for any $u \in W^{s,p}(\mathbb{R}^n)$,*

$$\|u\|^p_{L^{p_s^*}(\mathbb{R}^n)} \leq C \int_{\mathbb{R}^n \times \mathbb{R}^n} \frac{|u(x)-u(y)|^p}{|x-y|^{n+ps}}\, dx\, dy,$$

where the constant

$$p_s^* := \frac{pn}{n-sp}$$

is the so-called fractional critical exponent. Consequently, the space $W^{s,p}(\mathbb{R}^n)$ is continuously embedded in $L^q(\mathbb{R}^n)$ for any $q \in [p, p_s^*]$. Moreover, the embedding $W^{s,p}(\mathbb{R}^n) \hookrightarrow L^q_{\mathrm{loc}}(\mathbb{R}^n)$ is compact for every $q \in [p, p_s^*)$.

In an extension domain Ω, the following embedding result holds:

Theorem 1.5 *Let $s \in (0,1)$ and $p \in [1,+\infty)$ such that $sp < n$. Let $\Omega \subset \mathbb{R}^n$ be an extension domain for $W^{s,p}$. Then there exists a positive constant $C := C(n,p,s,\Omega)$ such that, for any $u \in W^{s,p}(\Omega)$,*

$$\|u\|_{L^q(\Omega)} \leq C\|u\|_{W^{s,p}(\Omega)},$$

for any $q \in [p, p_s^*]$; that is, the space $W^{s,p}(\Omega)$ is continuously embedded in $L^q(\Omega)$ for any $q \in [p, p_s^*]$. If, in addition, Ω is bounded, then the space $W^{s,p}(\Omega)$ is compactly embedded in $L^q(\Omega)$ for any $q \in [1, p_s^*)$.

Case 2: $sp = n$.

Theorem 1.6 *Let $s \in (0,1)$ and $p \in [1,+\infty)$ such that $sp = n$. Then there exists a positive constant $C := C(n,p,s)$ such that, for any $u \in W^{s,p}(\mathbb{R}^n)$,*

$$\|u\|_{L^q(\mathbb{R}^n)} \leq C\|u\|_{W^{s,p}(\mathbb{R}^n)},$$

for any $q \in [p,+\infty)$; that is, the space $W^{s,p}(\mathbb{R}^n)$ is continuously embedded in $L^q(\mathbb{R}^n)$ for any $q \in [p,+\infty)$.

For an extension domain Ω, we have the following embedding result:

Theorem 1.7 *Let $s \in (0,1)$ and $p \in [1,+\infty)$ such that $sp = n$. Let $\Omega \subset \mathbb{R}^n$ be an extension domain for $W^{s,p}$. Then there exists a positive constant $C := C(n,p,s,\Omega)$ such that, for any $u \in W^{s,p}(\Omega)$,*

$$\|u\|_{L^q(\Omega)} \leq C\|u\|_{W^{s,p}(\Omega)},$$

for any $q \in [p,+\infty)$; that is, the space $W^{s,p}(\Omega)$ is continuously embedded in $L^q(\mathbb{R}^n)$ for any $q \in [p,+\infty)$. If, in addition, Ω is bounded, then the space $W^{s,p}(\Omega)$ is continuously embedded in $L^q(\Omega)$ for any $q \in [1,+\infty)$.

Case 3: $sp > n$. Here $C^{0,\alpha}(\Omega)$ denotes the space of Hölder continuous functions, endowed with the standard norm

$$\|u\|_{C^{0,\alpha}(\Omega)} := \|u\|_{L^\infty(\Omega)} + \sup_{\substack{x,y\in\Omega \\ x\neq y}} \frac{|u(x)-u(y)|}{|x-y|^\alpha}.$$

Theorem 1.8 *Let $s \in (0,1)$ and $p \in [1,+\infty)$ such that $sp > n$. Let Ω be a $C^{0,1}$ domain of $\mathbb{R}^n$. Then there exists a positive constant $C := C(n,p,s,\Omega)$ such that, for any $u \in W^{s,p}(\Omega)$,*

$$\|u\|_{C^{0,\alpha}(\Omega)} \leq C\|u\|_{W^{s,p}(\Omega)},$$

with $\alpha := (sp-n)/p$; that is, the space $W^{s,p}(\Omega)$ is continuously embedded in $C^{0,\alpha}(\Omega)$.

The preceding regularity property remains valid for functions in $W^{s,p}$ when $sp > n$ and Ω is an extension domain for $W^{s,p}$ with no external cusps (see [83, theorem 8.2]). As a consequence of Theorem 1.8, we have the following result:

Corollary 1.9 Let $s \in (0,1)$ and $p \in [1,+\infty)$ such that $sp > n$. Let Ω be a $C^{0,1}$ bounded domain of $\mathbb{R}^n$. Then the embedding

$$W^{s,p}(\Omega) \hookrightarrow C^{0,\beta}(\Omega)$$

is compact for every $\beta < \alpha$, with $\alpha := (sp-n)/p$.

Proof Let $\{u_j\}_{j\in\mathbb{N}}$ be a bounded sequence in $W^{s,p}(\Omega)$. Theorem 1.8 ensures that $\{u_j\}_{j\in\mathbb{N}}$ is bounded in $C^{0,\alpha}(\Omega)$. Hence, there exists $C > 0$ such that

$$\|u_j\|_{L^\infty(\Omega)} + \sup_{\substack{x,y\in\Omega \\ x\neq y}} \frac{|u_j(x) - u_j(y)|}{|x-y|^\alpha} \leq C \quad \forall\, j \in \mathbb{N}. \tag{1.7}$$

By using (1.7), the Ascoli–Arzelà theorem gives that

$$u_j \to u_\infty \quad \text{uniformly in } \Omega \tag{1.8}$$

as $j \to +\infty$, for some $u_\infty \in C(\Omega)$. Moreover, (1.7) and (1.8) give

$$|u_\infty(x) - u_\infty(y)| = \lim_{j\to+\infty} |u_j(x) - u_j(y)| \leq C|x-y|^\alpha, \tag{1.9}$$

for any $x, y \in \Omega$. Thus, the function u_∞ belongs to $C^{0,\alpha}(\Omega)$.

Let us prove that $u_j \to u_\infty$ in $C^{0,\beta}(\Omega)$ as $j \to +\infty$, for every $\beta < \alpha$. Taking into account (1.8), we have to show that

$$\sup_{\substack{x,y\in\Omega \\ x\neq y}} \frac{|(u_j - u_\infty)(x) - (u_j - u_\infty)(y)|}{|x-y|^\beta} \to 0 \tag{1.10}$$

as $j \to +\infty$. Since

$$\|u_j - u_\infty\|_{L^\infty(\Omega)} \to 0 \quad \text{as } j \to +\infty,$$

for every $\varepsilon > 0$, there exists $j_\varepsilon \in \mathbb{N}$ such that

$$\|u_j - u_\infty\|_{L^\infty(\Omega)} \leq \frac{\varepsilon}{2}\left(\frac{\varepsilon}{2C}\right)^{\beta/(\alpha-\beta)} \qquad \forall\, j \geq j_\varepsilon. \tag{1.11}$$

Now inequalities (1.7) and (1.9) give

$$|(u_j - u_\infty)(x) - (u_j - u_\infty)(y)| \leq 2C|x-y|^{\alpha-\beta}|x-y|^\beta \quad \forall\, x, y \in \Omega. \tag{1.12}$$

If $2C|x-y|^{\alpha-\beta} < \varepsilon$, inequality (1.12) ensures that

$$|(u_j - u_\infty)(x) - (u_j - u_\infty)(y)| \leq \varepsilon|x-y|^\beta \quad \forall\, x, y \in \Omega. \tag{1.13}$$

Here we use the fact that $\beta < \alpha$.

However, whenever $2C|x-y|^{\alpha-\beta} \geq \varepsilon$, by using (1.11), for every $j \geq j_\varepsilon$, one has

$$|(u_j - u_\infty)(x) - (u_j - u_\infty)(y)| \leq 2\|u_j - u_\infty\|_{L^\infty(\Omega)} \leq \varepsilon\left(\frac{\varepsilon}{2C}\right)^{\beta/(\alpha-\beta)} \leq \varepsilon|x-y|^\beta. \tag{1.14}$$

By (1.13) and (1.14), relation (1.10) is clearly verified, and this ends the proof. □

1.2.2 The Sobolev space $H^s(\Omega)$

In this subsection, we focus our attention on the Hilbert case $p = 2$, dealing with its relation with the fractional Laplacian. Let Ω be an open subset in $\mathbb{R}^n$, and denote

$$H^s(\Omega) := W^{s,2}(\Omega),$$

for any $s \in (0,1)$. This is an important case because the preceding fractional Sobolev space turns out to be a Hilbert space. Indeed, the inner product on $H^s(\Omega)$ defined by

$$\langle u, v\rangle_{H^s(\Omega)} := \int_\Omega u(x)v(x)\,dx + \int_{\Omega\times\Omega} \frac{(u(x)-u(y))(v(x)-v(y))}{|x-y|^{n+2s}}\,dx\,dy,$$

for any $u, v \in H^s(\Omega)$, induces the norm given in (1.5) when $p = 2$.

Clearly, for every $s \in (0,1)$, one has

$$H^s(\mathbb{R}^n) := W^{s,2}(\mathbb{R}^n) = \left\{u \in L^2(\mathbb{R}^n) : [u]_{W^{s,2}(\mathbb{R}^n)} < +\infty\right\}, \tag{1.15}$$

where $[\,\cdot\,]_{W^{s,2}(\mathbb{R}^n)}$ is defined in formula (1.6).

The space $H^s(\mathbb{R}^n)$ can be defined in an alternative way via a Fourier transform. Precisely, let us define

$$\widehat{H}^s(\mathbb{R}^n) := \left\{u \in L^2(\mathbb{R}^n) : \int_{\mathbb{R}^n} (1 + |\xi|^{2s})|\mathscr{F}u(\xi)|^2\,dx < +\infty\right\}, \tag{1.16}$$

for any $s > 0$, and

$$\widehat{H}^s(\mathbb{R}^n) := \left\{u \in \mathscr{S}' : \int_{\mathbb{R}^n} (1 + |\xi|^2)^s|\mathscr{F}u(\xi)|^2\,dx < +\infty\right\},$$

for every $s < 0$. The equivalence between the space $\widehat{H}^s(\mathbb{R}^n)$ defined in (1.16) and the one defined via the Gagliardo norm in (1.15) is stated and proved for any $s \in (0,1)$ in what follows (see Corollary 1.15).

1.3 The fractional Laplacian operator

Nonlocal equations have attracted much attention in recent decades. The basic operator involved in this kind of problems is the so-called fractional Laplacian $(-\Delta)^s$ with $s \in (0,1)$. This operator and its generalization appear in many areas of mathematics, such as, for instance, harmonic analysis, probability theory, potential theory, quantum mechanics, statistical physics, and cosmology, as well as in many applications, as we highlighted at the beginning of this chapter. This section is devoted to the definition of this operator and to its properties.

Let $s \in (0,1)$, and define the operator $(-\Delta)^s : \mathscr{S} \to L^2(\mathbb{R}^n)$ by

$$(-\Delta)^s u(x) := C(n,s) \lim_{\varepsilon \to 0^+} \int_{\mathbb{R}^n \setminus B(x,\varepsilon)} \frac{u(x)-u(y)}{|x-y|^{n+2s}}\, dy \quad x \in \mathbb{R}^n, \tag{1.17}$$

where $B(x,\varepsilon)$ is the ball centered at $x \in \mathbb{R}^n$ with radius ε, and $C(n,s)$ is the following (positive) normalization constant:

$$C(n,s) := \left(\int_{\mathbb{R}^n} \frac{1-\cos(\zeta_1)}{|\zeta|^{n+2s}}\, d\zeta \right)^{-1}, \tag{1.18}$$

with $\zeta = (\zeta_1, \zeta')$, $\zeta' \in \mathbb{R}^{n-1}$.

The operator defined in (1.17) is the *fractional Laplacian*. Commonly, in defining $(-\Delta)^s$, the abbreviation for "in the principal-value sense" is adopted. Precisely, setting

$$\text{P.V.} \int_{\mathbb{R}^n} \frac{u(x)-u(y)}{|x-y|^{n+2s}}\, dy := \lim_{\varepsilon \to 0^+} \int_{\mathbb{R}^n \setminus B(x,\varepsilon)} \frac{u(x)-u(y)}{|x-y|^{n+2s}}\, dy,$$

we can write

$$(-\Delta)^s u(x) := C(n,s) \text{P.V.} \int_{\mathbb{R}^n} \frac{u(x)-u(y)}{|x-y|^{n+2s}}\, dy \quad x \in \mathbb{R}^n. \tag{1.19}$$

The singular integral given in (1.19) can be written as a weighted second-order differential quotient as follows (see [83, lemma 3.2]):

Proposition 1.10 Let $s \in (0,1)$. Then, for any $u \in \mathscr{S}$,

$$(-\Delta)^s u(x) = -\frac{1}{2} C(n,s) \int_{\mathbb{R}^n} \frac{u(x+y)+u(x-y)-2u(x)}{|y|^{n+2s}}\, dy \quad x \in \mathbb{R}^n. \tag{1.20}$$

Proof By (1.19), we have that

$$(-\Delta)^s u(x) = -C(n,s) \text{P.V.} \int_{\mathbb{R}^n} \frac{u(y)-u(x)}{|x-y|^{n+2s}}\, dy, \tag{1.21}$$

for every $x \in \mathbb{R}^n$. Hence, substituting $z = y - x$ in (1.21), it follows that

$$(-\Delta)^s u(x) = -C(n,s) \text{P.V.} \int_{\mathbb{R}^n} \frac{u(x+z)-u(x)}{|z|^{n+2s}}\, dz, \tag{1.22}$$

for every $x \in \mathbb{R}^n$. However, by putting $z^\star = -z$, one has

$$\mathrm{P.V.} \int_{\mathbb{R}^n} \frac{u(x+z)-u(x)}{|z|^{n+2s}}\,dz = \mathrm{P.V.} \int_{\mathbb{R}^n} \frac{u(x-z^\star)-u(x)}{|z^\star|^{n+2s}}\,dz^\star.$$

So, after relabeling $z^\star$ as z, the following equalities hold:

$$\begin{aligned} 2\mathrm{P.V.} \int_{\mathbb{R}^n} \frac{u(x+z)-u(x)}{|z|^{n+2s}}\,dz &= \mathrm{P.V.} \int_{\mathbb{R}^n} \frac{u(x+z)-u(x)}{|z|^{n+2s}}\,dz \\ &\quad + \mathrm{P.V.} \int_{\mathbb{R}^n} \frac{u(x-z)-u(x)}{|z|^{n+2s}}\,dz \\ &= \mathrm{P.V.} \int_{\mathbb{R}^n} \frac{u(x+y)+u(x-y)-2u(x)}{|y|^{n+2s}}\,dy. \end{aligned} \tag{1.23}$$

Finally, a second-order Taylor expansion yields

$$\frac{u(x+y)+u(x-y)-2u(x)}{|y|^{n+2s}} \le \frac{\|D^2u\|_{L^\infty(\mathbb{R}^n)}}{|y|^{n+2s-2}},$$

and since $s \in (0,1)$, one has

$$\frac{u(x+y)+u(x-y)-2u(x)}{|y|^{n+2s}} \in L^1(\mathbb{R}^n).$$

Thus, for any $u \in \mathscr{S}$, we have that

$$\mathrm{P.V.} \int_{\mathbb{R}^n} \frac{u(x+y)+u(x-y)-2u(x)}{|y|^{n+2s}}\,dy = \int_{\mathbb{R}^n} \frac{u(x+y)+u(x-y)-2u(x)}{|y|^{n+2s}}\,dy. \tag{1.24}$$

In conclusion, relation (1.20) holds thanks to (1.22)–(1.24). □

Remark 1.11 Let $s \in (0,1/2)$. Observe that for any $u \in \mathscr{S}$ and for a fixed $x \in \mathbb{R}^n$, we have that

$$\begin{aligned} \int_{\mathbb{R}^n} \frac{u(x)-u(y)}{|x-y|^{n+2s}}\,dy &\le C \int_{B(x,R)} \frac{|x-y|}{|x-y|^{n+2s}}\,dy \\ &\quad + \|u\|_{L^\infty(\mathbb{R}^n)} \int_{\mathbb{R}^n \setminus B(x,R)} \frac{1}{|x-y|^{n+2s}}\,dy \\ &\le C\left(\int_0^R \frac{1}{\rho^{2s}}\,d\rho + \int_R^{+\infty} \frac{1}{\rho^{2s+1}}\,d\rho\right) < +\infty, \end{aligned}$$

where C is a positive constant depending only on the dimension n and the L^∞-norm of the function u. Hence, in the case $s \in (0, 1/2)$, the integral

$$\int_{\mathbb{R}^n} \frac{u(x) - u(y)}{|x-y|^{n+2s}}\, dy$$

is not singular near the point x, so one can get rid of the P.V. in (1.19).

Some recent results on fractional Laplacian equations can be found in [50, 55, 56, 57, 83, 99, 102, 184, 185, 186, 206] and references therein. Moreover, very recently, a new nonlocal and nonlinear operator (the *fractional p-Laplacian* $(-\Delta)^s_p$) was considered (see, e.g., the papers [165, 168] and references therein). Namely, for $p \in (1, +\infty)$, $s \in (0,1)$, and u smooth enough, it is defined as

$$(-\Delta)^s_p u(x) = \mathrm{P.V.} \int_{\mathbb{R}^n} \frac{|u(x)-u(y)|^{p-2}(u(x)-u(y))}{|x-y|^{n+sp}}\, dy \quad x \in \mathbb{R}^n. \tag{1.25}$$

Up to some normalization constant depending on n, p, and s, this definition is consistent with the one of the fractional Laplacian $(-\Delta)^s$ in the case $p = 2$.

For the motivations that lead to the study of such operators, we refer the reader to the seminal paper of Caffarelli [52] (see also the recent papers [37, 107, 133]).

1.3.1 The constant $C(n, s)$: some properties

Here we recall some properties of the constant $C(n,s)$ proved in [83].

Lemma 1.12 *Let* $s \in (0,1)$ *and* $C(n,s)$ *be the constant defined in* (1.18), *and let* $A(n,s)$ *and* $B(n,s)$ *be as follows:*

$$A(n,s) := \begin{cases} 1 & \text{if } n = 1 \\ \displaystyle\int_{\mathbb{R}^{n-1}} \frac{1}{(1+|\eta'|^2)^{n+2s/2}}\, d\eta' & \text{if } n \geq 2, \end{cases}$$

and

$$B(s) := s(1-s) \int_{\mathbb{R}} \frac{1-\cos t}{|t|^{1+2s}}\, dt.$$

Then

$$C(n,s) = \frac{s(1-s)}{A(n,s)B(s)}. \tag{1.26}$$

Proof The proof is based on direct computations. Let $n \geq 2$ and $\zeta = (\zeta_1, \zeta')$, with $\zeta' \in \mathbb{R}^{n-1}$. By using the change of variables $\eta' = \zeta'/|\zeta_1|$, one has

$$\begin{aligned}
\int_{\mathbb{R}^n} \frac{1-\cos(\zeta_1)}{|\zeta|^{n+2s}}\, d\zeta &= \int_{\mathbb{R}} \left(\int_{\mathbb{R}^{n-1}} \frac{1-\cos(\zeta_1)}{|\zeta_1|^{n+2s}} \frac{1}{(1+|\zeta'|^2/|\zeta_1|^2)^{\frac{n+2s}{2}}}\, d\zeta' \right) d\zeta_1 \\
&= \int_{\mathbb{R}} \left(\int_{\mathbb{R}^{n-1}} \frac{1-\cos(\zeta_1)}{|\zeta_1|^{1+2s}} \frac{1}{(1+|\eta'|^2)^{\frac{n+2s}{2}}}\, d\eta' \right) d\zeta_1 \\
&= \frac{A(n,s)B(s)}{s(1-s)}.
\end{aligned}$$

This completes the proof. □

Remark 1.13 Note that by (1.18) and (1.26) it follows that

$$C(n,s)^{-1} = \int_{\mathbb{R}^n} \frac{1-\cos(\zeta_1)}{|\zeta|^{n+2s}}\, d\zeta = A(n,s) \int_{\mathbb{R}} \frac{1-\cos t}{|t|^{1+2s}}\, dt.$$

For $n \geq 3$, denoting by S_{n-2} the Lebesgue measure of the unit sphere in $\mathbb{R}^{n-1}$, one has

$$A(n,s) = \int_{\mathbb{R}^{n-1}} \frac{1}{(1+|\eta'|^2)^{\frac{n+2s}{2}}}\, d\eta' < S_{n-2}\left(\frac{\pi}{4} + \frac{1}{1+2s} \right)$$

and

$$\int_{\mathbb{R}} \frac{1-\cos t}{|t|^{1+2s}}\, dt < \frac{1}{2(1-s)} + \frac{2}{s}.$$

Consequently, in this setting it is easy to see that

$$C(n,s)^{-1} < S_{n-2}\left(\frac{\pi}{4} + \frac{1}{1+2s} \right)\left(\frac{1}{2(1-s)} + \frac{2}{s} \right). \tag{1.27}$$

However, we also have

$$C(n,s)^{-1} < \begin{cases} \dfrac{1}{2(1-s)} + \dfrac{2}{s} & \text{if } n = 1 \\ \left(\dfrac{\pi}{2} + \dfrac{2}{1+2s} \right)\left(\dfrac{1}{2(1-s)} + \dfrac{2}{s} \right) & \text{if } n = 2. \end{cases} \tag{1.28}$$

The following property will be used in what follows:

Proposition 1.14 Let $s \in (0,1)$ and $C(n,s)$ be the constant defined in (1.18). Then, for any $\xi \in \mathbb{R}^n$, the equality

$$\int_{\mathbb{R}^n} \frac{1-\cos(\xi \cdot y)}{|y|^{n+2s}}\, dy = C(n,s)^{-1}|\xi|^{2s} \tag{1.29}$$

holds.

Proof First, we observe that, for $\zeta = (\zeta_1, \dots, \zeta_n)$, we have

$$\frac{1-\cos\zeta_1}{|\zeta|^{n+2s}} \le \frac{|\zeta_1|^2}{|\zeta|^{n+2s}} \le \frac{1}{|\zeta|^{n-2+2s}}$$

near the origin. As a consequence,

$$\int_{\mathbb{R}^n} \frac{1-\cos\zeta_1}{|\zeta|^{n+2s}}\, d\zeta \quad \text{is finite and positive,}$$

also thanks to the choice of s.

Now let us define the map $J : \mathbb{R}^n \to \mathbb{R}$ as follows:

$$J(\xi) := \int_{\mathbb{R}^n} \frac{1-\cos(\xi \cdot y)}{|y|^{n+2s}}\, dy,$$

for every $\xi \in \mathbb{R}^n$. We claim that J is rotationally invariant; that is,

$$J(\xi) = J(|\xi| e_1) \quad \xi \in \mathbb{R}^n, \tag{1.30}$$

where e_1 stands for the first direction vector on the space $\mathbb{R}^n$.

For $n = 1$, equality (1.30) is trivial because J is an odd function. When $n \ge 2$, we consider a rotation $\mathscr{R}$ for which $\mathscr{R}(|\xi| e_1) = \xi$, and we denote by $\mathscr{R}^T$ its transpose. Hence, by substituting $\widetilde{y} = \mathscr{R}^T y$, we obtain

$$\begin{aligned}
J(\xi) &= \int_{\mathbb{R}^n} \frac{1-\cos((\mathscr{R}(|\xi| e_1)) \cdot y)}{|y|^{n+2s}}\, dy \\
&= \int_{\mathbb{R}^n} \frac{1-\cos((|\xi| e_1) \cdot (\mathscr{R}^T \cdot y))}{|y|^{n+2s}}\, dy \\
&= \int_{\mathbb{R}^n} \frac{1-\cos((|\xi| e_1) \cdot \widetilde{y})}{|\widetilde{y}|^{n+2s}}\, d\widetilde{y} \\
&= J(|\xi| e_1),
\end{aligned}$$

so claim (1.30) is proved.

Then, by (1.30), the substitution $\zeta = |\xi| y$ gives that

$$\begin{aligned}
J(\xi) &= J(|\xi| e_1) \\
&= \int_{\mathbb{R}^n} \frac{1-\cos(|\xi| y_1)}{|y|^{n+2s}}\, dy \\
&= \frac{1}{|\xi|^n} \int_{\mathbb{R}^n} \frac{1-\cos\zeta_1}{|\zeta/|\xi||^{n+2s}}\, d\zeta \\
&= C(n,s)^{-1} |\xi|^{2s},
\end{aligned}$$

thanks to (1.18). In conclusion, relation (1.29) is proved. □

Corollary 1.15 Let $s \in (0,1)$ and let $C(n,s)$ be the constant defined in (1.18). Then, for any $u \in H^s(\mathbb{R}^n)$,

$$[u]^2_{H^s(\mathbb{R}^n)} = 2C(n,s)^{-1} \int_{\mathbb{R}^n} |\xi|^{2s} |\mathscr{F}u(\xi)|^2 \, d\xi. \tag{1.31}$$

Moreover, $H^s(\mathbb{R}^n) = \widehat{H}^s(\mathbb{R}^n)$.

Proof Let us fix $y \in \mathbb{R}^n$. Using the change of variable $z = x - y$, and applying the Parseval–Plancherel formula given in (1.4), it follows that

$$\begin{aligned}
\int_{\mathbb{R}^n} \left(\int_{\mathbb{R}^n} \frac{|u(x)-u(y)|^2}{|x-y|^{n+2s}} \, dx \right) dy &= \int_{\mathbb{R}^n} \left(\int_{\mathbb{R}^n} \frac{u(z+y)-u(y)}{|z|^{n+2s}} \, dz \right) dy \\
&= \int_{\mathbb{R}^n} \left(\int_{\mathbb{R}^n} \left| \frac{u(z+y)-u(y)}{|z|^{n/2+s}} \right|^2 dy \right) dz \\
&= \int_{\mathbb{R}^n} \left\| \frac{u(z+\cdot)-u(\cdot)}{|z|^{n/2+s}} \right\|^2_{L^2(\mathbb{R}^n)} dz \\
&= \int_{\mathbb{R}^n} \left\| \mathscr{F}\left(\frac{u(z+\cdot)-u(\cdot)}{|z|^{n/2+s}} \right) \right\|^2_{L^2(\mathbb{R}^n)} dz.
\end{aligned} \tag{1.32}$$

Elementary computations ensure that

$$\begin{aligned}
\int_{\mathbb{R}^n} \left\| \mathscr{F}\left(\frac{u(z+\cdot)-u(\cdot)}{|z|^{n/2+s}} \right) \right\|^2_{L^2(\mathbb{R}^n)} dz &= \int_{\mathbb{R}^n} \left(\int_{\mathbb{R}^n} \frac{|e^{i\xi\cdot z}-1|^2}{|z|^{n+2s}} |\mathscr{F}u(\xi)|^2 \, d\xi \right) dz \\
&= 2\int_{\mathbb{R}^n \times \mathbb{R}^n} \frac{1-\cos(\xi\cdot z)}{|z|^{n+2s}} |\mathscr{F}u(\xi)|^2 \, dz \, d\xi,
\end{aligned}$$

so, by (1.29), we can write

$$\int_{\mathbb{R}^n} \left\| \mathscr{F}\left(\frac{u(z+\cdot)-u(\cdot)}{|z|^{n/2+s}} \right) \right\|^2_{L^2(\mathbb{R}^n)} dz = 2C(n,s)^{-1} \int_{\mathbb{R}^n} |\xi|^{2s} |\mathscr{F}u(\xi)|^2 \, d\xi. \tag{1.33}$$

Hence, relation (1.31) follows by (1.32) and (1.33).

Finally, the equivalence between the fractional spaces $H^s(\mathbb{R}^n)$ and $\widehat{H}^s(\mathbb{R}^n)$ comes from (1.3) and (1.31). □

A class of truncated functions. The computations in Remark 1.13 and Corollary 1.15 are useful in proving some estimates on the norm of some truncated (fractional) functions, which we will manage in Chapter 10 when we study existence and multiplicity results for sublinear fractional problems.

Precisely, fix an element $x_0 \in \Omega$ (where $\Omega \subset \mathbb{R}^n$ is of class $C^{0,1}$), and choose $\tau > 0$ in such a way that

$$\bar{B}(x_0, \tau) := \{x \in \mathbb{R}^n : |x - x_0| \leq \tau\} \subset \Omega, \tag{1.34}$$

where $|\cdot|$ denotes the usual Euclidean distance in $\mathbb{R}^n$.

Now let $\sigma \in (0,1)$ and $t_0 \in \mathbb{R}$, and define $u_\sigma^{t_0} \in H^s(\mathbb{R}^n)$ as follows:

$$u_\sigma^{t_0}(x) := \begin{cases} 0 & \text{if } x \in \mathbb{R}^n \setminus B(x_0, \tau) \\ \dfrac{t_0}{(1-\sigma)\tau}(\tau - |x - x_0|) & \text{if } x \in B(x_0, \tau) \setminus B(x_0, \sigma\tau) \\ t_0 & \text{if } x \in B(x_0, \sigma\tau), \end{cases} \tag{1.35}$$

where $B(x_0, r)$ denotes the n-dimensional ball with center $x_0 \in \Omega$ and radius $r > 0$. Set

$$\nu_0 := \left(1 + \frac{1}{\widetilde{\lambda}_1}\right), \tag{1.36}$$

where

$$\widetilde{\lambda}_1 := \inf_{u \in H_0^1(\Omega) \setminus \{0\}} \frac{\|\nabla u\|^2_{L^2(\Omega)}}{\|u\|^2_{L^2(\Omega)}}. \tag{1.37}$$

The following result holds:

Proposition 1.16 Let $\sigma, s \in (0,1)$, $t_0 \in \mathbb{R}$, and τ be such that (1.34) is verified. Let $u_\sigma^{t_0}$ be the function given in (1.35), S_{n-2} be the Lebesgue measure of the unit sphere in $\mathbb{R}^{n-1}$, and

$$\Gamma(t) := \int_0^{+\infty} z^{t-1} e^{-z} dz \quad t > 0,$$

be the usual Gamma function. Then $u_\sigma^{t_0} \in H^s(\mathbb{R}^n)$, and one has

$$\left(\int_{\mathbb{R}^n \times \mathbb{R}^n} \frac{|u_\sigma^{t_0}(x) - u_\sigma^{t_0}(y)|^2}{|x-y|^{n+2s}} dx\, dy\right)^{1/2} < \frac{|t_0|}{(1-\sigma)} \sqrt{\frac{\pi^{\frac{n}{2}} \tau^{n-2}(1-\sigma^n)}{\Gamma\left(1 + \frac{n}{2}\right)} \kappa_1 \kappa_2}, \tag{1.38}$$

where

$$\kappa_1 := \begin{cases} 2\nu_0 & \text{if } n = 1 \\ \left(\pi + \dfrac{4}{1+2s}\right)\nu_0 & \text{if } n = 2 \\ S_{n-2}\left(\dfrac{\pi}{2} + \dfrac{2}{1+2s}\right)\nu_0 & \text{if } n \geq 3 \end{cases} \quad \text{and} \quad \kappa_2 := \frac{1}{2(1-s)} + \frac{2}{s} \tag{1.39}$$

with ν_0 given in (1.36).

Proof Computing the standard seminorm of the function $u_\sigma^{t_0}$ in $H^1(\mathbb{R}^n)$, we easily have

$$\begin{aligned}
[u_\sigma^{t_0}]^2_{H^1(\mathbb{R}^n)} &= \int_{\mathbb{R}^n} |\nabla u_\sigma^{t_0}(x)|^2\,dx \\
&= \int_{B(x_0,\tau)\setminus B(x_0,\sigma\tau)} \frac{t_0^2}{(1-\sigma)^2\tau^2}\,dx \\
&= \frac{t_0^2}{(1-\sigma)^2\tau^2}\big(|B(x_0,\tau)| - |B(x_0,\sigma\tau)|\big) \\
&= \frac{t_0^2}{(1-\sigma)^2}\,\frac{\pi^{\frac{n}{2}}\tau^{n-2}(1-\sigma^n)}{\Gamma\left(1+\frac{n}{2}\right)}.
\end{aligned} \tag{1.40}$$

Since $u_\sigma^{t_0} \in H_0^1(\Omega)$, by Proposition 1.1(b), it follows that $u_\sigma^{t_0} \in W^{s,2}(\Omega)$. Moreover, the boundary $\partial\Omega$ is Lipschitz, $\bar{B}(x_0,\tau) \subset \Omega$, and $u_\sigma^{t_0} = 0$ in $\Omega \setminus \bar{B}(x_0,\tau)$. Then, by Lemma 1.3, one has that $u_\sigma^{t_0} \in H^s(\mathbb{R}^n)$.

Hence, since $s \in (0,1)$, Corollary 1.15 yields

$$\begin{aligned}
[u_\sigma^{t_0}]^2_{H^s(\mathbb{R}^n)} &\le 2\left(\int_{\mathbb{R}^n} \frac{1-\cos x_1}{|x|^{n+2s}}\,dx\right)\int_{\mathbb{R}^n} |\xi|^{2s}|\mathscr{F}u_\sigma^{t_0}(\xi)|^2\,d\xi \\
&< 2\left(\int_{\mathbb{R}^n} \frac{1-\cos x_1}{|x|^{n+2s}}\,dx\right)\int_{\mathbb{R}^n} (1+|\xi|^2)|\mathscr{F}u_\sigma^{t_0}(\xi)|^2\,d\xi.
\end{aligned} \tag{1.41}$$

Now, by standard arguments on the Fourier transform and Poincaré inequality, one has

$$\int_{\mathbb{R}^n} (1+|\xi|^2)|\mathscr{F}u_\sigma^{t_0}(\xi)|^2\,d\xi \le \nu_0[u_\sigma^{t_0}]^2_{H^1(\mathbb{R}^n)}. \tag{1.42}$$

Indeed, by Parseval–Plancherel identity (1.4), it follows that

$$u_\sigma^{t_0} \in L^2(\mathbb{R}^n) \quad \text{if and only if} \quad \mathscr{F}u_\sigma^{t_0} \in L^2(\mathbb{R}^n)$$

and

$$\|u_\sigma^{t_0}\|^2_{L^2(\mathbb{R}^n)} = \|\mathscr{F}u_\sigma^{t_0}\|^2_{L^2(\mathbb{R}^n)}. \tag{1.43}$$

Furthermore,

$$|\nabla u_\sigma^{t_0}| \in L^2(\mathbb{R}^n) \quad \text{if and only if} \quad |\xi|\mathscr{F}u_\sigma^{t_0} \in L^2(\mathbb{R}^n)$$

and

$$\|\nabla u_\sigma^{t_0}\|^2_{L^2(\mathbb{R}^n)} = \||\xi|\mathscr{F}u_\sigma^{t_0}\|^2_{L^2(\mathbb{R}^n)}. \tag{1.44}$$

Relations (1.43) and (1.44) give

$$\int_{\mathbb{R}^n} (1+|\xi|^2)|\mathscr{F}u_\sigma^{t_0}(\xi)|^2\,d\xi = \left(\|u_\sigma^{t_0}\|^2_{L^2(\mathbb{R}^n)} + \|\nabla u_\sigma^{t_0}\|^2_{L^2(\mathbb{R}^n)}\right). \tag{1.45}$$

Hence, by (1.36) and (1.37), one has that inequality (1.42) is a direct consequence of (1.45), taking into account that $u_\sigma^{t_0} \in W_0^{1,2}(\Omega)$.

Then, by (1.41) and (1.42), il follows that

$$\int_{\mathbb{R}^n\times\mathbb{R}^n} \frac{|u_\sigma^{t_0}(x)-u_\sigma^{t_0}(y)|^2}{|x-y|^{n+2s}}\,dx\,dy < 2\nu_0\left(\int_{\mathbb{R}^n}\frac{1-\cos x_1}{|x|^{n+2s}}\,dx\right)[u_\sigma^{t_0}]^2_{H^1(\mathbb{R}^n)}.$$

Finally, inequality (1.38) follows by (1.40) and Remark 1.13 (see relations (1.27) and (1.28)). □

1.3.2 The fractional Laplacian via Fourier transform

Here we prove that the fractional Laplacian $(-\Delta)^s$ can be viewed as a pseudo-differential operator of symbol $|\xi|^{2s}$ (see [83, section 3]).

Proposition 1.17 Let $s \in (0,1)$. Then, for any $u \in \mathscr{S}$,

$$(-\Delta)^s u(x) = \mathscr{F}^{-1}(|\xi|^{2s}(\mathscr{F}u)(\xi))(x), \quad x \in \mathbb{R}^n, \tag{1.46}$$

where $\mathscr{F}^{-1}$ is the inverse Fourier transform defined in (1.2).

Proof Denote by

$$\mathscr{L}u(x) := -\frac{1}{2}C(n,s)\int_{\mathbb{R}^n}\frac{u(x+y)+u(x-y)-2u(x)}{|y|^{n+2s}}\,dy \quad x\in\mathbb{R}^n,$$

where $C(n,s)$ is the constant given in (1.18). We look for a function $S:\mathbb{R}^n\to\mathbb{R}$ such that

$$\mathscr{L}u = \mathscr{F}^{-1}(S(\mathscr{F}u)). \tag{1.47}$$

We will prove that for every $\xi\in\mathbb{R}^n$,

$$S(\xi) = |\xi|^{2s}. \tag{1.48}$$

Since

$$\frac{u(x+y)+u(x-y)-2u(x)}{|y|^{n+2s}} \in L^1(\mathbb{R}^n\times\mathbb{R}^n),$$

by Fubini–Tonelli's result we can exchange the integral in y with the Fourier transform in x. Applying the Fourier transform in the variable x in (1.47), we obtain

$$\begin{aligned} S(\xi)(\mathscr{F}u)(\xi) &= \mathscr{F}(\mathscr{L}u) \\ &= -\frac{1}{2}C(n,s)\int_{\mathbb{R}^n}\frac{\mathscr{F}(u(x+y)+u(x-y)-2u(x))}{|y|^{n+2s}}\,dy \\ &= -\frac{1}{2}C(n,s)\int_{\mathbb{R}^n}\frac{e^{i\xi\cdot y}+e^{-i\xi\cdot y}-2}{|y|^{n+2s}}\,dy(\mathscr{F}u)(\xi) \\ &= C(n,s)\int_{\mathbb{R}^n}\frac{1-\cos(\xi\cdot y)}{|y|^{n+2s}}\,dy(\mathscr{F}u)(\xi). \end{aligned} \tag{1.49}$$

By substituting (1.29) in (1.49), necessarily the function S has the form given in (1.48), and this ends the proof. □

Different definitions of the fractional Laplacian consider different normalizing constants. The constant $C(n,s)$ chosen here (see (1.18)) is the one that ensures the equivalence of the integral definition of $(-\Delta)^s$ with that given by the Fourier transform (see, e.g., [55] and [83, (3.1)–(3.3) and (3.8)]). Moreover, $C(n,s)$ has the following additional properties:

$$\lim_{s\to 1^-}(-\Delta)^s u = -\Delta u$$

and

$$\lim_{s\to 0^+}(-\Delta)^s u = u$$

(see [83, proposition 4.4]). Here $-\Delta$ denotes the classical Laplacian operator.

Finally, we are able to prove the relation between the fractional Laplacian operator $(-\Delta)^s$ and the fractional Sobolev space $H^s(\mathbb{R}^n)$ (see [83]).

Proposition 1.18 Let $s\in(0,1)$ and let $C(n,s)$ be the constant defined in (1.18). Then, for any $u\in H^s(\mathbb{R}^n)$,

$$[u]^2_{H^s(\mathbb{R}^n)} = 2C(n,s)^{-1}\|(-\Delta)^{s/2}u\|^2_{L^2(\mathbb{R}^n)}. \tag{1.50}$$

Proof By (1.4), it follows that

$$\|(-\Delta)^{s/2}u\|^2_{L^2(\mathbb{R}^n)} = \|\mathscr{F}(-\Delta)^{s/2}u\|^2_{L^2(\mathbb{R}^n)}. \tag{1.51}$$

Moreover, Proposition 1.17 yields

$$\|\mathscr{F}(-\Delta)^{s/2}u\|^2_{L^2(\mathbb{R}^n)} = \||\xi|\mathscr{F}u\|^2_{L^2(\mathbb{R}^n)}. \tag{1.52}$$

Finally, by Corollary 1.15, one has

$$\||\xi|\mathscr{F}u\|^2_{L^2(\mathbb{R}^n)} = \frac{1}{2}C(n,s)\|(-\Delta)^{s/2}u\|^2_{L^2(\mathbb{R}^n)}. \tag{1.53}$$

Relations (1.51)–(1.53) immediately prove (1.50). □

As consequence of the Parseval–Plancherel formula, as well as Propositions 1.17 and 1.18 and Corollary 1.15, the norms in $H^s(\mathbb{R}^n)$ given by

$$u \mapsto \left(\int_{\mathbb{R}^n} |u(x)|^2\,dx + \int_{\mathbb{R}^n\times\mathbb{R}^n} \frac{|u(x)-u(y)|^2}{|x-y|^{n+2s}}\,dx\,dy\right)^{1/2},$$

$$u \mapsto \left(\int_{\mathbb{R}^n} (1+|\xi|^2)^s |\mathscr{F}u(\xi)|^2\,d\xi\right)^{1/2},$$

$$u \mapsto \left(\int_{\mathbb{R}^n} |u(x)|^2\,dx + \int_{\mathbb{R}^n} |\xi|^{2s} |\mathscr{F}u(\xi)|^2\,d\xi\right)^{1/2},$$

and

$$u \mapsto \left(\int_{\mathbb{R}^n} |u(x)|^2\,dx + \|(-\Delta)^{s/2}u\|^2_{L^2(\mathbb{R}^n)}\right)^{1/2}$$

are all equivalent. This will be useful in studying fractional Schrödinger equations (see Chapter 13).

1.3.3 *A generalization of* $(-\Delta)^s$

Here we introduce a general integrodifferential operator of nonlocal type that generalizes $(-\Delta)^s$. For any fixed $s \in (0,1)$, the operator $\mathscr{L}_K$ is given by

$$\mathscr{L}_K u(x) := \int_{\mathbb{R}^n} \big(u(x+y)+u(x-y)-2u(x)\big)K(y)\,dy, \tag{1.54}$$

for every $x \in \mathbb{R}^n$, where the kernel $K : \mathbb{R}^n \setminus \{0\} \to (0,+\infty)$ is a function with the properties that

$$mK \in L^1(\mathbb{R}^n), \quad \textit{where } m(x) := \min\{|x|^2, 1\}; \tag{1.55}$$

$$\textit{there exists } \theta > 0 \quad \textit{such that} \quad K(x) \geq \theta|x|^{-(n+2s)}, \quad \textit{for any } x \in \mathbb{R}^n \setminus \{0\}. \tag{1.56}$$

Of course, as a model for K, we can take the function

$$K(x) = |x|^{-(n+2s)}, \quad x \in \mathbb{R}^n \setminus \{0\}. \tag{1.57}$$

In this case, up to some normalization constant, $\mathscr{L}_K = -(-\Delta)^s$.

1.4 The fractional Sobolev space $\widetilde{H}^s_0(\Omega)$

There is another way to define the fractional Laplace operator given by the inverse Fourier transform (1.46). In fact, in the case $s = 1/2$, the half-Laplacian acting on a function u in the whole space $\mathbb{R}^n$ can be computed as the normal derivative on the boundary of its harmonic extension to the upper half-space $\mathbb{R}_+^{n+1} := \mathbb{R}^n \times (0, +\infty)$, the so-called Dirichlet to Neumann operator (see, e.g., [55]).

The s-derivative $(-\Delta)^s u$ can be characterized in a similar way, defining its s-harmonic extension to the upper half-space. This extension is used commonly in the recent literature because it allows one to write nonlocal problems in a local way, and this permits the use of variational techniques for studying these kinds of problems.

In the case of the operator defined in bounded domains, the preceding characterization has to be adapted. More precisely, the powers of the Laplace operator $-\Delta$, in a bounded domain Ω with zero boundary conditions, are defined through its spectral decomposition using the powers of its eigenvalues. Hence, according to classical results on positive operators in Ω, if $\{\widetilde{e}_j, \widetilde{\lambda}_j\}_{j\in\mathbb{N}}$ are the eigenfunctions and eigenvalues of the following Dirichlet problem

$$\begin{cases} -\Delta u = \lambda u & \text{in } \Omega \\ u = 0 & \text{on } \partial\Omega, \end{cases}$$

then $\{\widetilde{e}_j, \widetilde{\lambda}_j^s\}_{j\in\mathbb{N}}$ are eigenfunctions and eigenvalues of the corresponding fractional problem

$$\begin{cases} A_s u = \lambda u & \text{in } \Omega \\ u = 0 & \text{on } \partial\Omega, \end{cases}$$

where, for $u = \sum_{j\in\mathbb{N}} a_j \widetilde{e}_j \in L^2(\Omega)$, we set

$$A_s u = \sum_{j\in\mathbb{N}} a_j \widetilde{\lambda}_j^s \widetilde{e}_j.$$

The operator A_s (the *spectral Laplacian*) is well defined on the Sobolev space

$$\widetilde{H}_0^s(\Omega) := \left\{ u = \sum_{j\in\mathbb{N}} a_j \widetilde{e}_j \in L^2(\Omega) : \sum_{j\in\mathbb{N}} a_j^2 \widetilde{\lambda}_j^s < +\infty \right\},$$

endowed with the norm

$$\|u\|_{\widetilde{H}_0^s(\Omega)} := \left(\sum_{j\in\mathbb{N}} a_j^2 \widetilde{\lambda}_j^s \right)^{1/2}.$$

1.4.1 The extension problem and the space $X_0^s(\mathscr{C}_\Omega)$

Associated with the bounded domain Ω, let us consider the cylinder

$$\mathscr{C}_\Omega := \{(x, y) : x \in \Omega,\ y \in \mathbb{R}_+\} \subset \mathbb{R}_+^{n+1}$$

and denote by $\partial_L \mathscr{C}_\Omega := \partial\Omega \times \mathbb{R}_+$ its lateral boundary. For a function $u \in \widetilde{H}_0^s(\Omega)$, define the s-harmonic extension $\mathbb{E}_s(u)$ to the cylinder $\mathscr{C}_\Omega$ as the (unique) solution of the problem

$$\begin{cases} -\operatorname{div}(y^{1-2s}\nabla \mathbb{E}_s(u)) = 0 & \text{in } \mathscr{C}_\Omega \\ \mathbb{E}_s(u) = 0 & \text{on } \partial_L \mathscr{C}_\Omega \\ \operatorname{Tr}(\mathbb{E}_s(u)) = u & \text{in } \Omega, \end{cases}$$

where the trace operator $\operatorname{Tr} : X_0^s(\mathscr{C}_\Omega) \to L^2(\Omega)$ is given by

$$\operatorname{Tr}(\mathbb{E}_s(u)) := \mathbb{E}_s(u)(\cdot, 0).$$

The extension function $\mathbb{E}_s(u)$ belongs to the Hilbert space

$$X_0^s(\mathscr{C}_\Omega) := \left\{ w \in L^2(\mathscr{C}_\Omega) : w = 0 \text{ on } \partial_L \mathscr{C}_\Omega,\ \int_{\mathscr{C}_\Omega} y^{1-2s} |\nabla w(x,y)|^2 \, dx\, dy < +\infty \right\},$$

with the standard norm

$$\|w\|_{X_0^s(\mathscr{C}_\Omega)} := \left(\kappa_s \int_{\mathscr{C}_\Omega} y^{1-2s} |\nabla w(x,y)|^2 \, dx\, dy \right)^{1/2},$$

where κ_s is a suitable normalization factor. We can choose this constant in such a way that the extension operator $\mathbb{E}_s : \widetilde{H}_0^s(\Omega) \to X_0^s(\mathscr{C}_\Omega)$ is an isometry; that is, for any $u \in \widetilde{H}_0^s(\Omega)$,

$$\|\mathbb{E}_s(u)\|_{X_0^s(\mathscr{C}_\Omega)} = \|u\|_{\widetilde{H}_0^s(\Omega)},$$

and the trace inequality

$$\|\operatorname{Tr}(w)\|_{\widetilde{H}_0^s(\Omega)} \le \|w\|_{X_0^s(\mathscr{C}_\Omega)}$$

holds true for every $w \in X_0^s(\mathscr{C}_\Omega)$.

By using the s-extension $\mathbb{E}_s(u) \in X_0^s(\mathscr{C}_\Omega)$ of the function $u \in \widetilde{H}_0^s(\Omega)$, the fractional operator A_s in Ω can be defined as follows:

$$A_s u(x) := -\kappa_s \lim_{y \to 0^+} y^{1-2s} \frac{\partial \mathbb{E}_s(u)}{\partial y}(x,y).$$

Note that when $\Omega = \mathbb{R}^n$, an analogous Dirichlet–Neumann procedure provides a formula for the fractional Laplacian in the whole space equivalent to that obtained using the Fourier transform. In that case, the s-harmonic extension and the fractional Laplacian have explicit expressions in terms of the Poisson and Riesz kernels, which, up to a normalization constant, are given by

$$\mathbb{E}_s(u)(x,y) = y^{2s} \int_{\mathbb{R}^n} \frac{u(\zeta)}{(|x-\zeta|^2 + y^2)^{(n+2s)/2}} \, d\zeta$$

and

$$(-\Delta)^s u(x) = \text{P.V.} \int_{\mathbb{R}^n} \frac{u(x) - u(y)}{|x-y|^{n+2s}} \, dy.$$

Hence, the fractional Laplacian also can be considered as a local operator in an augmented space.

We will not use this characterization directly in this book. However, regularity theorems for the fractional Laplacian are often easier to prove with this characterization. For instance, Cabré and Sire in [48] establish necessary conditions on the nonlinearity in order to admit special classes of solutions (e.g., bounded increasing solutions in the whole real line). These necessary conditions are derived in the spirit of a classical result due to Modica (see [143]) and are valid for the Laplacian operator. In addition, the authors studied regularity issues, as well as maximum and Harnack principles associated with the treated equation (see also the related paper [49]).

To end this section, we will provide a recent result obtained in [147, 148], where the extension procedure is used for studying nonlocal problems via variational techniques. Let $f : \mathbb{R} \to \mathbb{R}$ be a continuous function, and let us consider the existence of weak solutions of the following problem:

$$\begin{cases} A_s u = f(u) & \text{in } \Omega \\ u = 0 & \text{on } \partial\Omega. \end{cases} \tag{1.58}$$

By using the preceding extension arguments, we can formulate problem (1.58) as

$$\begin{cases} -\operatorname{div}(y^{1-2s}\nabla \mathbb{E}_s(u)) = 0 & \text{in } \mathscr{C}_\Omega \\ \mathbb{E}_s(u) = 0 & \text{on } \partial_L \mathscr{C}_\Omega \\ \operatorname{Tr}(\mathbb{E}_s(u)) = f(u) & \text{in } \Omega. \end{cases} \tag{1.59}$$

A weak (or energy) solution to (1.59) is a function $w \in X_0^s(\mathscr{C}_\Omega)$ such that

$$\kappa_s \int_{\mathscr{C}_\Omega} y^{1-2s} \langle \nabla w, \nabla \varphi \rangle \, dx\, dy = \int_\Omega f(\operatorname{Tr}(w)(x)) \operatorname{Tr}(\varphi)(x) dx,$$

for every $\varphi \in X_0^s(\mathscr{C}_\Omega)$. Moreover, for any energy solution $w \in X_0^s(\mathscr{C}_\Omega)$ of (1.59), the function $u = w(\cdot, 0)$, defined in the sense of traces, belongs to the space $\widetilde{H}_0^s(\Omega)$, and it is an energy solution of (1.58); that is,

$$\int_\Omega A_{s/2} u(x) A_{s/2} \varphi(x) dx = \int_\Omega f(u(x)) \varphi(x) dx,$$

for every $\varphi \in \widetilde{H}_0^s(\Omega)$. The converse is also true. Therefore, both the variational formulations of problems (1.58) and (1.59) are equivalent.

Owing to the preceding abstract framework and by using variational methods, a nonlinear elliptic problem defined in a bounded domain $\Omega \subset \mathbb{R}^n$, with smooth boundary $\partial\Omega$, and involving fractional powers of the Laplacian operator, together with a suitable nonlinear term f, has been studied in [148]. More precisely, the following characterization theorem on the existence of one weak solution for the Dirichlet problem

$$\begin{cases} A_s u = \lambda f(u) & \text{in } \Omega \\ u > 0 & \text{in } \Omega \\ u = 0 & \text{on } \partial\Omega, \end{cases} \tag{1.60}$$

has been proved:

Theorem 1.19 *Let $f : [0,+\infty) \to [0,+\infty)$ be a continuous function with $f(0) = 0$ and such that, for some $a > 0$, the map $h : (0,+\infty) \to [0,+\infty)$ defined by*

$$h(\xi) := \frac{\int_0^\xi f(t)dt}{\xi^2}$$

is nonincreasing in the real interval $(0,a]$. Then the following assertions are equivalent:

(a) h is not constant in $(0,b]$ for each $b > 0$; and

(b) For each $r > 0$ there exists an open interval $J_r \subseteq (0,+\infty)$ such that, for every $\lambda \in J_r$, the nonlocal problem (1.60) has a weak solution in $\widetilde{H}_0^s(\Omega)$ whose norm is less than r.

This result extends to the nonlocal setting recent theorems for ordinary and classical elliptic equations, as well as a characterization for elliptic problems on certain nonsmooth domains. To make the nonlinear methods work, some careful analysis of the fractional spaces involved is necessary.

For other interesting contributions devoted to the existence, multiplicity, and regularity of solutions for fractional equations involving the spectral Laplacian operator, we refer to [22, 42, 50, 51, 58, 59, 139, 208, 214] and references therein.

1.5 Other fractional Sobolev-type spaces

One of the aims of this book is to study nonlocal problems driven by $(-\Delta)^s$ (or its generalization) and with Dirichlet boundary data via variational methods. For this purpose, we need to work in a suitable fractional Sobolev space: for this, we consider a functional analytical setting that is inspired by (but not equivalent to) the fractional Sobolev spaces in order to correctly encode the Dirichlet boundary datum in the variational formulation. This section is devoted to the definition of this space as well as to its properties. Other fractional Sobolev-type spaces have been considered in the literature (see [41, 137, 138] and the references therein).

Let $s \in (0,1)$ be fixed, let Ω be an open-bounded subset of $\mathbb{R}^n$, $n > 2s$, and denote by Q the set

$$Q := (\mathbb{R}^n \times \mathbb{R}^n) \setminus (\mathscr{C}\Omega \times \mathscr{C}\Omega), \tag{1.61}$$

where $\mathscr{C}\Omega := \mathbb{R}^n \setminus \Omega$. Moreover, let $K : \mathbb{R}^n \setminus \{0\} \to (0,+\infty)$ be a function satisfying (1.55) and (1.56).

The space $X^s(\Omega)$ is the linear space of Lebesgue measurable functions from $\mathbb{R}^n$ to $\mathbb{R}$ such that the restriction to Ω of any function u in $X^s(\Omega)$ belongs to $L^2(\Omega)$, and

$$\textit{the map } (x,y) \mapsto (u(x)-u(y))\sqrt{K(x-y)} \quad \textit{is in} \quad L^2(Q, dx\,dy).$$

The norm in $X^s(\Omega)$ can be defined as follows:

$$\|u\|_{X^s(\Omega)} = \|u\|_{L^2(\Omega)} + \left(\int_Q |u(x)-u(y)|^2 K(x-y)dx\,dy\right)^{1/2}. \tag{1.62}$$

It is easy to check that $\|\cdot\|_{X^s(\Omega)}$ is a norm on $X^s(\Omega)$. We only show that if $\|u\|_{X^s(\Omega)} = 0$, then $u=0$ a.e. in $\mathbb{R}^n$. Indeed, by $\|u\|_{X^s(\Omega)}=0$, we get $\|u\|_{L^2(\Omega)}=0$, which implies that

$$u=0 \quad \text{a.e. in } \Omega, \tag{1.63}$$

and

$$\int_Q |u(x)-u(y)|^2 K(x-y)dx\,dy = 0. \tag{1.64}$$

By (1.64), we deduce that $u(x)=u(y)$ a.e. $(x,y)\in Q$; that is, u is constant a.e. in $\mathbb{R}^n$, say, $u=c\in\mathbb{R}$ a.e. in $\mathbb{R}^n$. By (1.63), it easily follows that $c=0$, so $u=0$ a.e. in $\mathbb{R}^n$.

We remark that even in the model case in which K is as in (1.57), the norms in (1.5) and (1.62) are not the same because $\Omega\times\Omega$ is strictly contained in Q: this makes the classical fractional Sobolev space approach not sufficient for studying the nonlocal problem we consider in this book.

We also define

$$X_0^s(\Omega) := \left\{u\in X^s(\Omega) : u=0 \text{ a.e. in } \mathscr{C}\Omega\right\}. \tag{1.65}$$

If K is the prototype given in (1.57), the Sobolev-type spaces $X^s(\Omega)$ and $X_0^s(\Omega)$ will be denoted by $\mathbb{H}^s(\Omega)$ and $\mathbb{H}_0^s(\Omega)$, respectively.

Now we recall some properties of the functional spaces $X^s(\Omega)$ and $X_0^s(\Omega)$.

Lemma 1.20 *Let $\varphi\in C_0^2(\Omega)$. Then*

$$\textit{the map } \mathbb{R}^n\times\mathbb{R}^n \ni (x,y)\mapsto |\varphi(x)-\varphi(y)|^2 K(x-y) \quad \textit{belongs to} \quad L^1(\mathbb{R}^n\times\mathbb{R}^n).$$

Proof Since φ vanishes outside Ω,

$$\begin{aligned}\int_{\mathbb{R}^n\times\mathbb{R}^n} |\varphi(x)-\varphi(y)|^2K(x-y)dx\,dy &= \int_{\Omega\times\Omega}|\varphi(x)-\varphi(y)|^2K(x-y)dx\,dy\\ &\quad +2\int_{\Omega\times\mathscr{C}\Omega}|\varphi(x)-\varphi(y)|^2K(x-y)dx\,dy\\ &\le 2\int_{\Omega\times\mathbb{R}^n}|\varphi(x)-\varphi(y)|^2K(x-y)dx\,dy.\end{aligned} \tag{1.66}$$

Now we notice that

$$|\varphi(x)-\varphi(y)| \leq \|\nabla\varphi\|_{L^\infty(\mathbb{R}^n)}|x-y|$$

and

$$|\varphi(x)-\varphi(y)| \leq 2\|\varphi\|_{L^\infty(\mathbb{R}^n)}.$$

Accordingly, we get

$$|\varphi(x)-\varphi(y)| \leq 2\|\varphi\|_{C^1(\mathbb{R}^n)}\min\{|x-y|,1\} = 2\|\varphi\|_{C^1(\mathbb{R}^n)}\sqrt{m(x-y)},$$

where m is defined in (1.55). Therefore, from (1.66), we deduce that

$$\begin{aligned}\int_{\mathbb{R}^n\times\mathbb{R}^n}|\varphi(x)-\varphi(y)|^2K(x-y)\,dx\,dy &\leq 8\|\varphi\|^2_{C^1(\mathbb{R}^n)}\int_{\Omega\times\mathbb{R}^n} m(x-y)K(x-y)\,dx\,dy\\ &= 8|\Omega|\,\|\varphi\|^2_{C^1(\mathbb{R}^n)}\int_{\mathbb{R}^n} m(\xi)K(\xi)\,d\xi.\end{aligned}$$

Thus, Lemma 1.20 follows by (1.55) and by the fact that Ω is bounded. □

As a trivial consequence of Lemma 1.20, we get that $C^2_0(\Omega) \subseteq X^s(\Omega)$. In what follows, we denote by g^+ the positive part of a function g; that is,

$$g^+(x) := \max\{g(x),0\}.$$

Lemma 1.21 *Let $\alpha,\beta : \mathbb{R}^n \to \mathbb{R}$, $n \geq 1$. Then, for any $x,y \in \mathbb{R}^n$, the following inequality holds:*

$$|\alpha^+(x)-\beta^+(y)| \leq |\alpha(x)-\beta(y)|.$$

Proof Since the map

$$\mathbb{R} \ni \tau \mapsto \tau^+ := \max\{\tau,0\}$$

is Lipschitz continuous with constant 1 and (sub)linear at infinity, the conclusion easily follows. □

With Lemma 1.21, we can prove the following result:

Lemma 1.22 *Let u be a function in $X^s(\Omega)$. Then its positive part $u^+ \in X^s(\Omega)$.*

Proof By Lemma 1.21 with $\beta \equiv 0$, we have

$$\int_\Omega |u^+(x)|^2dx \leq \int_\Omega |u(x)|^2dx < +\infty,$$

while, taking $\beta = \alpha$, we get

$$\int_Q |u^+(x)-u^+(y)|^2K(x-y)\,dx\,dy \leq \int_Q |u(x)-u(y)|^2K(x-y)\,dx\,dy < +\infty;$$

that is, $u^+ \in X^s(\Omega)$. □

The following summability property can be easily proved:

Lemma 1.23 *Let $u, v \in X^s(\Omega)$. Then*

$$\text{the map } Q \ni (x,y) \mapsto (u(x)-u(y))(v(x)-v(y))K(x-y) \quad \text{belongs to} \quad L^1(Q, dx\,dy).$$

Proof By the Cauchy–Schwarz inequality, we have

$$\begin{aligned} &2\,|u(x)-u(y)|\,|v(x)-v(y)|\,\sqrt{K(x-y)}\sqrt{K(x-y)} \\ &\quad \le |u(x)-u(y)|^2K(x-y)+|v(x)-v(y)|^2K(x-y). \end{aligned}$$

This and the fact that $u, v \in X^s(\Omega)$ give the assertion. □

Now we compare the spaces $X_0^s(\Omega)$ and H^s.

Lemma 1.24 *Let $s \in (0,1)$ and $K : \mathbb{R}^n \setminus \{0\} \to (0,+\infty)$ satisfy assumptions* (1.55) *and* (1.56). *Then the following assertions hold true:*

(*a*) *If $u \in X^s(\Omega)$, then $u \in H^s(\Omega)$. Moreover,*

$$\|u\|_{H^s(\Omega)} \le c(\theta)\|u\|_{X^s(\Omega)},$$

where $c(\theta) := \max\{1, 1/\sqrt{\theta}\}$;

(*b*) *If $u \in X_0^s(\Omega)$, then $u \in H^s(\mathbb{R}^n)$. Moreover,*

$$\|u\|_{H^s(\Omega)} \le \|u\|_{H^s(\mathbb{R}^n)} \le c(\theta)\|u\|_{X^s(\Omega)},$$

where $c(\theta)$ is the constant given in a); and

(*c*) *Let $K(x) = |x|^{-(n+2s)}$. Then*

$$X_0^s(\Omega) = \{u \in H^s(\mathbb{R}^n) : u = 0 \text{ a.e. in } \mathscr{C}\Omega\}.$$

Proof Let us prove part (a). By (1.56), we get

$$\begin{aligned} \int_{\Omega\times\Omega} \frac{|u(x)-u(y)|^2}{|x-y|^{n+2s}}\,dx\,dy &\le \frac{1}{\theta}\int_{\Omega\times\Omega} |u(x)-u(y)|^2K(x-y)\,dx\,dy \\ &\le \frac{1}{\theta}\int_Q |u(x)-u(y)|^2K(x-y)\,dx\,dy < +\infty, \end{aligned}$$

because $u \in X^s(\Omega)$. The first assertion is proved.

For part (b), note that $u \in X^s(\Omega)$ and $u = 0$ a.e. in $\mathscr{C}\Omega$. As a consequence,

$$\|u\|_{L^2(\mathbb{R}^n)} = \|u\|_{L^2(\Omega)} < +\infty$$

and

$$\int_{\mathbb{R}^n\times\mathbb{R}^n} \frac{|u(x)-u(y)|^2}{|x-y|^{n+2s}}\,dx\,dy = \int_Q \frac{|u(x)-u(y)|^2}{|x-y|^{n+2s}}\,dx\,dy$$

$$\leq \frac{1}{\theta}\int_Q |u(x)-u(y)|^2 K(x-y)\,dx\,dy < +\infty.$$

Hence, $u \in H^s(\mathbb{R}^n)$. The estimate on the norm easily follows.

Finally, let us show part (c). Clearly, $X_0^s(\Omega) \subseteq \{u \in H^s(\mathbb{R}^n) : u = 0 \text{ a.e. in } \mathscr{C}\Omega\}$, thanks to part (b). Conversely, let $u \in H^s(\mathbb{R}^n)$ be such that $u = 0$ a.e. in $\mathscr{C}\Omega$. Since $u \in H^s(\mathbb{R}^n)$, we have that

$$\int_{\mathbb{R}^n} |u(x)|^2\,dx < +\infty$$

and

$$\int_{\mathbb{R}^n\times\mathbb{R}^n} \frac{|u(x)-u(y)|^2}{|x-y|^{n+2s}}\,dx\,dy < +\infty,$$

from which, using the property that $u = 0$ a.e. in $\mathscr{C}\Omega$, we deduce

$$\int_\Omega |u(x)|^2\,dx = \int_{\mathbb{R}^n} |u(x)|^2\,dx < +\infty$$

and

$$\int_Q \frac{|u(x)-u(y)|^2}{|x-y|^{n+2s}}\,dx\,dy = \int_{\mathbb{R}^n\times\mathbb{R}^n} \frac{|u(x)-u(y)|^2}{|x-y|^{n+2s}}\,dx\,dy < +\infty.$$

Hence, $u \in X_0^s(\Omega)$, and this ends the proof. □

In the spaces $X^s(\Omega)$ and $X_0^s(\Omega)$, the following convergence property is satisfied:

Lemma 1.25 *Let $\{u_j\}_{j\in\mathbb{N}}$ be a sequence in $X^s(\Omega)$ such that $u_j \to u_\infty$ a.e. in $\mathbb{R}^n$ as $j \to +\infty$ and*

$$\sup_{j\in\mathbb{N}} \|u_j\|_{X^s(\Omega)} < +\infty. \tag{1.67}$$

Then $u_\infty \in X^s(\Omega)$. If, in addition, $u_j \in X_0^s(\Omega)$, for any $j \in \mathbb{N}$, then $u_\infty \in X_0^s(\Omega)$.

Proof By (1.67) and the Fatou lemma, we have

$$+\infty > \liminf_{j\to+\infty} \int_\Omega |u_j(x)|^2 dx \geq \int_\Omega |u_\infty(x)|^2 dx;$$

that is, $u_\infty \in L^2(\Omega)$, and

$$+\infty > \liminf_{j\to+\infty} \int_Q |u_j(x)-u_j(y)|^2 K(x-y)\,dx\,dy$$

$$\geq \int_Q |u_\infty(x)-u_\infty(y)|^2 K(x-y)\,dx\,dy.$$

Thus, $u_\infty \in X^s(\Omega)$.

Now suppose that $u_j = 0$ a.e. in $\mathscr{C}\Omega$ for any $j \in \mathbb{N}$. Then it is easy to see that $u_\infty = 0$ a.e. in $\mathscr{C}\Omega$. Hence, $u_\infty \in X_0^s(\Omega)$, and the assertion is proved. □

In the following lemma we prove a sort of formula of integration by parts in $X^s(\Omega)$, which will be useful in what follows.

Lemma 1.26 *Let $\varphi \in C_0^2(\Omega)$ and $u \in X^s(\Omega) \cap L^\infty(\mathscr{C}\Omega)$. Then the following equalities hold true:*

$$\begin{aligned}
\int_{\mathbb{R}^n \times \mathbb{R}^n} & (u(x)-u(y))\,(\varphi(x)-\varphi(y))\,K(x-y)\,dx\,dy \\
&= \int_Q (u(x)-u(y))\,(\varphi(x)-\varphi(y))\,K(x-y)\,dx\,dy \\
&= \int_{\mathbb{R}^n \times \mathbb{R}^n} u(x)(2\varphi(x)-\varphi(x+\xi)-\varphi(x-\xi))\,K(\xi)\,dx\,d\xi.
\end{aligned}$$

Proof In what follows, for notation consistency, it will be convenient to denote

$$\mathscr{D}_0 := Q = \mathbb{R}^n \times \mathbb{R}^n \setminus (\mathscr{C}\Omega) \times (\mathscr{C}\Omega). \tag{1.68}$$

Also, given $\varepsilon \in [0,1)$, we define

$$\mathscr{D}_\varepsilon := \big\{(x,y) \in \mathscr{D}_0 : \ |x-y| \geq \varepsilon\big\}$$

and

$$\mathscr{D}_\varepsilon^\pm := \big\{(x,\xi) \in \mathbb{R}^n \times \mathbb{R}^n : \ (x, x \pm \xi) \in \mathscr{D}_0 \text{ and } |\xi| \geq \varepsilon\big\}.$$

We remark that the preceding notation for $\mathscr{D}_\varepsilon$ is consistent with that given in (1.68) for $\varepsilon = 0$.

Recalling that φ vanishes outside Ω, direct computations ensure that

$$\begin{aligned}
\int_{\mathscr{D}_\varepsilon} & |u(x)||\varphi(x)-\varphi(y)|\,K(x-y)\,dx\,dy \\
&= \int_{(\Omega\times\Omega)\cap\{|x-y|\geq\varepsilon\}} |u(x)||\varphi(x)-\varphi(y)|\,K(x-y)\,dx\,dy \\
&\quad + \int_{(\Omega\times\mathscr{C}\Omega)\cap\{|x-y|\geq\varepsilon\}} |u(x)||\varphi(x)-\varphi(y)|\,K(x-y)\,dx\,dy \\
&\quad + \int_{(\mathscr{C}\Omega\times\Omega)\cap\{|x-y|\geq\varepsilon\}} |u(x)||\varphi(x)-\varphi(y)|\,K(x-y)\,dx\,dy \\
&= \int_{(\Omega\times\mathbb{R}^n)\cap\{|x-y|\geq\varepsilon\}} |u(x)||\varphi(x)-\varphi(y)|\,K(x-y)\,dx\,dy
\end{aligned}$$

$$+\int_{(\Omega\times\mathscr{C}\Omega)\cap\{|x-y|\geq\varepsilon\}} |u(y)||\varphi(x)-\varphi(y)|\,K(y-x)\,dx\,dy$$

$$\leq 2\|\varphi\|_{L^\infty(\mathbb{R}^n)}\left[\int_{(\Omega\times\mathbb{R}^n)\cap\{|x-y|\geq\varepsilon\}} |u(x)|K(x-y)\,dx\,dy\right.$$

$$\left.+\int_{(\Omega\times\mathscr{C}\Omega)\cap\{|x-y|\geq\varepsilon\}} |u(y)|K(y-x)\,dx\,dy\right]$$

$$=\frac{2\|\varphi\|_{L^\infty(\mathbb{R}^n)}}{\varepsilon^2}\left[\int_{(\Omega\times\mathbb{R}^n)\cap\{|x-y|\geq\varepsilon\}} |u(x)|m(x-y)K(x-y)\,dx\,dy\right.$$

$$\left.+\int_{(\Omega\times\mathscr{C}\Omega)\cap\{|x-y|\geq\varepsilon\}} |u(y)|m(y-x)K(y-x)\,dx\,dy\right]$$

$$\leq\frac{2\|\varphi\|_{L^\infty(\mathbb{R}^n)}}{\varepsilon^2}\left[\int_{\Omega\times\mathbb{R}^n} |u(x)|m(\xi)K(\xi)\,dx\,d\xi\right.$$

$$\left.+\|u\|_{L^\infty(\mathscr{C}\Omega)}\int_{\Omega\times\mathbb{R}^n} m(\xi)K(\xi)\,dx\,d\xi\right]$$

$$=\frac{2\|\varphi\|_{L^\infty(\mathbb{R}^n)}}{\varepsilon^2}\big(\|u\|_{L^1(\Omega)}+|\Omega|\,\|u\|_{L^\infty(\mathscr{C}\Omega)}\big)\int_{\mathbb{R}^n} m(\xi)K(\xi)\,d\xi,$$

which is finite, thanks to (1.55).

As a consequence,

$$\begin{aligned}&\text{the maps } (x,y)\mapsto u(x)(\varphi(x)-\varphi(y))\,K(x-y) \quad\text{and}\quad (x,y)\\ &\qquad\mapsto u(y)(\varphi(x)-\varphi(y))\,K(x-y) \quad\text{belong to}\quad L^1(\mathscr{D}_\varepsilon,dx\,dy).\end{aligned} \tag{1.69}$$

Thanks to Lemma 1.23 and (1.69), we can split the integrals

$$\int_{\mathscr{D}_\varepsilon} (u(x)-u(y))\,(\varphi(x)-\varphi(y))\,K(x-y)\,dx\,dy$$

$$=\int_{\mathscr{D}_\varepsilon} u(x)(\varphi(x)-\varphi(y))\,K(x-y)\,dx\,dy$$

$$+\int_{\mathscr{D}_\varepsilon} u(y)(\varphi(y)-\varphi(x))\,K(x-y)\,dx\,dy$$

$$=\int_{\mathscr{D}_\varepsilon} u(x)(\varphi(x)-\varphi(y))\,K(x-y)\,dx\,dy$$

$$+\int_{\mathscr{D}_\varepsilon} u(x)(\varphi(x)-\varphi(y))\,K(y-x)\,dx\,dy.$$

Now, with the change of variable $\xi = x - y$ in the first integral and $\xi = y - x$ in the second one and the definition of $\mathscr{D}_\varepsilon^\pm$, we get

$$\begin{aligned}
&\int_{\mathscr{D}_\varepsilon} (u(x)-u(y))\,(\varphi(x)-\varphi(y))\,K(x-y)\,dx\,dy \\
&\quad = \int_{\mathscr{D}_\varepsilon} u(x)(\varphi(x)-\varphi(y))\,K(x-y)\,dx\,dy \\
&\qquad + \int_{\mathscr{D}_\varepsilon} u(x)(\varphi(x)-\varphi(y))\,K(y-x)\,dx\,dy \\
&\quad = \int_{\mathscr{D}_\varepsilon^-} u(x)(\varphi(x)-\varphi(x-\xi))\,K(\xi)\,dx\,d\xi \\
&\qquad + \int_{\mathscr{D}_\varepsilon^+} u(x)(\varphi(x)-\varphi(x+\xi))\,K(\xi)\,dx\,d\xi.
\end{aligned} \tag{1.70}$$

We claim that

$$\int_{\mathscr{D}_\varepsilon^- \setminus \mathscr{D}_\varepsilon^+} u(x)(\varphi(x)-\varphi(x+\xi))\,K(\xi)\,dx\,d\xi = 0. \tag{1.71}$$

To check this, let $(x,\xi) \in \mathscr{D}_\varepsilon^- \setminus \mathscr{D}_\varepsilon^+$. By definitions of $\mathscr{D}_\varepsilon^\pm$, it follows that $(x, x-\xi) \in \mathscr{D}_0$, $(x, x+\xi) \notin \mathscr{D}_0$, and $|\xi| \geq \varepsilon$. Hence, $(x, x+\xi) \in \mathscr{O}$; that is,

$$x \in \mathscr{C}\Omega \tag{1.72}$$

and

$$x + \xi \in \mathscr{C}\Omega. \tag{1.73}$$

Then, by (1.72) and (1.73), we have that $\varphi(x) = 0 = \varphi(x+\xi)$, for any $(x,\xi) \in \mathscr{D}_\varepsilon^- \setminus \mathscr{D}_\varepsilon^+$, which implies (1.71). Analogously, we get that

$$\int_{\mathscr{D}_\varepsilon^+ \setminus \mathscr{D}_\varepsilon^-} u(x)(\varphi(x)-\varphi(x-\xi))\,K(\xi)\,dx\,d\xi = 0. \tag{1.74}$$

In the light of (1.70), (1.71) and (1.74), we obtain that

$$\begin{aligned}
&\int_{\mathscr{D}_\varepsilon} (u(x)-u(y))\,(\varphi(x)-\varphi(y))\,K(x-y)\,dx\,dy \\
&\quad = \int_{\mathscr{D}_\varepsilon^+ \cup \mathscr{D}_\varepsilon^-} u(x)(\varphi(x)-\varphi(x+\xi))\,K(\xi)\,dx\,d\xi \\
&\qquad + \int_{\mathscr{D}_\varepsilon^- \cup \mathscr{D}_\varepsilon^+} u(x)(\varphi(x)-\varphi(x-\xi))\,K(\xi)\,dx\,d\xi \\
&\quad = \int_{\mathscr{D}_\varepsilon^+ \cup \mathscr{D}_\varepsilon^-} u(x)(2\varphi(x)-\varphi(x+\xi)-\varphi(x-\xi))\,K(\xi)\,dx\,d\xi.
\end{aligned} \tag{1.75}$$

Now let us define

$$\Phi(x,\xi) := (2\varphi(x)-\varphi(x+\xi)-\varphi(x-\xi))\,K(\xi). \tag{1.76}$$

We claim that

$$\Phi \in L^1(\mathbb{R}^n \times \mathbb{R}^n, dx\,dy). \tag{1.77}$$

To establish this, we observe that

$$|\Phi(x,\xi)| \le 4\|\varphi\|_{L^\infty(\mathbb{R}^n)} K(\xi) \tag{1.78}$$

and, by a Taylor expansion, that

$$|\Phi(x,\xi)| \le \|D^2\varphi\|_{L^\infty(\mathbb{R}^n)} |\xi|^2 K(\xi).$$

Therefore,

$$|\Phi(x,\xi)| \le 4\|\varphi\|_{C^2(\mathbb{R}^n)} m(\xi) K(\xi). \tag{1.79}$$

Let

$$R > 1 \tag{1.80}$$

be such that $\Omega \subset B_R$. If $x \in \mathscr{C} B_{2R}$ and $x \pm \xi \in \Omega \subset B_R$, then

$$|\xi| \ge |x| - |x \pm \xi| \ge 2R - R = R > 1 \tag{1.81}$$

by (1.80). Now we define

$$\mathscr{G} := \big\{(x,\xi) \in \mathbb{R}^n \times \mathbb{R}^n : x \in \mathscr{C} B_{2R} \text{ and } \xi \in B_R(x) \cup B_R(-x)\big\}$$

and

$$\mathscr{G}_* := \big\{(x,\xi) \in \mathbb{R}^n \times \mathbb{R}^n : x \in B_R(\xi) \cup B_R(-\xi) \text{ and } \xi \in \mathscr{C} B_1\big\}.$$

Since (1.81) holds true, then

$$\mathscr{G} \subset \mathscr{G}_*. \tag{1.82}$$

Moreover, by definition of m (see (1.55)),

$$K(\xi) = m(\xi)K(\xi) \quad \text{if} \quad (x,\xi) \in \mathscr{G}_*. \tag{1.83}$$

Now let $(x,\xi) \in (\mathscr{C} B_{2R} \times \mathbb{R}^n) \setminus \mathscr{G}$. Then $x \in \mathscr{C} B_{2R}$ and $\xi \in \mathscr{C}(B_R(x) \cup B_R(-x))$; that is, $|x \pm \xi| > R$. As a consequence, since φ vanishes outside Ω,

$$\varphi(x) = 0 = \varphi(x \pm \xi) \quad \text{if} \quad (x,\xi) \in (\mathscr{C} B_{2R} \times \mathbb{R}^n) \setminus \mathscr{G}. \tag{1.84}$$

Thus, using (1.78), (1.82)–(1.84), and (1.55), we have

$$\begin{aligned}
\int_{(\mathscr{C} B_{2R}) \times \mathbb{R}^n} |\Phi(x,\xi)|\,dx\,d\xi &= \int_{\mathscr{G}} |2\varphi(x) - \varphi(x+\xi) - \varphi(x-\xi)|\,K(\xi)\,dx\,d\xi \\
&\le 4\|\varphi\|_{L^\infty(\mathbb{R}^n)} \int_{\mathscr{G}_*} K(\xi)\,dx\,d\xi \\
&= 4\|\varphi\|_{L^\infty(\mathbb{R}^n)} \int_{\mathscr{G}_*} m(\xi)K(\xi)\,dx\,d\xi \\
&\le 8\omega_n R^n \|\varphi\|_{L^\infty(\mathbb{R}^n)} \int_{\mathbb{R}^n} m(\xi)K(\xi)\,d\xi \\
&= C(n,R,K) < +\infty,
\end{aligned} \tag{1.85}$$

for a suitable positive constant $C(n,R,K)$, where ω_n is the volume of the unit ball in $\mathbb{R}^n$; that is, $\omega_n := 2\pi^{n/2}/(n\Gamma(n/2))$. By (1.79) and (1.85), we obtain

$$\begin{aligned}\int_{\mathbb{R}^n\times\mathbb{R}^n} |\Phi(x,\xi)|\,dx\,d\xi &\le C(n,R,K) + \int_{B_{2R}\times\mathbb{R}^n} |\Phi(x,\xi)|\,dx\,d\xi \\ &\le C(n,R,K) + 4\|\varphi\|_{C^2(\mathbb{R}^n)} \int_{B_{2R}\times\mathbb{R}^n} m(\xi)K(\xi)\,dx\,d\xi \\ &\le C(n,R,K) + 4\omega_n(2R)^n \|\varphi\|_{C^2(\mathbb{R}^n)} \int_{\mathbb{R}^n} m(\xi)K(\xi)\,d\xi,\end{aligned}$$

which, once more, is finite due to (1.55). This establishes (1.77).

Owing to Lemma 1.23 and (1.77), we can send $\varepsilon \to 0^+$ in (1.75), and we obtain

$$\begin{aligned}&\int_{\mathscr{D}_0} (u(x)-u(y))\,(\varphi(x)-\varphi(y))\,K(x-y)\,dx\,dy \\ &\qquad = \int_{\mathscr{D}_0^+\cup\mathscr{D}_0^-} u(x)(2\varphi(x)-\varphi(x+\xi)-\varphi(x-\xi))\,K(\xi)\,dx\,d\xi.\end{aligned}$$

By using once again the idea that φ vanishes outside Ω, we get the assertion of Lemma 1.26. $\square$

As a consequence of Lemma 1.26, the following result holds true:

Corollary 1.27 Let $\varphi \in C_0^2(\Omega)$. Let $\{u_j\}_{j\in\mathbb{N}} \subset X^s(\Omega)$ be a sequence of functions converging uniformly in $\mathbb{R}^n$ to $u_\infty \in X^s(\Omega)$ as $j \to +\infty$ and such that $u_j - u_\infty \in X_0^s(\Omega)$ for any $j \in \mathbb{N}$. Then

$$\begin{aligned}&\lim_{j\to+\infty} \int_Q (u_j(x)-u_j(y))\,(\varphi(x)-\varphi(y))\,K(x-y)\,dx\,dy \\ &\qquad = \int_Q (u_\infty(x)-u_\infty(y))\,(\varphi(x)-\varphi(y))\,K(x-y)\,dx\,dy.\end{aligned}$$

Proof We define $w_j := u_j - u_\infty$. Notice that $w_j \in X^s(\Omega)$, being $X^s(\Omega)$ a linear space, and $w_j = 0$ a.e. in $\mathscr{C}\Omega$, because $u_j = u_\infty$ a.e. in $\mathscr{C}\Omega$. Hence, $w_j \in X^s(\Omega) \cap L^\infty(\mathscr{C}\Omega)$, for any $j \in \mathbb{N}$. Then, by Lemma 1.26 (applied to w_j), we obtain that

$$\begin{aligned}&\int_Q (u_j(x)-u_j(y))(\varphi(x)-\varphi(y))K(x-y)\,dx\,dy \\ &\quad - \int_Q (u_\infty(x)-u_\infty(y))(\varphi(x)-\varphi(y))K(x-y)\,dx\,dy \\ &= \int_Q (w_j(x)-w_j(y))(\varphi(x)-\varphi(y))K(x-y)\,dx\,dy\end{aligned}$$

$$= \int_Q w_j(x)(2\varphi(x) - \varphi(x+\xi) - \varphi(x-\xi))K(\xi)\,dx\,d\xi$$

$$= \int_Q (u_j(x) - u_\infty(x))(2\varphi(x) - \varphi(x+\xi) - \varphi(x-\xi))K(\xi)\,dx\,d\xi, \qquad (1.86)$$

for any $\varphi \in C_0^2(\Omega)$ and any $j \in \mathbb{N}$.

Moreover, exploiting the notation in (1.76) and (1.77), we have

$$\lim_{j\to+\infty} \left| \int_{\mathbb{R}^n\times\mathbb{R}^n} (u_j(x) - v_\infty(x))(2\varphi(x) - \varphi(x+\xi) - \varphi(x-\xi))\, K(\xi)\,dx\,d\xi \right|$$

$$= \lim_{j\to+\infty} \left| \int_{\mathbb{R}^n\times\mathbb{R}^n} (u_j(x) - u_\infty(x))\Phi(x,\xi)\,dx\,d\xi \right|$$

$$\leq \lim_{j\to+\infty} \|u_j - u_\infty\|_{L^\infty(\mathbb{R}^n)} \|\Phi\|_{L^1(\mathbb{R}^n\times\mathbb{R}^n)} = 0,$$

because $u_j \to u_\infty$ uniformly in $\mathbb{R}^n$ as $j \to +\infty$. This and (1.86) imply the assertion. □

1.5.1 The space $X_0^s(\Omega)$

In this subsection, we discuss some properties of the space $X_0^s(\Omega)$, which will be the functional space in which we work in this book when studying nonlocal fractional problems. First of all, we observe that since $u = 0$ in $\mathscr{C}\Omega$, the integral in formula (1.62) can be extended to the whole of $\mathbb{R}^n$; that is, for any $u \in X_0^s(\Omega)$, we can write the norm as follows:

$$\|u\|_{X^s(\Omega)} = \|u\|_{L^2(\mathbb{R}^n)} + \left(\int_{\mathbb{R}^n\times\mathbb{R}^n} |u(x) - u(y)|^2 K(x-y)\,dx\,dy \right)^{1/2}.$$

Let us start by proving some properties of $X_0^s(\Omega)$.

Lemma 1.28 *Let $s \in (0,1)$, $n > 2s$, and $K : \mathbb{R}^n \setminus \{0\} \to (0,+\infty)$ satisfy assumptions* (1.55) *and* (1.56)*. Then*

(*a*) *There exists a positive constant c, depending only on n and s, such that, for any $u \in X_0^s(\Omega)$,*

$$\|u\|^2_{L^{2_s^*}(\Omega)} = \|u\|^2_{L^{2_s^*}(\mathbb{R}^n)} \leq c \int_{\mathbb{R}^n\times\mathbb{R}^n} \frac{|u(x) - u(y)|^2}{|x-y|^{n+2s}}\,dx\,dy,$$

where $2_s^* := 2n/(n-2s)$ is the fractional critical Sobolev exponent; and

(*b*) *There exists a constant $C > 1$, depending only on n, s, θ, and Ω, such that, for any $u \in X_0^s(\Omega)$,*

$$\int_Q |u(x) - u(y)|^2 K(x-y)\,dx\,dy \leq \|u\|^2_{X^s(\Omega)}$$

$$\leq C \int_Q |u(x) - u(y)|^2 K(x-y)\,dx\,dy;$$

that is

$$\|u\|_{X_0^s(\Omega)} = \left(\int_{\mathbb{R}^n\times\mathbb{R}^n} |u(x)-u(y)|^2 K(x-y)\,dx\,dy\right)^{1/2} \tag{1.87}$$

is a norm on $X_0^s(\Omega)$ *equivalent to the usual norm defined as in* (1.62).

Proof Let u be in $X_0^s(\Omega)$. By Lemma 1.24, we know that $u \in H^s(\mathbb{R}^n)$, so, using Theorem 1.4 (here with $p=2$), we get

$$\|u\|^2_{L^{2_s^*}(\mathbb{R}^n)} \le c\int_{\mathbb{R}^n\times\mathbb{R}^n} \frac{|u(x)-u(y)|^2}{|x-y|^{n+2s}}\,dx\,dy,$$

where c is a positive constant depending only on n and s. Since $u=0$ a.e. in $\mathscr{C}\Omega$, we get assertion (a).

For part (b), we note that by (1.62) it easily follows that

$$\|u\|^2_{X^s(\Omega)} \ge \int_Q |u(x)-u(y)|^2 K(x-y)\,dx\,dy.$$

Moreover, using the fact that $L^{2_s^*}(\Omega) \hookrightarrow L^2(\Omega)$ continuously (being Ω bounded and $2 < 2_s^* = 2n/(n-2s)$), part (a), and assumption (1.56), we get

$$\begin{aligned}
\|u\|^2_{X^s(\Omega)} &= \left(\|u\|_{L^2(\Omega)} + \left(\int_Q |u(x)-u(y)|^2 K(x-y)dx\,dy\right)^{1/2}\right)^2\\
&\le 2\|u\|^2_{L^2(\Omega)} + 2\int_Q |u(x)-u(y)|^2 K(x-y)dx\,dy\\
&\le 2|\Omega|^{(2_s^*-2)/2_s^*}\|u\|^2_{L^{2_s^*}(\Omega)} + 2\int_Q |u(x)-u(y)|^2 K(x-y)\,dx\,dy\\
&\le 2c|\Omega|^{(2_s^*-2)/2_s^*}\int_{\mathbb{R}^n\times\mathbb{R}^n}\frac{|u(x)-u(y)|^2}{|x-y|^{n+2s}}\,dx\,dy\\
&\quad + 2\int_Q |u(x)-u(y)|^2 K(x-y)\,dx\,dy\\
&\le 2\left(\frac{c|\Omega|^{(2_s^*-2)/2_s^*}}{\theta}+1\right)\int_Q |u(x)-u(y)|^2 K(x-y)\,dx\,dy.
\end{aligned}$$

Hence, assertion (b) follows by taking

$$C = 2\left(\frac{c}{\theta}|\Omega|^{(2_s^*-2)/2_s^*}+1\right) > 1.$$

In order to show that (1.87) is a norm on $X_0^s(\Omega)$, it is enough to prove that if $\|u\|_{X_0^s(\Omega)}=0$, then $u=0$ a.e. in $\mathbb{R}^n$. Indeed, by $\|u\|_{X_0^s(\Omega)}=0$, we get

$$\int_{\mathbb{R}^n\times\mathbb{R}^n} |u(x)-u(y)|^2 K(x-y)dx\,dy = 0,$$

so $u(x) = u(y)$ a.e. $(x, y) \in \mathbb{R}^n \times \mathbb{R}^n$; that is, u is constant a.e. in $\mathbb{R}^n$, say, $u = c \in \mathbb{R}$ a.e. in $\mathbb{R}^n$. Since $u = 0$ a.e. in $\mathscr{C}\Omega$, it easily follows that $c = 0$, so $u = 0$ a.e. in $\mathbb{R}^n$.

The preceding argument is indeed similar to the one used for (1.62), with a technical difference: there u was 0 in Ω, while here u is 0 in $\mathscr{C}\Omega$. An alternative proof could be the following: if $\|u\|_{X_0^s(\Omega)} = 0$, then $\|u\|_{X^s(\Omega)} = 0$ by Lemma 1.28(a). Hence, $u = 0$ a.e. in $\mathbb{R}^n$ because $\|\cdot\|_{X^s(\Omega)}$ is a norm. □

From now on, we take (1.87) as a norm on $X_0^s(\Omega)$. The following result is valid:

Lemma 1.29 $\left(X_0^s(\Omega), \|\cdot\|_{X_0^s(\Omega)}\right)$ *is a Hilbert space with scalar product*

$$\langle u, v\rangle_{X_0^s(\Omega)} := \int_{\mathbb{R}^n \times \mathbb{R}^n} (u(x) - u(y))(v(x) - v(y))K(x - y)\,dx\,dy. \tag{1.88}$$

Proof First of all, let us consider the map

$$X_0^s(\Omega) \times X_0^s(\Omega) \ni (u, v) \mapsto \langle u, v\rangle_{X_0^s(\Omega)}.$$

Thanks to the properties of the integrals and the positivity of K, it is easy to see that $\langle\cdot,\cdot\rangle_{X_0^s(\Omega)}$ is a scalar product in $X_0^s(\Omega) \times X_0^s(\Omega)$, which induces the norm defined in (1.87).

In order to show that $X_0^s(\Omega)$ is a Hilbert space, it remains to prove that $X_0^s(\Omega)$ is complete with respect to the norm $\|\cdot\|_{X_0^s(\Omega)}$. For this, let $\{u_j\}_{j\in\mathbb{N}}$ be a Cauchy sequence in $X_0^s(\Omega)$. Thus, for any $\varepsilon > 0$, there exists ν_ε such that if $i, j \geq \nu_\varepsilon$, then

$$\varepsilon \geq \|u_i - u_j\|^2_{X_0^s(\Omega)} \geq \frac{1}{C}\|u_i - u_j\|^2_{X^s(\Omega)} \geq \frac{1}{C}\|u_i - u_j\|^2_{L^2(\Omega)}, \tag{1.89}$$

thanks to Lemma 1.28(b).

Hence, being $L^2(\Omega)$ complete, there exists $u_\infty \in L^2(\Omega)$ such that $u_j \to u_\infty$ in $L^2(\Omega)$ as $j \to +\infty$. Since $u_j = 0$ a.e. in $\mathscr{C}\Omega$, we may define $u_\infty := 0$ in $\mathscr{C}\Omega$, and then $u_j \to u_\infty$ in $L^2(\mathbb{R}^n)$ as $j \to +\infty$. Thus, there exists a subsequence $\{u_{j_k}\}_{k\in\mathbb{N}}$ in $X_0^s(\Omega)$ such that $u_{j_k} \to u_\infty$ a.e. in $\mathbb{R}^n$ (see [43, theorem IV.9]).

Therefore, by the Fatou lemma and the first inequality in (1.89) with $\varepsilon = 1$, we have that

$$\begin{aligned}
&\int_{\mathbb{R}^n \times \mathbb{R}^n} |u_\infty(x) - u_\infty(y)|^2 K(x - y)\,dx\,dy \\
&\leq \liminf_{k\to+\infty} \int_{\mathbb{R}^n \times \mathbb{R}^n} |u_{j_k}(x) - u_{j_k}(y)|^2 K(x - y)\,dx\,dy \\
&= \liminf_{k\to+\infty} \|u_{j_k}\|^2_{X_0^s(\Omega)} \\
&\leq \liminf_{k\to+\infty} \left(\|u_{j_k} - u_{\nu_1}\|_{X_0^s(\Omega)} + \|u_{\nu_1}\|_{X_0^s(\Omega)}\right)^2 \\
&\leq (1 + \|u_{\nu_1}\|_{X_0^s(\Omega)})^2 \\
&< +\infty,
\end{aligned}$$

so $u_\infty \in X_0^s(\Omega)$.

Now it remains to show that the whole sequence converges to u_∞ in $X_0^s(\Omega)$. For this, let us take $i \geq \nu_\varepsilon$. By the first inequality in (1.89), Lemma 1.28(b), and the Fatou lemma, we get

$$\begin{aligned}\|u_i - u_\infty\|^2_{X_0^s(\Omega)} &\leq \|u_i - u_\infty\|^2_{X_0^s(\Omega)} \\ &\leq \liminf_{k\to+\infty} \|u_i - u_{j_k}\|^2_{X_0^s(\Omega)} \\ &\leq C \liminf_{k\to+\infty} \|u_i - u_{j_k}\|^2_{X_0^s(\Omega)} \\ &\leq C\varepsilon;\end{aligned}$$

that is, $u_i \to u_\infty$ in $X_0^s(\Omega)$ as $i \to +\infty$. Hence, Lemma 1.29 is proved. □

In what follows, we will denote by $\|\cdot\|_{\mathbb{H}_0^s(\Omega)}$ and $\langle\cdot,\cdot\rangle_{\mathbb{H}_0^s(\Omega)}$ the norm and the scalar product in $\mathbb{H}_0^s(\Omega)$; that is,

$$\|u\|_{\mathbb{H}_0^s(\Omega)} := \int_{\mathbb{R}^n\times\mathbb{R}^n} \frac{|u(x)-u(y)|^2}{|x-y|^{n+2s}}\,dx\,dy$$

$$\langle u,v\rangle_{\mathbb{H}_0^s(\Omega)} := \int_{\mathbb{R}^n\times\mathbb{R}^n} \frac{(u(x)-u(y))(v(x)-v(y))}{|x-y|^{n+2s}}\,dx\,dy.$$

Finally, we prove a convergence property for bounded sequences in $X_0^s(\Omega)$.

Lemma 1.30 *Let $s \in (0,1)$, $n > 2s$, Ω be an open, bounded subset of $\mathbb{R}^n$ with continuous boundary, and $K : \mathbb{R}^n \setminus \{0\} \to (0,+\infty)$ satisfy assumptions* (1.55) *and* (1.56). *Let $\{u_j\}_{j\in\mathbb{N}}$ be a bounded sequence in $X_0^s(\Omega)$. Then there exists $u_\infty \in L^\nu(\mathbb{R}^n)$ such that, up to a subsequence,*

$$u_j \to u_\infty \quad \textit{in } L^\nu(\mathbb{R}^n)$$

as $j \to +\infty$, for any $\nu \in [1, 2_s^*)$.

Proof By Lemma 1.24(b), $u_j \in H^s(\mathbb{R}^n)$, so $u_j \in H^s(\Omega)$. More precisely, by Lemma 1.24(b), Lemma 1.28(b), and the definition of $X_0^s(\Omega)$, we see that

$$\|u_j\|_{H^s(\Omega)} \leq \|u_j\|_{H^s(\mathbb{R}^n)} \leq c(\theta)\|u_j\|_{X^s(\Omega)} \leq c(\theta)\sqrt{C}\|u_j\|_{X_0^s(\Omega)},$$

with $c(\theta)$ and C depending only on n, s, θ, and Ω. Hence, $\{u_j\}_{j\in\mathbb{N}}$ is bounded in $H^s(\Omega)$ and so in $L^2(\Omega)$.

Then, by Theorem 1.5 and our assumptions on Ω, there exists $u_\infty \in L^\nu(\Omega)$ such that, up to a subsequence,

$$u_j \to u_\infty \quad \text{in } L^\nu(\Omega)$$

as $j \to +\infty$, for any $\nu \in [1, 2_s^*)$. Since u_j vanishes outside Ω, we can define $u_\infty := 0$ in $\mathscr{C}\Omega$ and see that the convergence occurs in $L^\nu(\mathbb{R}^n)$. □

1.5.2 Embedding properties of $X_0^s(\Omega)$

The results obtained in Lemmas 1.28(a) and 1.30 can be summarized as follows in term of embeddings of $X_0^s(\Omega)$ into the classical Lebesgue spaces:

Lemma 1.31 *Let $s \in (0,1)$, $n > 2s$, and $K : \mathbb{R}^n \setminus \{0\} \to (0,+\infty)$ satisfy assumptions* (1.55) *and* (1.56).

Then the following assertions hold true:

(a) *If Ω has a continuous*[1] *boundary, then the embedding $X_0^s(\Omega) \hookrightarrow L^\nu(\Omega)$ is compact for any $\nu \in [1, 2_s^*)$; and*

(b) *The embedding $X_0^s(\Omega) \hookrightarrow L^{2_s^*}(\Omega)$ is continuous.*

Proof Part (a) comes from Lemma 1.30.

Let us show part (b). For this, by Lemma 1.28(a) and assumption (1.56), we get that, for any $u \in X_0^s(\Omega)$,

$$\begin{aligned}\|u\|^2_{L^{2_s^*}(\Omega)} &= \|u\|^2_{L^{2_s^*}(\mathbb{R}^n)} \\ &\le c \int_{\mathbb{R}^n \times \mathbb{R}^n} \frac{|u(x)-u(y)|^2}{|x-y|^{n+2s}}\,dx\,dy \\ &\le \frac{c}{\theta} \int_{\mathbb{R}^n \times \mathbb{R}^n} |u(x)-u(y)|^2 K(x-y)\,dx\,dy \\ &= \frac{c}{\theta} \|u\|^2_{X_0^s(\Omega)},\end{aligned}$$

for a suitable positive constant c, depending only on n and s. Hence, assertion (b) is proved. □

Thanks to Lemma 1.31, we can define the positive constant S_K given by

$$S_K := \inf_{u \in X_0^s(\Omega) \setminus \{0\}} S_K(u), \tag{1.90}$$

where, for any $u \in X_0^s(\Omega) \setminus \{0\}$,

$$S_K(u) := \frac{\displaystyle\int_{\mathbb{R}^n \times \mathbb{R}^n} |u(x)-u(y)|^2 K(x-y)\,dx\,dy}{\left(\displaystyle\int_\Omega |u(x)|^{2_s^*}\,dx\right)^{2/2_s^*}}. \tag{1.91}$$

Moreover, note that since in formula (1.91) the integral over Ω can be extended to all $\mathbb{R}^n$ (being $u = 0$ a.e. in $\mathscr{C}\Omega$), then the function $u \mapsto S_K(u)$ does not depend on the domain Ω, while, in general, S_K does (indeed, $X_0^s(\Omega)$ depends on Ω).

The counterpart of Lemma 1.31 in the usual fractional Sobolev spaces is given by the following result, which is a special case of Theorem 1.4 when $p = 2$:

[1] The continuity of the boundary $\partial\Omega$ can be assumed here, as will be clear in Chapter 2.

Lemma 1.32 *Let $s \in (0,1)$ and $n > 2s$. Then the embedding $H^s(\mathbb{R}^n) \hookrightarrow L^\nu(\mathbb{R}^n)$ is continuous for any $\nu \in [2, 2_s^*]$.*

Lemma 1.32 allows us to define the positive constant

$$S_s := \inf_{u \in H^s(\mathbb{R}^n)\setminus\{0\}} S_s(u), \tag{1.92}$$

where, for any $u \in H^s(\mathbb{R}^n) \setminus \{0\}$,

$$S_s(u) := \frac{\displaystyle\int_{\mathbb{R}^n \times \mathbb{R}^n} \frac{|u(x)-u(y)|^2}{|x-y|^{n+2s}}\, dx\, dy}{\left(\displaystyle\int_{\mathbb{R}^n} |u(x)|^{2_s^*} dx\right)^{2/2_s^*}}. \tag{1.93}$$

Remark 1.33 In the particular case where $K(x) = |x|^{-(n+2s)}$, it is easily seen that

$$\inf_{u \in H^s(\mathbb{R}^n)\setminus\{0\}} S_s(u) \leq \inf_{u \in \mathbb{H}_0^s(\Omega)\setminus\{0\}} S_K(u)$$

because $\mathbb{H}_0^s(\Omega) \subseteq H^s(\mathbb{R}^n)$ by Lemma 1.24(a).

As a final observation, we note that with respect to embeddings in Lebesgue spaces, the fractional Sobolev space $H^s(\mathbb{R}^n)$ behaves like the usual Sobolev space $H^1(\mathbb{R}^n)$, while $X_0^s(\Omega)$, like $H_0^1(\Omega)$ (this is due to the fact that the functions $u \in X_0^s(\Omega)$ are such that $u = 0$ a.e. in $\mathscr{C}\Omega$ and so $X_0^s(\Omega)$, may be seen somehow as a space of functions defined in the bounded set Ω).

1.5.3 The general case

The definition of the Sobolev-type spaces $X^s(\Omega)$ and $X_0^s(\Omega)$, considered in the preceding subsections, can be generalized to the case p. This subsection is devoted to this generalization.

For any fixed $s \in (0,1)$ and $p \in [1, +\infty)$, let $K : \mathbb{R}^n \setminus \{0\} \to (0, +\infty)$ be a kernel such that

$$mK \in L^1(\mathbb{R}^n), \quad \textit{where } m(x) := \min\{|x|^p, 1\};$$

$$\textit{there exists } \theta > 0 \textit{ such that } K(x) \geq \theta |x|^{-(n+sp)}, \textit{ for any } x \in \mathbb{R}^n \setminus \{0\}.$$

It is easily seen that a model for K is given by the singular kernel $K(x) = |x|^{-(n+sp)}$, which gives rise (up to some normalization constant) to the fractional p-Laplace operator $-(-\Delta)_p^s$ defined; in (1.25) and to the fractional Laplace operator $-(-\Delta)^s$ defined in (1.20) when $p = 2$.

Furthermore, let Ω be an open, bounded subset of $\mathbb{R}^n$, $n > sp$. In this framework, $X^{s,p}(\Omega)$ is the linear space of Lebesgue measurable functions from $\mathbb{R}^n$ to $\mathbb{R}$ such that the restriction to Ω of any function u in $X^{s,p}(\Omega)$ belongs to $L^p(\Omega)$, and

$$\text{the map } (x,y) \mapsto (u(x)-u(y))\big(K(x-y)\big)^{1/p} \quad \text{is in} \quad L^p(Q, dx\, dy),$$

where Q is the set given in (1.61). The norm in $X^{s,p}(\Omega)$ can be defined as follows:

$$\|u\|_{X^{s,p}(\Omega)} = \|u\|_{L^p(\Omega)} + \left(\int_Q |u(x)-u(y)|^p K(x-y)dx\,dy\right)^{1/p}. \tag{1.94}$$

Also, we define

$$X_0^{s,p}(\Omega) := \left\{u \in X^{s,p}(\Omega) : u = 0 \text{ a.e. in } \mathscr{C}\Omega\right\}.$$

With this notation, we set

$$X^s(\Omega) := X^{s,2}(\Omega)$$

and

$$X_0^s(\Omega) := X_0^{s,2}(\Omega).$$

We observe that the space $X_0^{s,p}(\Omega)$ is similar to, but different from, the usual fractional Sobolev space $W^{s,p}(\Omega)$, which is endowed with the norm (1.5). Even in the model case in which $K(x) = |x|^{-(n+sp)}$, the norms in (1.5) and (1.94) are, of course, not the same.

All the properties proved in the preceding subsections for the Sobolev-type spaces $X^s(\Omega)$ and $X_0^s(\Omega)$ can be easily generalized to $X^{s,p}(\Omega)$ and $X_0^{s,p}(\Omega)$.

2

A density result for fractional Sobolev spaces

The aim of this chapter is to give the details of the proof of some density properties of smooth and compactly supported functions in fractional Sobolev spaces and suitable modifications thereof, which have recently found application in variational problems. The arguments are rather technical, but roughly speaking, they rely on a basic technique of convolution, joined with a cutoff, with some care needed not to exceed the original support.

As is wellknown, the natural functional space associated with the classical Laplace setting is the Sobolev space $H_0^1(\Omega)$. When $\partial\Omega$ is sufficiently regular, such a space may be equivalently defined as the set of functions that belongs to $L^2(\Omega)$, together with their weak derivatives of order 1 and that vanish along $\partial\Omega$ in the trace sense, or as the closure of $C_0^\infty(\Omega)$ in the norm of $H^1(\Omega)$ (see, e.g., [3]).

In this chapter, we study the density properties of the fractional space $X_0^{s,p}(\Omega)$ (see Section 2.1 for a definition) in order to establish a result similar to the one known in the classical case. Precisely, we investigate the relation between the spaces $X_0^{s,p}(\Omega)$ and $C_0^\infty(\Omega)$.

Sometimes density properties in the fractional framework are similar to those in the classical case, as happens for the results presented in this chapter. The reader should not infer that such fractional density properties are always well expected. For instance, it turns out that fractional harmonic functions are locally dense in the space of continuous and smooth functions (see [86]). This is in clear contrast with the classical case because harmonic functions can never approximate a function with a strict maximum (simply by the maximum principle). The results of this chapter are based on the paper [103].

2.1 The main theorems

Let $K : \mathbb{R}^n \setminus \{0\} \to (0, +\infty)$ satisfy assumptions (1.55) and (1.56), and let $X^{s,p}(\mathbb{R}^n)$ be the linear space of Lebesgue measurable functions u from $\mathbb{R}^n$ to $\mathbb{R}$ such that

$$\|u\|_{L^p(\mathbb{R}^n)} + \left(\int_{\mathbb{R}^n \times \mathbb{R}^n} |u(x) - u(y)|^p K(x-y) dx\, dy \right)^{1/p} < +\infty. \tag{2.1}$$

Then $X_0^{s,p}(\Omega)$ denotes the space of functions $u \in X^{s,p}(\mathbb{R}^n)$ that vanish a.e. in $\mathscr{C}\Omega$. Clearly, the quantity in (2.1) defines a norm on $X^{s,p}(\mathbb{R}^n)$ and $X_0^{s,p}(\Omega)$: such a norm will be denoted in this chapter by $\|\cdot\|$.

We observe that the space $X_0^{s,p}(\Omega)$ is similar to but different from the fractional Sobolev space $W^{s,p}(\Omega)$ defined in Chapter 1, which is endowed with the norm (1.5). Since $X_0^{s,p}(\Omega)$ is a space of functions defined in $\mathbb{R}^n$, in this context, we denote by $C_0^\infty(\Omega)$ the following space:

$$C_0^\infty(\Omega) := \{u : \mathbb{R}^n \to \mathbb{R} : u \in C^\infty(\mathbb{R}^n), \operatorname{Supp} u \text{ is compact and} \operatorname{Supp} u \subseteq \Omega\}, \tag{2.2}$$

where $\operatorname{Supp} u := \overline{\{x \in \mathbb{R}^n : u(x) \neq 0\}}$. Notice that if $u \in C_0^\infty(\Omega)$, then the support of the function u, that is, $\operatorname{Supp} u$, is always bounded, even if here and in what follows we do not assume that Ω is bounded.

Indeed, the first case we deal with is when $\partial\Omega$ is a graph of a continuous function (notice that, in this case, neither Ω nor $\partial\Omega$ is bounded). The following definition will be crucial in what follows:

Definition 2.1 The open set $\Omega \subseteq \mathbb{R}^n$ is a *hypograph* if there exists a continuous function $\xi : \mathbb{R}^{n-1} \to \mathbb{R}$ such that, up to a rigid motion,

$$\Omega = \left\{(x', x_n) \in \mathbb{R}^{n-1} \times \mathbb{R} : \, x_n < \xi(x')\right\}.$$

In this framework, our density result can be stated in the following form:

Theorem 2.2 *Let $\Omega \subseteq \mathbb{R}^n$ be a hypograph. Then, for any $u \in X_0^{s,p}(\Omega)$, there exists a sequence $\{\rho_\varepsilon\}_\varepsilon$ in $C_0^\infty(\Omega)$ such that*

$$\|\rho_\varepsilon - u\| \to 0 \quad as\ \varepsilon \to 0.$$

In other words, $C_0^\infty(\Omega)$ is a dense subspace of $X_0^{s,p}(\Omega)$.

Remark 2.3 The sequence of function ρ_ε in Theorem 2.2 also enjoys the additional property of being supported in the vicinity of the support of the original function u. More precisely, fixed any $\gamma > 0$, there exists $\varepsilon_\gamma > 0$ such that, for any $\varepsilon \in (0, \varepsilon_\gamma]$, one has that

$$\operatorname{Supp} \rho_\varepsilon \subseteq \operatorname{Supp} u + B_\gamma .$$

The next case that we consider is when Ω is a domain with continuous boundary, according to the following definition:

Definition 2.4 The open set $\Omega \subseteq \mathbb{R}^n$ is a domain with continuous boundary $\partial\Omega$ if

- $\partial\Omega$ is compact, and
- There exist $M \in \mathbb{N}$, open sets $W_1,\dots,W_M \subseteq \mathbb{R}^n$, sets $\Omega_1,\dots,\Omega_M \subseteq \mathbb{R}^n$, continuous functions $\xi_1,\dots,\xi_M : \mathbb{R}^{n-1} \to \mathbb{R}$, and rigid motions $\mathscr{R}_1,\dots,\mathscr{R}_M : \mathbb{R}^n \to \mathbb{R}^n$ such that the following conditions hold true:
    - $\partial\Omega \subseteq \bigcup_{j=1}^{M} W_j$,
    - $\mathscr{R}_j(\Omega_j) := \{(x',x_n) \in \mathbb{R}^{n-1} \times \mathbb{R} : x_n < \xi_j(x')\}$, for any $j \in \{1,\dots,M\}$, and
    - $W_j \cap \Omega = W_j \cap \Omega_j$.

See, for example, [141, definition 3.28].

Roughly speaking, a domain with continuous boundary is characterized by being locally the hypograph of a continuous function. Notice, however, that a hypograph is not a domain with a continuous boundary because its boundary is unbounded; hence the cases presented in Definitions 2.1 and 2.4 are similar but conceptually different. We also remark that Definitions 2.1 and 2.4 are also valid when $n = 1$ (using the standard notation for which $\mathbb{R}^0$ reduces to the point $\{0\}$).

Remark 2.5 Simple examples of domains with continuous boundaries are the ball and its complement. More generally, if Ω is a domain with a continuous boundary, then so is $\mathscr{C}\Omega$. We also point out that not all the continuous, oriented, closed surfaces are boundaries of a domain with a continuous boundary, not even continuous closed curves in the plane that are boundaries of open and bounded sets. For instance, the boundary of the Koch snowflake is a closed curve in the plane, but the Koch snowflake itself is not a domain with a continuous boundary.

The main density result for domains with continuous boundaries goes as follows:

Theorem 2.6 *Let Ω be an open subset of $\mathbb{R}^n$ with continuous boundary $\partial\Omega$. Then, for any $u \in X_0^{s,p}(\Omega)$, there exists a sequence $\{\rho_\varepsilon\}_\varepsilon$ in $C_0^\infty(\Omega)$ such that*

$$\|\rho_\varepsilon - u\| \to 0 \quad \text{as } \varepsilon \to 0.$$

In other words, $C_0^\infty(\Omega)$ is a dense subspace of $X_0^{s,p}(\Omega)$.

In the case of the usual fractional Sobolev space $W^{s,p}(\Omega)$, the result given in Theorem 2.6 has already been mentioned in [118, theorem 1.4.2.2]. The detailed proof that we present here has a simple structure: to prove Theorem 2.2, we first translate the function "below" to set its support a bit far from the boundary of the domain; then we mollify the function; and then we cutoff. Some care is needed to check that any of these actions (translations, mollifications, and cutoffs) can be performed by paying only a small error in terms of the norm in (2.1).

Then the proof of Theorem 2.6 is a modification of the proof of Theorem 2.2: near the boundary, we reduce to the case of the hypograph, thanks to Definition 2.4. Far from the boundary, one has space enough to mollify without making the support exit the domain. A suitable partition of unity then glues these two features together.

Remark 2.7 An interesting problem is to determine the "minimal" regularity assumptions on the domain Ω under which the density of the smooth functions, compactly supported in Ω and stated in Theorem 2.6, holds true. We remark that such a property does not hold for *any* domain Ω, not even when $n=1$, as the following counterexample shows:

Example 2.8 Let $\Omega := (-1,0)\cup(0,1)$, $s\in(1/2,1)$, $\psi:\mathbb{R}\to\mathbb{R}$ be any fixed smooth function supported in $(-1,1)$ with $\psi(0)=1$, and define

$$\phi(x) := \begin{cases} \psi(x) & \text{if } x\in\Omega \\ 0 & \text{if } x\notin\Omega. \end{cases}$$

Then, because integrals disregard sets of measure zero, we have that, for any $s\in(0,1)$,

$$\|\phi\|_{H^s(\mathbb{R})} = \|\psi\|_{H^s(\mathbb{R})} < +\infty;$$

hence $\phi\in H^s(\mathbb{R})$. Also, ϕ vanishes outside Ω; that is, $\phi\in X_0^{s,2}(\Omega)$.

Now let η be any smooth function supported in Ω. We have that $\eta(0)=0$, and so, denoting by $f:=\phi-\eta$, by the fractional Sobolev embedding, we obtain that

$$1 = \lim_{\Omega\ni x\to 0} f(x) \le \|f\|_{L^\infty(\Omega)} \le C\|f\|_{H^s(\Omega)} \le C\|\phi-\eta\|_{X_0^{s,2}(\Omega)},$$

where C is a positive constant. Therefore, smooth functions compactly supported in Ω cannot approximate ϕ in $X_0^{s,2}(\Omega)$ (and also in $H^s(\Omega)$).

This example easily generalizes in $W^{s,p}(\Omega)$, with $\Omega := B_1\setminus\{0\}$ and $sp>n$, and also in $H^s(\Omega)$, for $n\ge 2$ and $s>1/2$, where $\Omega := B_1\setminus S$, and S is a closed segment contained in Ω. To achieve this goal, one needs to replace the fractional Sobolev embedding with the trace theory (see, e.g., [83, proposition 3.8]).

2.2 Some preliminary lemmas

This section is devoted to the proof of some preliminary lemmas that will be used in what follows.

2.2.1 Properties of the support

We start with some elementary properties of the support of a function.

Lemma 2.9 *Let u and $v:\mathbb{R}^n\to\mathbb{R}$ be two functions. Then*

$$\operatorname{Supp}(uv)\subseteq(\operatorname{Supp}u)\cap(\operatorname{Supp}v).$$

Proof Let $x \in \operatorname{Supp}(uv)$. We show that $x \in \operatorname{Supp} u$ (the proof that $x \in \operatorname{Supp} v$ is identical). By construction, there exists a sequence $x_k \to x$ in $\mathbb{R}^n$ as $k \to +\infty$ with $u(x_k)v(x_k) \neq 0$. In particular, we have $u(x_k) \neq 0$, so $x_k \in \operatorname{Supp} u$. Since the support is a closed set, we obtain that $x \in \operatorname{Supp} u$, as desired. □

Lemma 2.10 *Let $v : \mathbb{R}^n \to \mathbb{R}$, and $V \subseteq \mathbb{R}^n$ be open. Assume that*

$$\operatorname{Supp} v \subseteq V. \tag{2.3}$$

Then, for any $R > 0$, there exists $a > 0$ such that

$$B_R \cap \big(\operatorname{Supp} v + B_a\big) \subseteq V.$$

The quantity a depends on n, v, R, and V.

Proof Fix $R > 0$. We argue by contradiction, assuming that, for any $k \in \mathbb{N}$, there exists

$$x_k \in B_R \cap \big(\operatorname{Supp} v + B_{1/k}\big) \cap (\mathscr{C}V).$$

Hence, for any $k \in \mathbb{N}$, there exists $b_k \in B_{1/k}$ and $y_k \in \operatorname{Supp} v$ such that $x_k = y_k + b_k$.

We observe that $|x_k| \leq R$, for any $k \in \mathbb{N}$; hence, up to a subsequence, we may suppose that

$$x_k \to P \quad \text{in } \mathbb{R}^n, \tag{2.4}$$

for some $P \in \mathbb{R}^n$, as $k \to +\infty$. Moreover,

$$|y_k - P| \leq |y_k - x_k| + |x_k - P| = |b_k| + |x_k - P| \to 0.$$

Hence,

$$y_k \to P \quad \text{in } \mathbb{R}^n, \tag{2.5}$$

as $k \to +\infty$. We stress that both $\operatorname{Supp} v$ and $\mathscr{C}V$ are closed sets. Hence, because $x_k \in \mathscr{C}V$ and $y_k \in \operatorname{Supp} v$, we deduce from (2.4) and (2.5) that

$$P \in (\operatorname{Supp} v) \cap (\mathscr{C}V).$$

Therefore, by (2.3), we obtain

$$P \in V \cap (\mathscr{C}V) = \emptyset,$$

which is a contradiction. □

The proofs of Theorems 2.2 and 2.6 are mainly based on a basic technique of convolution, joined with a cutoff. Here we will give some properties of these operations with respect to the norm in (2.1). In this section, Ω will denote an open subset of $\mathbb{R}^n$.

2.2.2 Acting by convolution

In what follows, we will denote by $|B_R|$ the Lebesgue measure of the ball B_R. Let $\eta \in C_0^\infty(\mathbb{R}^n)$ be such that $\eta \geq 0$ in $\mathbb{R}^n$ and $\operatorname{Supp} \eta \subseteq B_1$. We also assume that $\int_{B_1} \eta(x)\,dx = 1$. Also, let $\varepsilon > 0$, and let η_ε be the mollifier defined as

$$\eta_\varepsilon(x) := \varepsilon^{-n}\eta(x/\varepsilon) \quad x \in \mathbb{R}^n.$$

Finally, for any $u \in X_0^{s,p}(\Omega)$, we will denote by u_ε the function defined as the convolution between u and η_ε; that is,

$$u_\varepsilon(x) := (u * \eta_\varepsilon)(x) \quad x \in \mathbb{R}^n.$$

Of course, by construction, u_ε is a smooth function; that is, $u_\varepsilon \in C^\infty(\mathbb{R}^n)$. However, if u is supported in Ω, it is not possible, in general, to conclude that $u_\varepsilon \in X_0^{s,p}(\Omega)$ because the support of u_ε may exceed that of u, so it may exit Ω.

For small ε, convolutions do not change the norm (2.1) too much, according to the following result:

Lemma 2.11 *Let $u \in X^{s,p}(\mathbb{R}^n)$. Then*

$$\|u_\varepsilon - u\| \to 0 \quad \text{as } \varepsilon \to 0.$$

Proof Since $u \in L^p(\mathbb{R}^n)$, by [43, theorem IV.22], we know that $\|u_\varepsilon - u\|_{L^p(\mathbb{R}^n)} \to 0$ as $\varepsilon \to 0$. Hence, to get the assertion, it is sufficient to show that

$$\int_{\mathbb{R}^n \times \mathbb{R}^n} |u_\varepsilon(x) - u(x) - u_\varepsilon(y) + u(y)|^p K(x-y)\,dx\,dy \to 0 \tag{2.6}$$

as $\varepsilon \to 0$.

For this, note that, using the definition of u_ε and the Hölder inequality, we obtain

$$\begin{aligned}
&\int_{\mathbb{R}^n \times \mathbb{R}^n} |u_\varepsilon(x) - u(x) - u_\varepsilon(y) + u(y)|^p K(x-y)\,dx\,dy \\
&\quad = \int_{\mathbb{R}^n \times \mathbb{R}^n} K(x-y) \left| \int_{\mathbb{R}^n} \big(u(x-z) - u(y-z)\big)\,\eta_\varepsilon(z)\,dz - u(x) + u(y) \right|^p dx\,dy \\
&\quad = \int_{\mathbb{R}^n \times \mathbb{R}^n} K(x-y) \left| \varepsilon^{-n} \int_{B_\varepsilon} \big(u(x-z) - u(y-z)\big) \right. \qquad (2.7) \\
&\qquad \left. \eta(z/\varepsilon)\,dz - u(x) + u(y) \right|^p dx\,dy = \int_{\mathbb{R}^n \times \mathbb{R}^n} K(x-y) \\
&\qquad \left| \int_{B_1} \big(u(x-\varepsilon\tilde{z}) - u(y-\varepsilon\tilde{z}) - u(x) + u(y)\big)\,\eta(\tilde{z})\,d\tilde{z} \right|^p dx\,dy \leq
\end{aligned}$$

$$
\begin{aligned}
&|B_1|^{p-1}\int_{\mathbb{R}^{2n}}\left(\int_{B_1}|u(x-\varepsilon z)-u(y-\varepsilon z)-u(x)+u(y)|^p K(x-y)\,\eta^p(z)dz\right)dx\,dy\\
&\quad=|B_1|^{p-1}\int_{\mathbb{R}^n\times\mathbb{R}^n\times B_1}|u(x-\varepsilon z)-u(y-\varepsilon z)-u(x)+u(y)|^p\\
&\qquad K(x-y)\,\eta^p(z)dx\,dy\,dz.
\end{aligned}
$$

In the last equality, we used both Tonelli's and Fubini's theorems. Now, by using the continuity in $L^p(\mathbb{R}^n\times\mathbb{R}^n)$ of the translations, as in [106, proposition 8.5], for any $v\in L^p(\mathbb{R}^n\times\mathbb{R}^n)$ and $w\in\mathbb{R}^n\times\mathbb{R}^n$, we have

$$
\int_{\mathbb{R}^n\times\mathbb{R}^n}|v(\widetilde{x}-\varepsilon w)-v(\widetilde{x})|^p\,d\widetilde{x}\to 0
$$

as $\varepsilon\to 0$. Thus, by fixing $z\in B_1$ and choosing $w:=(z,z)\in\mathbb{R}^n\times\mathbb{R}^n$, $\widetilde{x}:=(x,y)\in\mathbb{R}^n\times\mathbb{R}^n$, and

$$
v(x,y):=\big(u(x)-u(y)\big)\,(K(x-y))^{1/p},
$$

which is in $L^p(\mathbb{R}^n\times\mathbb{R}^n)$, because $u\in X^{s,p}(\mathbb{R}^n)$, we obtain

$$
\int_{\mathbb{R}^n\times\mathbb{R}^n}|u(x-\varepsilon z)-u(y-\varepsilon z)-u(x)+u(y)|^p\,K(x-y)dx\,dy\to 0
$$

as $\varepsilon\to 0$. Hence,

$$
\psi_\varepsilon(z):=\eta^p(z)\int_{\mathbb{R}^n\times\mathbb{R}^n}|u(x-\varepsilon z)-u(y-\varepsilon z)-u(x)+u(y)|^p\,K(x-y)dx\,dy\to 0
$$

as $\varepsilon\to 0$.

Moreover, for a.e. $z\in B_1$, taking into account the regularity of η and the fact that $u\in X^{s,p}(\mathbb{R}^n)$, we get

$$
\begin{aligned}
|\psi_\varepsilon(z)|&\le 2^{p-1}\eta^p(z)\left(\int_{\mathbb{R}^n\times\mathbb{R}^n}|u(x-\varepsilon z)-u(y-\varepsilon z)|^p K(x-y)dx\,dy\right.\\
&\qquad\left.+\int_{\mathbb{R}^n\times\mathbb{R}^n}|u(x)-u(y)|^p K(x-y)dx\,dy\right)\\
&=2^p\eta^p(z)\int_{\mathbb{R}^n\times\mathbb{R}^n}|u(x)-u(y)|^p K(x-y)dx\,dy\in L^\infty(B_1),
\end{aligned}
$$

for any $\varepsilon>0$. Hence, by the dominated convergence theorem, we have

$$
\int_{B_1}\psi_\varepsilon(z)dz\to 0
$$

as $\varepsilon\to 0$. This fact, combined with (2.7), gives (2.6), so the assertion of Lemma 2.11 is proved. □

2.2.3 Cutoff technique

In this subsection we discuss the cutoff technique needed for the density argument. For this, for any $j \in \mathbb{N}$, let $\tau_j \in C^\infty(\mathbb{R}^n)$ be a function such that

$$0 \leq \tau_j(x) \leq 1 \quad \text{for any } x \in \mathbb{R}^n, \tag{2.8}$$

$$\tau_j(x) := \begin{cases} 1 & \text{if } x \in B_j \\ 0 & \text{if } x \in \mathscr{C} B_{j+1}, \end{cases} \tag{2.9}$$

and

$$\left|\nabla \tau_j(x)\right| \leq C \quad \text{for any } x \in \mathbb{R}^n, \tag{2.10}$$

where C is a positive constant not depending on j. Here B_j denotes the ball centered at 0 with radius j.

We have the following result:

Lemma 2.12 *Let $u \in X^{s,p}(\mathbb{R}^n)$. Then* $\operatorname{Supp} \tau_j u \subseteq \overline{B_{j+1}} \cap \operatorname{Supp} u$, *and*

$$\|\tau_j u - u\| \to 0 \quad \text{as } j \to 0.$$

Proof The claim on the supports follows from Lemma 2.9 and (2.9). Now let us check that

$$\int_{\mathbb{R}^n} |\tau_j(x)u(x) - u(x)|^p \, dx \to 0 \tag{2.11}$$

as $j \to +\infty$. To this end, we observe that

$$|\tau_j(x)u(x) - u(x)|^p = |\tau_j(x) - 1|^p |u(x)|^p \leq 2^p |u(x)|^p \in L^1(\mathbb{R}^n) \tag{2.12}$$

by (2.8) and the fact that $u \in L^p(\mathbb{R}^n)$. Moreover, thanks to (2.9),

$$|\tau_j(x)u(x) - u(x)|^p \to 0 \quad \text{a.e. in} \quad \mathbb{R}^n \tag{2.13}$$

as $j \to +\infty$. Then, by (2.12), (2.13), and the dominated convergence theorem, we get (2.11).

Now let us show that

$$\int_{\mathbb{R}^n \times \mathbb{R}^n} |\tau_j(x)u(x) - u(x) - \tau_j(y)u(y) + u(y)|^p K(x-y) \, dx \, dy \to 0 \tag{2.14}$$

as $j \to +\infty$. For this, note that

$$\begin{aligned} &|\tau_j(x)u(x) - u(x) - \tau_j(y)u(y) + u(y)| \\ &\quad = |(\tau_j(x) - 1)\,(u(x) - u(y)) - (\tau_j(y) - \tau_j(x))\,u(y)|; \end{aligned}$$

thus,

$$\begin{aligned}\int_{\mathbb{R}^n\times\mathbb{R}^n} & |\tau_j(x)u(x)-u(x)-\tau_j(y)u(y)+u(y)|^p K(x-y)\,dx\,dy \\ & \le 2^{p-1}\left(\int_{\mathbb{R}^n\times\mathbb{R}^n} \left|\tau_j(x)-1\right|^p |u(x)-u(y)|^p K(x-y)\,dx\,dy \right. \\ & \quad \left. + \int_{\mathbb{R}^n\times\mathbb{R}^n} |\tau_j(x)-\tau_j(y)|^p\, |u(y)|^p\, K(x-y)\,dx\,dy\right).\end{aligned} \tag{2.15}$$

Since $u \in X^{s,p}(\mathbb{R}^n)$ and (2.8) holds true, it easily follows that, for any $j \in \mathbb{N}$,

$$\left|\tau_j(x)-1\right|^p |u(x)-u(y)|^p K(x-y) \le 2^p |u(x)-u(y)|^p K(x-y) \in L^1(\mathbb{R}^n\times\mathbb{R}^n),$$

while, by (2.9), we deduce that

$$\left|\tau_j(x)-1\right|^p |u(x)-u(y)|^p K(x-y) \to 0 \quad \text{a.e. in} \quad \mathbb{R}^n\times\mathbb{R}^n$$

as $j \to +\infty$. Hence, by the dominated convergence theorem, we get

$$\int_{\mathbb{R}^n\times\mathbb{R}^n} \left|\tau_j(x)-1\right|^p |u(x)-u(y)|^p K(x-y)\,dx\,dy \to 0 \tag{2.16}$$

as $j \to +\infty$.

Also, by (2.8) and (2.10), we have

$$\begin{aligned} & |\tau_j(x)-\tau_j(y)|^p\, |u(y)|^p\, K(x-y) \\ & \quad \le 2^p \|\tau_j\|^p_{C^1(\mathbb{R}^n)} \min\left\{1, |x-y|^p\right\} |u(y)|^p\, K(x-y) \\ & \quad = 2^p\,(C+1)^p\, m(x-y)\, |u(y)|^p\, K(x-y),\end{aligned} \tag{2.17}$$

where m is the function defined in (1.55). Since, by changing variables, by (1.55), and by the fact that $u \in X^{s,p}(\mathbb{R}^n)$,

$$\int_{\mathbb{R}^n\times\mathbb{R}^n} m(x-y)K(x-y)\,|u(y)|^p\,dx\,dy = \int_{\mathbb{R}^n} m(z)K(z)\,dz \int_{\mathbb{R}^n} |u(y)|^p\,dy < +\infty,$$

we get that the map

$$\mathbb{R}^n\times\mathbb{R}^n \ni (x,y) \mapsto m(x-y)K(x-y)\,|u(y)|^p \tag{2.18}$$

belongs to $L^1(\mathbb{R}^n\times\mathbb{R}^n)$. Thus, by this, (2.17), the fact that

$$|\tau_j(x)-\tau_j(y)|^p\, |u(y)|^p\, K(x-y) \to 0 \quad \text{a.e. in } \mathbb{R}^n\times\mathbb{R}^n$$

as $j \to +\infty$, and again by the dominated convergence theorem, we obtain

$$\int_{\mathbb{R}^n \times \mathbb{R}^n} |\tau_j(x) - \tau_j(y)|^p \, |u(y)|^p \, K(x-y)\, dx\, dy \to 0 \tag{2.19}$$

as $j \to +\infty$. By (2.15), (2.16), and (2.19), we get (2.14). This and (2.11) end the proof. □

2.2.4 *Effect of the translations*

Here we study the effect of the translations on the norm in (2.1). For any $\delta > 0$ and any function u, we set

$$u_\delta(x) := u(x', x_n + \delta) \quad x \in \mathbb{R}^n.$$

Then we have the following:

Lemma 2.13 *Let $u \in X^{s,p}(\mathbb{R}^n)$. Then*

$$\|u_\delta - u\| \to 0 \quad \text{as } \delta \to 0.$$

Proof We can argue as in the final part of the proof of Lemma 2.11: we prefer to give a brief explanation for the reader's convenience.

First of all, the continuity of the translations in $L^p(\mathbb{R}^n)$, as in [106, proposition 8.5], and the fact that $u \in L^p(\mathbb{R}^n)$ give that

$$\int_{\mathbb{R}^n} |u_\delta(x) - u(x)|^p \, dx = \int_{\mathbb{R}^n} |u(x', x_n + \delta) - u(x', x_n)|^p \, dx \to 0$$

as $\delta \to 0$; that is,

$$\|u_\delta - u\|_{L^p(\mathbb{R}^n)} \to 0 \quad \text{as } \delta \to 0. \tag{2.20}$$

Finally, let us show that

$$\int_{\mathbb{R}^n \times \mathbb{R}^n} |u_\delta(x) - u(x) - u_\delta(y) + u(y)|^p K(x-y)\, dx\, dy \to 0 \tag{2.21}$$

as $\delta \to 0$. Again, the continuity of the translations in $L^p(\mathbb{R}^n \times \mathbb{R}^n)$ yields that, for any $v \in L^p(\mathbb{R}^n \times \mathbb{R}^n)$ and $w \in \mathbb{R}^n \times \mathbb{R}^n$,

$$\int_{\mathbb{R}^n \times \mathbb{R}^n} |v(\widetilde{x} + \delta w) - v(\widetilde{x})|^p \, d\widetilde{x} \to 0$$

as $\delta \to 0$. Here we denote with $e_n = (\overline{0}, 1)$, where $\overline{0} = (0, \ldots, 0) \in \mathbb{R}^{n-1}$. Then, by choosing $w := (e_n, e_n) \in \mathbb{R}^n \times \mathbb{R}^n$, $\widetilde{x} := (x, y) \in \mathbb{R}^n \times \mathbb{R}^n$, and

$$v(x, y) := \big(u(x) - u(y)\big)\, (K(x-y))^{1/p},$$

which is in $L^p(\mathbb{R}^n \times \mathbb{R}^n)$ since $u \in X^{s,p}(\mathbb{R}^n)$, we obtain (2.21). Of course, (2.20) and (2.21) conclude the proof. □

The next lemma says that translations are useful to set the support away from the boundary of a hypograph.

Lemma 2.14 *Let Ω be a hypograph. Let $u : \mathbb{R}^n \to \mathbb{R}$ be such that*[1] $u = 0$ *in* $\mathscr{C}\Omega$. *Then*

$$\operatorname{Supp} u_\delta \subseteq \Omega. \tag{2.22}$$

More precisely, given any $R > 0$, there exists $a > 0$ such that

$$B_R \cap \left(\operatorname{Supp} u_\delta + B_a\right) \subseteq \Omega. \tag{2.23}$$

The quantity a depends only on n, u, δ, R, and Ω (say, $a := a(n,u,\delta,R,\Omega)$).

Proof The proof is an elementary set inclusion joined with a compactness argument. We provide full details for the convenience of the reader.

First, we prove (2.22). For this, we take $x \in \operatorname{Supp} u_\delta$. Then there exists a sequence of points x_k such that $x_k \to x$ in $\mathbb{R}^n$ as $k \to +\infty$ and

$$0 \neq u_\delta(x_k) = u(x'_k, x_{k,n} + \delta).$$

Therefore, $(x'_k, x_{k,n} + \delta) \in \Omega$ because $u = 0$ in $\mathscr{C}\Omega$ by assumption. Hence, by Definition 2.1, we have that $x_{k,n} + \delta < \xi(x'_k)$. Passing to the limit as $k \to +\infty$ and using the continuity of ξ, we obtain that $x_n + \delta \le \xi(x')$. In particular, since $\delta > 0$, we get that $x_n < \xi(x')$, so $x \in \Omega$. This proves (2.22).

Then (2.23) follows from (2.22) and Lemma 2.10. □

2.3 Proof of Theorem 2.2 (and of Remark 2.3)

Let Ω be as in Theorem 2.2, and let $u \in X_0^{s,p}(\Omega)$. By possibly changing u in a set of zero measure, we suppose that

$$u = 0 \quad \text{in } \mathscr{C}\Omega. \tag{2.24}$$

Let us fix $\sigma > 0$. By Lemma 2.13, we know that there exists a measurable function u_δ such that

$$\|u_\delta - u\| < \frac{\sigma}{3}, \tag{2.25}$$

for δ sufficiently small, say, $\delta \le \bar{\delta}$, with $\bar{\delta} = \bar{\delta}(\sigma) > 0$.

[1] We point out that if $u \in X_0^{s,p}(\Omega)$, one cannot say that $\operatorname{Supp} u \subseteq \overline{\Omega}$: this is only true by possibly modifying u in a set of measure zero. Indeed, as an example, one can take Ω to be an open set in $\mathbb{R}^n$ and u to be the characteristic function of $\mathbb{Q}^n$. In this case, $\operatorname{Supp} u = \mathbb{R}^n$, but, of course, u is equivalent to the zero function, up to sets of measure zero, so $u \in X_0^{s,p}(\Omega)$. This is the reason for which we assume that $u = 0$ in $\mathscr{C}\Omega$.

Now let us fix $\delta = \overline{\delta}$ and let τ_j be as in Subsection 2.2.3. By Lemma 2.12, we know that

$$\|\tau_j u_{\overline{\delta}} - u_{\overline{\delta}}\| < \frac{\sigma}{3}, \tag{2.26}$$

for j large enough, say, $j \geq \overline{\imath}$, with $\overline{\imath} = \overline{\imath}(\sigma) \in \mathbb{N}$.

For any $\varepsilon > 0$, let us consider

$$\rho_\varepsilon := \tau_{\overline{\imath}} u_{\overline{\delta}} * \eta_\varepsilon,$$

where η_ε is the function defined in Subsection 2.2.2. Of course, $\rho_\varepsilon \in C^\infty(\mathbb{R}^n)$ by construction. Furthermore, by the standard properties of the convolution (see, e.g., [43, proposition IV.18]), we have that

$$\operatorname{Supp} \rho_\varepsilon \subseteq \operatorname{Supp} \tau_{\overline{\imath}} u_{\overline{\delta}} + \overline{B}_\varepsilon. \tag{2.27}$$

However, by Lemma 2.12, we get that

$$\operatorname{Supp} \tau_{\overline{\imath}} u_{\overline{\delta}} \subseteq \overline{B_{\overline{\imath}+1}} \cap \operatorname{Supp} u_{\overline{\delta}}. \tag{2.28}$$

Now we claim that

$$\operatorname{Supp} \rho_\varepsilon \subseteq B_{\overline{\imath}+2} \cap \left(\operatorname{Supp} u_{\overline{\delta}} + B_{2\varepsilon}\right) \tag{2.29}$$

if ε is sufficiently small (possibly in dependence on σ). Indeed, let $P \in \operatorname{Supp} \rho_\varepsilon$. Then, by (2.27), there exists $Q \in \operatorname{Supp} \tau_{\overline{\imath}} u_{\overline{\delta}}$ such that $|P - Q| \leq \varepsilon < 2\varepsilon$. Hence, by (2.28), we know that $|Q| \leq \overline{\imath} + 1$ and $Q \in \operatorname{Supp} u_{\overline{\delta}}$. In particular, $|P| \leq |Q| + |P - Q| \leq \overline{\imath} + 1 + 2\varepsilon < \overline{\imath} + 2$, if ε is small enough, thus proving (2.29).

From (2.23) and (2.29), we deduce that $\operatorname{Supp} \rho_\varepsilon$ is compact and contained in Ω, as long as ε is small enough, say, $2\varepsilon < a(n, u, \overline{\delta}, \overline{\imath} + 2, \Omega)$ in the notation of Lemma 2.14 (we also remark that it is possible to use Lemma 2.14 here in virtue of the normalization performed in (2.24)). As a consequence of this,

$$\rho_\varepsilon \in C_0^\infty(\Omega),$$

for ε small enough.

Furthermore, by Lemma 2.11,

$$\|\rho_\varepsilon - \tau_{\overline{\imath}} u_{\overline{\delta}}\| < \frac{\sigma}{3}, \tag{2.30}$$

for ε small, say, $\varepsilon \leq \overline{\varepsilon}$, where $\overline{\varepsilon} = \overline{\varepsilon}(\sigma) > 0$.

Finally, thanks to (2.25), (2.26), and (2.30), we get

$$\begin{aligned}
\|u - \rho_\varepsilon\| &\leq \|u - u_{\overline{\delta}}\| + \|u_{\overline{\delta}} - \tau_{\overline{\imath}} u_{\overline{\delta}}\| + \|\tau_{\overline{\imath}} u_{\overline{\delta}} - \rho_\varepsilon\| \\
&< \frac{\sigma}{3} + \frac{\sigma}{3} + \frac{\sigma}{3} = \sigma.
\end{aligned}$$

The arbitrariness of σ concludes the proof of Theorem 2.2.

As for the proof of Remark 2.3, we argue as follows: by construction,

$$\operatorname{Supp} u_{\bar{\delta}} \subseteq \operatorname{Supp} u + B_{2\bar{\delta}}.$$

This and (2.29) yield that

$$\begin{aligned}\operatorname{Supp} \rho_\varepsilon &\subseteq \operatorname{Supp} u_{\bar{\delta}} + B_{2\varepsilon} \\ &\subseteq \operatorname{Supp} u + B_{2\bar{\delta}} + B_{2\varepsilon} \\ &\subseteq \operatorname{Supp} u + B_{2(\varepsilon+\bar{\delta})},\end{aligned}$$

thus checking Remark 2.3.

2.4 Proof of the main result

Roughly speaking, the idea for proving Theorem 2.6 is to use an appropriate partition of unity in order to reduce locally to the case of a hypograph and thus use Theorem 2.2. Here are the details of the argument.

Let Ω be as in Theorem 2.6. Since $\partial\Omega$ is compact, we may suppose that it is contained in B_R, for some $R > 0$. By possibly replacing W_j with $W_j \cap B_{R+1}$, we may also suppose that

$$\textit{each } W_j \textit{ is a bounded set contained in } B_{R+1}. \tag{2.31}$$

Hence, the set

$$W := W_1 \cup \cdots \cup W_M \tag{2.32}$$

is bounded and contained in B_{R+1}.

First of all, we claim that

$$\rho_0 := \inf_{\substack{x\in\mathbb{R}^n\setminus W\\ y\in\partial\Omega}} |x-y| > 0. \tag{2.33}$$

To prove this, we argue by contradiction, and we assume that there exist $\{x_j\}_{j\in\mathbb{N}}$ in $\mathbb{R}^n \setminus W$ and $\{y_j\}_{j\in\mathbb{N}}$ in $\partial\Omega$ such that $|x_j - y_j| \to 0$ in $\mathbb{R}$ as $j \to +\infty$. Since $\partial\Omega$ is compact, we may suppose that $y_j \to y_\infty \in \partial\Omega$ as $j \to +\infty$. Moreover, $|x_j| \leq |x_j - y_j| + |y_j| \leq 1 + R$, for large j; hence, we may also suppose that $x_j \to x_\infty$, for some $x_\infty \in \mathbb{R}^n$. By construction and thanks to the continuity of $|\cdot|$, we get

$$|x_\infty - y_\infty| = \lim_{j\to+\infty} |x_j - y_j| = 0,$$

so $x_\infty = y_\infty \in \partial\Omega$. Hence, up to renaming the covering sets, we may suppose that $x_\infty = y_\infty \in W_1$. Since W_1 is open, we have that both x_j and y_j must lie in W_1, for j sufficiently large. But this is in contradiction to the fact that $x_j \in \mathbb{R}^n \setminus W$, so (2.33) is proved.

Thus, we define

$$W_{M+1} := \{x \in \Omega \,:\, |x-y| > \rho_0/2 \text{ for any } y \in \partial\Omega\}. \tag{2.34}$$

Notice that W_{M+1} may be empty (in such case, one can simply neglect the index $M+1$ from now on). We claim that

$$\Omega \subseteq \bigcup_{j=1}^{M+1} W_j. \tag{2.35}$$

To prove it, we argue again by contradiction. Suppose that there exists $x \in \Omega \setminus \bigcup_{j=1}^{M+1} W_j$. In particular,

$$x \notin W \quad \text{and} \quad x \notin W_{M+1}.$$

Then, by the definition of W_{M+1}, there exists $y \in \partial\Omega$ such that

$$|x-y| \leq \rho_0/2.$$

But also $x \in \mathbb{R}^n \setminus W$, so we can use (2.33) and say that $|x-y| \geq \rho_0 > \rho_0/2$. This is a contradiction; hence, (2.35) is proved.

Now we take a partition of unity $\{\alpha_j\}_{j\in\{1,\dots,M+1\}}$ subordinate to the collection of sets $\{W_1,\dots,W_{M+1}\}$; that is, $\alpha_j : \mathbb{R}^n \to [0,1]$ is smooth, with

$$\operatorname{Supp} \alpha_j \subseteq W_j \quad \text{for } j \in \{1,\dots,M\}, \tag{2.36}$$

$$\operatorname{Supp} \alpha_{M+1} \subseteq W_{M+1} + B_{\rho_0/8}, \tag{2.37}$$

and

$$\sum_{j=1}^{M+1} \alpha_j(x) = 1 \quad \text{for any } x \in \Omega. \tag{2.38}$$

Of course, in this partition of unity, the index $M+1$ must be disregarded when $W_{M+1} = \emptyset$. Because this partition of unity is not completely standard, we give the details of its construction in Section 2.5.

Now we claim that there exists $c > 0$ such that, for any $j \in \{1,\dots,M\}$,

$$\operatorname{Supp} \alpha_j + B_c \subseteq W_j. \tag{2.39}$$

To this purpose, we use (2.36) and Lemma 2.10 to see that, for any $j \in \{1,\dots,M\}$, there exists $c_j \in (0,1)$ such that

$$B_{R+2} \cap \left(\operatorname{Supp} \alpha_j + B_{c_j}\right) \subseteq W_j. \tag{2.40}$$

Here R is the one given in (2.31). Thus, we notice that, by (2.36) and (2.31),

$$\operatorname{Supp} \alpha_j + B_{c_j} \subseteq W_j + B_{c_j} \subseteq B_{R+1} + B_1 \subseteq B_{R+2},$$

so

$$B_{R+2} \cap \left(\operatorname{Supp} \alpha_j + B_{c_j}\right) = \operatorname{Supp} \alpha_j + B_{c_j}.$$

Accordingly, (2.40) becomes

$$\operatorname{Supp}\alpha_j + B_{c_j} \subseteq W_j.$$

Because we are dealing with a finite number of indices j, we can set

$$c := \min\{c_1, \dots, c_M\} > 0$$

and complete the proof of (2.39).

Now let $u \in X_0^{s,p}(\Omega)$ be as in Theorem 2.6. For any $j \in \{1, \dots, M+1\}$, define

$$u_j := \alpha_j u.$$

By (2.38), one has

$$\sum_{j=1}^{M+1} u_j(x) = \sum_{j=1}^{M+1} \alpha_j(x) u(x) = u(x) \sum_{j=1}^{M+1} \alpha_j(x) = u(x),$$

for a.e. $x \in \Omega$. Moreover, we get

$$u(x) = 0 = u(x) \sum_{j=1}^{M+1} \alpha_j(x) = \sum_{j=1}^{M+1} u_j(x),$$

a.e. $x \in \mathscr{C}\Omega$. All in all,

$$u(x) = \sum_{j=1}^{M+1} u_j(x), \tag{2.41}$$

for a.e. $x \in \mathbb{R}^n$.

Now let us fix $\sigma > 0$ and $j \in \{1, \dots, M\}$. We note that

$$u_j \in X^{s,p}(\mathbb{R}^n) \tag{2.42}$$

because so is u (recall that α_j is smooth and compactly supported, by (2.36) and (2.31)). We check that

$$u_j = 0 \quad \text{a.e. in } \mathscr{C}\Omega_j, \tag{2.43}$$

where Ω_j is the set given in Definition 2.4. To this purpose, we notice that, by the properties of Ω_j and (2.36),

$$\Omega_j \supseteq \Omega_j \cap W_j = \Omega \cap W_j \supseteq \Omega \cap \operatorname{Supp}\alpha_j.$$

As a consequence,

$$\mathscr{C}\Omega_j \subseteq \mathscr{C}\big(\Omega \cap \operatorname{Supp}\alpha_j\big) = (\mathscr{C}\Omega) \cup (\mathscr{C}\operatorname{Supp}\alpha_j). \tag{2.44}$$

Now let $x \in \mathscr{C}\Omega_j$. By (2.44), we obtain that either $x \in \mathscr{C}\Omega$ (so $u(x) = 0$, up to neglecting a set of zero measure) or $x \in \mathscr{C}\operatorname{Supp}\alpha_j$ (so $\alpha_j(x) = 0$). This says that $u_j(x) = \alpha_j(x) u(x) = 0$ (up to a set of zero measure) and proves (2.43).

By (2.42) and (2.43), we have that $u_j \in X_0^{s,p}(\Omega_j)$, for any $j \in \{1,\dots,M\}$. Thus, we can use Theorem 2.2 and obtain a function

$$v_j \in C_0^\infty(\Omega_j) \tag{2.45}$$

such that

$$\|v_j - u_j\| \le \frac{\sigma}{2M}, \tag{2.46}$$

for any $j \in \{1,\dots,M\}$. Additionally, by Remark 2.3, we may choose v_j such that

$$\operatorname{Supp} v_j \subseteq \operatorname{Supp} u_j + B_c, \tag{2.47}$$

where $c > 0$ is the positive constant fixed in (2.39). Also, by Lemma 2.9, we have that $\operatorname{Supp} u_j \subseteq \operatorname{Supp} \alpha_j$. Using this, (2.39), and (2.47), we obtain

$$\operatorname{Supp} v_j \subseteq \operatorname{Supp} \alpha_j + B_c \subseteq W_j.$$

More precisely, recalling (2.45), we have

$$\operatorname{Supp} v_j \subseteq W_j \cap \Omega_j = W_j \cap \Omega \subseteq \Omega,$$

so

$$v_j \in C_0^\infty(\Omega). \tag{2.48}$$

The reader may appreciate the difference between (2.45) and (2.48).

Now let us consider u_{M+1}. We recall (2.34), and we obtain that

$$W_{M+1} + B_{\rho_0/4} \subseteq \Omega. \tag{2.49}$$

Thus, we define the mollification $u_{M+1,\varepsilon} := u_{M+1} * \eta_\varepsilon$. Using once more the standard properties of the convolution, Lemma 2.9, and (2.37), we conclude that

$$\begin{aligned}
\operatorname{Supp} u_{M+1,\varepsilon} &\subseteq \operatorname{Supp} u_{M+1} + \overline{B}_\varepsilon \\
&= \operatorname{Supp}(\alpha_{M+1} u) + \overline{B}_\varepsilon \\
&\subseteq \big(\operatorname{Supp} \alpha_{M+1} \cap \operatorname{Supp} u\big) + \overline{B}_\varepsilon \\
&\subseteq \big((W_{M+1} + B_{\rho_0/8}) \cap \operatorname{Supp} u\big) + \overline{B}_\varepsilon \\
&\subseteq (W_{M+1} + B_{\rho_0/8}) + \overline{B}_\varepsilon.
\end{aligned}$$

Therefore, recalling (2.49), we obtain that

$$\operatorname{Supp} u_{M+1,\varepsilon} \subseteq \Omega \tag{2.50}$$

if ε is sufficiently small, say, $\varepsilon < \rho_0/8$. Furthermore, by Lemma 2.11, we have that

$$\|u_{M+1,\varepsilon} - u_{M+1}\| \le \frac{\sigma}{4} \tag{2.51}$$

as long as ε is sufficiently small (possibly in dependence of σ).

We stress that (2.50) is not enough to ensure that $u_{M+1,\varepsilon}$ is compactly supported because Ω is not necessarily bounded (and neither is W_{M+1}). Thus, now we need to perform a further cutoff argument by defining

$$u_{M+1,\varepsilon,J} := \tau_J\, u_{M+1,\varepsilon},$$

with τ_J as in (2.8). Thus, by Lemma 2.12 and (2.50), we obtain that

$$\operatorname{Supp} u_{M+1,\varepsilon,J} \subseteq \overline{B_{J+1}} \cap \operatorname{Supp} u_{M+1,\varepsilon} \subseteq \overline{B_{J+1}} \cap \Omega;$$

hence,

$$u_{M+1,\varepsilon,J} \in C_0^\infty(\Omega) \tag{2.52}$$

and

$$\|u_{M+1,\varepsilon,J} - u_{M+1,\varepsilon}\| \le \frac{\sigma}{4}$$

if J is large enough (possibly depending on σ). This and (2.51) give that

$$\|u_{M+1,\varepsilon,J} - u_{M+1}\| \le \frac{\sigma}{2}. \tag{2.53}$$

Now we define

$$\rho_\sigma := \sum_{j=1}^{M} v_j + u_{M+1,\varepsilon,J}.$$

By (2.48) and (2.52), we know that

$$\rho_\sigma \in C_0^\infty(\Omega).$$

Furthermore, using (2.41), (2.46), and (2.53), we conclude that

$$\begin{aligned}
\|u - \rho_\sigma\| &= \Big\| \sum_{j=1}^{M+1} u_j - \sum_{j=1}^{M} v_j - u_{M+1,\varepsilon,J} \Big\| \\
&\le \sum_{j=1}^{M} \|u_j - v_j\| + \|u_{M+1} - u_{M+1,\varepsilon,J}\| \\
&\le \sum_{j=1}^{M} \frac{\sigma}{2M} + \frac{\sigma}{2} \\
&= \sigma.
\end{aligned}$$

The arbitrariness of σ ends the proof of Theorem 2.6.

2.5 Note on the partition of unity

In the last part of this chapter we show how to construct, under the assumption of Theorem 2.6, the partition of unity $\{\alpha_j\}_{j\in\{1,\dots,M+1\}}$ subordinate to the collection of sets $\{W_1, \dots, W_{M+1}\}$ and satisfying conditions (2.36)–(2.38).

To this end, we set

$$F := \overline{(\partial\Omega) + B_{3\rho_0/4}},$$

where ρ_0 is the positive constant given in (2.33). Since $\partial\Omega$ is bounded, so is F, which turns out to be compact. Also, we note that

$$F \subseteq W_1 \cup \cdots \cup W_M. \tag{2.54}$$

To prove this, we argue by contradiction: if not, there exists $x_* \in F \setminus (W_1 \cup \cdots \cup W_M) = F \setminus W$. Thus, since $x_* \in F$, there exists a sequence $\{x_k\}_{k\in\mathbb{N}}$ in $(\partial\Omega) + B_{3\rho_0/4}$ such that $x_k \to x_*$ in $\mathbb{R}^n$ as $k \to +\infty$. Hence, there exist $y_k \in \partial\Omega$ and $b_k \in B_{3\rho_0/4}$ such that $x_k = y_k + b_k$, for any $k \in \mathbb{N}$. Up to subsequences, since Ω is compact, we may suppose that $y_k \to y_* \in \partial\Omega$ as $k \to +\infty$. Thus,

$$|x_* - y_*| = \lim_{k\to+\infty} |x_k - y_k| = \lim_{k\to+\infty} |b_k| \leq \frac{3\rho_0}{4}. \tag{2.55}$$

However, $x_* \in \mathbb{R}^n \setminus W$; hence, by (2.33), one has

$$|x_* - y_*| \geq \inf_{\substack{x\in\mathbb{R}^n\setminus W\\ y\in\partial\Omega}} |x - y| = \rho_0 > \frac{3\rho_0}{4}.$$

This is in contradiction with (2.55), so (2.54) is proved.

Thanks to (2.54) and the fact that F is compact, we can now perform the covering argument used in the proof of [4, theorem 1.9]. Accordingly, there exist compact sets $C_j \subseteq W_j$ and functions $\xi_j \in C_0^\infty(W_j)$, with values in $[0,1]$, such that

$$F \subseteq C_1 \cup \cdots \cup C_M \tag{2.56}$$

and

$$\xi_j(x) = 1 \quad \text{for all } x \in C_j, \tag{2.57}$$

for any $j = 1, \ldots, M$.

Now let us consider the set W_{M+1} defined in (2.34). We define

$$\tilde{W}_{M+1} := W_{M+1} + B_{\rho_0/16}, \tag{2.58}$$

and we consider its characteristic function $\chi_{\tilde{W}_{M+1}}$ and its mollification ξ_{M+1} as described in Section 2.2.2; that is,

$$\xi_{M+1} := \eta_\varepsilon * \chi_{\tilde{W}_{M+1}},$$

for a fixed ε that we now specify. For this, we notice that ξ_{M+1} is smooth and

$$\operatorname{Supp} \xi_{M+1} \subseteq \tilde{W}_{M+1} + \overline{B_\varepsilon} \subseteq W_{M+1} + B_{\rho_0/8}$$

as long as ε is fixed suitably small, say, $\varepsilon < \rho_0/16$. Furthermore, taking into account this choice of ε and (2.58), we get

$$\begin{aligned}\xi_{M+1}(x) &= \int_{B_\varepsilon} \chi_{\tilde{W}_{M+1}}(x-y)\,\eta_\varepsilon(y)\,dy \\ &= \int_{B_\varepsilon} \eta_\varepsilon(y)\,dy \\ &= 1,\end{aligned} \tag{2.59}$$

for any $x \in W_{M+1}$.

Now we set $\alpha_1 := \xi_1$ and, recursively,

$$\alpha_j := (1-\xi_1)\cdots(1-\xi_{j-1})\xi_j,$$

for any $j \in \{2,\dots,M+1\}$. By construction, for any $j = 1,\dots,M+1$, the functions α_j are smooth, with values in $[0,1]$, and, by Lemma 2.9,

$$\operatorname{Supp}\alpha_j \subseteq \operatorname{Supp}\xi_j,$$

which implies (2.36) and (2.37), thanks to the properties of ξ_j.

It remains to prove (2.38). For this, we claim that, for any $\ell \in \{1,\dots,M+1\}$,

$$\sum_{j=1}^{\ell} \alpha_j = 1-(1-\xi_1)\cdots(1-\xi_\ell). \tag{2.60}$$

This can be checked by induction. Indeed, (2.60) is obvious when $\ell = 1$. Now suppose that (2.60) holds for some ℓ; then

$$\begin{aligned}\sum_{j=1}^{\ell+1} \alpha_j &= 1-(1-\xi_1)\cdots(1-\xi_\ell)+\alpha_{\ell+1} \\ &= 1-(1-\xi_1)\cdots(1-\xi_\ell)+(1-\xi_1)\cdots(1-\xi_\ell)\xi_{\ell+1} \\ &= 1-(1-\xi_1)\cdots(1-\xi_\ell)(1-\xi_{\ell+1}),\end{aligned}$$

thus completing the inductive step and proving (2.60).

With this, we can check (2.38) by arguing as follows: fix $x \in \Omega$. If $x \in F$, we use (2.56) to see that $x \in C_{j_o}$, for some $j_0 \in \{1,\dots,M\}$. Thus, by (2.57), we get that $\xi_{j_0}(x) = 1$ and therefore

$$(1-\xi_1(x))\cdots(1-\xi_{M+1}(x)) = 0.$$

Consequently, by (2.60), we see that

$$\sum_{j=1}^{M+1} \alpha_j(x) = 1-(1-\xi_1(x))\cdots(1-\xi_{M+1}(x)) = 1.$$

This proves (2.38) when $x \in F$.

Finally, we check (2.38) when $x \in \Omega \setminus F$. For this, we notice that, in this case,

$$|x - y| \geq \frac{3\rho_0}{4} \quad \text{for any } y \in \partial\Omega. \tag{2.61}$$

Indeed, if (2.61) were false, there would exist $y \in \partial\Omega$ such that $|x - y| < 3\rho_0/4$. Thus, one sets $b := x - y \in B_{3\rho_0/4}$ and obtains that $x = y + b \in (\partial\Omega) + B_{3\rho_0/4} \subseteq F$, against our assumption. This proves (2.61).

In turn, (2.61) and (2.34) imply that $x \in W_{M+1}$, and therefore, by (2.59), we obtain that $\xi_{M+1}(x) = 1$. Hence, by (2.60), we conclude that

$$\sum_{j=1}^{M+1} \alpha_j(x) = 1 - (1 - \xi_1(x)) \cdots (1 - \xi_{M+1}(x)) = 1.$$

This proves (2.38) also when $x \in \Omega \setminus F$, so the construction of the desired partition of unity is complete.

3

An eigenvalue problem

Study of the eigenvalues of a linear operator is a classical topic, and many functional analytic tools of general flavor may be used to deal with it. Moving along in this direction, this chapter focuses on the following eigenvalue problem:

$$\begin{cases} -\mathscr{L}_K u = \lambda u & \text{in } \Omega \\ u = 0 & \text{in } \mathbb{R}^n \setminus \Omega, \end{cases} \tag{3.1}$$

where $s \in (0,1)$, $n > 2s$, Ω is an open, bounded subset of $\mathbb{R}^n$, $K : \mathbb{R}^n \setminus \{0\} \to (0, +\infty)$ is a function satisfying (1.55) and (1.56), and $\mathscr{L}_K$ is the integrodifferential operator defined in (1.54).

More precisely, here we discuss the weak formulation of (3.1), which consists of the following eigenvalue problem:

$$\begin{cases} \displaystyle\int_{\mathbb{R}^n \times \mathbb{R}^n} (u(x) - u(y))(\varphi(x) - \varphi(y))K(x-y)\,dx\,dy = \lambda \int_\Omega u(x)\varphi(x)\,dx \\ \qquad\qquad\qquad\qquad\qquad\qquad\qquad\qquad\qquad \forall\, \varphi \in X_0^s(\Omega) \\ u \in X_0^s(\Omega). \end{cases} \tag{3.2}$$

We recall that $\lambda \in \mathbb{R}$ is an eigenvalue of $-\mathscr{L}_K$ provided that there exists a non-trivial solution $u \in X_0^s(\Omega)$ of problem (3.1) – in fact, of its weak formulation (3.2) – and, in this case, any solution will be called an *eigenfunction* corresponding to the eigenvalue λ.

The results on the spectrum of $-\mathscr{L}_K$ that we give here are more general and more precise than what we need, strictly speaking, for the proofs performed in this book; nevertheless, we believe that it is good to have a result stated in detail with complete proofs, as well as for further reference. This chapter is based on the papers [104, 200, 201].

3.1 Eigenvalues and eigenfunctions of $-\mathscr{L}_K$

In this section, we give the following characterization on the eigenvalues and eigenfunctions of the nonlocal operator $-\mathscr{L}_K$ (see [200]):

Proposition 3.1 Let $s \in (0,1)$, $n > 2s$, Ω be an open, bounded subset of $\mathbb{R}^n$, and let $K : \mathbb{R}^n \setminus \{0\} \to (0,+\infty)$ be a function satisfying assumptions (1.55) and (1.56). Then

(a) Problem (3.2) admits an eigenvalue λ_1 that is positive and that can be characterized as follows:

$$\lambda_1 = \min_{\substack{u \in X_0^s(\Omega) \\ \|u\|_{L^2(\Omega)}=1}} \int_{\mathbb{R}^n \times \mathbb{R}^n} |u(x)-u(y)|^2 K(x-y)\,dx\,dy \tag{3.3}$$

or, equivalently,

$$\lambda_1 = \min_{u \in X_0^s(\Omega)\setminus\{0\}} \frac{\displaystyle\int_{\mathbb{R}^n \times \mathbb{R}^n} |u(x)-u(y)|^2 K(x-y)\,dx\,dy}{\displaystyle\int_{\Omega} |u(x)|^2\,dx}. \tag{3.4}$$

(b) There exists a nonnegative[1] function $e_1 \in X_0^s(\Omega)$ that is an eigenfunction corresponding to λ_1, attaining the minimum in (3.3); that is, $\|e_1\|_{L^2(\Omega)} = 1$ and

$$\lambda_1 = \int_{\mathbb{R}^n \times \mathbb{R}^n} |e_1(x)-e_1(y)|^2 K(x-y)\,dx\,dy. \tag{3.5}$$

(c) λ_1 is simple; that is, if $u \in X_0^s(\Omega)$ is a solution of the equation

$$\int_{\mathbb{R}^n \times \mathbb{R}^n} (u(x)-u(y))(\varphi(x)-\varphi(y))K(x-y)\,dx\,dy = \lambda_1 \int_{\Omega} u(x)\varphi(x)\,dx, \tag{3.6}$$

for every $\varphi \in X_0^s(\Omega)$, then $u = \zeta e_1$, with $\zeta \in \mathbb{R}$.

(d) The set of the eigenvalues of problem (3.2) consists of a sequence $\{\lambda_k\}_{k\in\mathbb{N}}$ with[2]

$$0 < \lambda_1 < \lambda_2 \le \cdots \le \lambda_k \le \lambda_{k+1} \le \ldots \tag{3.7}$$

and

$$\lambda_k \to +\infty \quad \text{as } k \to +\infty. \tag{3.8}$$

Moreover, for any $k \in \mathbb{N}$, the eigenvalues can be characterized as follows:

$$\lambda_{k+1} = \min_{\substack{u \in \mathbb{P}_{k+1} \\ \|u\|_{L^2(\Omega)}=1}} \int_{\mathbb{R}^n \times \mathbb{R}^n} |u(x)-u(y)|^2 K(x-y)\,dx\,dy \tag{3.9}$$

[1] In the forthcoming Corollary 4.8, we will prove that in the model case of the fractional Laplacian, the first eigenvalue of $(-\Delta)^s$ is strictly positive in Ω.

[2] As usual, here we call λ_1 the *first eigenvalue* of the operator $-\mathscr{L}_K$. This notation is justified by (3.7). Notice also that some of the eigenvalues in the sequence $\{\lambda_k\}_{k\in\mathbb{N}}$ may repeat; that is, the inequalities in (3.7) may be not always strict.

or, equivalently,

$$\lambda_{k+1} = \min_{u\in\mathbb{P}_{k+1}\setminus\{0\}} \frac{\displaystyle\int_{\mathbb{R}^n\times\mathbb{R}^n} |u(x)-u(y)|^2 K(x-y)\,dx\,dy}{\displaystyle\int_\Omega |u(x)|^2\,dx}, \tag{3.10}$$

where

$$\mathbb{P}_{k+1} := \big\{u \in X_0^s(\Omega) : \langle u, e_j\rangle_{X_0^s(\Omega)} = 0 \quad \forall j = 1,\dots,k\big\}. \tag{3.11}$$

(e) For any $k \in \mathbb{N}$, there exists a function $e_{k+1} \in \mathbb{P}_{k+1}$ that is an eigenfunction corresponding to λ_{k+1}, attaining the minimum in (3.9); that is, $\|e_{k+1}\|_{L^2(\Omega)} = 1$ and

$$\lambda_{k+1} = \int_{\mathbb{R}^n\times\mathbb{R}^n} |e_{k+1}(x) - e_{k+1}(y)|^2 K(x-y)\,dx\,dy. \tag{3.12}$$

(f) The sequence $\{e_k\}_{k\in\mathbb{N}}$ of eigenfunctions corresponding to λ_k is an orthonormal basis of $L^2(\Omega)$ and an orthogonal basis of $X_0^s(\Omega)$.

(g) Each eigenvalue λ_k has finite multiplicity[3]; more precisely, if λ_k is such that

$$\lambda_{k-1} < \lambda_k = \cdots = \lambda_{k+h} < \lambda_{k+h+1}, \tag{3.13}$$

for some $h \in \mathbb{N}_0$, then the set of all the eigenfunctions corresponding to λ_k agrees with

$$\operatorname{span}\{e_k,\dots,e_{k+h}\}.$$

The proof of Proposition 3.1 we present is rather long because it is given in full detail, but it is self-contained and elementary. For instance, no explicit background on the theory of linear operators or on harmonic analysis is required to read it, and we do not really make use of the nonlocal elliptic regularity theory.

3.2 A direct approach

We start by proving some preliminary observations. Let $\mathscr{J} : X_0^s(\Omega) \to \mathbb{R}$ be the functional defined as follows:

$$\mathscr{J}(u) := \frac{1}{2}\int_{\mathbb{R}^n\times\mathbb{R}^n} |u(x)-u(y)|^2\,K(x-y)\,dx\,dy = \frac{1}{2}\,\|u\|^2_{X_0^s(\Omega)},$$

for every $u \in X_0^s(\Omega)$. We remark that $\mathscr{J}$ is smooth, and its derivative is given by

$$\langle \mathscr{J}'(u), v\rangle = \int_{\mathbb{R}^n\times\mathbb{R}^n} (u(x)-u(y))(v(x)-v(y))\,K(x-y)\,dx\,dy, \tag{3.14}$$

for any $v \in X_0^s(\Omega)$.

[3] We observe that we already know that the eigenfunctions corresponding to λ_1 are $\operatorname{span}\{e_1\}$, thanks to (c), so (g) is interesting only when $k \geq 2$.

The following facts are valid:

Claim 3.1 *If $X_\star$ is a (nonempty) weakly closed subspace of $X_0^s(\Omega)$ and*

$$\mathcal{M}_\star := \{u \in X_\star : \|u\|_{L^2(\Omega)} = 1\},$$

then there exists $u_\star \in \mathcal{M}_\star$ such that

$$\min_{u\in\mathcal{M}_\star} \mathcal{J}(u) = \mathcal{J}(u_\star) \tag{3.15}$$

and

$$\int_{\mathbb{R}^n\times\mathbb{R}^n} (u_\star(x)-u_\star(y))(\varphi(x)-\varphi(y))\,K(x-y)\,dx\,dy = \lambda_\star \int_\Omega u_\star(x)\varphi(x)\,dx, \tag{3.16}$$

for every $\varphi \in X_\star$, where $\lambda_\star := 2\mathcal{J}(u_\star) > 0$.

Proof In order to prove (3.15), we use the direct method of minimization. Let us take a minimizing sequence $\{u_j\}_{j\in\mathbb{N}}$ for $\mathcal{J}$ on $\mathcal{M}_\star$, that is, a sequence $\{u_j\}_{j\in\mathbb{N}} \subset \mathcal{M}_\star$ such that

$$\mathcal{J}(u_j) \to \inf_{u\in\mathcal{M}_\star} \mathcal{J}(u) \geq 0 > -\infty \text{ as } j \to +\infty. \tag{3.17}$$

Then the sequence $\{\mathcal{J}(u_j)\}_{j\in\mathbb{N}}$ is bounded in $\mathbb{R}$, so, by definition of $\mathcal{J}$, we have that

$$\{\|u_j\|_{X_0^s(\Omega)}\}_{j\in\mathbb{N}} \quad \text{is also bounded.} \tag{3.18}$$

Since $X_0^s(\Omega)$ is a reflexive space (being a Hilbert space, by Lemma 1.29), up to a subsequence, still denoted by u_j, we have that $\{u_j\}_{j\in\mathbb{N}}$ converges weakly in $X_0^s(\Omega)$ to some $u_\star \in X_\star$ ($X_\star$ being weakly closed). The weak convergence gives that

$$\begin{aligned}&\int_{\mathbb{R}^n\times\mathbb{R}^n} (u_j(x)-u_j(y))(\varphi(x)-\varphi(y))K(x-y)\,dx\,dy\\ &\quad\to \int_{\mathbb{R}^n\times\mathbb{R}^n} (u_\star(x)-u_\star(y))(\varphi(x)-\varphi(y))K(x-y)\,dx\,dy \quad \text{for any } \varphi \in X_0^s(\Omega)\end{aligned}$$

as $j \to +\infty$. Moreover, by (3.18) and Lemma 1.30, up to a subsequence,

$$u_j \to u_\star \quad \text{in } L^2(\mathbb{R}^n) \tag{3.19}$$

as $j \to +\infty$, so $\|u_\star\|_{L^2(\Omega)} = 1$; that is, $u_\star \in \mathcal{M}_\star$. Using the weak lower semicontinuity of the norm in $X_0^s(\Omega)$ (or simply the Fatou lemma), we deduce that

$$\begin{aligned}\lim_{j\to+\infty} \mathcal{J}(u_j) &= \frac{1}{2}\lim_{j\to+\infty}\int_{\mathbb{R}^n\times\mathbb{R}^n} |u_j(x)-u_j(y)|^2 K(x-y)\,dx\,dy\\ &\geq \frac{1}{2}\int_{\mathbb{R}^n\times\mathbb{R}^n} |u_\star(x)-u_\star(y)|^2 K(x-y)\,dx\,dy\\ &= \mathcal{J}(u_\star)\\ &\geq \inf_{u\in\mathcal{M}_\star} \mathcal{J}(u)\end{aligned}$$

so that, by (3.17),

$$\mathscr{J}(u_\star) = \inf_{u \in \mathscr{M}_\star} \mathscr{J}(u).$$

This gives (3.15).

Now we prove (3.16). For this, let $\varepsilon \in (-1,1)$, $\varphi \in X_\star$, $c_\varepsilon := \|u_\star + \varepsilon\varphi\|_{L^2(\Omega)}$, and $u_\varepsilon := (u_\star + \varepsilon\varphi)/c_\varepsilon$. We observe that $u_\varepsilon \in \mathscr{M}_\star$,

$$c_\varepsilon^2 = \|u_\star\|^2_{L^2(\Omega)} + 2\varepsilon \int_\Omega u_\star(x)\varphi(x)\,dx + o(\varepsilon),$$

and

$$\|u_\star + \varepsilon\varphi\|^2_{X_0^s(\Omega)} = \|u_\star\|^2_{X_0^s(\Omega)} + 2\varepsilon\langle u_\star, \varphi\rangle_{X_0^s(\Omega)} + o(\varepsilon).$$

Consequently, because $\|u_\star\|_{L^2(\Omega)} = 1$,

$$\begin{aligned}
2\mathscr{J}(u_\varepsilon) &= \frac{\|u_\star\|^2_{X_0^s(\Omega)} + 2\varepsilon\langle u_\star, \varphi\rangle_{X_0^s(\Omega)} + o(\varepsilon)}{1 + 2\varepsilon \int_\Omega u_\star(x)\varphi(x)\,dx + o(\varepsilon)} \\
&= \big(2\mathscr{J}(u_\star) + 2\varepsilon\langle u_\star, \varphi\rangle_{X_0^s(\Omega)} + o(\varepsilon)\big)\left(1 - 2\varepsilon \int_\Omega u_\star(x)\varphi(x)\,dx + o(\varepsilon)\right) \\
&= 2\mathscr{J}(u_\star) + 2\varepsilon\left(\langle u_\star, \varphi\rangle_{X_0^s(\Omega)} - 2\mathscr{J}(u_\star)\int_\Omega u_\star(x)\varphi(x)\,dx\right) + o(\varepsilon).
\end{aligned}$$

This and the minimality of $u_\star$ imply (3.16) (for this, notice also that $\mathscr{J}(u_\star) > 0$, because otherwise we would have $u_\star \equiv 0$, but $0 \notin \mathscr{M}_\star$). Hence, Claim 3.1 is proved. □

Claim 3.2 *If $\lambda \neq \tilde{\lambda}$ are different eigenvalues of problem* (3.2), *with eigenfunctions e and $\tilde{e} \in X_0^s(\Omega)$, respectively, then*

$$\langle e, \tilde{e}\rangle_{X_0^s(\Omega)} = 0 = \int_\Omega e(x)\tilde{e}(x)\,dx.$$

Proof To check this, we may suppose that $e \not\equiv 0$ and $\tilde{e} \not\equiv 0$. We set $f := e/\|e\|_{L^2(\Omega)}$ and $\tilde{f} := \tilde{e}/\|\tilde{e}\|_{L^2(\Omega)}$, which are eigenfunctions as well, and we compute (3.2) for f with test function $\tilde{f}$ and vice versa. We obtain

$$\begin{aligned}
\lambda \int_\Omega f(x)\tilde{f}(x)\,dx &= \int_{\mathbb{R}^n \times \mathbb{R}^n} (f(x) - f(y))(\tilde{f}(x) - \tilde{f}(y))K(x-y)\,dx\,dy \\
&= \tilde{\lambda} \int_\Omega f(x)\tilde{f}(x)\,dx;
\end{aligned} \tag{3.20}$$

that is,

$$(\lambda - \tilde{\lambda}) \int_\Omega f(x)\tilde{f}(x)\,dx = 0.$$

Thus, since $\lambda \neq \tilde{\lambda}$,

$$\int_{\Omega} f(x)\tilde{f}(x)\,dx = 0. \tag{3.21}$$

By plugging (3.21) into (3.20), we obtain

$$\langle f, \tilde{f}\rangle_{X_0^s(\Omega)} = \int_{\mathbb{R}^n\times\mathbb{R}^n} (f(x)-f(y))(\tilde{f}(x)-\tilde{f}(y))K(x-y)\,dx\,dy = 0.$$

This and (3.21) complete the proof. □

Claim 3.3 *If e is an eigenfunction of problem* (3.2) *corresponding to an eigenvalue λ, then*

$$\int_{\mathbb{R}^n\times\mathbb{R}^n} |e(x)-e(y)|^2 K(x-y)\,dx\,dy = \lambda \|e\|^2_{L^2(\Omega)}.$$

Proof By (3.2), one has

$$\int_{\mathbb{R}^n\times\mathbb{R}^n} (e(x)-e(y))(\varphi(x)-\varphi(y))K(x-y)\,dx\,dy$$
$$= \lambda \int_{\Omega} e(x)\varphi(x)\,dx \quad \forall\, \varphi \in X_0^s(\Omega).$$

By choosing $\varphi := e$ here, we obtain Claim 3.3. □

3.2.1 Proof of Proposition 3.1

Proof of assertion (*a*). For this, we note that the minimum defining λ_1 (see formula (3.3)) exists and that λ_1 is an eigenvalue, thanks to (3.15) and (3.16), applied here with $X_\star := X_0^s(\Omega)$. □

Proof of assertion (*b*). Again by (3.15), the minimum defining λ_1 is attained at some $e_1 \in X_0^s(\Omega)$, with $\|e_1\|_{L^2(\Omega)} = 1$. The fact that e_1 is an eigenfunction corresponding to λ_1 and formula (3.5) follow from (3.16), again with $X_\star = X_0^s(\Omega)$.

Now we show that we may assume that $e_1 \geq 0$ in $\mathbb{R}^n$. First, we claim that

if e is an eigenfunction relative to λ_1, with $\|e\|_{L^2(\Omega)} = 1$,
then both e and $|e|$ attain the minimum in (3.3); (3.22)
also, either $e \geq 0$ or $e \leq 0$ a.e. in Ω.

To check this, we use Claim 3.3 and (3.5) (which has already been proved): we obtain

$$2\mathcal{J}(e) = \int_{\mathbb{R}^n\times\mathbb{R}^n} |e(x)-e(y)|^2 K(x-y)\,dx\,dy = \lambda_1 = 2\mathcal{J}(e_1). \tag{3.23}$$

Also, by triangle inequality, a.e. $x, y \in \mathbb{R}^n$,

$$\big||e(x)| - |e(y)|\big| \leq |e(x)-e(y)|.$$

But, if $x \in \{e > 0\}$ and $y \in \{e < 0\}$, we have that

$$\begin{aligned}\big||e(x)| - |e(y)|\big| &= |e(x) + e(y)| \\ &= \max\{e(x) + e(y), -e(x) - e(y)\} \\ &< e(x) - e(y) \\ &= |e(x) - e(y)|.\end{aligned}$$

This says that

$$\begin{aligned}&\mathscr{J}(|e|) \leq \mathscr{J}(e), \\ &\textit{and } \mathscr{J}(|e|) < \mathscr{J}(e) \textit{ if both } \{e > 0\} \textit{ and } \{e < 0\} \textit{ have positive measure.}\end{aligned} \tag{3.24}$$

Also, $|e| \in X_0^s(\Omega)$ (see Lemma 1.22), and $\| |e| \|_{L^2(\Omega)} = \|e\|_{L^2(\Omega)} = 1$. Hence, (3.23), (3.24), and the minimality of e_1 imply that $\mathscr{J}(|e|) = \mathscr{J}(e) = \mathscr{J}(e_1)$ and that either $\{e > 0\}$ or $\{e < 0\}$ has zero measure. This proves (3.22).

By (3.22), by possibly replacing e_1 with $|e_1|$, we may and do suppose that $e_1 \geq 0$ in $\mathbb{R}^n$. This completes the proof of (b). □

Proof of assertion (c). Suppose that λ_1 also corresponds to another eigenfunction f_1 in $X_0^s(\Omega)$ with $f_1 \not\equiv e_1$. We may suppose that $f_1 \not\equiv 0$; otherwise, we are done. By (3.22), we know that either $f_1 \geq 0$ or $f_1 \leq 0$ a.e. in Ω. Let us consider the case

$$f_1 \geq 0 \quad \text{a.e. in } \Omega, \tag{3.25}$$

the other being analogous. We set

$$\tilde{f}_1 := \frac{f_1}{\|f_1\|_{L^2(\Omega)}} \quad \text{and} \quad g_1 := e_1 - \tilde{f}_1.$$

We show that

$$g_1(x) = 0 \quad \text{a.e. } x \in \mathbb{R}^n. \tag{3.26}$$

For this, note also that g_1 is an eigenfunction relative to λ_1, so, by (3.22), we get that either $g_1 \geq 0$ or $g_1 \leq 0$ a.e. in Ω. Then either $e_1 \geq \tilde{f}_1$ or $e_1 \leq \tilde{f}_1$, and thus, by (3.25) and the nonnegativity of e_1,

$$\text{either } e_1^2 \geq \tilde{f}_1^2 \quad \text{or} \quad e_1^2 \leq \tilde{f}_1^2 \quad \text{a.e. in } \Omega. \tag{3.27}$$

However,

$$\int_\Omega \big(e_1^2(x) - \tilde{f}_1^2(x)\big)\,dx = \|e_1\|^2_{L^2(\Omega)} - \|\tilde{f}_1\|^2_{L^2(\Omega)} = 1 - 1 = 0.$$

This and (3.27) give that $e_1^2 - \tilde{f}_1^2 = 0$ and hence $e_1 = \tilde{f}_1$, so $g_1 = 0$ a.e. in Ω. Because g_1 vanishes outside Ω, we conclude that $g_1 = 0$ a.e. in $\mathbb{R}^n$, proving (3.26).

Then, as a consequence of (3.26), we obtain that f_1 is proportional to e_1, and this proves (c). □

Proof of assertion (*d*). We define λ_{k+1} as in (3.9); we notice indeed that the minimum in (3.9) exists, and it is attained at some $e_{k+1} \in \mathbb{P}_{k+1}$, thanks to (3.15) and (3.16), applied here with $X_\star := \mathbb{P}_{k+1}$, which, by construction, is weakly closed. This fact easily follows from (1.88) and (3.10).

Moreover, since $\mathbb{P}_{k+1} \subseteq \mathbb{P}_k \subseteq X_0^s(\Omega)$, we have that

$$0 < \lambda_1 \leq \lambda_2 \leq \cdots \leq \lambda_k \leq \lambda_{k+1} \leq \dots . \tag{3.28}$$

We claim that

$$\lambda_1 \neq \lambda_2. \tag{3.29}$$

Indeed, if not, $e_2 \in \mathbb{P}_2$ also would be an eigenfunction relative to λ_1, and therefore, by assertion (c), $e_2 = \zeta e_1$, with $\zeta \in \mathbb{R}$, and $\zeta \neq 0$ because $e_2 \not\equiv 0$. Since $e_2 \in \mathbb{P}_2$, we get

$$0 = \langle e_1, e_2 \rangle_{X_0^s(\Omega)} = \zeta \, \|e_1\|^2_{X_0^s(\Omega)}.$$

This would say that $e_1 \equiv 0$, which is a contradiction, thus proving (3.29). From (3.28) and (3.29), we obtain (3.7).

Also, (3.16) with $X_\star = \mathbb{P}_{k+1}$ says that

$$\begin{aligned} &\int_{\mathbb{R}^n \times \mathbb{R}^n} (e_{k+1}(x) - e_{k+1}(y))(\varphi(x) - \varphi(y)) \, K(x-y) \, dx \, dy \\ &\quad = \lambda_{k+1} \int_\Omega e_{k+1}(x) \varphi(x) \, dx \quad \forall \varphi \in \mathbb{P}_{k+1}. \end{aligned} \tag{3.30}$$

In order to show that λ_{k+1} is an eigenvalue with eigenfunction e_{k+1}, we need to show that

formula (3.30) *holds for any* $\varphi \in X_0^s(\Omega)$, *not only in* $\mathbb{P}_{k+1}$. (3.31)

For this, we argue recursively, assuming that the claim holds for $1, \dots, k$ and proving it for $k+1$ (the base of induction is given to the fact that λ_1 is an eigenvalue, as shown in assertion (a)). We use the direct sum decomposition

$$X_0^s(\Omega) = \operatorname{span}\{e_1, \dots, e_k\} \oplus \left(\operatorname{span}\{e_1, \dots, e_k\}\right)^\perp = \operatorname{span}\{e_1, \dots, e_k\} \oplus \mathbb{P}_{k+1},$$

where the orthogonal $\perp$ is intended with respect to the scalar product of $X_0^s(\Omega)$, namely, $\langle \cdot, \cdot \rangle_{X_0^s(\Omega)}$ (see (1.88)). Thus, given any $\varphi \in X_0^s(\Omega)$, we write $\varphi = \varphi_1 + \varphi_2$, with $\varphi_2 \in \mathbb{P}_{k+1}$ and

$$\varphi_1 = \sum_{i=1}^k c_i e_i,$$

for some $c_1,\dots,c_k \in \mathbb{R}$. Then, from (3.30) tested with $\varphi_2 = \varphi - \varphi_1$, we know that

$$\begin{aligned}
&\int_{\mathbb{R}^n\times\mathbb{R}^n} (e_{k+1}(x)-e_{k+1}(y))(\varphi(x)-\varphi(y))\,K(x-y)\,dx\,dy \\
&\qquad -\lambda_{k+1}\int_\Omega e_{k+1}(x)\varphi(x)\,dx \\
&\quad = \int_{\mathbb{R}^n\times\mathbb{R}^n} (e_{k+1}(x)-e_{k+1}(y))(\varphi_1(x)-\varphi_1(y))\,K(x-y)\,dx\,dy \\
&\qquad -\lambda_{k+1}\int_\Omega e_{k+1}(x)\varphi_1(x)\,dx \\
&\quad = \sum_{i=1}^k c_i\Bigg[\int_{\mathbb{R}^n\times\mathbb{R}^n} (e_{k+1}(x)-e_{k+1}(y))(e_i(x)-e_i(y))\,K(x-y)\,dx\,dy \\
&\qquad\qquad -\lambda_{k+1}\int_\Omega e_{k+1}(x)e_i(x)\,dx\Bigg].
\end{aligned} \tag{3.32}$$

Furthermore, testing the eigenvalue equation (3.2) for e_i against e_{k+1} for $i = 1,\dots,k$ (notice that this is allowed by inductive assumption) and recalling that $e_{k+1} \in \mathbb{P}_{k+1}$, we see that

$$\begin{aligned}
0 &= \int_{\mathbb{R}^n\times\mathbb{R}^n} (e_{k+1}(x)-e_{k+1}(y))(e_i(x)-e_i(y))\,K(x-y)\,dx\,dy \\
&= \lambda_i \int_\Omega e_{k+1}(x)e_i(x)\,dx,
\end{aligned}$$

so, by (3.28),

$$\int_{\mathbb{R}^n\times\mathbb{R}^n} (e_{k+1}(x)-e_{k+1}(y))\,(e_i(x)-e_i(y))\;K(x-y)\,dx\,dy = 0 = \int_\Omega e_{k+1}(x)e_i(x)\,dx,$$

for any $i = 1,\dots,k$. By plugging this into (3.32), we conclude that (3.30) holds true for any $\varphi \in X_0^s(\Omega)$; that is, λ_{k+1} is an eigenvalue with eigenfunction e_{k+1}.

Now we prove (3.8). For this, we start by showing that

$$\text{if } k,h \in \mathbb{N}, k \neq h, \text{ then } \langle e_k, e_h\rangle_{X_0^s(\Omega)} = 0 = \int_\Omega e_k(x)e_h(x)\,dx. \tag{3.33}$$

Indeed, let $k > h$; hence $k-1 \geq h$. Thus,

$$e_{k,K} \in \mathbb{P}_k = \big(\operatorname{span}\{e_1,\dots,e_{k-1}\}\big)^\perp \subseteq \big(\operatorname{span}\{e_h\}\big)^\perp,$$

and therefore,

$$\langle e_k, e_h\rangle_{X_0^s(\Omega)} = 0. \tag{3.34}$$

But e_k is an eigenfunction, so, using equation (3.2) for e_k tested with $\varphi = e_h$, we get

$$\int_{\mathbb{R}^n\times\mathbb{R}^n} (e_k(x)-e_k(y))(e_h(x)-e_h(y))\,K(x-y)\,dx\,dy = \lambda_k\int_\Omega e_k(x)e_h(x)\,dx.$$

This and (3.34) give (3.33).

To complete the proof of (3.8), suppose, by contradiction, that $\lambda_k \to c$ for some constant $c \in \mathbb{R}$. Then $\{\lambda_k\}_{k\in\mathbb{N}}$ is bounded in $\mathbb{R}$. Since $\|e_k\|^2_{X_0^s(\Omega)} = \lambda_k$ by Claim 3.3, we deduce by Lemma 1.30 that there is a subsequence for which

$$e_{k_j} \to e_\infty \quad \text{in } L^2(\Omega)$$

as $k_j \to +\infty$, for some $e_\infty \in L^2(\Omega)$. In particular,

$$\{e_{k_j}\}_{j\in\mathbb{N}} \textit{ is a Cauchy sequence in } L^2(\Omega). \tag{3.35}$$

But, from (3.33), e_{k_j} and e_{k_i} are orthogonal in $L^2(\Omega)$, so

$$\|e_{k_j} - e_{k_i}\|^2_{L^2(\Omega)} = \|e_{k_j}\|^2_{L^2(\Omega)} + \|e_{k_i}\|^2_{L^2(\Omega)} = 2.$$

Because this is in contradiction with (3.35), we have established the validity of (3.8).

Now, to complete the proof of (d), we need to show that the sequence of eigenvalues constructed in (3.9) exhausts all the eigenvalues of the problem, that is, that any eigenvalue of problem (3.2) can be written in the form (3.9). We show this by arguing, once more, by contradiction. Let us suppose that there exists an eigenvalue

$$\lambda \notin \{\lambda_k\}_{k\in\mathbb{N}}, \tag{3.36}$$

and let $e \in X_0^s(\Omega)$ be an eigenfunction relative to λ, normalized so that $\|e\|_{L^2(\Omega)} = 1$. Then, by Claim 3.3, we have that

$$2\mathcal{J}(e) = \int_{\mathbb{R}^n\times\mathbb{R}^n} |e(x) - e(y)|^2 K(x-y)\,dx\,dy = \lambda. \tag{3.37}$$

Thus, by the minimality of λ_1 given in (3.3) and (3.5), we have that

$$\lambda = 2\mathcal{J}(e) \geq 2\mathcal{J}(e_1) = \lambda_1.$$

This, (3.36), and (3.8) imply that there exists $k \in \mathbb{N}$ such that

$$\lambda_k < \lambda < \lambda_{k+1}. \tag{3.38}$$

We claim that

$$e \notin \mathbb{P}_{k+1}. \tag{3.39}$$

Indeed, if $e \in \mathbb{P}_{k+1}$, from (3.37) and (3.9), we deduce that

$$\lambda = 2\mathcal{J}(e) \geq \lambda_{k+1}.$$

This contradicts (3.38), so it proves (3.39).

As a consequence of (3.39), there exists $i \in \{1,\ldots,k\}$ such that $\langle e, e_i\rangle_{X_0^s(\Omega)} \neq 0$. But this is in contradiction with Claim 3.2, and therefore, it proves that (3.36) is false, so all the eigenvalues belong to the sequence $\{\lambda_k\}_{k\in\mathbb{N}}$. This completes the proof of (d). □

Proof of assertion (e). Again using (3.15) with $X_\star = \mathbb{P}_{k+1}$, the minimum defining λ_{k+1} is attained in some $e_{k+1} \in \mathbb{P}_{k+1}$. The fact that e_{k+1} is an eigenfunction corresponding to λ_{k+1} was checked in (3.31), and (3.12) follows from (3.16). □

Proof of assertion (f). The orthogonality claimed in (f) follows from (3.33). Thus, to end the proof of (f), we need to show that the sequence of eigenfunctions $\{e_k\}_{k\in\mathbb{N}}$ is a basis for both $L^2(\Omega)$ and $X_0^s(\Omega)$.

Let us start to prove that it is a basis of $X_0^s(\Omega)$. For this, we show that

$$\textit{if } v \in X_0^s(\Omega) \textit{ is such that } \langle v, e_k\rangle_{X_0^s(\Omega)} = 0 \textit{ for any } k \in \mathbb{N}, \textit{ then } v \equiv 0. \tag{3.40}$$

For this, we argue, once more by contradiction, and we suppose that there exists a nontrivial $v \in X_0^s(\Omega)$ satisfying

$$\langle v, e_k\rangle_{X_0^s(\Omega)} = 0 \quad \text{for any } k \in \mathbb{N}. \tag{3.41}$$

Then, up to normalization, we can assume that $\|v\|_{L^2(\Omega)} = 1$. Hence, from (3.8), there exists $k \in \mathbb{N}$ such that

$$2\mathscr{J}(v) < \lambda_{k+1} = \min_{\substack{u\in\mathbb{P}_{k+1}\\ \|u\|_{L^2(\Omega)}=1}} \int_{\mathbb{R}^n\times\mathbb{R}^n} |u(x)-u(y)|^2 K(x-y)\,dx\,dy.$$

Hence, $v \notin \mathbb{P}_{k+1}$, so there exists $j \in \mathbb{N}$ for which $\langle v, e_j\rangle_{X_0^s(\Omega)} \neq 0$. This contradicts (3.41), so it proves (3.40).

A standard Fourier analysis technique then shows that $\{e_k\}_{k\in\mathbb{N}}$ is a basis for $X_0^s(\Omega)$. We give the details for completeness (the expert reader is welcome to skip the argument): we define $E_i := e_i/\|e_i\|_{X_0^s(\Omega)}$ and, given $f \in X_0^s(\Omega)$,

$$f_j := \sum_{i=1}^{j} \langle f, E_i\rangle_{X_0^s(\Omega)} E_i.$$

We point out that, for any $j \in \mathbb{N}$,

$$f_j \quad \textit{belongs to} \quad \operatorname{span}\{e_1,\dots,e_j\}. \tag{3.42}$$

Let $v_j := f - f_j$. By the orthogonality of $\{e_k\}_{k\in\mathbb{N}}$ in $X_0^s(\Omega)$,

$$\begin{aligned}
0 &\le \|v_j\|^2_{X_0^s(\Omega)} \\
&= \langle v_j, v_j\rangle_{X_0^s(\Omega)} \\
&= \|f\|^2_{X_0^s(\Omega)} + \|f_j\|^2_{X_0^s(\Omega)} - 2\langle f, f_j\rangle_{X_0^s(\Omega)} \\
&= \|f\|^2_{X_0^s(\Omega)} + \langle f_j, f_j\rangle_{X_0^s(\Omega)} - 2\sum_{i=1}^{j} \langle f, E_i\rangle^2_{X_0^s(\Omega)} \\
&= \|f\|^2_{X_0^s(\Omega)} - \sum_{i=1}^{j} \langle f, E_i\rangle^2_{X_0^s(\Omega)}.
\end{aligned}$$

Therefore, for any $j \in \mathbb{N}$,

$$\sum_{i=1}^{j} \langle f, E_i \rangle^2_{X_0^s(\Omega)} \leq \| f \|^2_{X_0^s(\Omega)},$$

so

$$\sum_{i=1}^{+\infty} \langle f, E_i \rangle^2_{X_0^s(\Omega)} \quad \textit{is a convergent series.}$$

Thus, if we set

$$\tau_j := \sum_{i=1}^{j} \langle f, E_i \rangle^2_{X_0^s(\Omega)},$$

we have that

$$\{\tau_j\}_{j\in\mathbb{N}} \quad \textit{is a Cauchy sequence in } \mathbb{R}. \tag{3.43}$$

Moreover, using again the orthogonality of $\{e_k\}_{k\in\mathbb{N}}$ in $X_0^s(\Omega)$, we see that if $J > j$,

$$\|v_J - v_j\|^2_{X_0^s(\Omega)} = \left\| \sum_{i=j+1}^{J} \langle f, E_i \rangle_{X_0^s(\Omega)} E_i \right\|^2_{X_0^s(\Omega)} = \sum_{i=j+1}^{J} \langle f, E_i \rangle^2_{X_0^s(\Omega)} = \tau_J - \tau_j.$$

This and (3.43) say that $\{v_j\}_{j\in\mathbb{N}}$ is a Cauchy sequence in $X_0^s(\Omega)$; by the completeness of $X_0^s(\Omega)$ (recall Lemma 1.29), it follows that there exists $v \in X_0^s(\Omega)$ such that

$$v_j \to v \textit{ in } X_0^s(\Omega) \quad \textit{as } j \to +\infty. \tag{3.44}$$

Now we observe that if $j \geq k$,

$$\langle v_j, E_k \rangle_{X_0^s(\Omega)} = \langle f, E_k \rangle_{X_0^s(\Omega)} - \langle f_j, E_k \rangle_{X_0^s(\Omega)} = \langle f, E_k \rangle_{X_0^s(\Omega)} - \langle f, E_k \rangle_{X_0^s(\Omega)} = 0.$$

Hence, by (3.44), it easily follows that $\langle v, E_k \rangle_{X_0^s(\Omega)} = 0$, for any $k \in \mathbb{N}$, so, by (3.40), we have that $v = 0$. All in all, we have that as $j \to +\infty$,

$$f_j = f - v_j \to f - v = f \ \text{ in } \ X_0^s(\Omega).$$

This and (3.42) yield that $\{e_k\}_{k\in\mathbb{N}}$ is a basis in $X_0^s(\Omega)$.

To complete the proof of (f), we need to show that $\{e_k\}_{k\in\mathbb{N}}$ is a basis for $L^2(\Omega)$. For this, take $v \in L^2(\Omega)$ and let $v_j \in C_0^2(\Omega)$ be such that $\|v_j - v\|_{L^2(\Omega)} \leq 1/j$. Notice that $v_j \in X_0^s(\Omega)$, due to (1.27); therefore, since we know that $\{e_k\}_{k\in\mathbb{N}}$ is a basis for $X_0^s(\Omega)$, there exists $k_j \in \mathbb{N}$ and a function w_j belonging to $\operatorname{span}\{e_1, \dots, e_{k_j}\}$ such that

$$\|v_j - w_j\|_{X_0^s(\Omega)} \leq 1/j.$$

Thus, by Lemma 1.28(b),

$$\|v_j - w_j\|_{L^2(\Omega)} \leq \|v_j - w_j\|_X \leq C\|v_j - w_j\|_{X_0^s(\Omega)} \leq C/j.$$

Accordingly,

$$\|v - w_j\|_{L^2(\Omega)} \le \|v - v_j\|_{L^2(\Omega)} + \|v_j - w_j\|_{L^2(\Omega)} \le (C+1)/j.$$

This shows that the sequence $\{e_k\}_{k\in\mathbb{N}}$ of eigenfunctions of (3.2) is a basis in $L^2(\Omega)$. Thus, the proof of (f) is complete. □

Proof of assertion *(g).* Let $h \in \mathbb{N}_0$ be such that (3.13) holds true. We already know that each element of $\mathrm{span}\{e_k,\dots,e_{k+h}\}$ is an eigenfunction of problem (3.2) corresponding to $\lambda_k = \dots = \lambda_{k+h}$, due to (e). So we need to show that any eigenfunction $\psi \not\equiv 0$ corresponding to λ_k belongs to $\mathrm{span}\{e_k,\dots,e_{k+h}\}$. For this, we write

$$X_0^s(\Omega) = \mathrm{span}\{e_k,\dots,e_{k+h}\} \oplus \left(\mathrm{span}\{e_k,\dots,e_{k+h}\}\right)^\perp,$$

so $\psi = \psi_1 + \psi_2$, with

$$\psi_1 \in \mathrm{span}\{e_k,\dots,e_{k+h}\} \quad \text{and} \quad \psi_2 \in \left(\mathrm{span}\{e_k,\dots,e_{k+h}\}\right)^\perp. \tag{3.45}$$

In particular,

$$\langle \psi_1, \psi_2 \rangle_{X_0^s(\Omega)} = 0. \tag{3.46}$$

Because ψ is an eigenfunction corresponding to λ_k, we can write (3.2) and test it against ψ itself. We obtain

$$\lambda_k \|\psi\|^2_{L^2(\Omega)} = \|\psi\|^2_{X_0^s(\Omega)} = \|\psi_1\|^2_{X_0^s(\Omega)} + \|\psi_2\|^2_{X_0^s(\Omega)}, \tag{3.47}$$

thanks to (3.46).

Moreover, from (e), we know that $e_k,\dots,e_{k+h}$ are eigenfunctions corresponding to $\lambda_k = \dots = \lambda_{k+h}$, so

$$\psi_1 \quad \textit{is also an eigenfunction corresponding to} \quad \lambda_k. \tag{3.48}$$

As a consequence, we can write (3.2) for ψ_1 and test it against ψ_2. Thus, recalling (3.46), we obtain

$$\lambda_k \int_\Omega \psi_1(x)\psi_2(x)\,dx = \langle \psi_1, \psi_2 \rangle_{X_0^s(\Omega)} = 0;$$

that is

$$\int_\Omega \psi_1(x)\psi_2(x)\,dx = 0,$$

and therefore,

$$\|\psi\|^2_{L^2(\Omega)} = \|\psi_1 + \psi_2\|^2_{L^2(\Omega)} = \|\psi_1\|^2_{L^2(\Omega)} + \|\psi_2\|^2_{L^2(\Omega)}. \tag{3.49}$$

Now we write

$$\psi_1 = \sum_{i=k}^{k+h} c_i e_i,$$

with $c_i \in \mathbb{R}$. We use the orthogonality in (f) and (3.12) to obtain

$$\|\psi_1\|^2_{X_0^s(\Omega)} = \sum_{i=k}^{k+h} c_i^2 \|e_i\|^2_{X_0^s(\Omega)} = \sum_{i=k}^{k+h} c_i^2 \lambda_i = \lambda_k \sum_{i=k}^{k+h} c_i^2 = \lambda_k \|\psi_1\|^2_{L^2(\Omega)}. \tag{3.50}$$

Now we use (3.48) once more. From this and the fact that ψ is an eigenfunction corresponding to λ_k, we deduce that ψ_2 is also an eigenfunction corresponding to λ_k. Therefore, recalling (3.13) and Claim 3.2 , we conclude that

$$\langle \psi_2, e_1 \rangle_{X_0^s(\Omega)} = \cdots = \langle \psi_2, e_{k-1} \rangle_{X_0^s(\Omega)} = 0.$$

This and (3.45) imply that

$$\psi_2 \in \left(\operatorname{span}\{e_1, \ldots, e_{k+h}\}\right)^{\perp} = \mathbb{P}_{k+h+1}. \tag{3.51}$$

We claim that

$$\psi_2 \equiv 0. \tag{3.52}$$

We argue by contradiction: if not, by (3.10) and (3.51),

$$\begin{aligned}
\lambda_k < \lambda_{k+h+1} &= \min_{u \in \mathbb{P}_{k+h+1} \setminus \{0\}} \frac{\displaystyle\int_{\mathbb{R}^n \times \mathbb{R}^n} |u(x) - u(y)|^2 K(x-y)\,dx\,dy}{\displaystyle\int_\Omega |u(x)|^2\,dx} \\
&\le \frac{\displaystyle\int_{\mathbb{R}^n \times \mathbb{R}^n} |\psi_2(x) - \psi_2(y)|^2 K(x-y)\,dx\,dy}{\displaystyle\int_\Omega |\psi_2(x)|^2\,dx} \\
&= \frac{\|\psi_2\|^2_{X_0^s(\Omega)}}{\|\psi_2\|^2_{L^2(\Omega)}}.
\end{aligned} \tag{3.53}$$

Thus, we use (3.47), (3.49), (3.50), and (3.53) to compute

$$\begin{aligned}
\lambda_k \|\psi\|^2_{L^2(\Omega)} &= \|\psi_1\|^2_{X_0^s(\Omega)} + \|\psi_2\|^2_{X_0^s(\Omega)} \\
&> \lambda_k \|\psi_1\|^2_{L^2(\Omega)} + \lambda_k \|\psi_2\|^2_{L^2(\Omega)} \\
&= \lambda_k \|\psi\|^2_{L^2(\Omega)}.
\end{aligned}$$

This is a contradiction, so (3.52) is established.

From (3.45) and (3.52), we obtain that

$$\psi = \psi_1 \in \operatorname{span}\{e_k, \ldots, e_{k+h}\},$$

as desired. This completes the proof of (g), and it finishes the proof of Proposition 3.1. □

3.3 Another variational characterization of the eigenvalues

In Proposition 3.1, we have given a variational characterization of λ_k in terms of the space $\mathbb{P}_{k+1}$. In the following result (see [196]), we prove another characterization of the eigenvalues of $-\mathscr{L}_K$ in terms of a finite-dimensional space.

Proposition 3.2 Let $\{\lambda_k\}_{k\in\mathbb{N}}$ be the sequence of the eigenvalues of problem (3.2) with

$$0<\lambda_1<\lambda_2\leq\cdots\leq\lambda_k\leq\lambda_{k+1}\leq\ldots,$$

and let $\{e_k\}_{k\in\mathbb{N}}$ be the sequence of eigenfunctions corresponding to λ_k. Then, for any $k\in\mathbb{N}$, the eigenvalues can be characterized as follows:

$$\lambda_k=\max_{u\in\operatorname{span}\{e_1,\ldots,e_k\}\setminus\{0\}}\frac{\displaystyle\int_{\mathbb{R}^n\times\mathbb{R}^n}|u(x)-u(y)|^2K(x-y)\,dx\,dy}{\displaystyle\int_\Omega|u(x)|^2\,dx}.$$

Proof Let $k\in\mathbb{N}$. Because λ_k is the eigenvalue corresponding to the eigenfunction e_k, we have that

$$\begin{aligned}\lambda_k&=\frac{\displaystyle\int_{\mathbb{R}^n\times\mathbb{R}^n}|e_k(x)-e_k(y)|^2K(x-y)\,dx\,dy}{\displaystyle\int_\Omega|e_k(x)|^2\,dx}\\&\leq\max_{u\in\operatorname{span}\{e_1,\ldots,e_k\}\setminus\{0\}}\frac{\displaystyle\int_{\mathbb{R}^n\times\mathbb{R}^n}|u(x)-u(y)|^2K(x-y)\,dx\,dy}{\displaystyle\int_\Omega|u(x)|^2\,dx}.\end{aligned}\tag{3.54}$$

Furthermore, let $u\in\operatorname{span}\{e_1,\ldots,e_k\}\setminus\{0\}$. Then

$$u=\sum_{i=1}^k u_ie_i\quad\text{with}\ \sum_{i=1}^k u_i^2\neq0$$

and

$$\frac{\displaystyle\int_{\mathbb{R}^n\times\mathbb{R}^n}|u(x)-u(y)|^2K(x-y)\,dx\,dy}{\displaystyle\int_\Omega|u(x)|^2\,dx}=\frac{\displaystyle\sum_{i=1}^k u_i^2\|e_i\|_{X_0^s(\Omega)}}{\displaystyle\sum_{i=1}^k u_i^2}=\frac{\displaystyle\sum_{i=1}^k u_i^2\lambda_i}{\displaystyle\sum_{i=1}^k u_i^2}\leq\lambda_k$$

because $\{e_1,\ldots,e_k\}$ are orthonormal in $L^2(\Omega)$ and orthogonal in $X_0^s(\Omega)$ (see Proposition 3.1(f)) and $\lambda_k \geq \lambda_i$, for any $i = 1,\ldots,k$. As a consequence, passing to the maximum over $u \in \operatorname{span}\{e_1,\ldots,e_k\} \setminus \{0\}$, we get

$$\max_{u\in \operatorname{span}\{e_1,\ldots,e_k\}\setminus\{0\}} \frac{\displaystyle\int_{\mathbb{R}^n\times\mathbb{R}^n} |u(x)-u(y)|^2 K(x-y)\,dx\,dy}{\displaystyle\int_\Omega |u(x)|^2\,dx} \leq \lambda_k,$$

which together with (3.54) gives the assertion. □

3.4 A regularity result for the eigenfunctions

In this section we give a regularity result for the eigenfunctions of the integrodifferential operator $-\mathscr{L}_K$ proved in [196] (see also [201, proposition 4], where the case of the fractional Laplacian is considered).

First of all, we perform a simple observation on the positive part of a function $v \in X^s(\Omega)$; that is, $v^+ := \max\{v, 0\}$.

Lemma 3.3 *For any* $v \in X^s(\Omega)$,

$$(v(x)-v(y))\left(v^+(x)-v^+(y)\right) \geq \left|v^+(x)-v^+(y)\right|^2, \tag{3.55}$$

for any $x,\ y \in \mathbb{R}^n$.

Proof To check the claim in (3.55), because the role of x and y is symmetric, we can always suppose that $v(x) \geq v(y)$. Also, (3.55) is clearly an identity when x, $y \in \{v \geq 0\}$, and when x, $y \in \{v < 0\}$.

Thus, it only remains to check (3.55) when $x \in \{v \geq 0\}$ and $y \in \{v < 0\}$. In this case,

$$v^+(x)-v^+(y) = v(x) < v(x)+|v(y)| = v(x)-v(y),$$

so if we multiply both sides by $v^+(x)-v^+(y) \geq 0$, we obtain (3.55). □

Note that in the classical case, (3.55) boils down to $\nabla v \cdot \nabla v^+ \geq |\nabla v^+|^2$, which is obvious (and the equality holds).

With this, we can prove the following result:

Proposition 3.4 Let $e \in X_0^s(\Omega)$ and $\lambda > 0$ be such that

$$\langle e,\varphi\rangle_{X_0^s(\Omega)} = \lambda \int_\Omega e(x)\varphi(x)\,dx, \tag{3.56}$$

for any $\varphi \in X_0^s(\Omega)$. Then $e \in L^\infty(\Omega)$, and there exists $C > 0$, possibly depending on n, s, θ, and λ, such that

$$\|e\|_{L^\infty(\Omega)} \leq C\|e\|_{L^2(\Omega)}. \tag{3.57}$$

Proof We may assume that e does not vanish identically (otherwise, there is nothing to prove). Let $\delta > 0$, to be taken appropriately small in what follows (the choice of δ will be made in (3.66)). Up to multiplying e by a small constant, we may and do assume that

$$\|e\|_{L^2(\Omega)} = \sqrt{\delta}. \tag{3.58}$$

Now, for any $k \in \mathbb{N}$, we let $C_k := 1 - 2^{-k}$, $v_k := e - C_k$, $w_k := v_k^+$, and $U_k := \|w_k\|^2_{L^2(\Omega)}$. Notice that $w_k \in X$ because X is a linear space, and a.e. $x \in \mathscr{C}\Omega$, we have that

$$e(x) - C_k = -C_k \leq 0.$$

Hence, $w_k(x) = (-C_k)^+ = 0$ in $\mathscr{C}\Omega$, so $w_k \in X_0^s(\Omega)$. Moreover, we get that

$$0 \leq w_k \leq |e| + |C_k| \leq |e| + 1 \in L^2(\Omega),$$

with Ω bounded, and

$$\lim_{k\to+\infty} w_k = (e-1)^+.$$

Therefore, by the dominated convergence theorem,

$$\lim_{k\to+\infty} U_k = \int_\Omega \left[(e(x)-1)^+\right]^2 dx. \tag{3.59}$$

Moreover, for any $k \in \mathbb{N}$, we have $C_{k+1} > C_k$, so $v_{k+1} < v_k$ a.e. in $\mathbb{R}^n$, from which we deduce that

$$w_{k+1} \leq w_k \quad \text{a.e. in } \mathbb{R}^n. \tag{3.60}$$

Let also $A_k := C_{k+1}/(C_{k+1} - C_k) = 2^{k+1} - 1$, for any $k \in \mathbb{N}$. We claim that, for any $k \in \mathbb{N}$,

$$e < A_k w_k \quad \text{on} \quad \{w_{k+1} > 0\}. \tag{3.61}$$

To check this, let $x \in \{w_{k+1} > 0\}$. Then $e(x) - C_{k+1} > 0$, so, by the properties of C_k, we have $e(x) > C_{k+1} > C_k$. Hence, $w_k(x) = v_k(x) = e(x) - C_k$ and

$$\begin{aligned}
A_k w_k(x) &= A_k(e(x) - C_k) \\
&= \frac{C_{k+1}}{C_{k+1} - C_k} e(x) - \frac{C_k C_{k+1}}{C_{k+1} - C_k} \\
&= e(x) + \frac{C_k}{C_{k+1} - C_k}(e(x) - C_{k+1}) \\
&> e(x).
\end{aligned}$$

This proves (3.61).

Notice also that

$$v_{k+1}(x) - v_{k+1}(y) = e(x) - e(y),$$

for any $x, y \in \mathbb{R}^n$. Using this, (3.55), (3.56), and the fact that $v_{k+1}^+ = w_{k+1} \in X_0^s(\Omega)$, we obtain

$$
\begin{aligned}
&\int_{\mathbb{R}^n \times \mathbb{R}^n} |w_{k+1}(x) - w_{k+1}(y)|^2 K(x-y)\,dx\,dy \\
&= \int_{\mathbb{R}^n \times \mathbb{R}^n} |v_{k+1}^+(x) - v_{k+1}^+(y)|^2 K(x-y)\,dx\,dy \\
&\le \int_{\mathbb{R}^n \times \mathbb{R}^n} (v_{k+1}(x) - v_{k+1}(y))(v_{k+1}^+(x) - v_{k+1}^+(y)) K(x-y)\,dx\,dy \\
&= \int_{\mathbb{R}^n \times \mathbb{R}^n} (e(x) - e(y))(v_{k+1}^+(x) - v_{k+1}^+(y)) K(x-y)\,dx\,dy \\
&= \lambda \int_{\Omega} e(x) w_{k+1}(x)\,dx \\
&= \lambda \int_{\{w_{k+1} > 0\}} e(x) w_{k+1}(x)\,dx.
\end{aligned}
$$

Therefore, as a consequence of this, recalling (3.60) and (3.61) and using the fact that λ is positive, we deduce that

$$
\begin{aligned}
&\int_{\mathbb{R}^n \times \mathbb{R}^n} |w_{k+1}(x) - w_{k+1}(y)|^2 K(x-y)\,dx\,dy \\
&= \lambda \int_{\{w_{k+1} > 0\}} e(x) w_{k+1}(x)\,dx \\
&\le \lambda A_k \int_{\{w_{k+1} > 0\}} w_k(x) w_{k+1}(x)\,dx \\
&\le \lambda A_k \int_{\{w_{k+1} > 0\}} w_k^2(x)\,dx \le 2^{k+1} \lambda U_k.
\end{aligned}
\tag{3.62}
$$

Now we claim that

$$
\{w_{k+1} > 0\} \subseteq \{w_k > 2^{-(k+1)}\}. \tag{3.63}
$$

To establish this, we observe that if $x \in \{w_{k+1} > 0\}$, then

$$
0 < v_{k+1}^+(x) = \max\{e(x) - C_{k+1}, 0\}.
$$

Hence, $e(x) - C_{k+1} > 0$. Accordingly,

$$
v_k(x) = e(x) - C_k > C_{k+1} - C_k = 2^{-(k+1)},
$$

so, as a consequence, $w_k(x) = v_k(x) > 2^{-(k+1)}$, which proves (3.63).

As a consequence of (3.63), we obtain

$$\begin{aligned} U_k &= \|w_k\|^2_{L^2(\Omega)} \\ &\geq \int_{\{w_k > 2^{-(k+1)}\}} w_k^2(x)\,dx \\ &\geq 2^{-2(k+1)} |\{w_k \geq 2^{-(k+1)}\}| \\ &\geq 2^{-2(k+1)} |\{w_{k+1} > 0\}|. \end{aligned} \tag{3.64}$$

Now we use the Hölder inequality (with exponents $2^*_s/2$ and $n/(2s)$) and the fractional Sobolev inequality given in Lemma 1.31(b) to get that

$$\begin{aligned} U_{k+1} &\leq \left(\int_\Omega |w_{k+1}(x)|^{2^*_s}\,dx\right)^{2/2^*_s} |\{w_{k+1} > 0\}|^{2s/n} \\ &\leq \tilde{c} \int_{\mathbb{R}^n \times \mathbb{R}^n} |w_{k+1}(x) - w_{k+1}(y)|^2 K(x-y)\,dx\,dy\ |\{w_{k+1} > 0\}|^{2s/n}, \end{aligned}$$

for some positive constant $\tilde{c}$ depending only on n, s, and θ (see assumption (1.56)). Consequently, by (3.62) and (3.64), we see that

$$\begin{aligned} U_{k+1} &\leq (\tilde{c}\, 2^{k+1} \lambda U_k)(2^{2(k+1)} U_k)^{2s/n} \\ &= 2^{1+(4s/n)} \tilde{c} \lambda \cdot (2^{1+(4s/n)})^k U_k^{1+(2s/n)} \\ &\leq \left((1 + 2^{1+(4s/n)} \tilde{c} \lambda) \cdot 2^{1+(4s/n)}\right)^k U_k^{1+(2s/n)} \\ &= C^k U_k^\beta, \end{aligned} \tag{3.65}$$

where $\beta := 1 + (2s/n) > 1$ and $C > 1$ only depends on n, s, and λ. Now we are ready to perform our choice of δ: namely, we assume that $\delta > 0$ is so small that

$$\delta^{\beta-1} < \frac{1}{C^{1/(\beta-1)}}. \tag{3.66}$$

We also fix

$$\eta \in \left(\delta^{\beta-1}, \frac{1}{C^{1/(\beta-1)}}\right).$$

Notice that since $C > 1$ and $\beta > 1$,

$$\eta \in (0,1). \tag{3.67}$$

Moreover,

$$\delta^{\beta-1} \leq \eta \quad \text{and} \quad C\eta^{\beta-1} \leq 1. \tag{3.68}$$

We claim that

$$U_k \leq \delta \eta^k. \tag{3.69}$$

The proof is by induction. First of all, we have that

$$U_0 = \|e^+\|^2_{L^2(\Omega)} \leq \|e\|^2_{L^2(\Omega)} = \delta,$$

which is (3.69) when $k = 0$. Let us now suppose that (3.69) holds true for k, and let us prove it for $k+1$. For this, we use (3.65) and (3.68):

$$U_{k+1} \leq C^k U_k^\beta \leq C^k (\delta \eta^k)^\beta = \delta (C \eta^{\beta-1})^k \delta^{\beta-1} \eta^k \leq \delta \eta^{k+1}.$$

This proves (3.69). Then, by (3.67) and (3.69), we conclude that

$$\lim_{k \to +\infty} U_k = 0.$$

Hence, by (3.59), $(e-1)^+ = 0$ a.e. in Ω; that is, $e \leq 1$ a.e. in Ω. By replacing e with $-e$, we obtain $\|e\|_{L^\infty(\Omega)} \leq 1$. Then (3.57) follows by recalling the scaling in (3.58). This completes the proof. □

3.5 Nodal set of the eigenfunctions of $(-\Delta)^s$: 1D case

We conclude this chapter dealing with the model case of the fractional Laplace operator. The aim of this last section is to prove a property of the nodal set of the eigenfunctions of $(-\Delta)^s$, that is, the fact the nodal set of the eigenfunctions has zero Lebesgue measure, in the one-dimensional case (see [104]).

For this purpose, we will use some recent results about the regularity of the solutions of the fractional Laplacian equation (see [24, 201, 202]) and a unique continuation principle for nonlocal equations (see [94]).

First of all, let us prove the following proposition:

Proposition 3.5 Let $s \in (0, 1/2)$, $n = 1$, and Ω be an open, bounded subset of $\mathbb{R}^n$. Finally, let $u \in C^\infty(\Omega)$ be a function such that its nodal set

$$\mathcal{N}_u := \{x \in \Omega : u(x) = 0\}$$

has positive Lebesgue measure. Then, for almost any $\bar{x} \in \mathcal{N}_u$, we have that $u(x) = o(|x - \bar{x}|^k)$ as $x \to \bar{x}$, for any $k \in \mathbb{N}$.

Proof We take $\bar{x} \in \mathcal{N}_u$ to be a Lebesgue density point. Then there exists a sequence $\{x_j\}_{j \in \mathbb{N}}$ in $\mathcal{N}_u \setminus \{\bar{x}\}$ such that $x_j \to \bar{x}$ as $j \to +\infty$. Without loss of generality, we may suppose that such sequence is decreasing.

We claim that for any $k \in \mathbb{N}$ there exists a decreasing subsequence $\{x_{j_k}\}_{k \in \mathbb{N}}$ of $\{x_j\}_{j \in \mathbb{N}}$ such that $D^k u(x_{j_k}) = 0$, for any $j \in \mathbb{N}$. For this, we argue by induction. When $k = 0$, we can take $x_{j_k} = x_j$, since $u(x_j) = 0$, for any $j \in \mathbb{N}$, by construction of x_j.

Now suppose that the claim holds true for k, and let us show it for $k+1$. For purpose, we fix $j \in \mathbb{N}$, and we apply the Rolle theorem to the function $D^k u$ in the interval $[x_{(j+1)_k}, x_{j_k}]$. We get that there exists $x_{j_{k+1}} \in [x_{(j+1)_k}, x_{j_k}]$ such that

$$D^{k+1}u(x_{j_{(k+1)}}) = D^k u(x_{j_k}) - D^k u(x_{(j+1)_k}) = 0,$$

thanks to the induction assumption. Hence, the claim is proved.

By continuity and thanks to the fact that $x_j \to \bar{x}$ as $j \to +\infty$, we deduce that $D^k u(\bar{x}) = 0$, for any $k \in \mathbb{N}$. As a consequence of this, since $u \in C^\infty(\Omega)$, by Taylor expansion, we get

$$u(x) = \sum_{j=1}^{k-1} \frac{D^j u(\bar{x})}{j!} + o(|x-\bar{x}|^k) = o(|x-\bar{x}|^k),$$

for any $k \in \mathbb{N}$, which gives the assertion. □

Let us consider the following eigenvalue problem:

$$\begin{cases} (-\Delta)^s u + q(x)u = \lambda u & \text{in } \Omega \\ u = 0 & \text{in } \mathbb{R}^n \setminus \Omega. \end{cases} \tag{3.70}$$

The main result of this section can be stated as follows:

Theorem 3.6 *Let $s \in (0, 1/2)$, $n = 1$, and Ω be an open, bounded subset of $\mathbb{R}^n$ with Lipschitz boundary. Let $e \in H^s(\mathbb{R}^n)$ be a weak solution of problem* (3.70), *with $q \in C^\infty(\Omega) \cap L^\infty(\Omega)$. Then the nodal set of e has zero Lebesgue measure.*

Proof We argue by contradiction, and we suppose that the nodal set of e has positive Lebesgue measure. First of all, note that

$$e \in C^\infty(\Omega). \tag{3.71}$$

This follows via the following steps: first, we have that $e \in L^\infty(\Omega)$ by Proposition 3.4. Hence, by [103] (see also the results in Chapter 4), we obtain that e is uniformly continuous in $\overline{\Omega}$, and it is a viscosity solution of the equation. Then, by [24, theorem 5], we get that $e \in C^k(\Omega)$, for any $k \in \mathbb{N}$, thus establishing (3.71).

Hence, by Proposition 3.5, we easily get that, for any $\bar{x}$ such that $e(\bar{x}) = 0$,

$$e(x) = O(|x-\bar{x}|^k) \quad \text{for any } k \in \mathbb{N},$$

as $x \to \bar{x}$.

Now, applying [94, theorem 1.2] (here with $h(x) = \lambda - q(x) \in C^\infty(\Omega)$), we get that $e \equiv 0$ in Ω, which is a contradiction, e being an eigenfunction. Thus, the assertion is proved. □

The two preceding results have been proved in [104]. For the general case of Theorem 3.6, we refer to the paper [94].

4
Weak and viscosity solutions

In the literature, different notions of solutions are taken into account when dealing with elliptic equations, such as, for instance, the *weak* (also called *distributional*, or *variational*, or *energy*) solutions (i.e., the solutions that belong to a suitable Sobolev space and satisfy the equation in a distributional sense when integrated against a suitable set of test functions) and the *viscosity* solutions (i.e., all the smooth functions that touch either from above or below the continuous solution are required to be either subsolutions or supersolutions).

In this chapter, we point out that for the fractional Laplace equation, the notion of weak solution implies one of the viscosity solutions. More precisely, our aim is to show that weak solutions of the fractional Laplacian equation

$$\begin{cases} (-\Delta)^s u = f & \text{in } \Omega \\ u = g & \text{in } \mathbb{R}^n \setminus \Omega \end{cases}$$

are also continuous solutions (up to the boundary) of this problem in the viscosity sense. In order to get this regularity result, we first prove a maximum principle and then, using it, an interior and boundary regularity result for weak solutions of the problem. As a consequence, we also deduce that the first eigenfunction of $(-\Delta)^s$ is strictly positive in Ω.

We also discuss the boundedness of weak solutions for nonlocal fractional problems in both the linear and the nonlinear settings. The results of this chapter are based on the papers [23, 202].

4.1 Viscosity solutions

In what follows, we take Ω to be a bounded, open subset of $\mathbb{R}^n$ with C^2-boundary, and we consider the fractional boundary value problem

$$\begin{cases} (-\Delta)^s u = f & \text{in } \Omega \\ u = g & \text{in } \mathbb{R}^n \setminus \Omega, \end{cases} \tag{4.1}$$

where $s \in (0,1)$, and f and g are sufficiently smooth. As is customary, the operator $(-\Delta)^s$ is the fractional Laplacian, which may be written as

$$(-\Delta)^s u(x) := C(n,s) \int_{\mathbb{R}^n} \frac{2u(x) - u(x+y) - u(x-y)}{|y|^{n+2s}}\, dy, \tag{4.2}$$

where the positive constant $C(n,s)$ is given in (1.18) up to a constant factor.

The definition in (4.2) is well posed if u is smooth enough and it does not grow too much at infinity (e.g., if $u \in C^2(\mathbb{R}^n) \cap L^\infty(\mathbb{R}^n)$). However, it is convenient to consider weak formulations of (4.1) both to construct solutions using variational methods and to develop a regularity theory based on comparison principles.

These two different scopes indeed reflect two different attitudes and methodologies (namely, test functions and energy estimates versus touching the solution with a barrier), and they are reminiscent of the dichotomy between operators in divergence and nondivergence form in the classical elliptic framework. These two different points of view naturally translate into two different notions of solutions, that is, weak versus viscosity solutions.

As a matter of fact, in the recent literature, some interest has been developed for weak solutions of (4.1), that is, for functions $u \in \mathbb{H}^s(\Omega) \cap L^\infty(\mathbb{R}^n)$ that satisfy

$$\begin{cases} C(n,s) \displaystyle\int_{\mathbb{R}^n \times \mathbb{R}^n} \frac{(u(x)-u(y))(\varphi(x)-\varphi(y))}{|x-y|^{n+2s}}\, dx\, dy = \int_{\mathbb{R}^n} f(x)\varphi(x)\, dx \\ \qquad\qquad\qquad\qquad\qquad\qquad\qquad \text{for any } \varphi \in \mathbb{H}^s_0(\Omega), \\ u = g \quad \text{a.e. in } \mathbb{R}^n \setminus \Omega. \end{cases} \tag{4.3}$$

We recall here that $\mathbb{H}^s(\Omega)$ denotes the set of all measurable functions $u : \mathbb{R}^n \to \mathbb{R}$ such that $u \in L^2(\Omega)$ and

$$\int_{Q} \frac{|u(x)-u(y)|^2}{|x-y|^{n+2s}}\, dx\, dy < +\infty. \tag{4.4}$$

We also denoted by $\mathbb{H}^s_0(\Omega)$ the functions in $\mathbb{H}^s(\Omega)$ that vanish outside Ω (see also Subsection 1.5.1).

The interest of these weak solutions mainly comes from the calculus of variations because the expression in (4.3) arises from the Euler–Lagrange equation of a fractional-type energy functional.

However, when dealing with comparison principles, it is also natural to consider viscosity solutions of (4.1) (see, e.g., [206]). For this, we recall that $u \in C(\mathbb{R}^n)$ is a *viscosity subsolution*[1] for (4.1) if, for any open set $U \subset \Omega$, any $x_0 \in U$, and any $\phi \in$

[1] More generally, as in [56], one can define the notion of viscosity subsolution and supersolution for semicontinuous functions so that this class is closed under sup and inf.

$C^2(U)$ such that $\phi(x_0) = u(x_0)$ and $\phi \geq u$ in U, we let

$$v(x) := \begin{cases} \phi(x) & \text{in } U \\ u(x) & \text{outside } U, \end{cases} \tag{4.5}$$

we have that $-(-\Delta)^s v(x_0) \geq -f(x_0)$. Similarly, we say that $u \in C(\mathbb{R}^n)$ is a *viscosity supersolution* for (4.1) if, for any $\psi \in C^2(U)$ such that $\psi(x_0) = u(x_0)$ and $\psi \leq u$ in U, we let

$$w(x) := \begin{cases} \psi(x) & \text{in } U \\ u(x) & \text{outside } U, \end{cases}$$

we have that $-(-\Delta)^s w(x_0) \leq -f(x_0)$. Then u is a *viscosity solution* of problem (4.1) if it is both a viscosity subsolution and a viscosity supersolution.

Here we prove a relation between weak and viscosity solutions of problem (4.1). With respect to this, the main observation that we make in this chapter is the following:

Theorem 4.1 *Let $f \in C(\mathbb{R}^n)$ and $g \in C(\mathbb{R}^n) \cap L^\infty(\mathbb{R}^n)$. Let $u \in \mathbb{H}^s(\Omega) \cap L^\infty(\mathbb{R}^n)$ be a weak solution of* (4.1). *Then u is a viscosity solution of* (4.1).

The converse of Theorem 4.1 holds true if f is smooth enough, thanks to the regularity theory for viscosity solutions (see [184, 186]). In order to prove Theorem 4.1, we first need to show that every weak solution of (4.1) is continuous in $\mathbb{R}^n$, as stated in the following theorem:

Theorem 4.2 *Let $f \in L^\infty(\Omega)$, and let $u \in \mathbb{H}^s(\Omega) \cap L^\infty(\mathbb{R}^n) \cap C(\mathscr{C}\Omega)$ be a weak solution of*

$$(-\Delta)^s u = f \quad \text{in } \Omega.$$

Then $u \in C(\mathbb{R}^n)$.

For this, we first give an interior regularity result for weak solutions (see Proposition 4.5 for the details), and later, we show that the weak solutions of our problem are continuous up to the boundary (hence in the whole $\mathbb{R}^n$) if they are outside the domain where the problem is set.

As a by-product of Theorem 4.2, in Corollary 4.8 we show that the first eigenfunction of $(-\Delta)^s$ with Dirichlet data is strictly positive in Ω.

We would like to note that in the definition of weak solution and in all the previous results we assume that the weak solutions of problem (4.1) are in $L^\infty(\Omega)$. We do this only for the sake of clarity and concreteness. In fact, this condition can be removed. Indeed, in Proposition 4.10, we show that every weak solution of (4.1) is in $L^\infty(\Omega)$, provided that the data are smooth enough.

4.2 Perron method and existence theory for viscosity solutions

Here we recall that the classical Perron method (see e.g. [117]) can be suitably adapted to fractional problems, giving an existence theory in the viscosity setting.

One way to construct solutions somehow explicitly consists of using the fractional Poisson kernel in the ball to reconstruct harmonic functions from their values outside (see e.g. [129]) and then taking the supremum of all possible subsolutions. The notion of half-relaxed limit makes this construction compatible with the notion of viscosity solutions. Suitable barriers are needed to check that the boundary data are attained, as in [57]: this is more complicated than in the classical framework of the Laplacian because the fundamental solution is not a barrier and because an inequality in the whole complement is necessary for the comparison principle in this case.

For instance, as a particular case of [21, theorem 1], we have the following result:

Proposition 4.3 Let Ω be open, bounded, and with C^2-boundary, and let $g \in C(\mathbb{R}^n) \cap L^\infty(\mathbb{R}^n)$. Then there exists a viscosity solution $u \in C(\mathbb{R}^n) \cap L^\infty(\mathbb{R}^n)$ of the problem

$$\begin{cases} (-\Delta)^s u = 0 & \text{in } \Omega \\ u = g & \text{in } \mathbb{R}^n \setminus \Omega. \end{cases}$$

As a consequence of Proposition 4.3 we have the following existence result:

Corollary 4.4 Let Ω be open, bounded, and with C^2-boundary. Let $f \in C^2(\mathbb{R}^n)$ and $g \in C(\mathbb{R}^n) \cap L^\infty(\mathbb{R}^n)$. Then there exists a viscosity solution $u \in C(\mathbb{R}^n) \cap L^\infty(\mathbb{R}^n)$ of the problem

$$\begin{cases} (-\Delta)^s u = f & \text{in } \Omega \\ u = g & \text{in } \mathbb{R}^n \setminus \Omega. \end{cases}$$

Proof We may modify f outside Ω in such a way that f is compactly supported in some B_R, $R > 0$, and therefore bounded and uniformly continuous. We take $\Gamma(x) := c(n,s)|x|^{2s-n}$, the fundamental solution of the fractional Laplacian (see, e.g., [129] and [55, subsection 2.2], for the basic properties of fractional fundamental solutions); that is, $(-\Delta)^s \Gamma = \delta_0$, where δ_0 is the Dirac delta function in 0.

Let $\tilde{g} := g - \Gamma * f$. Notice that

$$\Gamma * f(x) = c(n,s) \int_{B_R(x)} |y|^{2s-n} f(x-y)\,dy,$$

thanks to our further assumptions on f. Since $s < 1$,

$$\int_{B_R(x)} |y|^{2s-n}\,dy < +\infty,$$

so we obtain that $\Gamma * f \in C(\mathbb{R}^n) \cap L^\infty(\mathbb{R}^n)$, again by the assumptions on f. Accordingly, $\tilde{g} \in C(\mathbb{R}^n) \cap L^\infty(\mathbb{R}^n)$, so we may apply Proposition 4.3 and obtain a viscosity solution $\tilde{u}$ of $(-\Delta)^s \tilde{u} = 0$ in Ω and $\tilde{u} = \tilde{g}$ outside Ω.

Also, $\Gamma * f \in C^2(\mathbb{R}^n)$, and it solves

$$(-\Delta)^s(\Gamma * f) = ((-\Delta)^s \Gamma) * f = f.$$

Let $u := \tilde{u} + \Gamma * f$. Then u solves (4.4), and this ends the proof. □

4.3 Regularity theory for weak solutions

In this section we give some regularity results for weak solutions of problem (4.1). First of all, the next proposition deals with the interior regularity theory for weak solutions.

Proposition 4.5 Let $f \in L^\infty(B_1)$, and let $u \in \mathbb{H}^s(B_3) \cap L^\infty(\mathbb{R}^n)$ be a weak solution of

$$(-\Delta)^s u = f \quad \text{in } B_1. \tag{4.6}$$

Then $u \in C^\alpha(B_{1/4})$, for any $0 < \alpha < \min\{2s, 1\}$, and

$$\|u\|_{C^\alpha(B_{1/4})} \leq C\big(\|u\|_{L^\infty(\mathbb{R}^n)} + \|f\|_{L^\infty(B_1)}\big),$$

for a suitable $C > 0$ depending on n, s, and α.

Proof The case of smooth solutions was considered in [89], so it is enough to reduce to this case by a standard convolution argument. Namely, we take $\rho \in C_0^\infty(B_1; [0,1])$, and we consider the mollifier

$$\rho_\varepsilon(x) := \varepsilon^{-n} \rho(x/\varepsilon),$$

where $\varepsilon > 0$. We define $u_\varepsilon := u * \rho_\varepsilon$ and $f_\varepsilon := f * \rho_\varepsilon$.

Of course, u_ε is a smooth function, by construction. Let us show that u_ε solves the following equation:

$$(-\Delta)^s u_\varepsilon = f_\varepsilon \quad \text{in } B_{1/2}. \tag{4.7}$$

For this, first we observe that, for any test function ϕ supported in $B_{1/2}$,

$$\begin{aligned}
&\int_{\mathbb{R}^n} \left[\int_{\mathbb{R}^n \times \mathbb{R}^n} \frac{|u(x+z) - u(y+z)|\, |\phi(x) - \phi(y)|\, \rho_\varepsilon(z)}{|x-y|^{n+2s}} \, dx\, dy \right] dz \\
&\quad = \int_{B_\varepsilon} \left[\int_{Q(B_1)} \frac{|u(x+z) - u(y+z)|\, |\phi(x) - \phi(y)|\, \rho_\varepsilon(z)}{|x-y|^{n+2s}} \, dx\, dy \right] dz \\
&\quad \leq \frac{\varepsilon^{-n}}{2} \int_{B_\varepsilon} \left[\int_{Q(B_2)} \frac{|u(x) - u(y)|^2}{|x-y|^{n+2s}} \, dx\, dy + \int_{Q(B_1)} \frac{|\phi(x) - \phi(y)|^2}{|x-y|^{n+2s}} \, dx\, dy \right] dz \\
&\quad < +\infty,
\end{aligned}$$

because $u \in \mathbb{H}^s(B_3) \cap L^\infty(\mathbb{R}^n)$, where $Q(B_r)$ is defined as

$$Q(B_r) := \mathbb{R}^n \times \mathbb{R}^n \setminus \big((\mathbb{R}^n \setminus B_r) \times (\mathbb{R}^n \setminus B_r)\big) \quad r > 0.$$

Hence, by Tonelli's theorem, the function

$$\mathbb{R}^{2n} \times \mathbb{R}^n \ni (x, y, z) \mapsto \frac{(u(x+z) - u(y+z))(\phi(x) - \phi(y))\, \rho_\varepsilon(z)}{|x-y|^{n+2s}}$$

belongs to $L^1(\mathbb{R}^{2n} \times \mathbb{R}^n)$. Then, by Fubini's theorem,

$$
\begin{aligned}
&\int_{\mathbb{R}^n} \left[\int_{\mathbb{R}^{2n}} \frac{(u(x+z)-u(y+z))(\phi(x)-\phi(y))\,\rho_\varepsilon(z)}{|x-y|^{n+2s}}\,dx\,dy \right] dz \\
&= \int_{\mathbb{R}^{2n}} \left[\int_{\mathbb{R}^n} \frac{(u(x+z)-u(y+z))(\phi(x)-\phi(y))\,\rho_\varepsilon(z)}{|x-y|^{n+2s}}\,dz \right] dx\,dy \\
&= \int_{\mathbb{R}^{2n}} \frac{(u_\varepsilon(x)-u_\varepsilon(y))(\phi(x)-\phi(y))}{|x-y|^{n+2s}}\,dx\,dy,
\end{aligned}
$$

thanks to the definition of u_ε.

Therefore, we may use Fubini's theorem to obtain that, for any $\phi \in C_0^\infty(B_{1/2})$,

$$
\begin{aligned}
&\int_{\mathbb{R}^n} f_\varepsilon(x)\phi(x)\,dx \\
&= \int_{\mathbb{R}^n} \left[\int_{\mathbb{R}^n} f(x+z)\phi(x)\rho_\varepsilon(z)\,dx \right] dz \\
&= \int_{B_\varepsilon} \left[\int_{\mathbb{R}^n} f(\tilde{x})\phi(\tilde{x}-z)\rho_\varepsilon(z)\,d\tilde{x} \right] dz \\
&= \int_{B_\varepsilon} \left[\int_{\mathbb{R}^n\times\mathbb{R}^n} \frac{(u(\tilde{x})-u(\tilde{y}))(\phi(\tilde{x}-z)-\phi(\tilde{y}-z))\,\rho_\varepsilon(z)}{|\tilde{x}-\tilde{y}|^{n+2s}}\,d\tilde{x}\,d\tilde{y} \right] dz \\
&= \int_{\mathbb{R}^n} \left[\int_{\mathbb{R}^n\times\mathbb{R}^n} \frac{(u(x+z)-u(y+z))(\phi(x)-\phi(y))\,\rho_\varepsilon(z)}{|x-y|^{n+2s}}\,dx\,dy \right] dz \\
&= \int_{\mathbb{R}^n\times\mathbb{R}^n} \frac{(u_\varepsilon(x)-u_\varepsilon(y))(\phi(x)-\phi(y))}{|x-y|^{n+2s}}\,dx\,dy,
\end{aligned}
$$

thanks to the fact that u is a weak solution of (4.6). This means that u_ε is a smooth solution of (4.7).

As a consequence of this, we may apply [89, corollary 4.3] to obtain that $u_\varepsilon \in C^\alpha(B_{1/4})$, for any $0 < \alpha < \min\{2s, 1\}$, and

$$
\|u_\varepsilon\|_{C^\alpha(B_{1/4})} \leq C\big(\|u_\varepsilon\|_{L^\infty(\mathbb{R}^n)} + \|f_\varepsilon\|_{L^\infty(B_1)}\big),
$$

for a suitable $C > 0$ depending on n, s, and α. Hence, the desired result follows by sending $\varepsilon \searrow 0$. □

4.3.1 Maximum principle for weak solutions

Here we prove a maximum principle for weak solutions in $\mathbb{H}^s(\Omega)$, which is the natural extension of the classical one in H^1.

Lemma 4.6 *Let $v \in \mathbb{H}^s(\Omega)$ satisfy $(-\Delta)^s v \leq 0$ in the weak sense, with $v \leq 0$ in $\mathbb{R}^n \setminus \Omega$. Then $v \leq 0$ in $\mathbb{R}^n$.*

Proof By assumption, $v^+ = 0$ in $\mathbb{R}^n \setminus \Omega$; hence, $v^+ \in \mathbb{H}^s_0(\Omega)$, so we can use it as a test function in the weak formulation of the problem. Therefore, setting for simplicity

$$\langle \phi, \psi \rangle_s := C(n,s) \int_{\mathbb{R}^n \times \mathbb{R}^n} \frac{(\phi(x)-\phi(y))(\psi(x)-\psi(y))}{|x-y|^{n+2s}}\, dx\, dy$$

and

$$\|\phi\|_s := \sqrt{\langle \phi, \phi \rangle_s},$$

from the problem we obtain that

$$\langle v^+, v \rangle_s \leq 0.$$

As a consequence of this, writing $v^- := v^+ - v \geq 0$ and using that $v^+(x)v^-(x) = 0$ and $v^+(x)v^-(y) \geq 0$ a.e. $x, y \in \mathbb{R}^n$, we get

$$\begin{aligned} 0 \geq \langle v^+, v \rangle_s &= \langle v^+, v^+ - v^- \rangle_s \\ &= \|v^+\|_s^2 - \langle v^+, v^- \rangle_s \\ &= \|v^+\|_s^2 - \int_{\mathbb{R}^n \times \mathbb{R}^n} \frac{(v^+(x)-v^+(y))(v^-(x)-v^-(y))}{|x-y|^{n+2s}}\, dx\, dy \\ &= \|v^+\|_s^2 + \int_{\mathbb{R}^n \times \mathbb{R}^n} \frac{(v^+(x)v^-(y)+v^-(x)v^+(y))}{|x-y|^{n+2s}}\, dx\, dy \\ &\geq \|v^+\|_s^2. \end{aligned}$$

Therefore, $v^+ = 0$ in $\mathbb{R}^n$, so the claim is proved. □

Now we are ready to show that the weak solutions of the fractional Laplace equation are continuous up to the boundary (hence in the whole of the space) if they are outside the domain where the problem is set: with respect to this, our result is stated in Theorem 4.2. Here we provide a proof of it.

4.3.2 Proof of Theorem 4.2

The idea of the proof is to consider a smooth neighborhood Ω_ε of Ω and a continuous function β_ε that agrees with u outside Ω_ε and with $(-\Delta)^s \beta_\varepsilon$ very large (by the smoothness of β_ε in Ω_ε, the notion of viscosity and weak solution are the same). By possibly translating β_ε a little bit up, one checks that β_ε must be above u. Then the continuity of β_ε gives a control from above of the values of u near $\partial\Omega$. By a reverse argument, one obtains a control from below and hence the continuity of u near $\partial\Omega$.

We argue by contradiction by supposing that there exist $x_0 \in \mathbb{R}^n$ and a sequence $\{x_j\}_{j\in\mathbb{N}}$ in $\mathbb{R}^n$ such that

$$\lim_{j\to+\infty} x_j = x_0 \quad \text{but} \quad |u(x_j) - u(x_0)| \geq a,$$

for some $a > 0$. For definiteness, we suppose that

$$u(x_j) - u(x_0) \geq a \tag{4.8}$$

because the case $u(x_0) - u(x_j) \geq a$ can be treated in a similar way.

First of all, we claim that

$$x_0 \in \partial\Omega. \tag{4.9}$$

Indeed, $x_0 \in \overline{\Omega}$ because u is continuous in the open set $\mathbb{R}^n \setminus \overline{\Omega}$ by assumption. Moreover, it cannot be that x_0 lies in the interior of Ω because u is also continuous in any domain compactly contained in Ω, thanks to the local regularity theory stated in Proposition 4.5. This proves (4.9).

Now let $\varepsilon \in (0,2]$, and let Ω_ε be a smooth ε-neighborhood of Ω, that is, a set with C^2-boundary and such that

$$\bigcup_{x\in\Omega} B_{\varepsilon/2}(x) \subseteq \Omega_\varepsilon \subseteq \bigcup_{x\in\Omega} B_\varepsilon(x).$$

Moreover, let β_ε be the viscosity solution of

$$\begin{cases} (-\Delta)^s \beta_\varepsilon = M & \text{in } \Omega_\varepsilon \\ \beta_\varepsilon = u & \text{in } \mathbb{R}^n \setminus \Omega_\varepsilon, \end{cases} \tag{4.10}$$

where $M := \|f\|_{L^\infty(\Omega)}$. The existence of such a viscosity solution is warranted by Corollary 4.4 and by the regularity assumptions[2] on u.

Note that $\beta_\varepsilon \in C(\overline{\Omega}_\varepsilon)$ (again by Corollary 4.4) with modulus of continuity of β_ε in Ω_ε independent of ε: see, for instance, [57, theorem 32] for the continuity up to the boundary. Also, by interior regularity (see [24, theorem 5]), we have that β_ε is as smooth as we like in any domain compactly contained in Ω_ε (of course, the smooth

[2] To be precise, in order to use Corollary 4.4 to obtain the desired solution β_ε in (4.10), one needs to extend the datum u continuously inside Ω_ε. This can be accomplished by taking $\phi_\varepsilon \in C^\infty(\mathbb{R}^n)$ such that $\phi_\varepsilon = 1$ in $\mathbb{R}^n \setminus \Omega_{\varepsilon/2}$ and $\phi_\varepsilon = 0$ in $\Omega_{\varepsilon/4}$. Then one defines $g_\varepsilon := \phi_\varepsilon u + (1 - \phi_\varepsilon)$. By construction, $g_\varepsilon \in C(\mathbb{R}^n)$, so Corollary 4.4 gives the existence of a viscosity solution β_ε of $(-\Delta)^s \beta_\varepsilon = M$ in Ω_ε, with $\beta_\varepsilon = g_\varepsilon = u$ in $\mathbb{R}^n \setminus \Omega_\varepsilon$.

norm involved depends on the distance of the domain from $\partial\Omega_\varepsilon$). All in all, we obtain that

$$\begin{aligned} (-\Delta)^s \beta_\varepsilon = M \ in \ \Omega_\varepsilon \ in \ the \ viscosity \ sense, \\ \beta_\varepsilon = u \ in \ \mathbb{R}^n \setminus \Omega_\varepsilon, \ and \\ \beta_\varepsilon \in C^2(\overline{\Omega}) \cap C(\overline{\Omega}_\varepsilon), \end{aligned} \tag{4.11}$$

with modulus of continuity ρ of β_ε in Ω_ε defined as

$$\rho(t) = \sup_{\substack{x,y \in \Omega_\varepsilon \\ |x-y|<t}} |\beta_\varepsilon(x) - \beta_\varepsilon(y)|.$$

Note that ρ is independent of ε, while $\|\beta_\varepsilon\|_{C^2(\overline{\Omega})}$ may depend on ε.

Now let σ also be the modulus of continuity of u in the compact[3] set $\overline{\Omega}_2 \setminus \Omega \subset \mathbb{R}^n \setminus \Omega$; that is,

$$\sigma(t) = \sup_{\substack{x,y \in \overline{\Omega}_2 \setminus \Omega \\ |x-y|<t}} |u(x) - u(y)|.$$

We set

$$\eta_\varepsilon := \varepsilon + \rho(\varepsilon) + \sigma(\varepsilon). \tag{4.12}$$

First of all, we point out that, thanks to the continuity properties of β_ε and u,

$$\lim_{\varepsilon \searrow 0} \eta_\varepsilon = 0. \tag{4.13}$$

Moreover, we claim that

$$\beta_\varepsilon + \eta_\varepsilon > u \quad \text{in } \mathbb{R}^n \setminus \Omega. \tag{4.14}$$

To check this, let $x \in \mathbb{R}^n \setminus \Omega$. First, if $x \in \mathbb{R}^n \setminus \Omega_\varepsilon$, then $\beta_\varepsilon(x) = u(x)$, due to (4.11), so (4.14) holds true. If, however, $x \in \Omega_\varepsilon \setminus \Omega$, then there exists $y \in \partial\Omega_\varepsilon \subseteq \mathbb{R}^n \setminus \Omega_\varepsilon$ such that $|x - y| \le \varepsilon$. Therefore, $\beta_\varepsilon(y) = u(y)$, $|\beta_\varepsilon(x) - \beta_\varepsilon(y)| \le \rho(\varepsilon)$, and $|u(x) - u(y)| \le \sigma(\varepsilon)$. As a consequence,

$$\beta_\varepsilon(x) - u(x) \ge \beta_\varepsilon(y) - u(y) - (\rho(\varepsilon) + \sigma(\varepsilon)) = -(\rho(\varepsilon) + \sigma(\varepsilon)) = -\eta_\varepsilon + \varepsilon > -\eta_\varepsilon,$$

which proves (4.14).

Now let $v_\varepsilon := u - \beta_\varepsilon - \eta_\varepsilon$. We claim that

$$v_\varepsilon \le 0 \quad \text{in } \mathbb{R}^n. \tag{4.15}$$

For this, note that, by (4.11) and the assumptions on u, the function v_ε is a weak solution of

$$(-\Delta)^s v_\varepsilon = f - M \le 0 \quad \text{in } \Omega,$$

thanks to the choice of M. Moreover, by (4.14), we know that $v_\varepsilon \le 0$ in $\mathbb{R}^n \setminus \Omega$. Also, $v_\varepsilon \in \mathbb{H}^s(\Omega)$ because so are u and β_ε, by the assumptions on u and (4.11). Accordingly, we can use Lemma 4.6 and obtain (4.15).

[3] Of course, Ω_2 is simply Ω_ε, for $\varepsilon = 2$.

This gives that $u \leq \beta_\varepsilon + \eta_\varepsilon$ in $\mathbb{R}^n$ by the definition of v_ε. Thus, from this, (4.8), and the continuity of β_ε, we see that, fixed $\varepsilon \in (0,2)$, for large j,

$$\begin{aligned} a + u(x_0) &\leq u(x_j) \\ &\leq \beta_\varepsilon(x_j) + \eta_\varepsilon \\ &\leq \beta_\varepsilon(x_0) + \rho(|x_j - x_0|) + \eta_\varepsilon \\ &\leq \beta_\varepsilon(x_0) + 2\eta_\varepsilon \end{aligned} \tag{4.16}$$

because $\rho(|x_j - x_0|) = \eta_{|x_j - x_0|} - \sigma(|x_j - x_0|) - |x_j - x_0| < \eta_{|x_j - x_0|} < \eta_\varepsilon$ by (4.12) and the choice of x_j.

Moreover, recalling (4.9), we know that there exists $y_\varepsilon \in \mathbb{R}^n \setminus \Omega_\varepsilon$ such that $|y_\varepsilon - x_0| \leq \varepsilon$. Notice that $\beta_\varepsilon(y_\varepsilon) = u(y_\varepsilon)$ by (4.11). Therefore, (4.16) gives

$$\begin{aligned} a + u(x_0) &\leq \beta_\varepsilon(x_0) + 2\eta_\varepsilon \\ &= u(y_\varepsilon) + \beta_\varepsilon(x_0) - \beta_\varepsilon(y_\varepsilon) + 2\eta_\varepsilon \\ &\leq u(y_\varepsilon) + 3\eta_\varepsilon, \end{aligned} \tag{4.17}$$

because

$$|\beta_\varepsilon(x_0) - \beta_\varepsilon(y_\varepsilon)| \leq \rho(|x_0 - y_\varepsilon|) = \eta_{|x_0 - y_\varepsilon|} - \sigma(|x_0 - y_\varepsilon|) - |x_0 - y_\varepsilon| < \eta_{|x_0 - y_\varepsilon|} < \eta_\varepsilon.$$

Also, $y_\varepsilon, x_0 \in \Omega_2 \setminus \Omega$ because, by construction, $\Omega_\varepsilon \subset \Omega_2$ (with $\varepsilon < 2$) and $|y_\varepsilon - x_0| \leq \varepsilon$. Thus, since $u \in C(\mathbb{R}^n \setminus \Omega)$,

$$u(y_\varepsilon) - u(x_0) \leq \sigma(|y_\varepsilon - x_0|) \leq \sigma(\varepsilon) \leq \eta_\varepsilon,$$

thanks to the fact that $\sigma(\varepsilon) = \eta_\varepsilon - \rho(\varepsilon) - \varepsilon < \eta_\varepsilon$. Thus, this and (4.17) imply that

$$a + u(x_0) \leq u(y_\varepsilon) + 3\eta_\varepsilon \leq u(y_\varepsilon) - u(x_0) + u(x_0) + 3\eta_\varepsilon \leq u(x_0) + 4\eta_\varepsilon;$$

that is,

$$a \leq 4\eta_\varepsilon.$$

By taking the limit as $\varepsilon \searrow 0$ and recalling (4.13), we obtain $a \leq 0$, which contradicts our assumption. Thus, $u \in C(\mathbb{R}^n)$, and this concludes the proof of Theorem 4.2.

4.4 Proof of Theorem 4.1

By extending f in a bounded way outside Ω, we may suppose that $f \in L^\infty(\mathbb{R}^n)$. Let u be a weak solution of problem (4.1). Since $u = g$ a.e. in $\mathbb{R}^n \setminus \Omega$, then $u \in C(\mathbb{R}^n \setminus \Omega)$, thanks to the assumptions on g. Then we can apply Theorem 4.2 and get that

$$u \in C(\mathbb{R}^n). \tag{4.18}$$

Now let us fix $x_0 \in \Omega$. By convolution (as done in Proposition 4.5) and recalling the stability of viscosity solutions (see, e.g., [56, lemma 4.5]), we may suppose that u is also smooth in a small neighborhood Ω' of x_0, and from (4.3), for any $\varphi \in C_0^\infty(\Omega')$,

$$\begin{aligned} C(n,s) \int_{\mathbb{R}^n \times \mathbb{R}^n} & \frac{(2u(x) - u(x+Y) - u(x-Y))\,\varphi(x)}{|Y|^{n+2s}}\,dx\,dY \\ &= C(n,s) \int_{\mathbb{R}^n \times \mathbb{R}^n} \frac{(u(x)-u(y))(\varphi(x)-\varphi(y))}{|x-y|^{n+2s}}\,dx\,dy \\ &= \int_{\mathbb{R}^n} f(x)\varphi(x)\,dx. \end{aligned}$$

Therefore,

$$C(n,s) \int_{\mathbb{R}^n} \frac{(2u(x) - u(x+Y) - u(x-Y))}{|Y|^{n+2s}}\,dY = f(x),$$

for almost any $x \in \Omega'$ and, in fact, for any $x \in \Omega'$, due to (4.18). In particular,

$$C(n,s) \int_{\mathbb{R}^n} \frac{(2u(x_0) - u(x_0+Y) - u(x_0-Y))}{|Y|^{n+2s}}\,dY = f(x_0).$$

Now let us take v as in (4.5): because $v \geq u$ in $\mathbb{R}^n$ and $v(x_0) = u(x_0)$, we obtain

$$\begin{aligned} (-\Delta)^s v(x_0) &= C(n,s) \int_{\mathbb{R}^n} \frac{(2v(x_0) - v(x_0+Y) - v(x_0-Y))}{|Y|^{n+2s}}\,dY \\ &\leq C(n,s) \int_{\mathbb{R}^n} \frac{(2u(x_0) - u(x_0+Y) - u(x_0-Y))}{|Y|^{n+2s}}\,dY \\ &= f(x_0). \end{aligned}$$

This shows that u is a viscosity subsolution of (4.1), and one can prove similarly that u is a viscosity supersolution. This ends the proof of Theorem 4.1. □

Remark 4.7 The result of Theorem 4.1 also holds when $f = f(x,u)$, for $f \in C(\mathbb{R}^{n+1})$. Indeed, one sets $\tilde{f}(x) := f(x,u(x)) \in L^\infty(\mathbb{R}^{n+1})$ and applies Theorem 4.2 to obtain that u is continuous and therefore so is $\tilde{f}$ (in this case it is necessary to assume that u is bounded from the beginning). Then one may apply Theorem 4.1 (with $\tilde{f}$ in the place of f) and obtain that u is a viscosity solution of (4.1).

As a consequence of Theorem 4.2, we can prove the following result about the first eigenfunction of the operator $(-\Delta)^s$:

Corollary 4.8 Let $\lambda_{1,s}$ be the first eigenvalue of the operator $(-\Delta)^s$ in Ω with Dirichlet boundary datum and $e_{1,s} \in \mathbb{H}^s_0(\Omega)$ be the corresponding eigenfunction; that is,

$$\begin{cases} (-\Delta)^s e_{1,s} = \lambda_{1,s} e_{1,s} & \text{in } \Omega \\ e_{1,s} = 0 & \text{in } \mathbb{R}^n \setminus \Omega. \end{cases}$$

Then $e_{1,s} > 0$ in Ω.

Proof By Proposition 3.4, we know that $e_{1,s} \in L^\infty(\Omega)$. Hence, by Theorem 4.2, it is easily seen that $e_{1,s} \in C(\mathbb{R}^n)$. Moreover, as a consequence of this and of Proposition 3.1, we have that

$$e_{1,s}(x) \geq 0 \quad \text{for any } x \in \Omega. \tag{4.19}$$

Also,

$$(-\Delta)^s e_{1,s}(x) \geq 0 \quad \text{for any } x \in \Omega, \tag{4.20}$$

thanks to the positivity of $\lambda_{1,s}$.

Assume by contradiction that there exists $\tilde{x} \in \Omega$ such that $e_{1,s}(\tilde{x}) = 0$. Then, by (4.2), (4.19), and (4.20), we get

$$0 \leq (-\Delta)^s e_{1,s}(\tilde{x}) = -C(n,s) \int_{\mathbb{R}^n} \frac{e_{1,s}(\tilde{x}+y) + e_{1,s}(\tilde{x}-y)}{|y|^{n+2s}} \, dy \leq 0,$$

so that, using also (4.19), $e_{1,s}(\tilde{x}+y) + e_{1,s}(\tilde{x}-y) = 0$ in the whole of $\mathbb{R}^n$. Hence, by this and again (4.19), we get that $e_{1,s} \equiv 0$ in $\mathbb{R}^n$. Of course, this is a contradiction, since $e_{1,s}$ is an eigenfunction.

Hence, $e_{1,s} > 0$ in Ω, and this completes the proof. □

4.5 On the boundedness of weak solutions

In this section we discuss the boundedness of weak solutions to nonlocal problems in both the linear and nonlinear cases.

4.5.1 The linear case

In all the previous results of this chapter, for the sake of simplicity, we have focused on the case of bounded solutions of (4.1), that is, on solutions $u \in L^\infty(\mathbb{R}^n)$. Here we show that, in fact, this assumption can be removed. Indeed, the boundedness of solutions may be proved directly using a barrier and the comparison principle for weak solutions (see the barrier β_ε and the argument in Section 4.3.2). However, we present here a different argument of classical flavor that is suited also for more general fractional operators in divergence form.

For the proof of it, we need some general energy estimates of fractional type, stated in Lemma 3.3 and in the forthcoming Lemma 4.9.

Lemma 4.9 *Let $f \in L^2(\Omega)$ and $v \in \mathbb{H}^s(\Omega)$. Suppose that $(-\Delta)^s v = f$ in Ω in the weak sense, and let $w := v^+$. Then*

$$\int_{\mathbb{R}^n \times \mathbb{R}^n} \frac{|w(x) - w(y)|^2}{|x - y|^{n+2s}} \, dx\, dy \le \int_{\Omega} f(x) w(x) \, dx. \tag{4.21}$$

Proof By construction, $v^+ = 0$ outside Ω, so $v^+ \in \mathbb{H}^s_0(\Omega)$; hence, we can use it as a test function. Using this and Lemma 3.3, we obtain

$$\begin{aligned}
\int_{\mathbb{R}^n \times \mathbb{R}^n} \frac{|w(x) - w(y)|^2}{|x - y|^{n+2s}} \, dx\, dy &= \int_{\mathbb{R}^n \times \mathbb{R}^n} \frac{|v^+(x) - v^+(y)|^2}{|x - y|^{n+2s}} \, dx\, dy \\
&\le \int_{\mathbb{R}^n \times \mathbb{R}^n} \frac{(v(x) - v(y))(v^+(x) - v^+(y))}{|x - y|^{n+2s}} \, dx\, dy \\
&= \int_{\Omega} f(x) v^+(x) \, dx,
\end{aligned}$$

which concludes the proof. □

The classical counterpart of (4.21) is that if $-\Delta v = f$ in Ω and $v \le 0$ on $\partial\Omega$, then one may test the equation against $w := v^+$ and obtain

$$\int_{\Omega} |\nabla w(x)|^2 \, dx = \int_{\Omega} \nabla v(x) \nabla w(x) \, dx = \int_{\Omega} (-\Delta v)(x) w(x) \, dx = \int_{\Omega} f(x) w(x) \, dx.$$

The main result of this section is the following:

Proposition 4.10 Let $u \in \mathbb{H}^s(\Omega)$ be a weak solution of problem (4.1) (i.e., suppose that (4.3) holds), with $f \in C(\mathbb{R}^n)$ and $g \in C(\mathbb{R}^n) \cap L^\infty(\mathbb{R}^n)$. Then $u \in L^\infty(\mathbb{R}^n)$.

Proof We use the same arguments considered in Proposition 3.4 to prove an L^∞-regularity result for the eigenfunctions of the operator $(-\Delta)^s$. The argument is a fractional version of the classical De Giorgi–Stampacchia iteration method.

By extending f in a bounded way outside Ω, we may suppose that $f \in L^\infty(\mathbb{R}^n)$. Moreover, we may assume that u does not vanish identically (otherwise, there is nothing to prove). Let $\delta > 0$, to be taken appropriately small in what follows. Up to multiplying u by a small constant, we may and do assume that

$$\|g\|_{L^\infty(\mathbb{R}^n)} \le \frac{1}{2}. \tag{4.22}$$

Here we use the fact that $u = g$ a.e. in $\mathbb{R}^n \setminus \Omega$) and

$$\|u\|_{L^2(\Omega)} \le \sqrt{\delta}. \tag{4.23}$$

Now, for any $k \in \mathbb{N}$, we let $C_k := 1 - 2^{-k}$, $v_k := u - C_k$, $w_k := v_k^+$, and $U_k := \|w_k\|^2_{L^2(\Omega)}$.

We remark that

$$0 \le w_k \le |u| + |C_k| \le |u| + 1 \in L^2(\Omega),$$

because Ω is bounded, and

$$w_k \to (u-1)^+ \quad \text{a.e. in } \mathbb{R}^n$$

as $k \to +\infty$. Therefore, by the dominated convergence theorem applied in $L^2(\Omega)$, we get

$$U_k := \|w_k\|^2_{L^2(\Omega)} \to \|(u-1)^+\|^2_{L^2(\Omega)} \tag{4.24}$$

as $k \to +\infty$.

Moreover, for any $k \in \mathbb{N}$, we have that $C_{k+1} > C_k$, so $v_{k+1} < v_k$ a.e. in $\mathbb{R}^n$, from which we deduce that

$$w_{k+1} \le w_k \quad \text{a.e. in } \mathbb{R}^n. \tag{4.25}$$

Furthermore, by (4.22), in $\mathbb{R}^n \setminus \Omega$, we have that

$$v_{k+1} = g - 1 + 2^{-k-1} \le -\frac{1}{2} + 2^{-k-1} \le 0$$

and $(-\Delta)^s v_{k+1} = (-\Delta)^s u = f$ in Ω in the weak sense. Therefore, we can apply Lemma 4.9 with $v := v_{k+1}$ and $w := v_{k+1}^+ = w_{k+1}$, which gives that

$$\begin{aligned}\int\limits_{\Omega\times\Omega} \frac{|w_{k+1}(x) - w_{k+1}(y)|^2}{|x-y|^{n+2s}}\,dx\,dy &\le \int\limits_{\mathbb{R}^n\times\mathbb{R}^n} \frac{|w_{k+1}(x) - w_{k+1}(y)|^2}{|x-y|^{n+2s}}\,dx\,dy \\ &\le \int\limits_{\Omega} f(x)\, w_{k+1}(x)\,dx.\end{aligned}$$

As a consequence of this,

$$\begin{aligned}\int\limits_{\Omega\times\Omega} \frac{|w_{k+1}(x) - w_{k+1}(y)|^2}{|x-y|^{n+2s}}\,dx\,dy &\le \int\limits_{\Omega\cap\{w_{k+1}>0\}} |f(x)|\, w_{k+1}(x)\,dx \\ &\le \|f\|_{L^\infty(\mathbb{R}^n)} \int\limits_{\Omega\cap\{w_{k+1}>0\}} w_{k+1}(x)\,dx \\ &\le \|f\|_{L^\infty(\mathbb{R}^n)} \int\limits_{\Omega\cap\{w_{k+1}>0\}} w_k(x)\,dx \\ &\le \|f\|_{L^\infty(\mathbb{R}^n)} |\Omega \cap \{w_{k+1} > 0\}|^{1/2} \|w_k\|_{L^2(\Omega)} \\ &\le \|f\|_{L^\infty(\mathbb{R}^n)} |\Omega \cap \{w_{k+1} > 0\}|^{1/2} U_k^{1/2},\end{aligned} \tag{4.26}$$

thanks (4.25), the regularity of f, the Hölder inequality, and the definition of U_k.

Now we claim that

$$\{w_{k+1} > 0\} \subseteq \{w_k > 2^{-(k+1)}\}. \tag{4.27}$$

To establish this, we observe that if $x \in \{w_{k+1} > 0\}$, then

$$0 < w_{k+1}(x) = v_{k+1}^+(x) = \max\{u(x) - C_{k+1}, 0\}.$$

Hence, $u(x) - C_{k+1} > 0$. Accordingly,

$$v_k(x) = u(x) - C_k > C_{k+1} - C_k = 2^{-(k+1)};$$

hence, $w_k(x) := v_k^+(x) = v_k(x) > 2^{-(k+1)}$, which proves (4.27).

As a consequence of (4.27), we obtain

$$\begin{aligned} U_k = \|w_k\|^2_{L^2(\Omega)} &\geq \int_{\Omega \cap \{w_k > 2^{-(k+1)}\}} w_k^2(x)\,dx \\ &\geq 2^{-2(k+1)} |\Omega \cap \{w_k \geq 2^{-(k+1)}\}| \\ &\geq 2^{-2(k+1)} |\Omega \cap \{w_{k+1} > 0\}|, \end{aligned}$$

so

$$|\Omega \cap \{w_{k+1} > 0\}| \leq 2^{2(k+1)}\, U_k. \tag{4.28}$$

Then, by (4.26) and (4.28), we get that

$$\int_{\Omega \times \Omega} \frac{|w_{k+1}(x) - w_{k+1}(y)|^2}{|x-y|^{n+2s}}\,dx\,dy \leq 2^{k+1}\, \|f\|_{L^\infty(\mathbb{R}^n)}\, U_k. \tag{4.29}$$

Now we use the Hölder inequality (with exponents $2_s^*/2$ and $n/(2s)$) and the fractional Sobolev inequality (see, e.g. [83, theorem 6.7]) to get that

$$\begin{aligned} U_{k+1} &\leq \left(\int_\Omega |w_{k+1}(x)|^{2_s^*}\,dx \right)^{2/2_s^*} |\Omega \cap \{w_{k+1} > 0\}|^{2s/n} \\ &\leq \tilde{c}\, |\Omega \cap \{w_{k+1} > 0\}|^{2s/n} \int_{\Omega \times \Omega} \frac{|w_{k+1}(x) - w_{k+1}(y)|^2}{|x-y|^{n+2s}}\,dx\,dy, \end{aligned}$$

for some positive constant $\tilde{c}$ depending only on n and s. Consequently, by (4.28) and (4.29), we see that

$$\begin{aligned} U_{k+1} &\le \tilde{c}\, 2^{k+1} \, \|f\|_{L^\infty(\mathbb{R}^n)} \, U_k \, (2^{2(k+1)} U_k)^{2s/n} \\ &= \tilde{c}\, \|f\|_{L^\infty(\mathbb{R}^n)} \, 2^{1+4s/n} (2^{1+4s/n})^k U_k^{1+2s/n} \\ &\le \left(1+\tilde{c}\, \|f\|_{L^\infty(\mathbb{R}^n)} \, 2^{1+4s/n}\right)(2^{1+4s/n})^k U_k^{1+2s/n} \\ &\le \left[\left(1+\tilde{c}\, \|f\|_{L^\infty(\mathbb{R}^n)} \, 2^{1+4s/n}\right)2^{1+4s/n}\right]^k U_k^{1+2s/n} \\ &= C^k U_k^\beta, \end{aligned} \tag{4.30}$$

where $\beta := 1 + (2s/n) > 1$ and $C > 1$ only depends on n, s, and f.

Now we are ready to perform our choice of δ: namely, we assume that $\delta > 0$ is so small that

$$\delta^{\beta-1} < \frac{1}{C^{1/(\beta-1)}} \tag{4.31}$$

We also fix

$$\eta \in \left(\delta^{\beta-1}, \frac{1}{C^{1/(\beta-1)}}\right).$$

Notice that since $C > 1$ and $\beta > 1$,

$$\eta \in (0,1). \tag{4.32}$$

Moreover,

$$\delta^{\beta-1} \le \eta \quad \text{and} \quad C\eta^{\beta-1} \le 1. \tag{4.33}$$

We claim that, for any $k \in \mathbb{N}$,

$$U_k \le \delta \eta^k. \tag{4.34}$$

The proof is by induction. First of all, by (4.23),

$$U_0 := \|w_0\|^2_{L^2(\Omega)} = \|u^+\|^2_{L^2(\Omega)} \le \|u\|^2_{L^2(\Omega)} \le \delta,$$

which is (4.34) when $k = 0$. Now let us suppose that (4.34) holds true for k, and let us prove it for $k+1$. For this, we use (4.30) and (4.33):

$$U_{k+1} \le C^k U_k^\beta \le C^k (\delta\eta^k)^\beta = \delta (C\eta^{\beta-1})^k \delta^{\beta-1} \eta^k \le \delta \eta^{k+1}.$$

This proves (4.34).

Then, by (4.32) and (4.34), we conclude that

$$\lim_{k\to+\infty} U_k = 0.$$

Hence, by (4.24), $(u-1)^+ = 0$ a.e. in Ω; that is, $u \le 1$ a.e. in Ω. By replacing u with $-u$, we obtain $\|u\|_{L^\infty(\Omega)} \le 1$. This ends the proof. □

4.5.2 The nonlinear case

We end this chapter with a regularity result for nonlocal nonlinear problems (see [23]), which will be useful in several applications later in this book.

Proposition 4.11 Let u be a positive weak solution of problem

$$\begin{cases} (-\Delta)^s u = f(x,u) & \text{in } \Omega \\ u = g & \text{in } \mathbb{R}^n \setminus \Omega, \end{cases}$$

and assume that

$$|f(x,t)| \le C(1+|t|^p),$$

for some $1 \le p \le 2_s^* - 1$ and $C > 0$, where $2_s^* := 2n/(n-2s)$ is the fractional critical Sobolev exponent. Then $u \in L^\infty(\Omega)$.

Proof For the sake of simplicity, we set

$$(-\Delta)^s u(x) = \int_{\mathbb{R}^n} \frac{u(x+y)+u(x-y)-2u(x)}{|y|^{n+2s}}\,dy.$$

The proof uses standard techniques for the fractional Laplacian, in particular, the following inequality: if φ is a convex and differentiable function, then

$$(-\Delta)^s \varphi(u) \le \varphi'(u)(-\Delta)^s u.$$

Let us define, for $\beta \ge 1$ and $T > 0$ large,

$$\varphi(t) = \varphi_{T,\beta}(t) := \begin{cases} 0 & \text{if } \; t \le 0 \\ t^\beta & \text{if } \; 0 < t < T \\ \beta T^{\beta-1}(t-T)+T^\beta & \text{if } \; t \ge T. \end{cases}$$

Observe that $\varphi(u) \in \mathbb{H}_0^s(\Omega)$ because φ is Lipschitz with constant $Z = \beta T^{\beta-1}$, and therefore,

$$\begin{aligned} \|\varphi(u)\|_{\mathbb{H}_0^s(\Omega)} &= \left(\int_{\mathbb{R}^n\times\mathbb{R}^n} \frac{|\varphi(u(x))-\varphi(u(y))|^2}{|x-y|^{n+2s}}\,dx\,dy\right)^{1/2} \\ &\le \left(\int_{\mathbb{R}^n\times\mathbb{R}^n} \frac{Z^2|u(x)-u(y)|^2}{|x-y|^{n+2s}}\,dx\,dy\right)^{1/2} \\ &= Z\|u\|_{\mathbb{H}_0^s(\Omega)}. \end{aligned}$$

Moreover, by Proposition 1.18, we have the following identity:

$$\|u\|_{\mathbb{H}_0^s(\Omega)} = \|(-\Delta)^{s/2}u\|_{L^2(\mathbb{R}^n)}. \tag{4.35}$$

Thus, by (4.35) and the Sobolev embedding theorem, we have

$$\int_\Omega \varphi(u(x))(-\Delta)^s\varphi(u(x))\,dx = \|\varphi(u)\|^2_{\mathbb{H}_0^s(\Omega)} \ge S_s\|\varphi(u)\|^2_{L^{2_s^*}(\Omega)}, \tag{4.36}$$

where S_s is the Sobolev constant defined in (1.92). However, since φ is convex and $\varphi(u)\varphi'(u) \in \mathbb{H}_0^s(\Omega)$, it follows that

$$\begin{aligned}\int_\Omega \varphi(u(x))(-\Delta)^s\varphi(u(x))dx &\leq \int_\Omega \varphi(u(x))\varphi'(u(x))(-\Delta)^s u(x)dx\\ &\leq C\int_\Omega \varphi(u(x))\varphi'(u(x))\left(1+u(x)^{2_s^*-1}\right)dx.\end{aligned}$$

From (4.36) and the preceding inequality, we get the following basic estimate:

$$\|\varphi(u)\|^2_{L^{2_s^*}(\Omega)} \leq C\int_\Omega \varphi(u(x))\varphi'(u(x))\left(1+u(x)^{2_s^*-1}\right)dx. \tag{4.37}$$

Since $u\varphi'(u) \leq \beta\varphi(u)$ and $\varphi'(u) \leq \beta(1+\varphi(u))$, the preceding estimate (4.37) becomes

$$\begin{aligned}\left(\int_\Omega (\varphi(u(x)))^{2_s^*}dx\right)^{2/2_s^*} &\leq C\beta\left(1+\int_\Omega (\varphi(u(x)))^2dx\right.\\ &\quad\left.+\int_\Omega (\varphi(u(x)))^2 u(x)^{2_s^*-2}dx\right).\end{aligned}\tag{4.38}$$

It is important to point out here that since $\varphi(u)$ grows linearly, both sides of the preceding expression are finite.

Claim 4.1 *Let β_1 be such that $2\beta_1 = 2_s^*$. Then $u \in L^{\beta_1 2_s^*}(\Omega)$.*

To see this, we take R large to be determined later. Then the Hölder inequality with $p=\beta_1 = 2_s^*/2$ and $p' = 2_s^*/(2_s^*-2)$ gives

$$\begin{aligned}&\int_\Omega (\varphi(u(x)))^2 u(x)^{2_s^*-2}dx\\ &\quad= \int_{\{u\leq R\}} (\varphi(u(x)))^2 u(x)^{2_s^*-2}dx + \int_{\{u>R\}} (\varphi(u(x)))^2 u(x)^{2_s^*-2}dx\\ &\quad\leq \int_{\{u\leq R\}} (\varphi(u(x)))^2 R^{2_s^*-2}dx\\ &\qquad+\left(\int_\Omega (\varphi(u(x)))^{2_s^*}dx\right)^{2/2_s^*}\left(\int_{\{u>R\}} u(x)^{2_s^*}dx\right)^{(2_s^*-2)/2_s^*}.\end{aligned}$$

By the monotone convergence theorem, we may take R so that

$$\left(\int_{\{u>R\}} u(x)^{2_s^*}dx\right)^{(2_s^*-2)/2_s^*} \leq \frac{1}{2C\beta_1}.$$

In this way, the second term in the preceding is absorbed by the left-hand side of (4.38) to get

$$\left(\int_\Omega (\varphi(u(x)))^{2_s^*}\,dx\right)^{2/2_s^*} \le 2C\beta_1\left(1+\int_\Omega (\varphi(u(x)))^2\,dx + \int_{\{u\le R\}} (\varphi(u(x)))^2 R^{2_s^*-2}\,dx\right). \tag{4.39}$$

Using that $\varphi_{T,\beta_1}(u) \le u^{\beta_1}$ on the right-hand side of (4.39) and then letting $T \to \infty$ on the left-hand side, since $2\beta_1 = 2_s^*$, we obtain

$$\left(\int_\Omega u(x)^{2_s^*\beta_1}\,dx\right)^{2/2_s^*} \le 2C\beta_1\left(1+\int_\Omega u(x)^{2_s^*}\,dx + R^{2_s^*-2}\int_\Omega u(x)^{2_s^*}\,dx\right) < +\infty.$$

This proves Claim 4.1.

We now go back to inequality (4.38), and we use as before that $\varphi_{T,\beta}(u) \le u^\beta$ on the right-hand side and then we take $T \to \infty$ on the left-hand side. Then

$$\left(\int_\Omega u(x)^{2_s^*\beta}\,dx\right)^{2/2_s^*} \le C\beta\left(1+\int_\Omega u(x)^{2\beta}\,dx + \int_\Omega u(x)^{2\beta+2_s^*-2}\,dx\right).$$

Since $\int_\Omega u(x)^{2\beta}\,dx \le |\Omega| + \int_\Omega u(x)^{2\beta+2_s^*-2}\,dx$, we get the following recurrence formula

$$\left(\int_\Omega u(x)^{2_s^*\beta}\,dx\right)^{2/2_s^*} \le 2C\beta(1+|\Omega|)\left(1+\int_\Omega u(x)^{2\beta+2_s^*-2}\,dx\right).$$

Therefore,

$$\left(1+\int_\Omega u(x)^{2_s^*\beta}\,dx\right)^{(1/2_s^*(\beta-1))} \le C_\beta^{(1/2(\beta-1))}\left(1+\int_\Omega u(x)^{2\beta+2_s^*-2}\,dx\right)^{(1/2(\beta-1))}, \tag{4.40}$$

where $C_\beta := 4C\beta(1+|\Omega|)$.

For $m \ge 1$, we define β_{m+1} inductively so that $2\beta_{m+1} + 2_s^* - 2 = 2_s^*\beta_m$; that is,

$$\beta_{m+1} - 1 = \frac{2_s^*}{2}(\beta_m - 1) = \left(\frac{2_s^*}{2}\right)^m (\beta_1 - 1).$$

Hence, from (4.40), it follows that

$$\left(1+\int_\Omega u(x)^{2_s^*\beta_{m+1}}\,dx\right)^{(1/2_s^*(\beta_{m+1}-1))} \le C_{\beta_{m+1}}^{(1/2(\beta_{m+1}-1))}\left(1+\int_\Omega u(x)^{2_s^*\beta_m}\,dx\right)^{(1/2_s^*(\beta_m-1))},$$

with $C_{m+1} := C_{\beta_{m+1}} = 4C\,\beta_{m+1}(1+|\Omega|)$. Then, defining for $m \geq 1$

$$A_m := \left(1 + \int_\Omega u(x)^{2_s^*\beta_m}\,dx\right)^{(1/2_s^*(\beta_m-1))},$$

by Claim 4.1 and using a limiting argument, we conclude that there exists $C_0 > 0$, independent of $m > 1$, such that

$$A_{m+1} \leq \prod_{k=2}^{m+1} C_k^{(1/2(\beta_k-1))} A_1 \leq C_0 A_1.$$

This implies that $\|u\|_{L^\infty(\Omega)} \leq C_0 A_1$, that is, $u \in L^\infty(\Omega)$. □

5

Spectral fractional Laplacian problems

Among others, two notions of fractional operators are well known and are widely studied in the literature in connection with elliptic problems of fractional type, namely, the *integral* operator (which reduces to the *classical fractional Laplacian*; see, e.g., [50, 55, 56, 57, 83, 99, 102, 184, 185, 186, 196, 197, 198, 200, 201, 202, 204, 205, 206] and references therein) and the *spectral* operator (sometimes called the *local fractional Laplacian*; see, e.g., [22, 42, 50, 51, 58, 59, 139, 208, 214] and references therein). This chapter is devoted to the study of these two fractional operators.

More precisely, for a fixed $s \in (0,1)$, in this chapter we consider the *integral* definition of the fractional Laplacian given by

$$(-\Delta)^s u(x) = -\frac{1}{2}C(n,s)\int_{\mathbb{R}^n} \frac{u(x+y)+u(x-y)-2u(x)}{|y|^{n+2s}}\,dy \quad x \in \mathbb{R}^n$$

and another fractional operator obtained via a *spectral* definition; that is,

$$A_s u = \sum_{i\in\mathbb{N}} a_i \widetilde{\lambda}_i^s \widetilde{e}_i,$$

where $\widetilde{e}_i, \widetilde{\lambda}_i$ are eigenfunctions and eigenvalues of the Laplace operator $-\Delta$ in Ω with homogeneous Dirichlet boundary data, while a_i represents the projection of u in the direction $\widetilde{e}_i$ (see Sections 1.3 and 1.4 for details).

Our aim here is to compare these two operators, with particular reference to their spectrum, in order to emphasize their differences. The result of this chapter have been obtained from [203].

5.1 Two different fractional operators

For any fixed $s \in (0,1)$, the fractional Laplace operator $(-\Delta)^s$ at the point x is defined by formula (1.20). A different operator, which is sometimes denoted by A_s, is defined as the power of the Laplace operator $-\Delta$, obtained by using the spectral decomposition of the Laplacian (see Chapter 1). Namely, let Ω be a smooth, bounded domain of $\mathbb{R}^n$, and let $\{\widetilde{\lambda}_k\}_{k\in\mathbb{N}}$, with

$$0<\widetilde{\lambda}_1<\widetilde{\lambda}_2\leq\cdots\leq\widetilde{\lambda}_k\leq\ldots,$$

be the divergent sequence of eigenvalues of the Laplacian operator $-\Delta$ in Ω with zero Dirichlet boundary data on $\partial\Omega$, and $\widetilde{e}_k$, $k\in\mathbb{N}$, the corresponding eigenfunctions, that is,

$$\begin{cases} -\Delta\widetilde{e}_k=\widetilde{\lambda}_k\widetilde{e}_k & \text{in } \Omega \\ \widetilde{e}_k=0 & \text{on } \partial\Omega, \end{cases}$$

normalized in such a way that $\|\widetilde{e}_k\|_{L^2(\Omega)}=1$.

For any $s\in(0,1)$ and any $u\in H_0^1(\Omega)$ with

$$u(x)=\sum_{i\in\mathbb{N}}a_i\,\widetilde{e}_i(x)\quad x\in\Omega,$$

one considers the operator

$$A_s u=\sum_{i\in\mathbb{N}}a_i\,\widetilde{\lambda}_i^s\,\widetilde{e}_i. \tag{5.1}$$

Our aim is to compare the two previous definitions of fractional Laplace operators. First of all, we note that these two fractional operators (i.e., the integral one and the spectral one) are *different* (despite some confusion that it is possible to find in some of the existing literature in which the two operators are somehow freely interchanged). Indeed, the spectral operator A_s depends on the domain Ω considered (because its eigenfunctions and eigenvalues depend on Ω), while the integral operator $(-\Delta)^s$ evaluated at some point is independent on the domain in which the equation is set.

Of course, by definition of A_s, it is easily seen that *the eigenvalues and the eigenfunctions of A_s are, respectively, λ_k^s and $\widetilde{e}_k$*, $k\in\mathbb{N}$, that is, the s-power of the eigenvalues of the Laplacian and the very same eigenfunctions of the Laplacian, respectively. However, we know that the spectrum of $(-\Delta)^s$ may be less explicit to be described (see Chapter 3).

A natural question is whether or not there is a relation between the spectrum of A_s and $(-\Delta)^s$ and, of course, between the respective eigenfunctions. By using the classical regularity theory for the eigenfunctions of the Laplace operator $-\Delta$ and some recent regularity results for the fractional Laplace equation (see [184, 185, 186, 202]), in this chapter we show that the eigenfunctions of A_s and $(-\Delta)^s$ are different. In particular, we prove that *the eigenfunctions of $(-\Delta)^s$ are, in general, no better than Hölder continuous up to the boundary*, different from the eigenfunctions of A_s (i.e., of the classical Laplacian), which are smooth up to the boundary (if the domain is too). Furthermore, with respect to the eigenvalues of A_s and $(-\Delta)^s$, we show that *the first eigenvalue of $(-\Delta)^s$ is strictly less than the first eigenvalue of A_s*. To this end, we will use some extension results for the fractional operators A_s and $(-\Delta)^s$ (see [55, 208]). Similar results for higher eigenvalues have been obtained in [161].

Summarizing, the main results of this chapter are as follows:

Theorem 5.1 *The operators* $(-\Delta)^s$ *and* A_s *are not the same because they have different eigenvalues and eigenfunctions (with respect to Dirichlet boundary conditions). In particular,*

- *The first eigenvalue of* $(-\Delta)^s$ *is strictly less than that of* A_s.
- *The eigenfunctions of* $(-\Delta)^s$ *are only Hölder continuous up to the boundary, different from the eigenfunctions of* A_s, *which are as smooth up to the boundary as the boundary allows.*

A more precise statement of Theorem 5.1 will be given in Theorem 5.2. For further comments on the similarities and differences between the operators A_s and $(-\Delta)^s$ for $s = 1/2$, see [93, remark 0.4].

5.2 A comparison between the eigenfunctions of A_s and $(-\Delta)^s$

This section is devoted to some remarks about the eigenfunctions of the operators A_s and $(-\Delta)^s$. Precisely, we will consider the following eigenvalue problems in a smooth, bounded domain $\Omega \subset \mathbb{R}^n$ with Dirichlet homogeneous boundary data driven, respectively, by A_s and $(-\Delta)^s$:

$$\begin{cases} A_s u = \lambda u & \text{in } \Omega \\ u = 0 & \text{on } \partial\Omega \end{cases} \tag{5.2}$$

and

$$\begin{cases} (-\Delta)^s u = \lambda u & \text{in } \Omega \\ u = 0 & \text{in } \mathbb{R}^n \setminus \Omega. \end{cases} \tag{5.3}$$

Note that in (5.3) the boundary condition is given in $\mathbb{R}^n \setminus \Omega$ and not simply on $\partial\Omega$ because of the nonlocal character of the operator $(-\Delta)^s$.

In what follows, we will denote by e_{k,A_s} and $e_{k,s}$, $k \in \mathbb{N}$, the kth eigenfunction of A_s and $(-\Delta)^s$, respectively, and by λ_{k,A_s} and $\lambda_{k,s}$, $k \in \mathbb{N}$, the corresponding eigenvalues, respectively.

With this setting, we can give a more precise statement of Theorem 5.1 as follows:

Theorem 5.2 *Problems* (5.2) *and* (5.3) *are not the same because they have different eigenvalues and eigenfunctions (with respect to Dirichlet boundary conditions). In particular,*

- *The first eigenvalue in* (5.3) *is strictly less than that in* (5.2).
- *The eigenfunctions of* (5.3) *are in general only Hölder continuous up to the boundary, different from those of* (5.2), *which are as smooth up to the boundary as the boundary allows.*

Taking into account the definition of A_s, it is easily seen that its eigenfunctions e_{k,A_s}, $k \in \mathbb{N}$, are exactly the eigenfunctions of the Laplace operator $-\Delta$; that is,

$$e_{k,A_s} = \widetilde{e}_k.$$

Also, since $\widetilde{e}_k \in C^\infty(\Omega)\cap C^m(\overline{\Omega})$, for any $m \in \mathbb{N}$ (see, e.g., [91]), then

$$e_{k,A_s} \in C^\infty(\Omega)\cap C^m(\overline{\Omega}). \tag{5.4}$$

Of course, constructing the eigenfunctions of $(-\Delta)^s$ is more difficult. Despite this, we have some regularity results for them. Precisely, by Proposition 3.1, $e_{k,s} \in H^s(\mathbb{R}^n)$, and by construction, $e_{k,s}$ vanishes outside Ω. Then, denoting by $\delta(x) := dist(x, \partial\Omega)$, $x \in \mathbb{R}^n$, by [186, theorems 1.1 and 1.3] and Proposition 3.4, we have that

$$e_{k,s}/\delta^s_{|\Omega} \in C^{0,\alpha}(\overline{\Omega}) \quad \text{for some } \alpha \in (0,1);$$

namely, $e_{k,s}/\delta^s_{|\Omega}$ has a continuous extension to $\overline{\Omega}$ that is $C^{0,\alpha}(\overline{\Omega})$. In particular, $e_{k,s}$ is Hölder continuous up to the boundary.

The aim of this section is to show that *the Hölder regularity is optimal* for the eigenfunctions $e_{k,s}$ of $(-\Delta)^s$. To this end, first of all, we recall the notion of a Poisson kernel of fractional type, and then we discuss the optimal regularity of the eigenfunctions $e_{k,s}$.

5.2.1 Poisson kernel of fractional type

Here we recall the notion of Poisson kernels of fractional type and their relation with the Dirichlet problem (see [129, chapter I]). First of all, for any $r > 0$, $x \in B_r$ (i.e., a ball of radius r centered at the origin), and $y \in \mathbb{R}^n \setminus B_r$, we define

$$P_r(x,y) := c_0(n,s)\left(\frac{r^2 - |x|^2}{|y|^2 - r^2}\right)^s \frac{1}{|x-y|^n},$$

with $c_0(n,s) > 0$. It is known (see [129, appendix]) that, for any fixed $x \in B_r$ the function

$$I(x) := \int_{\mathbb{R}^n\setminus B_r} P_r(x,y)\,dy$$

is constant in x. Therefore, we normalize $c_0(n,s)$ in such a way that[1]

$$\int_{\mathbb{R}^n\setminus B_r} P_r(x,y)\,dy = 1. \tag{5.5}$$

The function P_r plays the role of a fractional Poisson kernel; that is, if $g \in C(\mathbb{R}^n)\cap L^\infty(\mathbb{R}^n)$ and

$$u_g(x) := \begin{cases} \displaystyle\int_{\mathbb{R}^n\setminus B_r} P_r(x,y)g(y)\,dy & \text{if } x \in B_r \\ g(x) & \text{if } x \in \mathbb{R}^n \setminus B_r, \end{cases} \tag{5.6}$$

[1] More explicitly, one can choose $c_0(n,s) := \Gamma(n/2)\sin(\pi s)/\pi^{(n/2)+1}$ (see [129, pp. 399–400]).

then u_g is the unique solution of

$$\begin{cases} (-\Delta)^s u_g = 0 & \text{in } B_r \\ u_g = g & \text{outside } B_r. \end{cases} \tag{5.7}$$

For this, see [129, 206].

5.2.2 Optimal regularity for the eigenfunctions of $(-\Delta)^s$

In this subsection we prove that the $C^{0,\alpha}$-regularity of the eigenfunctions $e_{k,s}$ is optimal. Precisely, we show that, in general, *the eigenfunctions of* $(-\Delta)^s$ *need not to be Lipschitz continuous up to the boundary* (*i.e., the Hölder regularity is optimal*).

For concreteness, we consider the case[2]

$$n > 2s, \tag{5.8}$$

the domain $\Omega := B_r$ and the first eigenfunction $e_{1,s}$ (normalized in such a way that $\|e_{1,s}\|_{L^2(\mathbb{R}^n)} = 1$ and $e_{1,s} \geq 0$ in $\mathbb{R}^n$; see Proposition 3.1) of $(-\Delta)^s$ in B_r, that is,

$$\begin{cases} (-\Delta)^s e_{1,s} = \lambda_{1,s} e_{1,s} & \text{in } B_r \\ e_{1,s} = 0 & \text{in } \mathbb{R}^n \setminus B_r. \end{cases} \tag{5.9}$$

Our result is the following[3]:

Proposition 5.3 The function $e_{1,s}$ given in (5.9) is such that

$$e_{1,s} \notin W^{1,\infty}(B_r).$$

Proof The proof is by contradiction. We suppose that $e_{1,s} \in W^{1,\infty}(B_r)$, so $e_{1,s} \in W^{1,\infty}(\mathbb{R}^n)$; that is,

$$|e_{1,s}(x)| + |\nabla e_{1,s}(x)| \leq M \quad x \in \mathbb{R}^n, \tag{5.10}$$

for some $M > 0$.

From now on, we proceed by steps.

Step 5.1 *The function $e_{1,s}$ is spherically symmetric and radially decreasing in $\mathbb{R}^n$.*

Proof For this, because $e_{1,s} \geq 0$ in $\mathbb{R}^n$, we consider its symmetric, radially decreasing rearrangement $e^\star_{1,s}$ (see, e.g., [121, chapter 2] for the basics of such a rearrangement). We observe that $e^\star_{1,s}$ vanishes outside B_r because so does $e_{1,s}$.

Moreover, we recall that the L^2-norm is preserved by the rearrangement, while the fractional Gagliardo seminorm decreases (see, e.g. [6, 32, 163]).

[2] When condition (5.8) does not hold, that is, when $n = 1$ and $s \in [1/2, 1)$, one has to use the appropriate fundamental solution, that is, $Q(x) = -c_1(1,s)|x|^{2s-1}$, for $n = 1 < 2s$, and $Q(x) = -c_1(1, 1/2)\log|x|$, for $n = 1$ and $s = 1/2$. Since the constant $c_1(1,s)$ is positive in these cases too, we have that Q is always radially decreasing, and our argument goes through. For further comments, see also Remarks 5.4 and 5.5.

[3] As customary, $W^{1,\infty}(B_r)$ is the set of a.e. bounded functions with a.e. bounded distributional derivatives in B_r. Equivalently, it is the space of Lipschitz functions in B_r.

Then, by this and because $\lambda_{1,s}$ is obtained by minimizing the fractional Gagliardo seminorm under constraint on the L^2-norm for functions that vanish outside B_r (see Proposition 3.1), we conclude that the minimum is attained by $e^\star_{1,s}$ (as well as by $e_{1,s}$).

Since $\lambda_{1,s}$ is a simple eigenvalue (see again Proposition 3.1), it follows that $e^\star_{1,s} = e_{1,s}$, and Step 5.1 is proved. □

Now let Q be the fractional fundamental solution given by

$$Q(x) := c_1(n,s)|x|^{2s-n} \quad x \in \mathbb{R}^n \setminus \{0\}.$$

Here the constant $c_1(n,s) > 0$ is chosen in such a way that $(-\Delta)^s Q$ is the Dirac delta δ_0 centered at the origin (see, e.g., [129, p. 44] for the basic properties of fractional fundamental solutions).

We define

$$\tilde{v}(x) := \lambda_{1,s} Q * e_{1,s}(x) = \lambda_{1,s} c_1(n,s) \int_{\mathbb{R}^n} |y|^{2s-n} e_{1,s}(x-y)\,dy \quad x \in \mathbb{R}^n \tag{5.11}$$

and

$$v(x) := e_{1,s}(x) - \tilde{v}(x) \quad x \in \mathbb{R}^n. \tag{5.12}$$

First of all, notice that $\tilde{v} \geq 0$ in $\mathbb{R}^n$ because $\lambda_{1,s} > 0$, $Q > 0$, and $e_{1,s} \geq 0$ in $\mathbb{R}^n$.

Step 5.2 *The function $\tilde{v}$ is spherically symmetric and radially decreasing in $\mathbb{R}^n$.*

Proof Indeed, if $\mathscr{R}$ is a rotation, we use Step 5.1 and the substitution $\tilde{y} := \mathscr{R}y$ to obtain, for any $x \in \mathbb{R}^n$,

$$\begin{aligned}
\tilde{v}(x) &= \lambda_{1,s} c_1(n,s) \int_{\mathbb{R}^n} |y|^{2s-n} e_{1,s}(x-y)\,dy \\
&= \lambda_{1,s} c_1(n,s) \int_{\mathbb{R}^n} |y|^{2s-n} e_{1,s}(\mathscr{R}(x-y))\,dy \\
&= \lambda_{1,s} c_1(n,s) \int_{\mathbb{R}^n} |\tilde{y}|^{2s-n} e_{1,s}(\mathscr{R}x-\tilde{y})\,d\tilde{y} = \tilde{v}(\mathscr{R}x),
\end{aligned}$$

which shows the spherical symmetry of $\tilde{v}$.

As for the fact that $\tilde{v}$ is radially decreasing in $\mathbb{R}^n$, we take $\rho > 0$ and define

$$v_\star(\rho) := -(\lambda_{1,s} c_1(n,s))^{-1} \tilde{v}(0,\ldots,0,\rho) = -\int_{\mathbb{R}^n} |y|^{2s-n} e_{1,s}(-y', \rho - y_n)\,dy, \tag{5.13}$$

where we used the notation $y = (y', y_n) \in \mathbb{R}^{n-1} \times \mathbb{R}$ for the coordinates in $\mathbb{R}^n$.

The goal is to show that, for any $\rho > 0$,

$$v'_\star(\rho) \geq 0. \tag{5.14}$$

For this, first note that

$$\begin{aligned} v_\star(\rho) &= -\int_{\mathbb{R}^n\cap\{|\rho-y_n|\le r\}} |y|^{2s-n} e_{1,s}(-y',\rho-y_n)\,dy \\ &\quad -\int_{\mathbb{R}^n\cap\{|\rho-y_n|>r\}} |y|^{2s-n} e_{1,s}(-y',\rho-y_n)\,dy \\ &= -\int_{\mathbb{R}^n\cap\{|\rho-y_n|\le r\}} |y|^{2s-n} e_{1,s}(-y',\rho-y_n)\,dy \end{aligned}$$

because $\{|\rho - y_n| > r\} \subseteq \{|(-y',\rho-y_n)| > r\}$ and $e_{1,s}$ vanishes outside B_r. Also, since the function $e_{1,s}$ is spherically symmetric and radially decreasing in $\mathbb{R}^n$ by Step 5.1, we write $e_{1,s}(x) = -E(|x|)$, with $E' \ge 0$ in $\mathbb{R}^+$. Thus,

$$v_\star(\rho) = \int_{\mathbb{R}^n\cap\{|\rho-y_n|\le r\}} |y|^{2s-n} E(|(-y',\rho-y_n)|)\,dy,$$

so

$$v_\star'(\rho) = \int_{\mathbb{R}^n\cap\{|\rho-y_n|\le r\}} |y|^{2s-n} E'(|(-y',\rho-y_n)|)\frac{\rho-y_n}{|(-y',\rho-y_n)|}\,dy. \tag{5.15}$$

Now let us consider the following change of variables:

$$\begin{cases} \tilde{y}' := y' \\ \tilde{y}_n := 2\rho - y_n. \end{cases} \tag{5.16}$$

First of all, note that if $\tilde{y}_n - \rho \ge 0$, then $-\tilde{y}_n \le 2\rho - \tilde{y}_n \le \tilde{y}_n$, so

$$(2\rho-\tilde{y}_n)^2 \le \tilde{y}_n^2$$

and

$$|(\tilde{y}',2\rho-\tilde{y}_n)| = \sqrt{|\tilde{y}'|^2+(2\rho-\tilde{y}_n)^2} \le \sqrt{|\tilde{y}'|^2+\tilde{y}_n^2} = |\tilde{y}|.$$

As a consequence of this and recalling that $n \ge 2s$, we obtain that

$$|(\tilde{y}',2\rho-\tilde{y}_n)|^{2s-n} \ge |\tilde{y}|^{2s-n}. \tag{5.17}$$

Therefore, by (5.16) and (5.17), we get

$$\begin{aligned} &\int_{\mathbb{R}^n\cap\{0\le\rho-y_n\le r\}} |y|^{2s-n} E'(|(-y',\rho-y_n)|)\frac{\rho-y_n}{|(-y',\rho-y_n)|}\,dy \\ &= \int_{\mathbb{R}^n\cap\{0\le\tilde{y}_n-\rho\le r\}} |(\tilde{y}',2\rho-\tilde{y}_n)|^{2s-n} E'(|(\tilde{y}',\rho-\tilde{y}_n)|)\frac{\tilde{y}_n-\rho}{(|(\tilde{y}',\rho-\tilde{y}_n)|}\,d\tilde{y} \\ &\ge \int_{\mathbb{R}^n\cap\{0\le\tilde{y}_n-\rho\le r\}} |\tilde{y}|^{2s-n} E'(|(\tilde{y}',\rho-\tilde{y}_n)|)\frac{\tilde{y}_n-\rho}{(|(\tilde{y}',\rho-\tilde{y}_n)|}\,d\tilde{y} \end{aligned}$$

owing to the fact that $E' \ge 0$ in $\mathbb{R}^+$.

Hence, recalling (5.15), we get

$$\begin{aligned}
v_\star'(\rho) &= \int_{\mathbb{R}^n\cap\{|\rho-y_n|\le r\}} |y|^{2s-n} E'(|(-y',\rho-y_n)|)\frac{\rho-y_n}{|(-y',\rho-y_n)|}\,dy \\
&= \int_{\mathbb{R}^n\cap\{0\le\rho-y_n\le r\}} |y|^{2s-n} E'(|(-y',\rho-y_n)|)\frac{\rho-y_n}{|(-y',\rho-y_n)|}\,dy \\
&\quad + \int_{\mathbb{R}^n\cap\{0\le y_n-\rho\le r\}} |y|^{2s-n} E'(|(-y',\rho-y_n)|)\frac{\rho-y_n}{|(-y',\rho-y_n)|}\,dy \\
&\ge \int_{\mathbb{R}^n\cap\{0\le \tilde{y}_n-\rho\le r\}} |\tilde{y}|^{2s-n} E'(|(\tilde{y}',\rho-\tilde{y}_n)|)\frac{\tilde{y}_n-\rho}{(|(\tilde{y}',\rho-\tilde{y}_n)|}\,d\tilde{y} \\
&\quad + \int_{\mathbb{R}^n\cap\{0\le \tilde{y}_n-\rho\le r\}} |\tilde{y}|^{2s-n} E'(|(-\tilde{y}',\rho-\tilde{y}_n)|)\frac{\rho-\tilde{y}_n}{|(-\tilde{y}',\rho-\tilde{y}_n)|}\,d\tilde{y} \\
&= 0,
\end{aligned}$$

owing to the fact that $|(\tilde{y}',\rho-\tilde{y}_n)| = |(-\tilde{y}',\rho-\tilde{y}_n)|$. Hence, (5.14) is proved.

Then, by (5.13), the spherical symmetry of $\tilde{v}$, and the fact that $\lambda_{1,s}$ and $c_1(n,s)$ are positive constants, we get that $\tilde{v}$ is radially decreasing in $\mathbb{R}^n$. This concludes the proof of Step 5.2 . □

The next step will exploit assumption (5.10) taken for the argument by contradiction.

Step 5.3 *The function* $\tilde{v}$ *is such that*

$$\tilde{v} \in W^{1,\infty}(B_{2r}).$$

Proof To check this, we observe that, for any $x \in \mathbb{R}^n$,

$$\tilde{v}(x) = \lambda_{1,s}c_1(n,s)\int_{\mathbb{R}^n} |y|^{2s-n} e_{1,s}(x-y)\,dy = \lambda_{1,s}c_1(n,s)\int_{B_r(x)} |y|^{2s-n} e_{1,s}(x-y)\,dy$$

because $e_{1,s}$ vanishes outside B_r by (5.9). Here $B_r(x)$ denotes a ball of radius r centered at x.

Now we notice that if $x \in B_{2r}$, then $B_r(x) \subset B_{3r}$. As a consequence, recalling also (5.10), we obtain that, for any $x \in B_{2r}$,

$$\begin{aligned}
|\tilde{v}(x)| + |\nabla\tilde{v}(x)| &\le \lambda_{1,s}c_1(n,s)\int_{B_r(x)} |y|^{2s-n}\big(|e_{1,s}(x-y)| + |\nabla e_{1,s}(x-y)|\big)\,dy \\
&\le \lambda_{1,s}c_1(n,s)M\int_{B_{3r}} |y|^{2s-n}\,dy,
\end{aligned}$$

which is finite (because $s > 0$). Hence, Step 5.3 is established. □

Now we can conclude the proof of Proposition 5.3. For this, note that from (5.10) and Step 5.3, we get

$$v = e_{1,s} - \tilde{v} \in W^{1,\infty}(B_{2r}),$$

that is, there exists $\tilde{M} > 0$ such that

$$|v(x) - v(y)| \leq \tilde{M}|x - y|, \tag{5.18}$$

for any $x, y \in B_{2r}$.

Also, by (5.11) and the choice of Q,

$$(-\Delta)^s \tilde{v} = \lambda_{1,s} e_{1,s} * (-\Delta)^s Q = \lambda_{1,s} e_{1,s} * \delta_0 = \lambda_{1,s} e_{1,s},$$

so, by (5.9) and (5.12),

$$(-\Delta)^s v = (-\Delta)^s e_{1,s} - (-\Delta)^s \tilde{v} = \lambda_{1,s} e_{1,s} - \lambda_{1,s} e_{1,s} = 0$$

in B_r. Therefore, we can reconstruct v by its values outside B_r via the fractional Poisson kernel; that is, for any $x \in B_r$,

$$v(x) = \int_{\mathbb{R}^n \setminus B_r} P_r(x,y) v(y)\, dy. \tag{5.19}$$

For this, see (5.6) and (5.7).

Since (5.12) holds true and $e_{1,s} = 0$ outside B_r, by (5.19), we deduce that

$$\begin{aligned} v(x) &= \int_{\mathbb{R}^n \setminus B_r} P_r(x,y) v(y)\, dy \\ &= \int_{\mathbb{R}^n \setminus B_r} P_r(x,y) e_{1,s}(y)\, dy - \int_{\mathbb{R}^n \setminus B_r} P_r(x,y) \tilde{v}(y)\, dy \\ &= -\int_{\mathbb{R}^n \setminus B_r} P_r(x,y) \tilde{v}(y)\, dy. \end{aligned} \tag{5.20}$$

By (5.12) and (5.18)–(5.20), we get

$$\begin{aligned} &\left|\int_{\mathbb{R}^n \setminus B_r} P_r(x,y) \tilde{v}(y)\, dy - \tilde{v}(0,\ldots,0,r)\right| \\ &= \left|-\int_{\mathbb{R}^n \setminus B_r} P_r(x,y) v(y)\, dy + v(0,\ldots,0,r) - e_{1,s}(0,\ldots,0,r)\right| \\ &= \left|\int_{\mathbb{R}^n \setminus B_r} P_r(x,y) v(y)\, dy - v(0,\ldots,0,r)\right| \\ &= |v(0,\ldots,0,r) - v(x)| \\ &\leq \tilde{M}|(0,\ldots,0,r) - x|, \end{aligned} \tag{5.21}$$

for any $x \in B_r$.

If in (5.21) we take $x := (0,\dots,0,r-\varepsilon) \in B_r$ for a small $\varepsilon \in (0,r)$, recalling (5.5), we deduce that

$$\begin{aligned}
\tilde{M}\varepsilon &= \tilde{M}\,|(0,\dots,0,r)-x| \\
&\geq \left| \int_{\mathbb{R}^n\setminus B_r} P_r(x,y)\tilde{v}(y)\,dy - \tilde{v}(0,\dots,0,r) \right| \\
&= \left| \int_{\mathbb{R}^n\setminus B_r} P_r(x,y)\big(\tilde{v}(y) - \tilde{v}(0,\dots,0,r)\big)\,dy \right| \\
&= c_0(n,s) \int_{|y|>r} \left(\frac{r^2-|x|^2}{|y|^2-r^2} \right)^s \frac{\tilde{v}(0,\dots,0,r)-\tilde{v}(y)}{|x-y|^n}\,dy \qquad (5.22)\\
&= c_0(n,s)(r^2-|x|^2)^s \int_{|y|>r} \frac{\tilde{v}(0,\dots,0,r)-\tilde{v}(y)}{(|y|^2-r^2)^s|x-y|^n}\,dy \\
&\geq c_0(n,s) r^s\,(r-|x|)^s \int_{|y|>r} \frac{\tilde{v}(0,\dots,0,r)-\tilde{v}(y)}{(|y|^2-r^2)^s(|y'|^2+|y_n-r+\varepsilon|^2)^{n/2}}\,dy \\
&= \varepsilon^s \int_{|y|>r} f_\varepsilon(y)\,dy,
\end{aligned}$$

where

$$f_\varepsilon(y) := c_0(n,s) r^s \frac{\tilde{v}(0,\dots,0,r)-\tilde{v}(y)}{(|y|^2-r^2)^s(|y'|^2+|y_n-r+\varepsilon|^2)^{n/2}}.$$

We remark that $f_\varepsilon(y) \geq 0$ for any $|y| > r$ because

$$\tilde{v}(0,\dots,r) \geq \tilde{v}(y) \quad \text{for any } |y| > r, \tag{5.23}$$

thanks to Step 5.1. Moreover,

$$\lim_{\varepsilon\to 0^+} f_\varepsilon(y) = c_0(n,s) r^s \frac{\tilde{v}(0,\dots,0,r)-\tilde{v}(y)}{(|y|^2-r^2)^s(|y'|^2+|y_n-r|^2)^{n/2}}.$$

Thus, we divide the inequality obtained in (5.22) by ε^s, and we use Fatou's lemma, and we conclude that

$$\begin{aligned}
0 &= \liminf_{\varepsilon\to 0^+} \tilde{M}\varepsilon^{1-s} \\
&\geq \liminf_{\varepsilon\to 0^+} \int_{|y|>r} f_\varepsilon(y)\,dy \\
&= c_0(n,s) r^s \int_{|y|>r} \frac{\tilde{v}(0,\dots,0,r)-\tilde{v}(y)}{(|y|^2-r^2)^s(|y'|^2+|y_n-r|^2)^{n/2}}\,dy.
\end{aligned}$$

This and (5.23) yield that $\tilde{v}(y)$ is constantly equal to $\tilde{v}(0,\dots,0,r)$, for any $|y|>r$, so, in particular, if $x^\star := (0,\dots,2r)$, we have that

$$\partial_n \tilde{v}(x^\star) = 0. \tag{5.24}$$

However, by (5.11),

$$\begin{aligned}\frac{1}{\lambda_{1,s}c_1(n,s)}\partial_n \tilde{v}(x^\star) &= \frac{\partial}{\partial x_n}\int_{B_r}|x-z|^{2s-n}e_{1,s}(z)\,dz\Big|_{x=x^\star}\\ &= (2s-n)\int_{B_r}|x^\star - z|^{2s-n-2}(x_n^\star - z_n)e_{1,s}(z)\,dz\\ &= (2s-n)\int_{B_r}(|z'|^2+|2r-z_n|^2)^{(2s-n-2)/2}(2r-z_n)e_{1,s}(z)\,dz,\end{aligned}$$

which is strictly negative, by (5.8). This is a contradiction with (5.24), and hence Proposition 5.3 is proved. □

Remark 5.4 Concerning condition (5.8), we point out that there is an alternative approach that shows that A_s and $(-\Delta)^s$ are different operators when $n \le 2s$, that is, when $n=1$ and $s\in[1/2,1)$. In this case, one studies the boundary behavior of the solution of the equation with a constant right-hand side, noticing that such behavior is different for the two operators.

For this, we point out that if I is an interval and u is a solution of

$$A_s u = 1 \quad \text{in } I,$$

then

$$\begin{aligned} &u\in C^1(\overline{I}) \quad \textit{when } s\in(1/2,1)\\ &\textit{and } u\in C^\alpha(\overline{I}),\ \textit{for any } \alpha\in(0,1),\ \textit{when } s=1/2.\end{aligned} \tag{5.25}$$

To show this, without loss of generality, we take $I:=(0,\pi)$. In this case, the eigenfunctions of A_s (with respect to Dirichlet boundary conditions) are $e_k(t) = \sqrt{2/\pi}\,\sin(kt)$, with $k\in\mathbb{N}$, and the corresponding eigenvalues are $\lambda_{k,A_s} = \widetilde{\lambda}_k^s = k^{2s}$. Therefore, if

$$u = \sum_{i\in\mathbb{N}} a_i \widetilde{e}_i \quad \text{solves} \quad A_s u = 1,$$

then, by definition of A_s, we have

$$\sum_{i\in\mathbb{N}} a_i i^{2s}\widetilde{e}_i = 1.$$

By fixing $k\in\mathbb{N}$ and integrating against $\widetilde{e}_k$ in I, we obtain

$$a_k = \begin{cases} \sqrt{\dfrac{2}{\pi}}\,\dfrac{4}{k^{2s+1}} & \text{if } k \text{ is odd}\\[2ex] 0 & \text{if } k \text{ is even.}\end{cases}$$

Hence, when $s \in (1/2, 1)$, the function $u \in C^1([0,\pi])$ because it is the uniform limit in $C^1([0,\pi])$ of the function

$$u_M := \sum_{\substack{k\in\mathbb{N}\\ k\le M}} a_k \widetilde{e}_k.$$

Indeed, if $M > m$,

$$\begin{aligned}
|u_M - u_m| + |\dot{u}_M - \dot{u}_m| &\le \sum_{\substack{k\in\mathbb{N}\\ m<k\le M}} |a_k|\,(|e_k| + |\dot{\widetilde{e}}_k|) \\
&\le C \sum_{\substack{k\in\mathbb{N}\\ m<k\le M}} k^{-2s-1}(1+k) \\
&\le 2C \sum_{\substack{k\in\mathbb{N}\\ m<k\le M}} k^{-2s},
\end{aligned}$$

which becomes infinitesimal for large m when $s \in (1/2, 1)$.

Moreover, when $s = 1/2$, then $u \in C^\alpha([0,\pi])$, for any $\alpha \in (0,1)$, because the C^α-norm of u_M is bounded independently of M. To check this, it is enough to notice that, for any $x \neq y \in (0,\pi)$,

$$\begin{aligned}
\frac{|u_M(x) - u_M(y)|}{|x-y|^\alpha} &\le |x-y|^{-\alpha} \sum_{k\in\mathbb{N}} |a_k|\,|\widetilde{e}_k(x) - \widetilde{e}_k(y)| \\
&\le C\,|x-y|^{-\alpha} \sum_{k\in\mathbb{N}} k^{-2s-1} \min\{1, k|x-y|\} \\
&\le C\,|x-y|^{-\alpha} \left(\sum_{\substack{k\in\mathbb{N}\\ k\le 1/|x-y|}} k^{-2s}|x-y| + \sum_{\substack{k\in\mathbb{N}\\ k\ge 1/|x-y|}} k^{-2s-1} \right) \\
&\le C'|x-y|^{1-\alpha} \int_1^{1/|x-y|} \tau^{-2s}\,d\tau + C|x-y|^{-\alpha} \int_{1/|x-y|}^{+\infty} \tau^{-2s-1}\,d\tau \\
&\le C''|x-y|^{1-\alpha}\left(\log|x-y|^{-1} + 1\right)
\end{aligned}$$

when $s = 1/2$, and the latter quantity is bounded by a constant independent of M, x, and y (here C, C', and C'' are positive constants). This proves (5.25).

However, if I is an interval,

solutions of $(-\Delta)^s u = 1$ in I that vanish outside I are only $C^s(\overline{I})$. (5.26)

Indeed, without loss of generality, we can check (5.26) when $I = (-1, 1)$. In this case, by [115] (see in particular equations (A) and (1.3) there), we have that $u(t) = c(1-t^2)^s$, for a suitable normalizing constant c, so (5.26) is proved.

By comparing (5.25) and (5.26), we see that A_s and $(-\Delta)^s$ are also different operators when $n = 1$ and $s \in [1/2, 1)$.

Remark 5.5 We would like to sketch a different method for proving Proposition 5.3. This proof is valid for any $n \in \mathbb{N}$ and any $s \in (0,1)$, and it is based on the Green representation formula (see [38]), according to which, if $u \in H^s(\mathbb{R}^n)$ and $g \in L^2(B_1)$ satisfy

$$\begin{cases} (-\Delta)^s u = g & \text{in } B_1 \\ u = 0 & \text{in } \mathbb{R}^n \setminus B_1, \end{cases}$$

then, for any $x \in B_1$, we have that

$$u(x) = \int_{B_1} G(x,y) g(y) dy, \tag{5.27}$$

where

$$G(x,y) := C_{n,s} |x-y|^{2s-n} \int_0^{r(x,y)} r^{s-1} (r+1)^{-n/2} dr$$

and

$$r(x,y) := \frac{(1-|x|^2)(1-|y|^2)}{|x-y|^2},$$

for a suitable $C_{n,s} > 0$.

Notice that $G(x,y) > 0$ for all $x \neq y \in B_1$, and $r(x,y) = 0 = G(x,y)$ if $x \in \partial B_1$. We take $x_0 := (0,\ldots,0,1)$ and $x_\varepsilon := (0,\ldots,0,\sqrt{1-\varepsilon})$, $\varepsilon \in (0,1)$. Note that if $y \in B_{1/4}$, we have that $|x_\varepsilon - y| \le |x_\varepsilon| + |y| \le 5/4$, for any $\varepsilon \in (0,1)$, and

$$r(x_\varepsilon, y) = \frac{(1-|x_\varepsilon|^2)(1-|y|^2)}{|x_\varepsilon - y|^2} = \frac{\varepsilon(1-|y|^2)}{|x_\varepsilon - y|^2} \ge c_1 \varepsilon,$$

for some $c_1 > 0$. Consequently, taking also into account that $|x_\varepsilon - y| \ge |x_\varepsilon| - |y| \ge 1/4$ for small $\varepsilon > 0$, we get

$$G(x_\varepsilon, y) \ge c_2 \int_0^{c_1 \varepsilon} r^{s-1} (r+1)^{-n/2} dr \ge c_3 \int_0^{c_1 \varepsilon} r^{s-1} dr = c_4 \varepsilon^s,$$

for any $y \in B_{1/4}$, for some $c_4 > 0$.

Thus, using (5.27) for the first eigenfunction of $(-\Delta)^s$ with homogeneous Dirichlet boundary data (i.e., take $u := e_{1,s}$ and $g := \lambda_{1,s} e_{1,s}$), we obtain

$$\begin{aligned} |e_{1,s}(x_\varepsilon) - e_{1,s}(x_0)| &= e_{1,s}(x_\varepsilon) \\ &\ge \lambda_{1,s} \int_{B_{1/4}} G(x_\varepsilon, y) e_{1,s}(y) dy \\ &= c_5 \varepsilon^s \\ &\ge c_6 |x_0 - x_\varepsilon|^s, \end{aligned} \tag{5.28}$$

thanks to the fact that $|x_0 - x_\varepsilon| \ge \varepsilon/4$ for ε small enough. Here c_5 and c_6 are positive constants.

Since $|x_0 - x_\varepsilon|$ is small for ε small enough (because $x_\varepsilon \to x_0$ as $\varepsilon \to 0$), then (5.28) shows that $e_{1,s}$ is no better than Hölder continuous in B_1, providing an alternative proof of Proposition 5.3.

5.3 The spectrum of A_s and $(-\Delta)^s$

In this section we focus on the spectrum of the operators A_s and $(-\Delta)^s$.

By the definition of A_s, it easily follows that the eigenvalues λ_{k,A_s} are exactly the s–power of the ones of the Laplacian; that is,

$$\lambda_{k,A_s} = \widetilde{\lambda}_k^s \quad k \in \mathbb{N}. \tag{5.29}$$

Moreover, we recall that, for $k \in \mathbb{N}$,

$$\lambda_{k,s} = \frac{C(n,s)}{2} \min_{u \in \mathbb{P}_{k,s} \setminus \{0\}} \frac{\displaystyle\int_{\mathbb{R}^n \times \mathbb{R}^n} \frac{|u(x)-u(y)|^2}{|x-y|^{n+2s}}\,dx\,dy}{\displaystyle\int_\Omega |u(x)|^2\,dx}, \tag{5.30}$$

where

$$\mathbb{P}_{1,s} = X_0^s(\Omega) := \{u \in H^s(\mathbb{R}^n) \,:\, u = 0 \text{ a.e. in } \mathbb{R}^n \setminus \Omega\}$$

and

$$\mathbb{P}_{k,s} := \{u \in X_0^s(\Omega) \,:\, \langle u, e_{j,s}\rangle_{X_0^s(\Omega)} = 0,\ \forall j = 1, \ldots, k-1\}, \tag{5.31}$$

for any $k \geq 2$, with $\langle\cdot,\cdot\rangle_{X_0^s(\Omega)}$ defined as in (1.88).

In what follows, we will show that A_s and $(-\Delta)^s$ have different eigenvalues. Of course, for this purpose, we will use properties (5.29) and (5.30), but the main ingredient will be extension of the operator A_s, carried out in the Section 5.4.

5.4 One-dimensional analysis

In this section we perform an ordinary differential equations (ODE) analysis related to the extension of the operator A_s, as will be clear in Section 5.5. This analysis is not new in itself (see also [55, section 3.2] and [208, section 3.1]): similar results were obtained, for instance, in [208] by using a conjugate equation and suitable special functions such as different kinds of Bessel and Hankel functions. Here we use an elementary and self-contained approach.

Given $a \in (-1, 1)$, in what follows, we denote by $W_a^{1,2}(\mathbb{R}^+)$ the following Sobolev space:

$$W_a^{1,2}(\mathbb{R}^+) := \left\{ u \in W_{\text{loc}}^{1,1}(\mathbb{R}^+) \,:\, \int_{\mathbb{R}^+} t^a |u(t)|^2\,dt < +\infty \text{ and } \int_{\mathbb{R}^+} t^a |\dot{u}(t)|^2\,dt < +\infty \right\}$$

endowed with the norm

$$\|u\|_{W_a^{1,2}(\mathbb{R}^+)} := \left(\int_{\mathbb{R}^+} t^a |u(t)|^2\,dt + \int_{\mathbb{R}^+} t^a |\dot{g}(t)|^2\,dt \right)^{1/2}. \tag{5.32}$$

Here, as usual, we used the notation $\mathbb{R}^+ := (0, +\infty)$. We also denote by $W^{1,2}_{1,a}(\mathbb{R}^+)$ the closure, with respect to the norm in (5.32), of the set of all functions $g \in C^\infty(\mathbb{R}^+) \cap C^0(\overline{\mathbb{R}^+})$ with bounded support and $g(0) = 1$.

It is useful to point out that $W^{1,2}_a(\mathbb{R}^+)$ and $W^{1,2}_{1,a}(\mathbb{R}^+)$ are contained in a classical Sobolev space. Precisely, denoting by $W^{1,p}((0,\kappa))$, $p \geq 1$, and $\kappa > 0$, the classical Sobolev space endowed with the norm

$$\|g\|_{W^{1,p}((0,\kappa))} = \left(\|g\|^p_{L^p((0,\kappa))} + \|\dot{g}\|^p_{L^p((0,\kappa))}\right)^{1/p},$$

the following result holds true:

Lemma 5.6 *Fix $a \in (-1, 1)$ and $\kappa > 0$. Then*

$$W^{1,2}_a(\mathbb{R}^+) \subseteq W^{1,p}((0,\kappa)),$$

for any $p \in [1, a^*)$, with

$$a^* := \begin{cases} 2/(a+1) & \text{if } a \in (0,1) \\ 2 & \text{if } a \in (-1,0]. \end{cases}$$

Moreover, there exists $C_\kappa > 0$ such that

$$\|g\|_{W^{1,p}((0,\kappa))} \leq C_\kappa \|g\|_{W^{1,2}_a(\mathbb{R}^+)},$$

for any $g \in W^{1,2}_a(\mathbb{R}^+)$.

Proof Let $a \in (-1, 1)$, $g \in W^{1,2}_a(\mathbb{R}^+)$, and $p \in [1, a^*)$. We use the Hölder inequality with exponents $2/(2-p)$ and $2/p$ (note that both these exponents are greater than 1, thanks to the choice of p) to see that

$$\begin{aligned}
\|g\|^p_{L^p((0,\kappa))} &= \int_0^\kappa t^{-pa/2} t^{pa/2} |g(t)|^p \, dt \\
&\leq \left[\int_0^\kappa t^{-pa/(2-p)} \, dt\right]^{(2-p)/2} \left[\int_0^\kappa t^a |g(t)|^2 \, dt\right]^{p/2} \\
&= \left[\frac{2-p}{2-p(1+a)} \kappa^{(2-p(1+a))/(2-p)}\right]^{(2-p)/2} \left[\int_{\mathbb{R}^+} t^a |g(t)|^2 \, dt\right]^{p/2} < +\infty,
\end{aligned}$$

again because $p < a^*$.

A similar inequality holds if we replace g with $\dot{g}$, and this proves the desired result. □

Hence, as a consequence of Lemma 5.6, the functions in $W^{1,2}_{1,a}(\mathbb{R}^+)$ are uniformly continuous in any interval, by the standard Sobolev embedding, and have a distributional derivative that is well defined almost everywhere.

Now, for any $\lambda > 0$ and any $g \in W^{1,2}_{1,a}(\mathbb{R}^+)$, we consider the functional

$$G_\lambda(g) := \int_{\mathbb{R}^+} t^a\big(\lambda|g(t)|^2 + |\dot g(t)|^2\big)\,dt.$$

The minimization problem of G_λ is described in detail by the following result:

Theorem 5.7 *There exists a unique $g_\lambda \in W^{1,2}_{1,a}(\mathbb{R}^+)$ such that*

$$\mathrm{m}_\lambda := \inf_{g\in W^{1,2}_{1,a}(\mathbb{R}^+)} G_\lambda(g) = G_\lambda(g_\lambda); \tag{5.33}$$

that is, the preceding infimum is attained. Moreover, $g_\lambda \in C^\infty(\mathbb{R}^+)\cap C^0(\overline{\mathbb{R}^+})$, and it satisfies

$$\begin{cases} \ddot g_\lambda(t) + \dfrac{a}{t}\dot g_\lambda(t) - \lambda g_\lambda(t) = 0 \quad \text{for any } t\in\mathbb{R}^+ \\ \\ g_\lambda(0) = 1, \end{cases} \tag{5.34}$$

and

$$\lim_{t\to 0^+} t^a \dot g_\lambda(t) = -\mathrm{m}_1 \lambda^{(1-a)/2}. \tag{5.35}$$

Finally,

$$g_\lambda(t)\in[0,1] \quad \text{and} \quad \dot g_\lambda(t)\le 0, \text{for all } t\in\mathbb{R}^+ \quad \text{and} \quad \lim_{t\to+\infty} g_\lambda(t) = 0. \tag{5.36}$$

Proof By plugging a smooth and compactly supported function in G_λ, we see that $\mathrm{m}_\lambda \in [0,+\infty)$, so we can take a minimizing sequence $\{g_j\}_{j\in\mathbb{N}}$ in $W^{1,2}_{1,a}(\mathbb{R}^+)$, that is, a sequence $\{g_j\}_{j\in\mathbb{N}}$ such that

$$G_\lambda(g_j) \to \mathrm{m}_\lambda$$

as $j\to+\infty$.

In particular, $G_\lambda(g_j) \le \mathrm{m}_\lambda + 1$. As a consequence of this, $\|g_j\|_{W^{1,2}_a(\mathbb{R}^+)}$ is bounded uniformly in j. Hence, there exists $g_\lambda \in W^{1,2}_{1,a}(\mathbb{R}^+)$ such that $g_j \to g_\lambda$ weakly in $W^{1,2}_{1,a}(\mathbb{R}^+)$ as $j\to+\infty$. Also, for any $k\in\mathbb{N}$, $k\ge 2$, we have that

$$\tilde C_k \int_{1/k}^{k} \big(\lambda|g_j(t)|^2 + |\dot g_j(t)|^2\big)\,dt \le \int_{1/k}^{k} t^a\big(\lambda|g_j(t)|^2 + |\dot g_j(t)|^2\big)\,dt \le \mathrm{m}_\lambda + 1,$$

where $\tilde C_k = (1/k)^a$ if $a\ge 0$, while $\tilde C_k = k^a$ if $a<0$. Thus, $\|g_j\|_{W^{1,2}([1/k,k])}$ is bounded uniformly in j.

Now we perform a diagonal compactness argument over the index k. That is, we take an increasing function $\phi_k : \mathbb{N}\to\mathbb{N}$, and we use it to extract subsequences. We have a subsequence $\{g_{\phi_2(j)}\}_{j\in\mathbb{N}}$ that converges a.e. in $[1/2,2]$ to g_λ with $\dot g_{\phi_2(j)}$ converging to $\dot g_\lambda$ weakly in $L^2([1/2,2])$ as $j\to+\infty$. Then, we take a further subsequence $\{g_{\phi_3(\phi_2(j))}\}_{j\in\mathbb{N}}$ that converges a.e. in $[1/3,3]$ to g_λ with $\dot g_{\phi_3(\phi_2(j))}$ converging to $\dot g_\lambda$ weakly in $L^2([1/3,3])$ as $j\to+\infty$. Iteratively, for any $k\in\mathbb{N}$, we get a subsequence $\{g_{\phi_k\circ\ldots\phi_2(j)}\}_{j\in\mathbb{N}}$ that converges a.e. in $[1/k,k]$ to g_λ with $\dot g_{\phi_k\circ\cdots\circ\phi_2(j)}$ converging to $\dot g_\lambda$ weakly in $L^2([1/k,k])$ as $j\to+\infty$.

Hence, we look at the diagonal sequence $\overline{g}_j := g_{\phi_j\circ\cdots\circ\phi_2(j)}$. By construction, $\overline{g}_j$ converges to g_λ a.e. in $\mathbb{R}^+$ as $j\to+\infty$, and therefore, by the Fatou lemma,

$$\liminf_{j\to+\infty}\int_{\mathbb{R}^+} t^a|\overline{g}_j(t)|^2\,dt \geq \int_{\mathbb{R}^+} t^a|g_\lambda(t)|^2\,dt. \tag{5.37}$$

However, by the weak convergence of $\dot{\overline{g}}_j$ to $\dot g_\lambda$ in $L^2([1/k,k])$ as $j\to+\infty$, we have that $\dot g_\lambda\in L^2([1/k,k])$, so $\psi(t):=t^a\dot g_\lambda(t)$ is also in $L^2([1/k,k])$, which gives

$$\lim_{j\to+\infty}\int_{1/k}^k \dot{\overline{g}}_j(t)\psi(t)\,dt = \int_{1/k}^k \dot g_\lambda(t)\psi(t)\,dt;$$

that is,

$$\lim_{j\to+\infty}\int_{1/k}^k t^a\dot{\overline{g}}_j(t)\dot g_\lambda(t)\,dt = \int_{1/k}^k t^a|\dot g_\lambda(t)|^2\,dt,$$

for any $k\in\mathbb{N}$, $k\geq 2$. As a consequence of this, we obtain that

$$\begin{aligned}
0 &\leq \liminf_{j\to+\infty}\int_{1/k}^k t^a|\dot{\overline{g}}_j(t)-\dot g_\lambda(t)|^2\,dt\\
&= \liminf_{j\to+\infty}\left(\int_{1/k}^k t^a|\dot{\overline{g}}_j(t)|^2\,dt + \int_{1/k}^k t^a|\dot g_\lambda(t)|^2\,dt - 2\int_{1/k}^k t^a\dot{\overline{g}}_j(t)\cdot\dot g_\lambda(t)\right)dt \qquad (5.38)\\
&= \liminf_{j\to+\infty}\int_{1/k}^k t^a|\dot{\overline{g}}_j(t)|^2\,dt - \int_{1/k}^k t^a|\dot g_\lambda(t)|^2\,dt,
\end{aligned}$$

for any $k\in\mathbb{N}$, $k\geq 2$.

By (5.37), (5.38), and the positivity of λ, we get

$$\begin{aligned}
\mathrm{m}_\lambda &= \lim_{j\to+\infty} G_\lambda(g_j)\\
&= \lim_{j\to+\infty}\left(\lambda\int_{\mathbb{R}^+} t^a|\overline{g}_j(t)|^2\,dt + \int_{\mathbb{R}^+} t^a|\dot{\overline{g}}_j(t)|^2\,dt\right)\\
&\geq \liminf_{j\to+\infty}\left(\lambda\int_{\mathbb{R}^+} t^a|\overline{g}_j(t)|^2\,dt + \int_{1/k}^k t^a|\dot{\overline{g}}_j(t)|^2\,dt\right)\\
&\geq \lambda\int_{\mathbb{R}^+} t^a|g_\lambda(t)|^2\,dt + \int_{1/k}^k t^a|\dot g_\lambda(t)|^2\,dt,
\end{aligned}$$

for any $k\in\mathbb{N}$, $k\geq 2$. By taking $k\to+\infty$, we deduce that $\mathrm{m}_\lambda\geq G_\lambda(g_\lambda)$. This proves that the infimum in (5.33) is attained at g_λ. The uniqueness of the minimizer is due to the convexity of the functional G_λ. This completes the proof of (5.33).

Now notice that since $g_\lambda \in W^{1,2}_{1,a}(\mathbb{R}^+)$, then $g_\lambda(0) = 1$ and $g_\lambda \in W^{1,p}((0,\kappa))$, for any $p \in [1,a^*)$ and any $\kappa > 0$, by Lemma 5.6. Hence, it is uniformly continuous on $(0,\kappa)$ for any $\kappa > 0$ by the standard Sobolev embedding, so it can be extended with continuity at 0, that is, the function $g_\lambda \in C^0(\overline{\mathbb{R}^+})$.

Moreover, by taking standard perturbation of the functional G_λ at $g_\lambda + \varepsilon\phi$, with $\phi \in C_0^\infty(\mathbb{R}^+)$ and $\varepsilon \in \mathbb{R}$ small, one obtains

$$\int_{\mathbb{R}^+} t^a \big(\lambda g_\lambda(t)\phi(t) + \dot{g}_\lambda(t)\dot{\phi}(t)\big)\, dt = 0. \tag{5.39}$$

Hence, g_λ satisfies weakly an ODE, so $g_\lambda \in C^\infty(\mathbb{R}^+)$ by uniformly elliptic regularity theory (see, e.g., [117, theorem 8.10][4]). Moreover, integrating by parts in (5.39), it easily follows that g_λ solves problem (5.34).

Now we prove (5.35). For this, it is convenient to reduce to the case $\lambda = 1$ by noticing that if $g^{(\lambda)}(t) := g(t\sqrt{\lambda})$, we have that

$$G_\lambda(g^{(\lambda)}) = \lambda^{(1-a)/2} G_1(g)$$

and, therefore,

$$\mathrm{m}_\lambda = \lambda^{(1-a)/2}\mathrm{m}_1 \quad \text{and} \quad g_\lambda(t) = g_1(t\sqrt{\lambda}). \tag{5.40}$$

Let us fix $\phi \in C_0^\infty([-1,1])$ with $\phi(0) = 1$, and let

$$\gamma(t) := t^a\big(g_1(t)\phi(t) + \dot{g}_1(t)\dot{\phi}(t)\big).$$

By the Cauchy–Schwarz inequality, we have that

$$\int_{\mathbb{R}^+} \gamma(t)\, dt \le G_1(g_1)\, G_1(\phi) < +\infty,$$

so $\gamma \in L^1(\mathbb{R}^+)$. Therefore, by the absolute continuity of the Lebesgue integral, for any fixed $\varepsilon > 0$, there exists $\delta_\varepsilon > 0$ such that if $0 < t_1 < t_2 < \delta_\varepsilon$, then

$$\int_{t_1}^{t_2} \gamma(\tau)\, d\tau < \varepsilon.$$

As a consequence, the function

$$\Gamma(t) := \int_t^{+\infty} \gamma(\tau)\, d\tau$$

is uniformly continuous in $(0,1)$, and therefore, it may be extended with continuity at 0 as follows:

$$\Gamma(0) = \int_0^{+\infty} \gamma(\tau)\, d\tau = \int_0^{+\infty} \tau^a\big(g_1(\tau)\phi(\tau) + \dot{g}_1(\tau)\dot{\phi}(\tau)\big)\, d\tau. \tag{5.41}$$

[4] In further detail, g_λ satisfies [117, equation (8.2)] with $n = 1$, $a^{ij} = a^{11} = t^a$, $b^i = b^1 = 0$, $c^i = c^1 = 0$, and $d = \lambda t^a$ and this equation is uniformly elliptic in bounded domains separated from 0, so we can apply [117, theorem 8.10] with $f = 0$ and obtain that $g_\lambda \in W^{k,2}(b_1, b_2)$, for any $b_2 > b_1 > 0$ and any $k \in \mathbb{N}$.

By (5.34) with $\lambda = 1$, it is easy to see that, for any $t \in \mathbb{R}^+$,

$$t^a g_1(t) = \frac{d}{dt}(t^a \dot{g}_1(t)).$$

Thus, by this and recalling that $\phi(0) = 1$ and $\phi(t) = 0$ if $t \geq 1$, we get

$$\begin{aligned}
\Gamma(0) &= \int_0^1 \tau^a \big(g_1(\tau)\phi(\tau) + \dot{g}_1(\tau)\dot{\phi}(\tau)\big)\, d\tau \\
&= \int_0^1 \left[\frac{d}{d\tau}\big(\tau^a \dot{g}_1(\tau)\big)\phi(\tau) + \tau^a \dot{g}_1(\tau)\dot{\phi}(\tau)\right] d\tau \\
&= \lim_{t\to 0^+} \int_t^1 \frac{d}{d\tau}\big(\tau^a \dot{g}_1(\tau)\phi(\tau)\big)\, d\tau \\
&= -\lim_{t\to 0^+} t^a \dot{g}_1(t)\phi(t) \\
&= -\lim_{t\to 0^+} t^a \dot{g}_1(t).
\end{aligned} \tag{5.42}$$

Note that the computation carried out in (5.42) also has shown that the preceding limit exists.

Now, to compute explicitly such a limit, we consider the perturbation

$$g_{1,\varepsilon} := (g_1 + \varepsilon\phi)/(1+\varepsilon)$$

with $\varepsilon \in \mathbb{R}$ small. First of all, notice that $g_{1,\varepsilon} = g_1 + \varepsilon\phi - \varepsilon g_1 + o(\varepsilon)$, so

$$|g_{1,\varepsilon}|^2 = |g_1|^2 + 2\varepsilon g_1\phi - 2\varepsilon |g_1|^2 + o(\varepsilon),$$

and, similarly, if we replace $g_{1,\varepsilon}$ with $\dot{g}_{1,\varepsilon}$. It follows that

$$\begin{aligned}
G_1(g_{1,\varepsilon}) = G_1(g_1) + 2\varepsilon \int_{\mathbb{R}^+} \tau^a \big(&g_1(\tau)\phi(\tau) \\
&- |g_1(\tau)|^2 + \dot{g}_1(\tau)\dot{\phi}(\tau) - |\dot{g}_1(t\tau)|^2\big)\, dt + o(\varepsilon).
\end{aligned}$$

Then the minimality condition implies that

$$\int_{\mathbb{R}^+} \tau^a \big(g_1(\tau)\phi(\tau) - |g_1(\tau)|^2 + \dot{g}_1(\tau)\dot{\phi}(\tau) - |\dot{g}_1(\tau)|^2\big)\, d\tau = 0.$$

Hence, by this, (5.41), and the definition of m_1, we deduce

$$\begin{aligned} 0 &= \int_{\mathbb{R}^+} \tau^a \big(g_1(\tau)\phi(\tau) - |g_1(\tau)|^2 + \dot{g}_1(\tau)\dot{\phi}(\tau) - |\dot{g}_1(\tau)|^2\big)\, d\tau \\ &= \int_{\mathbb{R}^+} \tau^a \big(g_1(\tau)\phi(\tau) + \dot{g}_1(\tau)\dot{\phi}(\tau)\big)\, d\tau - \int_{\mathbb{R}^+} \tau^a \big(|g_1(\tau)|^2 + |\dot{g}_1(\tau)|^2\big)\, d\tau \\ &= \Gamma(0) - m_1. \end{aligned}$$

This and (5.42) prove (5.35) for $\lambda = 1$. In general, recalling (5.40), we obtain

$$\lim_{t\to 0^+} t^a \dot{g}_\lambda(t) = \sqrt{\lambda} \lim_{t\to 0^+} t^a \dot{g}_1(t\sqrt{\lambda}) = \lambda^{(1-a)/2} \lim_{t\to 0^+} (t\sqrt{\lambda})^a \dot{g}_1(t\sqrt{\lambda}) = -m_1 \lambda^{(1-a)/2},$$

thus establishing (5.35) for any $\lambda > 0$.

Now let us prove (5.36) For this, we first observe that $G_1(|g_1|) = G_1(g_1)$, which implies, by the uniqueness of the minimizer, that $g_1 = |g_1|$ and so $g_1 \geq 0$ in $\mathbb{R}^+$. We start by showing that

$$\dot{g}_1 \leq 0 \quad \text{in the whole of } \mathbb{R}^+. \tag{5.43}$$

By contradiction, if g_1 were increasing somewhere, there would exist $t_2 > t_1 \geq 0$ such that $0 \leq g_1(t_1) < g_1(t_2)$. Let $b := (g_1(t_1) + g_1(t_2))/2 \in (g_1(t_1), g_1(t_2))$. Notice that there exists $t_3 > t_2$ such that $g(t_3) = b$; otherwise, by continuity, we would have that $g(t) > b > 0$, for any $t > t_2$, so, using that $a \in (-1, 1)$,

$$G_1(g_1) \geq \int_{t_2}^{+\infty} t^a |g_1(t)|^2\, dt \geq b^2 \int_{t_2}^{+\infty} t^a\, dt = +\infty,$$

which is against our contraction.

Having established the existence of the desired t_3, we use the Weierstrass theorem to obtain $t_\star \in [t_1, t_3]$ in such a way that

$$g_1(t_\star) = \max_{t\in[t_1,t_3]} g_1(t).$$

Note that, by definition of b,

$$g_1(t_\star) \geq g_1(t_2) > b > g_1(t_1).$$

Hence, $t_\star \neq t_1$ and also $t_\star \neq t_3$, because $g_1(t_3) = b$. Thus, $t_\star$ is an interior maximum for g_1. Accordingly, $\dot{g}_1(t_\star) = 0$ and $\ddot{g}_1(t_\star) \leq 0$. Thus, by (5.34),

$$0 = \ddot{g}_1(t_\star) + \frac{a}{t_\star} \dot{g}_1(t_\star) - g_1(t_\star) \leq 0 + 0 - b = -b < 0.$$

This is a contradiction, and it proves (5.43).

Another consequence of (5.43) is that $g_1(t) \leq g_1(0) = 1$, for any $t \in \mathbb{R}^+$. Moreover, it implies that the limit

$$\ell := \lim_{t\to+\infty} g_1(t) \in [0, 1]$$

exists. Necessarily, it must be

$$\ell = 0.$$

Otherwise, if $\ell > 0$, it would follow that $g(t) \geq \ell/2$, for any $t \geq t_0$, for a suitable $t_0 > 0$. This yields that (using also that $a \in (-1,1)$)

$$G_1(g_1) \geq \int_{t_0}^{+\infty} t^a |g_1(t)|^2 \, dt \geq (\ell/2)^2 \int_{t_0}^{+\infty} t^a \, dt = +\infty,$$

which is against our contraction. All these considerations imply (5.36) for $\lambda = 1$ and thus for any $\lambda > 0$, thanks to (5.40). □

5.5 The first eigenvalue of A_s and that of $(-\Delta)^s$

This section is devoted to the study of the relation between the first eigenvalues of A_s and of $(-\Delta)^s$, that is, between λ_{1,A_s} and $\lambda_{1,s}$. Precisely, in this framework, our main result is the following:

Proposition 5.8 The relation between the first eigenvalue of $(-\Delta)^s$ and that of A_s is given by

$$\lambda_{1,s} < \lambda_{1,A_s}.$$

Proof Let us take $a := 1 - 2s \in (-1,1)$ and, for any $(x,t) \in \Omega \times \mathbb{R}^+$, set

$$E_1(x,t) := g_{\widetilde{\lambda}_1}(t)\widetilde{e}_1(x),$$

where the setting of Theorem 5.7 is in use, $\widetilde{\lambda}_1$ is the first eigenvalue of the Laplacian $-\Delta$, and $\widetilde{e}_1 = e_{1,A_s}$ is the first eigenfunction of the operator A_s (see Section 5.2).

Notice that E_1 may be thought of as an extension of $\widetilde{e}_1$ in the half-space $\mathbb{R}^n \times \mathbb{R}^+$ that vanishes in $(\mathbb{R}^n \setminus \Omega) \times \mathbb{R}^+$. However, we point out that E_1 does not verify $\operatorname{div}(\nabla E_1) = 0$ in the whole of $\mathbb{R}^n \times \mathbb{R}^+$.

Also note that the function $E_1 \in C^\infty(\Omega \times \mathbb{R}^+) \cap C(\overline{\Omega \times \mathbb{R}^+})$ because $\widetilde{e}_1 \in C^\infty(\Omega) \cap C^m(\overline{\Omega})$, for any $m \in \mathbb{N}$ (see formula (5.4)), and $g_{\widetilde{\lambda}_1} \in C^\infty(\mathbb{R}^+) \cap C^0(\overline{\mathbb{R}^+})$ by Theorem 5.7. Also,

$$\lim_{t \to 0^+} t^a \partial_t E_1(x,t) = \lim_{t \to 0^+} t^a \dot{g}_{\widetilde{\lambda}_1}(t)\widetilde{e}_1(x) = -\mathrm{m}_1 \widetilde{\lambda}_1^{(1-a)/2} \widetilde{e}_1(x),$$

thanks to (5.35).

Furthermore, since $G_{\lambda_1}(g_{\widetilde{\lambda}_1})$ is finite by Theorem 5.7, we have that

$$L^1(\mathbb{R}^+) \ni t^a \widetilde{\lambda}_1 |g_{\widetilde{\lambda}_1}(t)|^2 + t^a |\dot{g}_{\widetilde{\lambda}_1}(t)|^2 \geq 2\sqrt{\widetilde{\lambda}_1} t^a |g_{\widetilde{\lambda}_1}(t) \dot{g}_{\widetilde{\lambda}_1}(t)|,$$

and therefore, there exists a diverging sequence of R for which

$$\lim_{R \to +\infty} R^a |g_{\widetilde{\lambda}_1}(R) \dot{g}_{\widetilde{\lambda}_1}(R)| = 0. \tag{5.44}$$

Now note that using[5] the definition of E_1, the fact that $\widetilde{e}_1$ is the first eigenfunction of $-\Delta_x$ (for this, see Section 5.2), for any $(x,t) \in \Omega \times \mathbb{R}^+$, we have

$$\begin{aligned} t^a|\nabla E_1(x,t)|^2 &= \operatorname{div}\big(t^a E_1(x,t)\nabla E_1(x,t)\big) - at^{a-1}E_1(x,t)\partial_t E_1(x,t) \\ &\quad - t^a E_1(x,t)\Delta E_1(x,t) \\ &= \operatorname{div}\big(t^a E_1(x,t)\nabla E_1(x,t)\big) - at^{a-1}E_1(x,t)\dot{g}_{\widetilde{\lambda}_1}(t)\widetilde{e}_1(x) \\ &\quad - t^a E_1(x,t)g_{\widetilde{\lambda}_1}(t)\Delta_x\widetilde{e}_1(x) - t^a E_1(x,t)\ddot{g}_{\widetilde{\lambda}_1}(t)\widetilde{e}_1(x) \\ &= \operatorname{div}\big(t^a E_1(x,t)\nabla E_1(x,t)\big) - at^{a-1}E_1(x,t)\dot{g}_{\widetilde{\lambda}_1}(t)\widetilde{e}_1(x) \\ &\quad + \widetilde{\lambda}_1 t^a E_1(x,t)g_{\widetilde{\lambda}_1}(t)\widetilde{e}_1(x) - t^a E_1(x,t)\ddot{g}_{\widetilde{\lambda}_1}(t)\widetilde{e}_1(x) \\ &= \operatorname{div}\big(t^a E_1(x,t)\nabla E_1(x,t)\big) \\ &\quad + t^a E_1(x,t)\widetilde{e}_1(x)\big(-at^{-1}\dot{g}_{\widetilde{\lambda}_1}(t) + \widetilde{\lambda}_1 g_{\widetilde{\lambda}_1}(t) - \ddot{g}_{\widetilde{\lambda}_1}(t)\big) \\ &= \operatorname{div}\big(t^a E_1(x,t)\nabla E_1(x,t)\big), \end{aligned} \tag{5.45}$$

thanks to (5.34).

By (5.45) and the divergence theorem, we have that

$$\begin{aligned} &\int_{\Omega\times\mathbb{R}^+} t^a|\nabla E_1(x,t)|^2\,dx\,dt \\ &= \lim_{R\to+\infty}\int_{\Omega\times(0,R)} t^a|\nabla E_1(x,t)|^2\,dx\,dt \\ &= \lim_{R\to+\infty}\int_{\Omega\times(0,R)} \operatorname{div}\big(t^a E_1(x,t)\nabla E_1(x,t)\big)\,dx\,dt \\ &= \lim_{R\to+\infty}\int_{\Omega} \big(t^a E_1(x,t)\partial_t E_1(x,t)\big)|_{t=R} - \big(t^a E_1(x,t)\partial_t E_1(x,t)\big)|_{t=0}\,dx \end{aligned} \tag{5.46}$$

because, for any $t \in \mathbb{R}^+$, $E_1(\cdot,t) = 0$ on $\partial\Omega$, $\widetilde{e}_1 = 0$ being outside Ω. Hence, by (5.46) and using again the definition of E_1, we deduce that

$$\begin{aligned} &\int_{\Omega\times\mathbb{R}^+} t^a|\nabla E_1(x,t)|^2\,dx\,dt \\ &= \lim_{R\to+\infty}\int_{\Omega}\Big[\big(t^a E_1(x,t)\partial_t E_1(x,t)\big)|_{t=R} - \big(t^a E_1(x,t)\partial_t E_1(x,t)\big)|_{t=0}\Big]\,dx \\ &= \lim_{R\to+\infty}\int_{\Omega}\Big[R^a E_1(x,R)\partial_t E_1(x,R) - \big(t^a E_1(x,t)\partial_t E_1(x,t)\big)|_{t=0}\Big]\,dx \end{aligned}$$

[5] We remark that here ∇ is the vector collecting all the derivatives, both in x and it t. Similarly, $\Delta = \Delta_x + \partial_t^2$.

$$= \lim_{R\to+\infty} \int_{\Omega} \left[R^a g_{\widetilde{\lambda}_1}(R)\dot{g}_{\widetilde{\lambda}_1}(R)|\widetilde{e}_1(x)|^2 - \left(t^a g_{\widetilde{\lambda}_1}(t)\dot{g}_{\widetilde{\lambda}_1}(t)|\widetilde{e}_1(x)|^2\right)|_{t=0}\right] dx$$

$$= \lim_{R\to+\infty} \int_{\Omega} \left(R^a g_{\widetilde{\lambda}_1}(R)\dot{g}_{\widetilde{\lambda}_1}(R)|\widetilde{e}_1(x)|^2 + \mathrm{m}_1 \widetilde{\lambda}_1^{(1-a)/2} |\widetilde{e}_1(x)|^2\right) dx$$

$$= \lim_{R\to+\infty} \left(R^a g_{\widetilde{\lambda}_1}(R)\dot{g}_{\widetilde{\lambda}_1}(R) + \mathrm{m}_1 \widetilde{\lambda}_1^{(1-a)/2}\right) \|\widetilde{e}_1\|^2_{L^2(\Omega)}$$

$$= \mathrm{m}_1 \widetilde{\lambda}_1^{(1-a)/2}$$

$$= \mathrm{m}_1 \widetilde{\lambda}_1^s,$$

thanks to (5.35), the fact that $g_\lambda(0) = 1$, (5.44), and the choice of a. As a consequence,

$$\begin{aligned} \inf_{\substack{U\in C(\mathbb{R}^n\times\mathbb{R}^+) \\ \|U(\cdot,0)\|_{L^2(\Omega)}=1 \\ U(\cdot,0)=0 \text{ outside } \Omega}} \int_{\mathbb{R}^n\times\mathbb{R}^+} t^a |\nabla U(x,t)|^2\,dx\,dt &\le \int_{\mathbb{R}^n\times\mathbb{R}^+} t^a |\nabla E_1(x,t)|^2\,dx\,dt \\ &= \int_{\Omega\times\mathbb{R}^+} t^a |\nabla E_1(x,t)|^2\,dx\,dt \\ &= \mathrm{m}_1 \widetilde{\lambda}_1^s \end{aligned} \tag{5.47}$$

because $E_1(\cdot,t) = \widetilde{e}_1(\cdot)g_\lambda(t) = 0$ in $\mathbb{R}^n \setminus \Omega$, for any $t \in \mathbb{R}^+$.

Now we use a result in [55] to relate the first term in (5.47) to $\lambda_{1,s}$ (which, roughly speaking, says that the optimal U is realized by the so-called a-harmonic extension of $u := U(\cdot,0)$). That is, by [55, formula (3.7) and its proof on p. 1250] and [83, proposition 3.4], we get

$$\inf_{\substack{U\in C(\mathbb{R}^n\times\mathbb{R}^+) \\ \|U(\cdot,0)\|_{L^2(\Omega)}=1 \\ U(\cdot,0)=0 \text{ outside } \Omega}} \int_{\mathbb{R}^n\times\mathbb{R}^+} t^a |\nabla U(x,t)|^2\,dx\,dt = \mathrm{m}_1 \inf_{\substack{u\in C(\mathbb{R}^n) \\ \|u\|_{L^2(\Omega)}=1 \\ u=0 \text{ outside } \Omega}} \int_{\mathbb{R}^n} |\xi|^{2s} |\mathscr{F}u(\xi)|^2\,d\xi$$

$$= \mathrm{m}_1 \frac{C(n,s)}{2} \inf_{\substack{u\in C(\mathbb{R}^n) \\ \|u\|_{L^2(\Omega)}=1 \\ u=0 \text{ outside } \Omega}} \int_{\mathbb{R}^n\times\mathbb{R}^n} \frac{|u(x)-u(y)|^2}{|x-y|^{n+2s}}\,dx\,dy,$$

$$\ge \mathrm{m}_1 \frac{C(n,s)}{2} \min_{\substack{u\in X_0^s(\Omega) \\ \|u\|_{L^2(\Omega)}=1}} \int_{\mathbb{R}^n\times\mathbb{R}^n} \frac{|u(x)-u(y)|^2}{|x-y|^{n+2s}}\,dx\,dy$$

$$= \mathrm{m}_1\, \lambda_{1,s},$$

thanks also to the variational characterization of $\lambda_{1,s}$ given in (3.4). Thus,

$$\inf_{\substack{U\in C(\mathbb{R}^n\times\mathbb{R}^+) \\ \|U(\cdot,0)\|_{L^2(\Omega)}=1 \\ U(\cdot,0)=0 \text{ outside } \Omega}} \int_{\mathbb{R}^n\times\mathbb{R}^+} t^a |\nabla U(x,t)|^2\,dx\,dt \ge \mathrm{m}_1 \lambda_{1,s}.$$

This and (5.47) give that

$$\lambda_{1,s} \leq \widetilde{\lambda}_1^s.$$

We claim that the strict inequality occurs. If, by contradiction, equality holds here, then it does in (5.47), namely,

$$E_1 \in \arg\min \left\{ \inf_{\substack{U \in C(\overline{\mathbb{R}^n \times \mathbb{R}^+}) \\ \|U(\cdot,0)\|_{L^2(\Omega)}=1 \\ U(\cdot,0)=0 \text{ outside } \Omega}} \int_{\mathbb{R}^n \times \mathbb{R}^+} t^a |\nabla U(x,t)|^2 \, dx \, dt \right\}.$$

We remark that such minimizers are continuous up to $\overline{\mathbb{R}^n \times \mathbb{R}^+}$, and they solve the associated elliptic partial differential equation in $\mathbb{R}^n \times \mathbb{R}^+$ (see [92]): in particular, E_1 would solve an elliptic partial differential equation in $\mathbb{R}^n \times \mathbb{R}^+$, and it vanishes in a nontrivial open set (just take a ball B outside Ω and consider $B \times (1,2)$).

As a consequence of this and of the unique continuation principle (see [119]), E_1 has to vanish identically in $\Omega \times \mathbb{R}^+$, so by taking $t \to 0^+$, we would have that $\widetilde{e}_1(x) = 0$, for any $x \in \Omega$ (here we also use the fact that $g_{\widetilde{\lambda}_1}(0) = 1$ by (5.34)). This is a contradiction and it establishes that $\lambda_{1,s} < \widetilde{\lambda}_1^s = \lambda_{1,A_s}$. □

We end this chapter with the following:

Proof of Theorem 5.1 It is a direct consequences of Propositions 5.3 and 5.8. □

Part II

Nonlocal subcritical problems

6

Mountain pass and linking results

In this chapter we study the existence of nontrivial solutions for equations driven by the nonlocal integrodifferential operator $\mathscr{L}_K$ (see (1.54)) with homogeneous Dirichlet boundary conditions. More precisely, we consider the problem

$$\begin{cases} \mathscr{L}_K u + \lambda u + f(x,u) = 0 & \text{in } \Omega \\ u = 0 & \text{in } \mathbb{R}^n \setminus \Omega, \end{cases}$$

where λ is a real parameter, and the nonlinear term f satisfies superlinear and subcritical growth conditions at zero and at infinity.

This equation has a variational nature, so its solutions can be found as critical points of the energy functional $\mathscr{J}_{K,\lambda}$ associated with the problem. Here we get such critical points using both the mountain pass theorem and the linking theorem, respectively, when $\lambda < \lambda_1$ and $\lambda \geq \lambda_1$, where λ_1 denotes the first eigenvalue of the operator $-\mathscr{L}_K$.

The results presented here may be seen as the extension of some classical nonlinear analysis theorems to the case of fractional operators: they are based on the papers [96, 198, 200]. For the sake of completeness, we would recall that nonexistence results for nonlocal problems involving the fractional Laplacian operator have been proved by Ros-Oton and Serra in [185] via a suitable Pohožaev identity.

6.1 Hypotheses and statements

In this chapter, we study the following equation

$$\begin{cases} \mathscr{L}_K u + \lambda u + f(x,u) = 0 & \text{in } \Omega \\ u = 0 & \text{in } \mathbb{R}^n \setminus \Omega, \end{cases} \tag{6.1}$$

where $\mathscr{L}_K$ is the nonlocal operator defined in (1.54), the parameter $s \in (0,1)$ is fixed, and the set $\Omega \subset \mathbb{R}^n$, $n > 2s$, is open, bounded, and with continuous boundary. The

Dirichlet datum is given in $\mathbb{R}^n \setminus \Omega$ and not simply on $\partial\Omega$, consistent with the nonlocal character of the operator $\mathcal{L}_K$.

The weak formulation of (6.1) is given by the following problem:

$$\begin{cases} \displaystyle\int_{\mathbb{R}^n \times \mathbb{R}^n} (u(x)-u(y))(\varphi(x)-\varphi(y))K(x-y)\,dx\,dy - \lambda \int_\Omega u(x)\varphi(x)\,dx \\ \displaystyle\qquad\qquad = \int_\Omega f(x,u(x))\varphi(x)\,dx \quad \forall\, \varphi \in X_0^s(\Omega) \\ u \in X_0^s(\Omega). \end{cases} \tag{6.2}$$

Here the functional space $X_0^s(\Omega)$ denotes the linear space introduced in (1.65).

Finally, we suppose that the nonlinear term in equation (6.1) is a function $f : \overline{\Omega} \times \mathbb{R} \to \mathbb{R}$ verifying the following conditions:

$$f \text{ is continuous in } \overline{\Omega} \times \mathbb{R}; \tag{6.3}$$

$$\begin{gathered} \textit{there exist } a_1, a_2 > 0, \textit{ and } q \in (2, 2_s^*), 2_s^* := 2n/(n-2s), \textit{ such that} \\ |f(x,t)| \le a_1 + a_2|t|^{q-1} \textit{ for any } x \in \Omega, t \in \mathbb{R}; \end{gathered} \tag{6.4}$$

$$\lim_{t\to 0} \frac{f(x,t)}{t} = 0 \textit{ uniformly in } x \in \Omega; \tag{6.5}$$

$$tf(x,t) \ge 0, \textit{ for any } x \in \Omega, t \in \mathbb{R}; \tag{6.6}$$

$$\begin{gathered} \textit{there exist } \mu > 2 \textit{ and } r > 0 \textit{ such that, for any } x \in \overline{\Omega}, t \in \mathbb{R}, |t| \ge r, \\ 0 < \mu F(x,t) \le tf(x,t), \end{gathered} \tag{6.7}$$

where the function F is the primitive of f with respect to its second variable; that is,

$$F(x,t) := \int_0^t f(x,\tau)\,d\tau. \tag{6.8}$$

As a model for f, we can take the odd nonlinearity

$$f(x,t) := a(x)|t|^{q-2}t, \tag{6.9}$$

with $a \in C(\overline{\Omega})$, $a > 0$ in $\overline{\Omega}$, and $q \in (2, 2_s^*)$.

We remark that $f(x,0) = 0$, thanks to (6.5); therefore, the function $u \equiv 0$ is a (trivial) solution of (6.1): our scope is then to construct nontrivial solutions for (6.1). For this, we exploit two different variational techniques: when $\lambda < \lambda_1$ (where, as usual, we denoted by λ_1 the first eigenvalue of $-\mathcal{L}_K$; see Section 3.1), we construct a nontrivial solution via the mountain pass theorem; however, when $\lambda \ge \lambda_1$, we accomplish our purposes by using the linking theorem. These two different approaches are indeed the nonlocal counterparts of the famous theory developed for the Laplace operator (see, e.g., [11, 172, 174]).

The main result of this chapter is an existence theorem for equations driven by general integrodifferential operators of nonlocal fractional type, as stated here:

Theorem 6.1 *Let $s \in (0,1)$, $n > 2s$, and Ω be an open, bounded subset of $\mathbb{R}^n$ with continuous boundary. Let $K : \mathbb{R}^n \setminus \{0\} \to (0,+\infty)$ be a function satisfying conditions* (1.55) *and* (1.56), *and let $f : \overline{\Omega} \times \mathbb{R} \to \mathbb{R}$ verify* (6.3)–(6.7).

Then, for any $\lambda \in \mathbb{R}$, problem (6.1) *admits a nontrivial weak solution $u \in X_0^s(\Omega)$.*

In fact, if λ is small (i.e., $\lambda < \lambda_1$), we can find a nonnegative (nonpositive) solution of problem (6.2), as stated in Corollary 6.16. When $\lambda < \lambda_1$, the thesis of Theorem 6.1 is still valid with weaker assumptions on f (see [198], where the case $\lambda = 0$ was considered).

In the nonlocal framework, the simplest example we can deal with is given by the fractional Laplacian $(-\Delta)^s$, according to the following result:

Theorem 6.2 *Let $s \in (0,1)$, $n > 2s$, and Ω be an open, bounded subset of $\mathbb{R}^n$ with continuous boundary. Let $f : \overline{\Omega} \times \mathbb{R} \to \mathbb{R}$ be a function verifying* (6.3)–(6.7). *Then, for any $\lambda \in \mathbb{R}$, the problem*

$$\begin{cases} (-\Delta)^s u - \lambda u = f(x,u) & \text{in } \Omega \\ u = 0 & \text{in } \mathbb{R}^n \setminus \Omega \end{cases} \tag{6.10}$$

admits a nontrivial weak solution $u \in \mathbb{H}_0^s(\Omega)$.

When $s = 1$, equation (6.10) reduces to a standard semilinear Laplace partial differential equation: in this sense, Theorem 6.2 may be seen as the fractional version of the classical existence result in [174, theorem 5.16] (see also [11, 172, 212, 220]).

This classical result is an application of two critical point theorems (the mountain pass theorem and the linking theorem) to elliptic partial differential equations. We prove here that the geometry of these classical minimax theorems is respected by the nonlocal framework: for this, we develop a functional analytical setting that is inspired by (but not equivalent to) the fractional Sobolev spaces in order to correctly encode the Dirichlet boundary datum in the variational formulation.

Of course, the compactness property required by these minimax theorems is also satisfied in the nonlocal setting, thanks to the classical Ambrosetti–Rabinowitz condition (6.7) and to the choice of the functional setting we work in.

6.2 Estimates on the nonlinearity and its primitive

Here we collect some elementary results that will be useful in the main estimates of this chapter. We use the growth conditions on f to deduce some bounds from above and below for the nonlinear term and its primitive. This part is quite standard (see, e.g., [174, 212]) and does not take into account the nonlocal nature of the problem: the reader familiar with these topics may skip it.

Lemma 6.3 *Assume that* $f : \overline{\Omega} \times \mathbb{R} \to \mathbb{R}$ *is a function satisfying conditions* (6.3)–(6.5). *Then, for any* $\varepsilon > 0$, *there exists* $\delta = \delta(\varepsilon)$ *such that, for any* $x \in \Omega$ *and* $t \in \mathbb{R}$,

$$|f(x,t)| \leq 2\varepsilon|t| + q\delta(\varepsilon)|t|^{q-1}, \tag{6.11}$$

and so, as a consequence,

$$|F(x,t)| \leq \varepsilon\,|t|^2 + \delta(\varepsilon)\,|t|^q, \tag{6.12}$$

where F is defined as in (6.8).

Proof By assumption (6.5) for any $\varepsilon > 0$, there exists $\sigma = \sigma(\varepsilon) > 0$ such that, for any $t \in \mathbb{R}$ with $|t| < \sigma$ and a.e. $x \in \Omega$, we get

$$|f(x,t)| \leq 2\varepsilon|t|. \tag{6.13}$$

Moreover, by (6.4), there exists $\delta = \delta(\sigma) > 0$ such that a.e. $x \in \Omega$ and, for any $t \in \mathbb{R}$ with $|t| \geq \sigma$, we have

$$|f(x,t)| \leq q\delta(\sigma)|t|^{q-1}. \tag{6.14}$$

Combining (6.13) and (6.14), (6.11) easily follows. Using the definition of F (see (6.8)), we also get (6.12). □

Lemma 6.4 *Let* $f : \overline{\Omega} \times \mathbb{R} \to \mathbb{R}$ *be a function satisfying conditions* (6.3) *and* (6.7). *Then there exist two positive constants* a_3 *and* a_4 *such that, for any* $x \in \overline{\Omega}$ *and* $t \in \mathbb{R}$,

$$F(x,t) \geq a_3|t|^\mu - a_4. \tag{6.15}$$

Proof Let $r > 0$ be as in (6.7). Then, for any $x \in \overline{\Omega}$ and $t \in \mathbb{R}$ with $|t| \geq r > 0$,

$$\frac{t\,f(x,t)}{F(x,t)} \geq \mu.$$

Suppose that $t > r$. Dividing by t and integrating both terms in $[r,t]$, we obtain

$$F(x,t) \geq \frac{F(x,r)}{r^\mu} t^\mu \quad \text{for any } x \in \overline{\Omega}, t \in \mathbb{R}, t > r. \tag{6.16}$$

Since $x \mapsto F(x,r)$ is continuous in $\overline{\Omega}$, by the Weierstrass theorem, there exists

$$\min_{x \in \overline{\Omega}} F(x,r).$$

Hence, by (6.16), we get

$$F(x,t) \geq \min_{x \in \overline{\Omega}} F(x,r) r^{-\mu} t^\mu \quad \text{for any } x \in \overline{\Omega}, t \in \mathbb{R}, t > r. \tag{6.17}$$

With the same arguments, it is easy to consider the case $t < -r$ and to prove that

$$F(x,t) \geq \frac{F(x,-r)}{r^\mu}|t|^\mu \geq \min_{x \in \overline{\Omega}} F(x,-r) r^{-\mu} t^\mu \quad \text{for any } x \in \overline{\Omega}, t \in \mathbb{R}, t < -r,$$

so, for any $x \in \overline{\Omega}$ and $t \in \mathbb{R}$ with $|t| \geq r$, we get

$$F(x,t) \geq a_3 |t|^{\mu}, \tag{6.18}$$

where

$$a_3 := r^{-\mu} \min \left\{ \min_{x \in \overline{\Omega}} F(x,r), \min_{x \in \overline{\Omega}} F(x,-r) \right\}.$$

Note that a_3 is a positive constant, because $F(x,t) > 0$, for any $x \in \overline{\Omega}$ and $t \in \mathbb{R}$, such that $|t| \geq r$ (see assumption (6.7)).

Because the function F is continuous in $\overline{\Omega} \times \mathbb{R}$, by the Weierstrass theorem, it is bounded for any $x \in \overline{\Omega}$ and $t \in \mathbb{R}$ such that $|t| \leq r$, say,

$$|F(x,t)| \leq \tilde{a}_4 \quad \text{in } \overline{\Omega} \times \{|t| \leq r\}, \tag{6.19}$$

for some positive constant $\tilde{a}_4$. Hence, the estimate (6.15) follows from (6.18) and (6.19), taking $a_4 = \tilde{a}_4 + a_3 r^{\mu} > 0$. □

6.3 Proofs of the main theorems

To prove Theorem 6.1, first we observe that problem (6.2) is the Euler–Lagrange equation of the energy functional $\mathcal{J}_{K,\lambda} : X_0^s(\Omega) \to \mathbb{R}$ defined as follows:

$$\begin{aligned} \mathcal{J}_{K,\lambda}(u) := \frac{1}{2} \int_{\mathbb{R}^n \times \mathbb{R}^n} |u(x) - u(y)|^2 K(x-y)\,dx\,dy - \frac{\lambda}{2} \int_{\Omega} |u(x)|^2\,dx \\ - \int_{\Omega} F(x,u(x))\,dx, \end{aligned} \tag{6.20}$$

for every $u \in X_0^s(\Omega)$.

Notice that the functional $\mathcal{J}_{K,\lambda}$ is well defined, thanks to assumptions (6.3) and (6.4), to the fact that $L^{2_s^*}(\Omega) \hookrightarrow L^q(\Omega)$ continuously (Ω being bounded), and to Lemma 1.28(a). Moreover, $\mathcal{J}_{K,\lambda}$ is Fréchet differentiable in $u \in X_0^s(\Omega)$, and

$$\begin{aligned} \langle \mathcal{J}'_{K,\lambda}(u), \varphi \rangle = \int_{\mathbb{R}^n \times \mathbb{R}^n} (u(x) - u(y))(\varphi(x) - \varphi(y)) K(x-y)\,dx\,dy \\ - \lambda \int_{\Omega} u(x)\varphi(x)\,dx - \int_{\Omega} f(x,u(x))\varphi(x)\,dx, \end{aligned}$$

for any $\varphi \in X_0^s(\Omega)$.

Thus, critical points of $\mathcal{J}_{K,\lambda}$ are solutions of problem (6.2). To find these critical points, we will apply two classical variational results, namely, the mountain pass theorem and the linking theorem (see [11, 172, 174, 212]), respectively, in the case where $\lambda < \lambda_1$ and $\lambda \geq \lambda_1$, where λ_1 is the first eigenvalue of the nonlocal operator $-\mathcal{L}_K$ (as introduced in Proposition 3.1).

Both these minimax theorems require that the functional $\mathcal{J}_{K,\lambda}$

1. Has a suitable *geometric structure* (as stated, e.g., for the mountain pass theorem in conditions (1°)–(3°) of [212, theorem 6.1] and for the linking theorem in (I_1') and (I_5) of [174, theorem 5.3]);

2. Satisfies the *Palais–Smale compactness condition* at any level $c \in \mathbb{R}$ (see, e.g., [212, p. 70]).

We recall that in a real Banach space E, a C^1-functional $\mathscr{J} : E \to \mathbb{R}$ satisfies the *Palais–Smale condition* at level $c \in \mathbb{R}$ if

$$\begin{gathered} \textit{any sequence } \{u_j\}_{j\in\mathbb{N}} \textit{ in } E \textit{ such that} \\ \mathscr{J}(u_j) \to c \textit{ and } \sup\big\{|\langle \mathscr{J}'(u_j), \varphi\rangle| : \varphi \in E, \|\varphi\|_E = 1\big\} \to 0 \\ \textit{as } j \to +\infty \textit{ admits a subsequence strongly convergent in } E. \end{gathered} \tag{6.21}$$

We will show indeed that our functional possesses this geometric structure (according to the different values of λ) and that it satisfies the Palais–Smale condition. We start by considering the case where the parameter λ is less than λ_1.

6.3.1 The case $\lambda < \lambda_1$: mountain pass–type solutions

In this setting, we assume that the nonlinearity f satisfies[1] conditions (6.3)–(6.5) and (6.7). In order to verify that the functional $\mathscr{J}_{K,\lambda}$ verifies the assumptions of the mountain pass theorem, we need the following lemma:

Lemma 6.5 *Let $K : \mathbb{R}^n \setminus \{0\} \to (0, +\infty)$ satisfy assumptions* (1.55) *and* (1.56), *and let $\lambda < \lambda_1$. Then there exist two positive constants m_1^λ and M_1^λ, depending only on λ, such that, for any $v \in X_0^s(\Omega)$,*

$$m_1^\lambda \|v\|^2_{X_0^s(\Omega)} \le \|v\|^2_{X_0^s(\Omega),\lambda} \le M_1^\lambda \|v\|^2_{X_0^s(\Omega)}, \tag{6.22}$$

where

$$\|v\|_{X_0^s(\Omega),\lambda} := \left(\int_{\mathbb{R}^n\times\mathbb{R}^n} |v(x)-v(y)|^2 K(x-y)\,dx\,dy - \lambda\int_\Omega |v(x)|^2\,dx\right)^{1/2}. \tag{6.23}$$

By (6.22), *it follows that $\|\cdot\|_{X_0,\lambda}$ is a norm on $X_0^s(\Omega)$ equivalent to the ones in* (1.62) *and* (1.87). *The constants m_1^λ and M_1^λ are given by*

$$m_1^\lambda := \min\{1, 1-\lambda/\lambda_1\} \quad \textit{and} \quad M_1^\lambda := \max\{1, 1-\lambda/\lambda_1\}.$$

Proof Of course, if $v \equiv 0$, then (6.22) is trivially verified, so we take a function $v \in X_0^s(\Omega) \setminus \{0\}$. For the following computation, it is convenient to distinguish the case $0 \le \lambda < \lambda_1$ from the case in which $\lambda < 0$. We remark that $\lambda_1 > 0$, by Proposition 3.1(a).

[1] For completeness, we observe that when $\lambda < \lambda_1$, we do not assume hypothesis (6.6), which is needed just when $\lambda \ge \lambda_1$. In fact, the result stated in Theorem 6.1 holds true under weaker assumptions on the nonlinearity f. For instance, when $\lambda < \lambda_1$, one could assume the conditions of [198, theorem 1], where the case $\lambda = 0$ was dealt with.

Thus, first, we assume that $0 \leq \lambda < \lambda_1$: then it is easily seen that

$$\int_{\mathbb{R}^n\times\mathbb{R}^n} |v(x)-v(y)|^2 K(x-y)\,dx\,dy - \lambda\int_\Omega |v(x)|^2\,dx \leq \int_{\mathbb{R}^n\times\mathbb{R}^n} |v(x)-v(y)|^2 K(x-y)\,dx\,dy.$$

Moreover, using the variational characterization of λ_1 (see formula (3.4)), we get

$$\int_{\mathbb{R}^n\times\mathbb{R}^n} |v(x)-v(y)|^2 K(x-y)\,dx\,dy - \lambda\int_\Omega |v(x)|^2\,dx \geq \left(1-\frac{\lambda}{\lambda_1}\right)\int_{\mathbb{R}^n\times\mathbb{R}^n} |v(x)-v(y)|^2 K(x-y)\,dx\,dy.$$

The preceding estimates imply (6.22) when $0 \leq \lambda < \lambda_1$.

When $\lambda < 0$, arguing in the same way, we obtain

$$\begin{aligned}\int_{\mathbb{R}^n\times\mathbb{R}^n} &|v(x)-v(y)|^2 K(x-y)\,dx\,dy\\ &\leq \int_{\mathbb{R}^n\times\mathbb{R}^n} |v(x)-v(y)|^2 K(x-y)\,dx\,dy - \lambda\int_\Omega |v(x)|^2\,dx\\ &\leq \left(1-\frac{\lambda}{\lambda_1}\right)\int_{\mathbb{R}^n\times\mathbb{R}^n} |v(x)-v(y)|^2 K(x-y)\,dx\,dy,\end{aligned}$$

which proves (6.22) in this case too.

Then we have to show that formula (6.23) defines a norm on $X_0^s(\Omega)$. For this, we claim that

$$\begin{aligned}\langle u,v\rangle_{X_0,\lambda} := &\int_{\mathbb{R}^n\times\mathbb{R}^n} (u(x)-u(y))(v(x)-v(y))K(x-y)\,dx\,dy\\ &-\lambda\int_\Omega u(x)v(x)\,dx\end{aligned} \tag{6.24}$$

is a scalar product on $X_0^s(\Omega)$. Indeed, by (6.22) and the fact that $\|\cdot\|_{X_0^s(\Omega)}$ is a norm (see Lemma 1.28(b)), it easily follows that $\langle v,v\rangle_{X_0,\lambda} \geq 0$, for any $v \in X_0^s(\Omega)$, and that $\langle v,v\rangle_{X_0,\lambda} = 0$ if and only if $v \equiv 0$, while the properties of the integrals give easily that $(u,v) \mapsto \langle u,v\rangle_{X_0,\lambda}$ is linear with respect both variables and symmetric. Hence, the claim is proved. Since

$$\|v\|_{X_0^s(\Omega),\lambda} := \sqrt{\langle v,v\rangle_{X_0,\lambda}},$$

formula (6.23) defines a norm on $X_0^s(\Omega)$.

Finally, the equivalency of the norms follows from Lemma 1.28(b). □

Now we can prove that the functional $\mathcal{J}_{K,\lambda}$ has the geometric features required by the mountain pass theorem.

Proposition 6.6 Let $\lambda < \lambda_1$, and let f be a function satisfying conditions (6.3)–(6.5). Then there exist $\rho > 0$ and $\beta > 0$ such that, for any $u \in X_0^s(\Omega)$ with $\|u\|_{X_0} = \rho$, it results that $\mathcal{J}_{K,\lambda}(u) \geq \beta$.

Proof Let u be a function in $X_0^s(\Omega)$. By (6.12), we get that, for any $\varepsilon > 0$,

$$\begin{aligned}
\mathcal{J}_{K,\lambda}(u) &\geq \frac{1}{2}\int_{\mathbb{R}^n\times\mathbb{R}^n} |u(x)-u(y)|^2 K(x-y)\,dx\,dy - \frac{\lambda}{2}\int_\Omega |u(x)|^2\,dx \\
&\quad - \varepsilon\int_\Omega |u(x)|^2 dx - \delta(\varepsilon)\int_\Omega |u(x)|^q\,dx \\
&\geq \frac{m_1^\lambda}{2}\int_{\mathbb{R}^n\times\mathbb{R}^n} |u(x)-u(y)|^2 K(x-y)\,dx\,dy - \varepsilon\|u\|^2_{L^2(\Omega)} - \delta(\varepsilon)\|u\|^q_{L^q(\Omega)} \\
&\geq \frac{m_1^\lambda}{2}\int_{\mathbb{R}^n\times\mathbb{R}^n} |u(x)-u(y)|^2 K(x-y)\,dx\,dy - \varepsilon|\Omega|^{(2_s^*-2)/2_s^*}\|u\|^2_{L^{2_s^*}(\Omega)} \\
&\quad - |\Omega|^{(2_s^*-q)/2_s^*}\delta(\varepsilon)\|u\|^q_{L^{2_s^*}(\Omega)},
\end{aligned} \tag{6.25}$$

thanks to Lemma 6.5 (here we need $\lambda < \lambda_1$) and the fact that $L^{2_s^*}(\Omega) \hookrightarrow L^2(\Omega)$ and $L^{2_s^*}(\Omega) \hookrightarrow L^q(\Omega)$ continuously (being Ω bounded and $\max\{2,q\} = q < 2_s^*$).

Using (1.56) and Lemma 1.28, we deduce from (6.25) that, for any $\varepsilon > 0$,

$$\begin{aligned}
\mathcal{J}_{K,\lambda}(u) &\geq \frac{m_1^\lambda}{2}\int_{\mathbb{R}^n\times\mathbb{R}^n} |u(x)-u(y)|^2 K(x-y)\,dx\,dy \\
&\quad - \varepsilon c|\Omega|^{(2_s^*-2)/2_s^*}\int_{\mathbb{R}^n\times\mathbb{R}^n} \frac{|u(x)-u(y)|^2}{|x-y|^{n+2s}}\,dx\,dy \\
&\quad - \delta(\varepsilon)c^{q/2}|\Omega|^{(2_s^*-q)/2_s^*}\left(\int_{\mathbb{R}^n\times\mathbb{R}^n} \frac{|u(x)-u(y)|^2}{|x-y|^{n+2s}}\,dx\,dy\right)^{q/2} \\
&\geq \left(\frac{m_1^\lambda}{2} - \frac{\varepsilon c|\Omega|^{(2_s^*-2)/2_s^*}}{\theta}\right)\int_{\mathbb{R}^n\times\mathbb{R}^n} |u(x)-u(y)|^2 K(x-y)\,dx\,dy \\
&\quad - \frac{\delta(\varepsilon)c^{q/2}|\Omega|^{(2_s^*-q)/2_s^*}}{\theta}\left(\int_{\mathbb{R}^n\times\mathbb{R}^n} |u(x)-u(y)|^2 K(x-y)\,dx\,dy\right)^{q/2} \\
&= \left(\frac{m_1^\lambda}{2} - \frac{\varepsilon c|\Omega|^{(2_s^*-2)/2_s^*}}{\theta}\right)\|u\|^2_{X_0^s(\Omega)} \\
&\quad - \frac{\delta(\varepsilon)c^{q/2}|\Omega|^{(2_s^*-q)/2_s^*}}{\theta}\|u\|^q_{X_0^s(\Omega)}.
\end{aligned} \tag{6.26}$$

Choosing $\varepsilon > 0$ such that $2\varepsilon c|\Omega|^{(2_s^*-2)/2_s^*} < m_1^\lambda \theta$, by (6.26), it easily follows that

$$\mathcal{J}_{K,\lambda}(u) \geq \alpha \|u\|^2_{X_0^s(\Omega)} \left(1 - \kappa \|u\|^{q-2}_{X_0^s(\Omega)}\right),$$

for suitable positive constants α and κ.

Now let $u \in X_0^s(\Omega)$ be such that $\|u\|_{X_0^s(\Omega)} = \rho > 0$. Because $q > 2$ by assumption, we can choose ρ sufficiently small (i.e., ρ such that $1 - \kappa\rho^{q-2} > 0$) so that

$$\inf_{\substack{u \in X_0^s(\Omega) \\ \|u\|_{X_0^s(\Omega)} = \rho}} \mathcal{J}_{K,\lambda}(u) \geq \alpha\rho^2(1 - \kappa\rho^{q-2}) =: \beta > 0.$$

Hence, Proposition 6.6 is proved. □

Proposition 6.7 Let $\lambda < \lambda_1$, and let f be a function satisfying conditions (6.3)–(6.5) and (6.7). Then there exists $e \in X_0^s(\Omega)$ such that $e \geq 0$ a.e. in $\mathbb{R}^n$, $\|e\|_{X_0^s(\Omega)} > \rho$, and $\mathcal{J}_{K,\lambda}(e) < \beta$, where ρ and β are given in Proposition 6.6.

Proof We fix $u \in X_0^s(\Omega)$ such that $\|u\|_{X_0^s(\Omega)} = 1$ and $u \geq 0$ a.e. in $\mathbb{R}^n$; we remark that this choice is possible thanks to (1.27) (alternatively, one can replace any $u \in X_0^s(\Omega)$ with its positive part, which belongs to $X_0^s(\Omega)$ too, thanks to Lemma 1.22).

Now let $\zeta > 0$. By Lemmas 6.4 and 6.5 (here we again use the fact that $\lambda < \lambda_1$), we have

$$\begin{aligned}
\mathcal{J}_{K,\lambda}(\zeta u) &= \frac{1}{2}\int_{\mathbb{R}^n \times \mathbb{R}^n} |\zeta u(x) - \zeta u(y)|^2 K(x-y)\,dx\,dy - \frac{\lambda}{2}\zeta^2 \int_\Omega |u(x)|^2\,dx \\
&\quad - \int_\Omega F(x, \zeta u(x))\,dx \\
&\leq \frac{M_1^\lambda}{2}\zeta^2 - a_3\zeta^\mu \int_\Omega |u(x)|^\mu\,dx + a_4|\Omega|.
\end{aligned}$$

Since $\mu > 2$, passing to the limit as $\zeta \to +\infty$, we get that $\mathcal{J}_{K,\lambda}(\zeta u) \to -\infty$, so the assertion follows taking $e = \zeta u$, with ζ sufficiently large. □

Propositions 6.6 and 6.7 give that the geometry of the mountain pass theorem is fulfilled by $\mathcal{J}_{K,\lambda}$. Therefore, to apply the mountain pass theorem, we have to check the validity of the Palais–Smale condition. This will be accomplished in Propositions 6.8 and 6.9.

Proposition 6.8 Let $\lambda < \lambda_1$, and let f be a function satisfying conditions (6.3)–(6.5) and (6.7). Let $c \in \mathbb{R}$, and let $\{u_j\}_{j\in\mathbb{N}}$ be a sequence in $X_0^s(\Omega)$ such that

$$\mathcal{J}_{K,\lambda}(u_j) \to c \tag{6.27}$$

and

$$\sup\left\{|\langle \mathcal{J}'_{K,\lambda}(u_j), \varphi\rangle| \,:\, \varphi \in X_0^s(\Omega), \|\varphi\|_{X_0^s(\Omega)} = 1\right\} \to 0 \tag{6.28}$$

as $j \to +\infty$. Then $\{u_j\}_{j\in\mathbb{N}}$ is bounded in $X_0^s(\Omega)$.

Proof For any $j \in \mathbb{N}$, by (6.27) and (6.28), it easily follows that there exists $\kappa > 0$ such that

$$|\mathscr{J}_{K,\lambda}(u_j)| \le \kappa \tag{6.29}$$

and

$$\Big|\langle \mathscr{J}'_{K,\lambda}(u_j), \frac{u_j}{\|u_j\|_{X_0^s(\Omega)}}\rangle\Big| \le \kappa. \tag{6.30}$$

Moreover, by Lemma 6.3 applied with $\varepsilon = 1$, we have that

$$\begin{aligned}&\Big|\int_{\Omega\cap\{|u_j|\le r\}} \Big(F(x,u_j(x)) - \frac{1}{\mu} f(x,u_j(x))u_j(x)\Big)\,dx\Big|\\ &\quad\le \Big(r^2 + \delta(1)r^q + \frac{2}{\mu}r + \frac{q}{\mu}\delta(1)r^{q-1}\Big)|\Omega| =: \tilde{\kappa}.\end{aligned} \tag{6.31}$$

Also, by Lemma 6.5 (which holds true because $\lambda < \lambda_1$), (6.7), and (6.31), we get

$$\begin{aligned}\mathscr{J}_{K,\lambda}(u_j) - \frac{1}{\mu}\langle \mathscr{J}'_{K,\lambda}(u_j), u_j\rangle &= \Big(\frac{1}{2} - \frac{1}{\mu}\Big)\Big(\|u_j\|^2_{X_0^s(\Omega)} - \lambda\|u\|^2_{L^2(\Omega)}\Big)\\ &\quad - \frac{1}{\mu}\int_\Omega \big(\mu F(x,u_j(x)) - f(x,u_j(x))u_j(x)\big)\,dx\\ &\ge \Big(\frac{1}{2} - \frac{1}{\mu}\Big) m_1^\lambda \|u_j\|^2_{X_0^s(\Omega)}\\ &\quad - \int_{\Omega\cap\{|u_j|\le r\}} \Big(F(x,u_j(x)) - \frac{1}{\mu} f(x,u_j(x))u_j(x)\Big)\,dx\\ &\ge \Big(\frac{1}{2} - \frac{1}{\mu}\Big) m_1^\lambda \|u_j\|^2_{X_0^s(\Omega)} - \tilde{\kappa}.\end{aligned} \tag{6.32}$$

As a consequence of (6.29) and (6.30), we also have

$$\mathscr{J}_{K,\lambda}(u_j) - \frac{1}{\mu}\langle \mathscr{J}'_{K,\lambda}(u_j), u_j\rangle \le \kappa\Big(1 + \|u_j\|_{X_0^s(\Omega)}\Big).$$

This and (6.32) imply that, for any $j \in \mathbb{N}$,

$$\|u_j\|^2_{X_0^s(\Omega)} \le \kappa_*\Big(1 + \|u_j\|_{X_0^s(\Omega)}\Big),$$

for a suitable positive constant κ_*. Hence, the assertion of Proposition 6.8 is proved. □

Proposition 6.9 Let f be a function satisfying conditions (6.3)–(6.5) and (6.7). Let $\{u_j\}_{j\in\mathbb{N}}$ be a sequence in $X_0^s(\Omega)$ such that $\{u_j\}_{j\in\mathbb{N}}$ is bounded in $X_0^s(\Omega)$ and (6.28) holds true. Then there exists $u_\infty \in X_0^s(\Omega)$ such that, up to a subsequence, $\|u_j - u_\infty\|_{X_0^s(\Omega)} \to 0$ as $j \to +\infty$.

Proof Since $\{u_j\}_{j\in\mathbb{N}}$ is bounded in $X_0^s(\Omega)$ and $X_0^s(\Omega)$ is a reflexive space (being a Hilbert space, by Lemma 1.29), up to a subsequence, still denoted by u_j, there exists $u_\infty \in X_0^s(\Omega)$ such that $u_j \to u_\infty$ weakly in $X_0^s(\Omega)$; that is,

$$\begin{aligned}&\int_{\mathbb{R}^n\times\mathbb{R}^n}(u_j(x)-u_j(y))(\varphi(x)-\varphi(y))K(x-y)\,dx\,dy\\&\quad\to \int_{\mathbb{R}^n\times\mathbb{R}^n}(u_\infty(x)-u_\infty(y))(\varphi(x)-\varphi(y))K(x-y)\,dx\,dy,\end{aligned}\tag{6.33}$$

for any $\varphi \in X_0^s(\Omega)$, as $j\to+\infty$. Moreover, by Lemma 1.30, up to a subsequence,

$$\begin{aligned}u_j &\to u_\infty \quad \text{in } L^2(\mathbb{R}^n)\\ u_j &\to u_\infty \quad \text{in } L^q(\mathbb{R}^n)\\ u_j &\to u_\infty \quad \text{a.e. in } \mathbb{R}^n\end{aligned}\tag{6.34}$$

as $j\to+\infty$, and there exists $\ell \in L^q(\mathbb{R}^n)$ such that

$$|u_j(x)| \le \ell(x) \quad \text{a.e. in } \mathbb{R}^n, \text{ for any } j\in\mathbb{N}\tag{6.35}$$

(see, e.g., [43, theorem IV.9]).

By (6.4), (6.33)–(6.35), the fact that the map $t\mapsto f(\cdot,t)$ is continuous in $t\in\mathbb{R}$, and the dominated convergence theorem, we get

$$\int_\Omega f(x,u_j(x))u_j(x)\,dx \to \int_\Omega f(x,u_\infty(x))u_\infty(x)\,dx\tag{6.36}$$

and

$$\int_\Omega f(x,u_j(x))u_\infty(x)\,dx \to \int_\Omega f(x,u_\infty(x))u_\infty(x)\,dx\tag{6.37}$$

as $j\to+\infty$. Moreover, by (6.28), we have that

$$\begin{aligned}0 \leftarrow \langle \mathcal{J}'_{K,\lambda}(u_j),u_j\rangle &= \int_{\mathbb{R}^n\times\mathbb{R}^n}|u_j(x)-u_j(y)|^2K(x-y)\,dx\,dy\\&\quad-\lambda\int_\Omega |u_j(x)|^2\,dx-\int_\Omega f(x,u_j(x))u_j(x)\,dx.\end{aligned}$$

Consequently, recalling also (6.34) and (6.36), we deduce that

$$\begin{aligned}&\int_{\mathbb{R}^n\times\mathbb{R}^n}|u_j(x)-u_j(y)|^2K(x-y)\,dx\,dy\\&\quad\to\lambda\int_\Omega |u_\infty(x)|^2\,dx+\int_\Omega f(x,u_\infty(x))u_\infty(x)\,dx\end{aligned}\tag{6.38}$$

as $j \to +\infty$. Furthermore, using again (6.28),

$$\langle \mathcal{J}'_{K,\lambda}(u_j), u_\infty \rangle \to 0 \tag{6.39}$$

as $j \to +\infty$. By (6.33), (6.34), (6.37), and (6.39), we obtain

$$\begin{aligned}\int_{\mathbb{R}^n \times \mathbb{R}^n} |u_\infty(x) - u_\infty(y)|^2 K(x-y)\, dx\, dy = \lambda \int_\Omega |u_\infty(x)|^2\, dx \\ + \int_\Omega f(x, u_\infty(x)) u_\infty(x)\, dx.\end{aligned} \tag{6.40}$$

Thus, (6.38) and (6.40) give that

$$\int_{\mathbb{R}^n \times \mathbb{R}^n} |u_j(x) - u_j(y)|^2 K(x-y)\, dx\, dy \to \int_{\mathbb{R}^n \times \mathbb{R}^n} |u_\infty(x) - u_\infty(y)|^2 K(x-y)\, dx\, dy,$$

so

$$\|u_j\|_{X_0^s(\Omega)} \to \|u_\infty\|_{X_0^s(\Omega)} \tag{6.41}$$

as $j \to +\infty$.

Finally, we have that

$$\begin{aligned}&\|u_j - u_\infty\|^2_{X_0^s(\Omega)} \\ &= \|u_j\|^2_{X_0^s(\Omega)} + \|u_\infty\|^2_{X_0^s(\Omega)} - 2\int_{\mathbb{R}^n \times \mathbb{R}^n} \big(u_j(x) \\ &\quad - u_j(y)\big)(u_\infty(x) - u_\infty(y)) K(x-y)\, dx\, dy \\ &\to 2\|u_\infty\|^2_{X_0^s(\Omega)} - 2\int_{\mathbb{R}^{2n}} |u_\infty(x) - u_\infty(y)|^2 K(x-y)\, dx\, dy = 0\end{aligned}$$

as $j \to +\infty$, thanks to (6.33) and (6.41). Then the assertion is proved. □

Remark 6.10 Note that Proposition 6.9 holds true for any value of the parameter λ, so we can also use such a result for $\lambda \geq \lambda_1$.

End of the proof of Theorem 6.1 when $\lambda < \lambda_1$. When $\lambda < \lambda_1$, the geometry of the mountain pass theorem for the functional $\mathcal{J}_{K,\lambda}$ is provided by Propositions 6.6 and 6.7, while the Palais–Smale condition is a consequence of Propositions 6.8 and 6.9.

Thus, we can make use of the mountain pass theorem (e.g., in the form given by [212, theorem 6.1] (see also [11, 174]): we conclude that there exists a critical point $u \in X_0^s(\Omega)$ of $\mathscr{J}_{K,\lambda}$ such that

$$\mathscr{J}_{K,\lambda}(u) \geq \beta > 0 = \mathscr{J}_{K,\lambda}(0)$$

so that $u \not\equiv 0$. Hence, problem (6.1) has a nontrivial weak solution.

6.3.2 The case $\lambda \geq \lambda_1$: linking-type solutions

Since $\lambda \geq \lambda_1$, we can suppose that

$$\lambda \in [\lambda_k, \lambda_{k+1}) \quad \text{for some } k \in \mathbb{N},$$

where λ_k is the kth eigenvalue of the operator $-\mathscr{L}_K$, as defined in Section 3.1. We start with the following preliminary result:

Lemma 6.11 *Let $K : \mathbb{R}^n \setminus \{0\} \to (0, +\infty)$ satisfy assumptions* (1.55) *and* (1.56), *and let $\lambda \in [\lambda_k, \lambda_{k+1})$, for some $k \in \mathbb{N}$. Then for any $v \in \mathbb{P}_{k+1}$,*

$$\int_{\mathbb{R}^n \times \mathbb{R}^n} |v(x) - v(y)|^2 K(x-y)\, dx\, dy - \lambda \int_\Omega |v(x)|^2\, dx \geq m_{k+1}^\lambda \|v\|^2_{X_0^s(\Omega)}, \tag{6.42}$$

where

$$m_{k+1}^\lambda = 1 - \lambda/\lambda_{k+1} > 0. \tag{6.43}$$

Proof First of all, note that $\lambda \geq \lambda_k \geq \lambda_1 > 0$, thanks to Proposition 3.1(a).

Let $v \in \mathbb{P}_{k+1}$. If $v \equiv 0$, then (6.42) is trivially verified. Now assume that $v \not\equiv 0$. The variational characterization of λ_{k+1} (see formula (3.10)) gives that

$$\int_{\mathbb{R}^n \times \mathbb{R}^n} |v(x) - v(y)|^2 K(x-y)\, dx\, dy - \lambda \int_\Omega |v(x)|^2\, dx$$
$$\geq \left(1 - \frac{\lambda}{\lambda_{k+1}}\right) \int_{\mathbb{R}^{2n}} |v(x) - v(y)|^2 K(x-y)\, dx\, dy.$$

Since $\lambda < \lambda_{k+1}$, Lemma 6.11 is proved. □

Now we prove that the functional $\mathscr{J}_{K,\lambda}$ has the geometric structure required by the linking theorem.

Proposition 6.12 Let $\lambda \in [\lambda_k, \lambda_{k+1})$, for some $k \in \mathbb{N}$, and let f be a function satisfying conditions (6.3)–(6.5). Then there exist $\rho > 0$ and $\beta > 0$ such that, for any $u \in \mathbb{P}_{k+1}$ with $\|u\|_{X_0^s(\Omega)} = \rho$, it results that $\mathscr{J}_{K,\lambda}(u) \geq \beta$.

Proof This proof is very similar to the one of Proposition 6.6: the only difference is that Lemma 6.5 is not available in this case, and we need to replace it with Lemma 6.11 (this will change m_1^λ in the proof of Proposition 6.6 with m_{k+1}^λ, and the rest is pretty much the same). We give full details for the facility of the reader.

Let $u \in \mathbb{P}_{k+1}$. By (6.12), we get that, for any $\varepsilon > 0$,

$$\begin{aligned}\mathscr{I}_{K,\lambda}(u) &\geq \frac{1}{2}\int_{\mathbb{R}^n\times\mathbb{R}^n}|u(x)-u(y)|^2K(x-y)dx\,dy - \frac{\lambda}{2}\int_\Omega |u(x)|^2dx\\ &\quad -\varepsilon\int_\Omega |u(x)|^2dx - \delta(\varepsilon)\int_\Omega |u(x)|^q\,dx\\ &\geq \frac{m^\lambda_{k+1}}{2}\int_{\mathbb{R}^n\times\mathbb{R}^n}|u(x)-u(y)|^2K(x-y)dx\,dy - \varepsilon\|u\|^2_{L^2(\Omega)} - \delta(\varepsilon)\|u\|^q_{L^q(\Omega)}\\ &\geq \frac{m^\lambda_{k+1}}{2}\int_{\mathbb{R}^n\times\mathbb{R}^n}|u(x)-u(y)|^2K(x-y)dx\,dy - \varepsilon|\Omega|^{(2^*_s-2)/2^*_s}\|u\|^2_{L^{2^*_s}(\Omega)}\\ &\quad -|\Omega|^{(2^*_s-q)/2^*_s}\delta(\varepsilon)\|u\|^q_{L^{2^*_s}(\Omega)},\end{aligned} \tag{6.44}$$

thanks to Lemma 6.11 (being $\lambda_k \leq \lambda < \lambda_{k+1}$) and to the fact that and $L^{2^*_s}(\Omega) \hookrightarrow L^2(\Omega)$ and $L^{2^*_s}(\Omega) \hookrightarrow L^q(\Omega)$ continuously (being Ω bounded and $\max\{2,q\} = q < 2^*_s$).

Using (1.56) and Lemma 1.28(a), we deduce from (6.44) that, for any $\varepsilon > 0$,

$$\begin{aligned}\mathscr{I}_{K,\lambda}(u) &\geq \frac{m^\lambda_{k+1}}{2}\int_{\mathbb{R}^n\times\mathbb{R}^n}|u(x)-u(y)|^2K(x-y)dx\,dy\\ &\quad -\varepsilon c|\Omega|^{(2^*_s-2)/2^*_s}\int_{\mathbb{R}^n\times\mathbb{R}^n}\frac{|u(x)-u(y)|^2}{|x-y|^{n+2s}}dx\,dy\\ &\quad -\delta(\varepsilon)c^{q/2}|\Omega|^{(2^*_s-q)/2^*_s}\left(\int_{\mathbb{R}^n\times\mathbb{R}^n}\frac{|u(x)-u(y)|^2}{|x-y|^{n+2s}}dx\,dy\right)^{q/2}\\ &\geq \left(\frac{m^\lambda_{k+1}}{2} - \frac{\varepsilon c|\Omega|^{(2^*_s-2)/2^*_s}}{\theta}\right)\int_{\mathbb{R}^n\times\mathbb{R}^n}|u(x)-u(y)|^2K(x-y)dx\,dy\\ &\quad -\frac{\delta(\varepsilon)c^{q/2}|\Omega|^{(2^*_s-q)/2^*_s}}{\theta}\left(\int_{\mathbb{R}^n\times\mathbb{R}^n}|u(x)-u(y)|^2K(x-y)dx\,dy\right)^{q/2}\\ &= \left(\frac{m^\lambda_{k+1}}{2} - \frac{\varepsilon c|\Omega|^{(2^*_s-2)/2^*_s}}{\theta}\right)\|u\|^2_{X^s_0(\Omega)} - \frac{\delta(\varepsilon)c^{q/2}|\Omega|^{(2^*_s-q)/2^*_s}}{\theta}\|u\|^q_{X^s_0(\Omega)},\end{aligned} \tag{6.45}$$

by Lemma 1.28(b).

Choosing $\varepsilon > 0$ such that $2\varepsilon c|\Omega|^{(2^*_s-2)/2^*_s} < m^\lambda_{k+1}\theta$, by (6.45), it easily follows that

$$\mathscr{I}_{K,\lambda}(u) \geq \alpha\|u\|^2_{X^s_0(\Omega)}\left(1-\kappa\|u\|^{q-2}_{X^s_0(\Omega)}\right),$$

for suitable positive constants α and κ.

Now let $u \in \mathbb{P}_{k+1}$ be such that $\|u\|_{X_0^s(\Omega)} = \rho > 0$. Because $q > 2$ by assumption, we can choose ρ sufficiently small (i.e., ρ such that $1 - \kappa\rho^{q-2} > 0$) so that

$$\inf_{\substack{u \in \mathbb{P}_{k+1} \\ \|u\|_{X_0^s(\Omega)} = \rho}} \mathscr{J}_{K,\lambda}(u) \geq \alpha\rho^2(1 - \kappa\rho^{q-2}) =: \beta > 0.$$

Hence, the assertion follows. □

Proposition 6.13 Let $\lambda \in [\lambda_k, \lambda_{k+1})$, for some $k \in \mathbb{N}$, and let f be a function satisfying conditions (6.3), (6.4), and (6.6). Then $\mathscr{J}_{K,\lambda}(u) \leq 0$, for any $u \in \operatorname{span}\{e_1, \ldots, e_k\}$.

Proof Let $u \in \operatorname{span}\{e_1, \ldots, e_k\}$. Then

$$u(x) = \sum_{i=1}^{k} u_i e_i(x),$$

with $u_i \in \mathbb{R}$, $i = 1, \ldots, k$. Because $\{e_1, \ldots, e_k, \ldots\}$ is an orthonormal basis of $L^2(\Omega)$ and an orthogonal basis of $X_0^s(\Omega)$, by Proposition 3.1(f), we get

$$\int_{\Omega} |u(x)|^2 \, dx = \sum_{i=1}^{k} u_i^2 \tag{6.46}$$

and

$$\int_{\mathbb{R}^{2n}} |u(x) - u(y)|^2 K(x-y) \, dx \, dy = \sum_{i=1}^{k} u_i^2 \|e_i\|^2_{X_0^s(\Omega)}. \tag{6.47}$$

Moreover, by (6.6) and (6.8), it is easily seen that

$$F(x,t) \geq 0 \quad \text{for any } x \in \Omega, t \in \mathbb{R}. \tag{6.48}$$

Then, by (6.46)–(6.48) and using (3.5) and (3.12), we get

$$\begin{aligned}
\mathscr{J}_{K,\lambda}(u) &= \frac{1}{2} \sum_{i=1}^{k} u_i^2 \left(\|e_i\|^2_{X_0^s(\Omega)} - \lambda \right) - \int_{\Omega} F(x, u(x)) \, dx \\
&\leq \frac{1}{2} \sum_{i=1}^{k} u_i^2 \left(\|e_i\|^2_{X_0^s(\Omega)} - \lambda \right) \\
&= \frac{1}{2} \sum_{i=1}^{k} u_i^2 (\lambda_i - \lambda) \leq 0,
\end{aligned}$$

thanks to the fact that $\lambda_i \leq \lambda_k \leq \lambda$, for any $i = 1, \ldots, k$. □

Proposition 6.14 Let $\lambda \geq 0$, and let f be a function satisfying (6.3)–(6.5) and (6.7). Moreover, let $\mathbb{F}$ be a finite-dimensional subspace of $X_0^s(\Omega)$. Then there exist $R > \rho$ such that $\mathscr{J}_{K,\lambda}(u) \leq 0$, for any $u \in \mathbb{F}$ with $\|u\|_{X_0^s(\Omega)} \geq R$, where ρ is given in Proposition 6.12.

Proof Let $u \in \mathbb{F}$. Then the nonnegativity of λ and Lemma 6.4 give

$$\begin{aligned}\mathcal{J}_{K,\lambda}(u) &\le \frac{1}{2}\|u\|^2_{X_0^s(\Omega)} - \frac{\lambda}{2}\|u\|^2_{L^2(\Omega)} - a_3\int_\Omega |u(x)|^\mu\,dx + a_4|\Omega| \\ &\le \frac{1}{2}\|u\|^2_{X_0^s(\Omega)} - a_3\|u\|^\mu_{L^\mu(\Omega)} + a_4|\Omega| \\ &\le \frac{1}{2}\|u\|^2_{X_0^s(\Omega)} - \tilde{a}_3\|u\|^\mu_{X_0^s(\Omega)} + a_4|\Omega|,\end{aligned} \tag{6.49}$$

for some positive constant $\tilde{a}_3$, thanks to the fact that in any finite-dimensional space all the norms are equivalent.

Hence, if $\|u\|_{X_0^s(\Omega)} \to +\infty$, then

$$\mathcal{J}_{K,\lambda}(u) \to -\infty$$

because $\mu > 2$ by assumption, so the assertion follows. □

Propositions 6.12–6.14 and [174, remark 5.5(iii)] give[2] that $\mathcal{J}_{K,\lambda}$ has the geometric structure required by the linking theorem. Therefore, it remains to check the validity of the Palais–Smale condition: this will be done in Proposition 6.15.

To prove the Palais–Smale compactness condition, we argue essentially as in the case of the mountain pass theorem, but some nontrivial technical differences arise (especially when dealing with the boundedness of the Palais–Smale sequence), so we prefer to give full details for the reader's convenience.

Proposition 6.15 Let $\lambda \ge \lambda_1$, and let f be a function satisfying conditions (6.3)–(6.7). Let $c \in \mathbb{R}$, and let $\{u_j\}_{j\in\mathbb{N}}$ be a sequence in $X_0^s(\Omega)$ satisfying (6.27) and (6.28). Then $\{u_j\}_{j\in\mathbb{N}}$ is bounded in $X_0^s(\Omega)$.

Proof The spirit of this proof is similar to that of Proposition 6.8. Nevertheless, the use of Lemma 6.5 is not possible in this case, and this causes some technical difficulties that require the introduction of an additional parameter γ. Here are the details of the proof.

Let us fix $\gamma \in (2,\mu)$, where $\mu > 2$ is given in assumption (6.7). By (6.7), (6.31), and Lemma 6.4,

$$\begin{aligned}\mathcal{J}_{K,\lambda}(u_j) - \frac{1}{\gamma}\langle \mathcal{J}'_{K,\lambda}(u_j), u_j\rangle &= \left(\frac{1}{2}-\frac{1}{\gamma}\right)\left(\|u_j\|^2_{X_0^s(\Omega)} - \lambda\|u_j\|^2_{L^2(\Omega)}\right) \\ &\quad - \int_\Omega \left(F(x,u_j(x)) - \frac{1}{\gamma}f(x,u_j(x))u_j(x)\right)dx\end{aligned}$$

[2] In particular, we use Proposition 6.14 with $\lambda \in [\lambda_k, \lambda_{k+1})$ and

$$\mathbb{F} := \mathrm{span}\{e_1,\dots,e_{k+1}\} = \mathrm{span}\{e_1,\dots,e_k\} \oplus \mathrm{span}\{e_{k+1}\},$$

while [174, remark 5.5(iii)] is used here with $V := \mathrm{span}\{e_1,\dots,e_k\}$ and $e := e_{k+1}$. With this choice, $\mathbb{F} = V \oplus \mathrm{span}\{e\}$.

$$\geq \left(\frac{1}{2}-\frac{1}{\gamma}\right)\left(\|u_j\|^2_{X_0^s(\Omega)} - \lambda\|u_j\|^2_{L^2(\Omega)}\right)$$
$$+\left(\frac{\mu}{\gamma}-1\right)\int_{\Omega\cap\{|u_j|\geq r\}} F(x,u_j(x))dx$$
$$-\int_{\Omega\cap\{|u_j|\leq r\}}\left(F(x,u_j(x)) - \frac{1}{\gamma}f(x,u_j(x))u_j(x)\right)dx$$
$$\geq \left(\frac{1}{2}-\frac{1}{\gamma}\right)\left(\|u_j\|^2_{X_0^s(\Omega)} - \lambda\|u_j\|^2_{L^2(\Omega)}\right)$$
$$+\left(\frac{\mu}{\gamma}-1\right)\int_{\Omega\cap\{|u_j|\geq r\}} F(x,u_j(x))dx - \tilde{\kappa}$$
$$\geq \left(\frac{1}{2}-\frac{1}{\gamma}\right)\left(\|u_j\|^2_{X_0^s(\Omega)} - \lambda\|u_j\|^2_{L^2(\Omega)}\right)$$
$$+a_3\left(\frac{\mu}{\gamma}-1\right)\|u_j\|^{\mu}_{L^{\mu}(\Omega)} - a_4\left(1-\frac{\mu}{\gamma}\right)|\Omega| - \tilde{\kappa}. \tag{6.50}$$

Moreover, for any $\varepsilon > 0$, the Young inequality (with conjugate exponents $\mu/2 > 1$ and $\mu/(\mu-2)$) gives

$$\|u_j\|^2_{L^2(\Omega)} \leq \frac{2\varepsilon}{\mu}\|u_j\|^{\mu}_{L^{\mu}(\Omega)} + \frac{\mu-2}{\mu}\varepsilon^{-2/(\mu-2)}|\Omega|. \tag{6.51}$$

Hence, by (6.50) and (6.51), we deduce that

$$\mathscr{J}_{K,\lambda}(u_j) - \frac{1}{\gamma}\langle \mathscr{J}'_{K,\lambda}(u_j), u_j\rangle \geq \left(\frac{1}{2}-\frac{1}{\gamma}\right)\|u_j\|^2_{X_0^s(\Omega)} - \lambda\left(\frac{1}{2}-\frac{1}{\gamma}\right)\frac{2\varepsilon}{\mu}\|u_j\|^{\mu}_{L^{\mu}(\Omega)}$$
$$-\lambda\left(\frac{1}{2}-\frac{1}{\gamma}\right)\frac{\mu-2}{\mu}\varepsilon^{-2/(\mu-2)}|\Omega|$$
$$+a_3\left(\frac{\mu}{\gamma}-1\right)\|u_j\|^{\mu}_{L^{\mu}(\Omega)} - a_4\left(1-\frac{\mu}{\gamma}\right)|\Omega| - \tilde{\kappa}$$
$$= \left(\frac{1}{2}-\frac{1}{\gamma}\right)\|u_j\|^2_{X_0^s(\Omega)}$$
$$+\left[a_3\left(\frac{\mu}{\gamma}-1\right) - \lambda\left(\frac{1}{2}-\frac{1}{\gamma}\right)\frac{2\varepsilon}{\mu}\right]\|u_j\|^{\mu}_{L^{\mu}(\Omega)} - C_{\varepsilon}, \tag{6.52}$$

where C_ε is a constant such that $C_\varepsilon \to +\infty$ as $\varepsilon \to 0$, being $\mu > \gamma > 2$.

Now, choosing ε so small that

$$a_3\left(\frac{\mu}{\gamma}-1\right)-\lambda\left(\frac{1}{2}-\frac{1}{\gamma}\right)\frac{2\varepsilon}{\mu}>0,$$

by (6.52), we get

$$\mathcal{J}_{K,\lambda}(u_j)-\frac{1}{\gamma}\langle\mathcal{J}'_{K,\lambda}(u_j),u_j\rangle\geq\left(\frac{1}{2}-\frac{1}{\gamma}\right)\|u_j\|^2_{X_0^s(\Omega)}-C_\varepsilon. \tag{6.53}$$

As a consequence of (6.29) and (6.30), we also have

$$\mathcal{J}_{K,\lambda}(u_j)-\frac{1}{\gamma}\langle\mathcal{J}'_{K,\lambda}(u_j),u_j\rangle\leq\kappa\left(1+\|u_j\|_{X_0^s(\Omega)}\right),$$

so, by (6.53), for any $j\in\mathbb{N}$,

$$\|u_j\|^2_{X_0^s(\Omega)}\leq\kappa_*\left(1+\|u_j\|_{X_0^s(\Omega)}\right),$$

for a suitable positive constant κ_*. Hence, the assertion is proved. □

By Proposition 6.15 and Remark 6.10, we deduce the validity of the Palais–Smale condition for the functional $\mathcal{J}_{K,\lambda}$, when $\lambda\geq\lambda_1$.

End of the proof of Theorem 6.1 when $\lambda\geq\lambda_1$. If $\lambda\geq\lambda_1$, we can assume that $\lambda\in[\lambda_k,\lambda_{k+1})$, for some $k\in\mathbb{N}$. In this setting, the geometry of the linking theorem is assured by Propositions 6.12–6.14. The Palais–Smale condition is given by Propositions 6.9 and 6.15 (recall also Remark 6.10), so we can exploit the linking theorem (e.g., in the form given by [174, theorem 5.3]): we conclude that there exists a critical point $u\in X_0^s(\Omega)$ of the functional $\mathcal{J}_{K,\lambda}$. Furthermore,

$$\mathcal{J}_{K,\lambda}(u)\geq\beta>0=\mathcal{J}_{K,\lambda}(0),$$

so $u\not\equiv 0$. Of course, u is a weak solution of problem (6.1), and this ends the proof of Theorem 6.1.

We conclude this section with the following:

Proof of Theorem 6.2 It is a consequence of Theorem 6.1 by choosing

$$K(x):=|x|^{-(n+2s)}$$

and by recalling that $\mathbb{H}_0^s(\Omega)\subseteq H^s(\mathbb{R}^n)$, due to Lemma 1.24(b). □

6.4 Comments on the sign of the solutions

In this section we discuss some properties about the sign of the solutions of equation (6.1). As in the classical case of the Laplacian (see [174, remark 5.19]), one can determine the sign of mountain pass–type solutions. Indeed, about problem (6.2), the following result holds true:

Corollary 6.16 Let all the assumptions of Theorem 6.1 be satisfied. Then, for any $\lambda < \lambda_1$, problem (6.1) admits a nonnegative weak solution u_+ and a nonpositive weak solution u_- in $X_0^s(\Omega)$ that are of mountain pass type and that are not identically zero.

Proof To prove the existence of a nonnegative (nonpositive) solution of problem (6.2), it is enough to introduce the functions

$$F_{\pm}(x,t) := \int_0^t f_{\pm}(x,\tau)d\tau,$$

with

$$f_+(x,t) := \begin{cases} f(x,t) & \text{if } t \geq 0 \\ 0 & \text{if } t < 0 \end{cases} \quad \text{and} \quad f_-(x,t) := \begin{cases} 0 & \text{if } t > 0 \\ f(x,t) & \text{if } t \leq 0. \end{cases}$$

Note that $f_{\pm}$ satisfy conditions (6.3)–(6.6), while assumption (6.7) is verified by f_+ and F_+ in Ω and for any $t > r$ and by f_- and F_- in Ω and for any $t < -r$.

Let $\mathcal{J}_{K,\lambda}^{\pm} : X_0^s(\Omega) \to \mathbb{R}$ be the functional defined as follows:

$$\begin{aligned}\mathcal{J}_{K,\lambda}^{\pm}(u) := &\frac{1}{2}\int_{\mathbb{R}^n\times\mathbb{R}^n} |u(x)-u(y)|^2 K(x-y)\,dx\,dy - \frac{\lambda}{2}\int_\Omega |u(x)|^2 dx \\ &- \int_\Omega F_{\pm}(x,u(x))dx,\end{aligned}$$

for every $u \in X_0^s(\Omega)$. It is easy to see that the functional $\mathcal{J}_{K,\lambda}^{\pm}$ is well defined, is Fréchet differentiable in $u \in X_0^s(\Omega)$, and for any $\varphi \in X_0^s(\Omega)$,

$$\begin{aligned}\langle(\mathcal{J}_{K,\lambda}^{\pm})'(u),\varphi\rangle = &\int_{\mathbb{R}^n\times\mathbb{R}^n} (u(x)-u(y))(\varphi(x)-\varphi(y))K(x-y)\,dx\,dy \\ &- \lambda\int_\Omega u(x)\varphi(x)dx - \int_\Omega f_{\pm}(x,u(x))\varphi(x)dx.\end{aligned} \tag{6.54}$$

Moreover, $\mathcal{J}_{K,\lambda}^{\pm}$ satisfies Propositions 6.6–6.9 (because we can choose the sign of e in Proposition 6.7) and $\mathcal{J}_{K,\lambda}^{\pm}(0) = 0$. Hence, by the mountain pass theorem, there exists a nontrivial critical point $u_{\pm} \in X_0^s(\Omega)$ of $\mathcal{J}_{K,\lambda}^{\pm}$.

Let us show that u_+ is nonnegative in $\mathbb{R}^n$. For this, first of all, we claim that, for any $w \in X_0^s(\Omega)$, the following relation holds true a.e. $x, y \in \mathbb{R}^n$:

$$(w(x)-w(y))(w^-(x)-w^-(y)) \leq -|w^-(x)-w^-(y)|^2, \tag{6.55}$$

where w^- is the negative part of w; that is, $w^- := \max\{-w, 0\}$. Indeed, writing $w = w^+ - w^-$ and taking into account that

$$w^+(x)w^-(x) = 0 \quad \text{and} \quad w^+(x)w^-(y) \geq 0 \quad \text{a.e. } x,y \in \mathbb{R}^n,$$

we get

$$\begin{aligned}
&(w(x)-w(y))(w^-(x)-w^-(y))\\
&\quad=(w^+(x)-w^+(y))(w^-(x)-w^-(y))-(w^-(x)-w^-(y))^2\\
&\quad=-w^+(x)w^-(y)-w^+(y)w^-(x)-(w^-(x)-w^-(y))^2\\
&\quad\le-|w^-(x)-w^-(y)|^2
\end{aligned}$$

a.e. $x,y\in\mathbb{R}^n$. Hence, the claim (6.55) is proved.

Now we define $\varphi:=(u_+)^-$, and we remark that because $u_+\in X_0^s(\Omega)$, we have that $(u_+)^-\in X_0^s(\Omega)$, by Lemma 1.22, and therefore, we can use φ in (6.54). In this way, we get

$$\begin{aligned}
0&=\langle(\mathscr{J}^{\pm}_{K,\lambda})'(u_+),(u_+)^-\rangle\\
&=\int_{\mathbb{R}^n\times\mathbb{R}^n}(u_+(x)-u_+(y))((u_+)^-(x)-(u_+)^-(y))K(x-y)\,dx\,dy\\
&\quad-\lambda\int_\Omega u_+(x)(u_+)^-(x)\,dx-\int_\Omega f_+(x,u_+(x))(u_+)^-(x)\,dx\\
&=\int_{\mathbb{R}^n\times\mathbb{R}^n}(u_+(x)-u_+(y))((u_+)^-(x)-(u_+)^-(y))K(x-y)\,dx\,dy\\
&\quad+\lambda\int_\Omega\left|(u_+)^-(x)\right|^2dx\\
&\le-\|(u_+)^-\|^2_{X_0^s(\Omega)}+\lambda\|(u_+)^-\|^2_{L^2(\Omega)}\\
&\le-m_1^\lambda\,\|(u_+)^-\|^2_{X_0^s(\Omega)}\le0,
\end{aligned}$$

thanks to (6.55), Lemma 6.5, the choice of λ, and the definition of f_+ and of the negative part.

Thus, again, because $\lambda<\lambda_1$, it follows that $\|(u_+)^-\|_{X_0^s(\Omega)}=0$, so $u_+\ge0$ a.e. in $\mathbb{R}^n$, which is the assertion. With the same arguments, it is easy to show that u_- is nonpositive in $\mathbb{R}^n$. This ends the proof. □

6.5 A remark on the case $\lambda=0$

Note that in the case $\lambda=0$, a direct application of the celebrated Pucci–Serrin theorem [169] and Ricceri's variational principle [176, Theorem 2.5] gives the following result for parametric nonlocal problems:

Theorem 6.17 *Let $s\in(0,1)$, $n>2s$, and Ω be an open, bounded subset of $\mathbb{R}^n$ with continuous boundary. Let $K:\mathbb{R}^n\setminus\{0\}\to(0,+\infty)$ be a function satisfying conditions* (1.55) *and* (1.56), *and let $f:\overline{\Omega}\times\mathbb{R}\to\mathbb{R}$ verify* (6.3), (6.4), *and* (6.7). *Then, for every $\rho>0$ and α,*

$$\alpha > \inf_{u \in B_\rho} \frac{\sup_{v \in B_\rho} \int_\Omega F(x, v(x))dx - \int_\Omega F(x, u(x))dx}{\rho - \|u\|^2_{X_0^s(\Omega)}},$$

where

$$B_\rho := \left\{ u \in X_0^s(\Omega) : \|u\|^2_{X_0^s(\Omega)} < \rho \right\},$$

the problem

$$\begin{cases} \mathscr{L}_K u + \frac{1}{2\alpha} f(x,u) = 0 & \text{in } \Omega \\ u = 0 & \text{in } \mathbb{R}^n \setminus \Omega \end{cases} \tag{6.56}$$

admits at least two weak solutions, one of which lies in B_ρ.

Remark 6.18 Let us fix $\rho > 0$, and let $\chi(\rho)$ be the following function

$$\chi(\rho) := \inf_{u \in B_\rho} \frac{\sup_{v \in B_\rho} \int_\Omega F(x, v(x))dx - \int_\Omega F(x, u(x))dx}{\rho - \|u\|^2_{X_0^s(\Omega)}}.$$

An explicit bound for $\chi(\rho)$ can be obtained. Indeed, because $0 \in B_\rho$, it follows that

$$\chi(\rho) \le \frac{\sup_{v \in B_\rho} \int_\Omega F(x, v(x))dx}{\rho}. \tag{6.57}$$

However, by using the growth condition (6.4), it is easy to see that

$$\frac{\sup_{v \in B_\rho} \int_\Omega F(x, v(x))dx}{\rho} \le \frac{c_1}{\sqrt{\rho}} a_1 + \frac{c_q^q a_2}{q} \rho^{q/2-1}, \tag{6.58}$$

where

$$c_t := \sup_{u \in X_0^s(\Omega) \setminus \{0\}} \frac{\|u\|_{L^t(\Omega)}}{\|u\|_{X_0^s(\Omega)}} \quad t \in \{1, q\}. \tag{6.59}$$

Conditions (6.57) and (6.58) immediately yield

$$\chi(\rho) \le \frac{c_1}{\sqrt{\rho}} a_1 + \frac{c_q^q a_2}{q} \rho^{q/2-1}.$$

Then Theorem 6.17 ensures that, for every $\rho > 0$ and α such that

$$\alpha > \frac{c_1}{\sqrt{\rho}} a_1 + \frac{c_q^q a_2}{q} \rho^{q/2-1},$$

problem (6.56) admits at least two weak solutions, one of which lies in B_ρ.

We emphasize that in Theorem 6.17, on the contrary of Theorem 6.1, no condition at zero is requested on the nonlinearity.

7

Existence and localization of solutions

In this chapter, we give some existence results for fractional Laplacian problems whose prototype is

$$\begin{cases} (-\Delta)^s u = h(x)f(u) & \text{in } \Omega \\ u = 0 & \text{in } \mathbb{R}^n \setminus \Omega, \end{cases} \tag{7.1}$$

where $(-\Delta)^s$ denotes the fractional Laplacian operator with $s \in (0,1)$, and $\Omega \subset \mathbb{R}^n$, $n > 2s$, is open, bounded, and with continuous boundary. Moreover, $h \in L^\infty(\Omega) \setminus \{0\}$ is a nonnegative map, and $f : \mathbb{R} \to \mathbb{R}$ represents a subcritical continuous function.

More precisely, we prove here some existence results (see Theorem 7.3 and its consequences) for fractional equations assuming that f has a suitable behavior at zero, together with some global properties formulated by means of an auxiliary function ψ. Problem (7.1) has a variational nature; hence, its weak solutions can be found as critical points of a suitable functional. In order to prove the existence of these critical points, the main tool used in this chapter is a recent existence and localization theorem by Ricceri [183], which is a consequence of the variational methods contained in [176]. The applicability of this variational principle which guarantees the existence and localization of at least one weak solution for quasi-linear problems, is only known in low dimension.

The main novelty of this new framework is that instead of the usual assumptions on functionals, it requires some hypotheses on the nonlinearity that allow us to better understand the existence phenomena. This allows us to enlarge the set of applications of [176, 183], exploiting these abstract methods without continuous embedding of the ambient space in $C(\overline{\Omega})$.

In this chapter, we also consider nonlocal problems depending on parameters. In this framework, we obtain some bifurcation theorems, exploiting in a suitable way the abstract methods developed in [176]. We would emphasize the fact that these results cannot be achieved by direct minimization. The results of this chapter have been obtained in the papers [150, 155].

7.1 Existence of one weak solution

In this section, we study the following nonlocal fractional problem

$$\begin{cases} \mathscr{L}_K u + h(x)f(u) = 0 & \text{in } \Omega \\ u = 0 & \text{in } \mathbb{R}^n \setminus \Omega, \end{cases} \tag{7.2}$$

where $\mathscr{L}_K$ is the operator given in (1.54), the parameter $s \in (0,1)$ is fixed, $\Omega \subset \mathbb{R}^n$, $n > 2s$, is open, bounded, and with continuous boundary, $h \in L^\infty(\Omega)$, and $f : \mathbb{R} \to \mathbb{R}$ satisfies the following assumptions:

$$f \text{ is continuous in } \mathbb{R}; \tag{7.3}$$

$$\text{there exists } \gamma \in [1, 2_s^*), 2_s^* := \frac{2n}{n-2s} \text{ such that } \sup_{t\in\mathbb{R}} \frac{|f(t)|}{1+|t|^{\gamma-1}} < +\infty. \tag{7.4}$$

In what follows, we set

$$F(t) := \int_0^t f(\tau)d\tau, \ t \in \mathbb{R}. \tag{7.5}$$

The weak formulation of problem (7.2) is given by

$$\begin{cases} \displaystyle\int_{\mathbb{R}^n\times\mathbb{R}^n} (u(x)-u(y))(\varphi(x)-\varphi(y))K(x-y)dx\,dy \\ \qquad\qquad\qquad = \displaystyle\int_\Omega h(x)f(u(x))\varphi(x)dx \quad \forall\, \varphi \in X_0^s(\Omega) \\ u \in X_0^s(\Omega), \end{cases}$$

which is the Euler–Lagrange equation of the functional $\mathscr{J}_{K,h} : X_0^s(\Omega) \to \mathbb{R}$ defined as follows:

$$\mathscr{J}_{K,h}(u) := \frac{1}{2}\int_{\mathbb{R}^n\times\mathbb{R}^n} |u(x)-u(y)|^2 K(x-y)dx\,dy - \int_\Omega h(x)F(u(x))dx,$$

for every $u \in X_0^s(\Omega)$.

Note that $\mathscr{J}_{K,h}$ is Fréchet differentiable in $u \in X_0^s(\Omega)$, and one has

$$\begin{aligned} \langle \mathscr{J}_K'(u), \varphi\rangle = &\int_{\mathbb{R}^n\times\mathbb{R}^n} (u(x)-u(y))(\varphi(x)-\varphi(y))K(x-y)dxdy \\ &- \int_\Omega h(x)f(u(x))\varphi(x)dx, \end{aligned}$$

for every $\varphi \in X_0^s(\Omega)$. Thus, critical points of $\mathscr{J}_{K,h}$ are solutions of problem (7.2). In order to find these critical points, we will make use of the following theorem:

Theorem 7.1 ([183]) *Let $(E, \|\cdot\|)$ be a reflexive real Banach space, and let $\Phi, \Psi : E \to \mathbb{R}$ be two sequentially weakly lower semicontinuous functionals, with Ψ coercive and $\Phi(0) = \Psi(0) = 0$. Further, set*

$$J_\mu := \mu\Psi + \Phi.$$

Then for each $\sigma > \inf_{u\in E} \Psi(u)$ *and each* μ *satisfying*

$$\mu > -\frac{\inf_{u\in\Psi^{-1}((-\infty,\sigma])} \Phi(u)}{\sigma},$$

the restriction of J_μ *to* $\Psi^{-1}((-\infty,\sigma))$ *has a global minimum.*

When exploiting this result, a key point is to prove the existence of a suitable $\sigma > 0$ such that

$$\sup_{\|u\|^2_{X_0^s(\Omega)}\le\sigma} \int_\Omega h(x)F(u(x))dx < \frac{\sigma}{2}.$$

One of the main novelties here is that, differently from several known results (see references contained in [180]), we obtain the preceding inequality even if the functional space we work in is not continuously embedded in $C^0(\overline{\Omega})$.

7.1.1 Notation

Before stating our existence result for problem (7.2), we need to introduce some notation. Let $0 \le a < b \le +\infty$. If $\lambda \in [a,b]$ and $\varphi,\psi : \mathbb{R} \to \mathbb{R}$ are two assigned functions, we set

$$g_\lambda^{\varphi,\psi} := \lambda\psi - \varphi.$$

Denote

$$M(\varphi,\psi,\lambda) := \begin{cases} \text{the set of global minima of } g_\lambda^{\varphi,\psi} & \text{if } \lambda < +\infty \\ \emptyset & \text{if } \lambda = +\infty, \end{cases}$$

$$\alpha(\varphi,\psi,b) := \max\left\{\inf_{t\in\mathbb{R}} \psi(t), \sup_{t\in M(\phi,\psi,b)} \psi(t)\right\},$$

and

$$\beta(\varphi,\psi,a) := \min\left\{\sup_{t\in\mathbb{R}} \psi(t), \inf_{t\in M(\varphi,\psi,b)} \psi(t)\right\},$$

where we assume that $\sup\emptyset = -\infty$ and $\inf\emptyset = +\infty$.

Furthermore, let $q \in [1,2_s^*)$, $2_s^* := 2n/(n-2s)$, and let $\mathfrak{F}_q$ be the family of all lower semicontinuous functions $\psi : \mathbb{R} \to \mathbb{R}$ such that the following conditions hold true:

$$\sup_{t\in\mathbb{R}} \psi(t) > 0,$$

$$\inf_{t\in\mathbb{R}} \frac{\psi(t)}{1+|t|^q} > -\infty,$$

$$\gamma_\psi := \sup_{t\in\mathbb{R}\setminus\{0\}} \frac{\psi(t)}{|t|^q} < +\infty.$$

The next result, due to Ricceri, will be crucial in the proof of the main theorem of this section.

Proposition 7.2 Let $\varphi, \psi : \mathbb{R} \to \mathbb{R}$ be such that, for each $\lambda \in (a,b)$, the function $g_\lambda^{\varphi,\psi}$ is lower semicontinuous, coercive, and has a global minimum in $\mathbb{R}$. Assume that

$$\alpha(\varphi, \psi, b) < \beta(\varphi, \psi, a).$$

Then, for any

$$r \in \big(\alpha(\varphi, \psi, b), \beta(\varphi, \psi, a)\big),$$

there exists $\lambda_r \in (a,b)$ such that the unique global minimum of $g_{\lambda_r}^{\varphi,\psi}$ lies in $\psi^{-1}(r)$.

7.1.2 Main results

With the preceding notation, an existence result for problem (7.2) reads as follows:

Theorem 7.3 *Let $s \in (0,1)$, $n > 2s$, and Ω be an open, bounded subset of $\mathbb{R}^n$ with continuous boundary. Let $K : \mathbb{R}^n \setminus \{0\} \to (0, +\infty)$ be a function satisfying (1.55) and (1.56). Moreover, let f satisfy (7.3) and (7.4) and $h \in L^\infty(\Omega) \setminus \{0\}$ with $h \geq 0$ in Ω. Assume that there exists $\psi \in \mathfrak{F}_q$ such that, for each $\lambda \in (a,b)$, the function $g_\lambda^{F,\psi}$ is coercive and has a unique global minimum in $\mathbb{R}$. Furthermore, suppose that there exists a number $r > 0$ satisfying*

$$r \in \big(\alpha(F, \psi, b), \beta(F, \psi, a)\big)$$

and

$$\sup_{\xi \in \psi^{-1}(r)} F(\xi) < \frac{r^{2/q}}{2c_q^2 \big(\gamma_\psi \operatorname*{esssup}_{x\in\Omega} h(x)\big)^{2/q} \|h\|_{L^1(\Omega)}^{(q-2)/q}}, \tag{7.6}$$

where c_q is given in (6.59). Then problem (7.2) admits at least one weak solution $u \in X_0^s(\Omega)$ that is a local minimum of the energy functional $\mathcal{J}_{K,h}$ and satisfies

$$\int_{\mathbb{R}^n \times \mathbb{R}^n} |u(x) - u(y)|^2 K(x-y)\, dx\, dy < \frac{1}{c_q^2} \left(\frac{r \|h\|_{L^1(\Omega)}}{\gamma_\psi \operatorname*{esssup}_{x\in\Omega} h(x)} \right)^{2/q}.$$

Proof Let us apply Theorem 7.1 by choosing $E := X_0^s(\Omega)$ and

$$\Phi(u) := -\int_\Omega h(x) F(u(x))\, dx, \qquad \Psi(u) := \|u\|_{X_0^s(\Omega)}^2,$$

for every $u \in E$. Set

$$\sigma := \frac{1}{c_q^2} \left(\frac{r \|h\|_{L^1(\Omega)}}{\gamma_\psi \operatorname*{esssup}_{x\in\Omega} h(x)} \right)^{2/q}, \tag{7.7}$$

where c_q is given in (6.59) (note that because $X_0^s(\Omega) \hookrightarrow L^q(\Omega)$ continuously, clearly, $c_q < +\infty$).

We claim that

$$\Psi^{-1}((-\infty,\sigma]) \subseteq D, \tag{7.8}$$

where

$$D := \left\{ u \in L^q(\Omega) : \int_\Omega h(x)\psi(u(x))dx \leq r\|h\|_{L^1(\Omega)} \right\}.$$

Indeed, because $X_0^s(\Omega) \hookrightarrow L^q(\Omega)$, it follows that

$$\Psi^{-1}((-\infty,\sigma]) \subseteq \left\{ u \in L^q(\Omega) : \|u\|_{L^q(\Omega)} \leq c_q\sqrt{\sigma} \right\}. \tag{7.9}$$

However, taking into account that $\psi \in \mathfrak{F}_q$, one also has

$$\int_\Omega h(x)\psi(u(x))dx \leq \gamma_\psi \|u\|^q_{L^q(\Omega)} \operatorname*{esssup}_{x\in\Omega} h(x). \tag{7.10}$$

Hence, inclusion (7.8) follows from inequalities (7.9) and (7.10).

Now, for each parameter $\lambda \in (a,b)$, denote by $\xi_\lambda^\star$ the unique global minimum (in $\mathbb{R}$) of the real function $g_\lambda^{F,\psi}$. By Lemma 7.2, since by assumption

$$r \in (\alpha(F,\psi,b), \beta(F,\psi,a)),$$

there exists $\lambda_r \in (a,b)$ such that $\psi(\xi^\star_{\lambda_r}) = r$. Hence, since

$$g_{\lambda_r}^{F,\psi}(\xi^\star_{\lambda_r}) \leq g_{\lambda_r}^{F,\psi}(\xi), \tag{7.11}$$

for every $\xi \in \mathbb{R}$, it follows that

$$F(\xi^\star_{\lambda_r}) = \sup_{\xi\in\psi^{-1}(r)} F(\xi). \tag{7.12}$$

Bearing in mind that h is nonnegative, by (7.11), one has

$$g_\lambda^{F,\psi}(\xi^\star_{\lambda_r})h(x) \leq h(x)g_{\lambda_r}^{F,\psi}(\xi), \tag{7.13}$$

for a.e. $x \in \Omega$. Hence, inequality (7.13) implies that

$$g_{\lambda_r}^{F,\psi}(\xi^\star_{\lambda_r})\|h\|_{L^1(\Omega)} \leq \int_\Omega h(x)g_{\lambda_r}^{F,\psi}(u(x))dx, \tag{7.14}$$

for every $u \in L^q(\Omega)$. Exploiting (7.14), for every $u \in D$, one has

$$\int_\Omega h(x)F(u(x))dx \leq F(\xi^\star_{\lambda_r})\|h\|_{L^1(\Omega)}.$$

Owing to (7.12), the preceding inequality assumes the form

$$\int_\Omega h(x)F(u(x))dx \le \sup_{\xi\in\psi^{-1}(r)} F(\xi)\|h\|_{L^1(\Omega)}, \tag{7.15}$$

for every $u \in D$.

Observing that

$$r = \frac{c_q^q \gamma_\psi \sigma^{q/2} \operatorname*{esssup}_{x\in\Omega} h(x)}{\|h\|_{L^1(\Omega)}},$$

and because inclusion (7.8) holds, it follows that

$$\sup_{u\in\Psi^{-1}((-\infty,\sigma])} \int_\Omega h(x)F(u(x))dx \le \sup_{\xi\in\psi^{-1}(r)} F(\xi)\|h\|_{L^1(\Omega)}. \tag{7.16}$$

Finally, relations (7.6) and (7.16) yield

$$\frac{\sup_{u\in\Psi^{-1}((-\infty,\sigma])} \int_\Omega h(x)F(u(x))dx}{\sigma} < \frac{1}{2};$$

that is,

$$\frac{1}{2} > -\frac{\inf_{u\in\Psi^{-1}((-\infty,\sigma])} \Phi(u)}{\sigma}.$$

Then the assertion of Theorem 7.1 follows, and the existence of one weak solution $u \in \Psi^{-1}((-\infty,\sigma))$ for problem (7.2) is established. □

Remark 7.4 The preceding existence theorem extends to the nonlocal setting some results already known in the literature in the case of the classical p-Laplace operator (see [183]).

7.1.3 The fractional Laplacian setting

As observed in Lemma 1.24(c), by taking $K(x) := |x|^{-(n+2s)}$, the space $X_0^s(\Omega)$ (which in this setting is denoted by $\mathbb{H}_0^s(\Omega)$) consists of all the functions of the usual fractional Sobolev space $H^s(\mathbb{R}^n)$ that vanish almost everywhere outside Ω. In this case, $\mathscr{L}_K$ is the fractional Laplace operator $-(-\Delta)^s$ that, up to a positive normalization factor, is defined as in (1.20), and problem (7.2) can be written as (7.1).

Let us consider the following assumptions on h

$$h \in L^\infty(\Omega)\setminus\{0\} \textit{ with } \operatorname*{ess\,inf}_{x\in\Omega} h(x) > 0 \tag{7.17}$$

and these assumptions on $f : [0,+\infty) \to [0,+\infty)$:

$$\begin{gathered} \textit{for some } q \in [1, 2_s^*) \textit{ the function } t \mapsto \frac{f(t)}{t^{q-1}} \\ \textit{is strictly decreasing in } (0,+\infty) \textit{ and } \lim_{t\to+\infty} \frac{f(t)}{t^{q-1}} = 0, \end{gathered} \tag{7.18}$$

$$\liminf_{\xi\to 0^+}\frac{F(\xi)}{\xi^2} > \frac{\lambda_{1,s}}{2\operatorname*{ess\,inf}_{x\in\Omega} h(x)}, \tag{7.19}$$

$$\textit{there exists } \xi_0 > 0 \textit{ such that } F(\xi_0) < \frac{\xi_0^2}{2c_q^2\left(\operatorname*{ess\,sup}_{x\in\Omega} h(x)\right)^{2/q}\|h\|_{L^1(\Omega)}^{(q-2)/q}}, \tag{7.20}$$

where $\lambda_{1,s}$ is the first eigenvalue of $(-\Delta)^s$ with homogeneous boundary data, and c_q is the constant given in (6.59).

In this setting, a meaningful consequence of Theorem 7.3 is as follows:

Theorem 7.5 *Let $s \in (0,1)$, $n > 2s$, and Ω be an open, bounded subset of $\mathbb{R}^n$ with continuous boundary. Moreover, let h satisfy* (7.17), *and let $f : [0,+\infty) \to [0,+\infty)$ be a continuous function satisfying* (7.18)–(7.20). *Then problem* (7.1) *admits at least one nonnegative and nontrivial weak solution $u_0 \in \mathbb{H}_0^s(\Omega)$. Moreover, u_0 is a local minimum of the energy functional*

$$X_0^s(\Omega) \ni u \mapsto \frac{1}{2}\int_{\mathbb{R}^n\times\mathbb{R}^n}\frac{|u(x)-u(y)|^2}{|x-y|^{n+2s}}\,dx\,dy - \int_\Omega h(x)F(u(x))dx$$

and satisfies

$$\int_{\mathbb{R}^n\times\mathbb{R}^n}\frac{|u_0(x)-u_0(y)|^2}{|x-y|^{n+2s}}\,dx\,dy < \left(\frac{\|h\|_{L^1(\Omega)}}{\operatorname*{ess\,sup}_{x\in\Omega} h(x)}\right)^{2/q}\left(\frac{\xi_0}{c_q}\right)^2. \tag{7.21}$$

Proof Let us define

$$\tilde{f}(t) := \begin{cases} f(t) & \text{if } t \geq 0 \\ f(0) & \text{if } t < 0 \end{cases}$$

and consider the following problem:

$$\begin{cases} (-\Delta)^s u = h(x)\tilde{f}(u) & \text{in } \Omega \\ u = 0 & \text{in } \mathbb{R}^n \setminus \Omega. \end{cases} \tag{7.22}$$

By Lemma 4.6, every weak solution of problem (7.22) is nonnegative in Ω. Furthermore, every nonnegative solution of problem (7.22) also solves problem (7.1). Taking $a := 0$ and $b := +\infty$, and by exploiting Theorem 7.1 with

$$\psi(t) := |t|^q \quad t \in \mathbb{R},$$

substantially arguing as in [183], the existence of one weak solution u_0 of problem (7.22) that is a local minimum of the associated energy functional

$$\tilde{\mathcal{J}}_{s,h}(u) := \frac{1}{2}\int_{\mathbb{R}^n\times\mathbb{R}^n}\frac{|u(x)-u(y)|^2}{|x-y|^{n+2s}}\,dx\,dy - \int_\Omega h(x)\tilde{F}(u(x))dx$$

and that satisfies (7.21) is established.

It remains to show that $u_0 \not\equiv 0$. For this, it is enough to prove that

$$0 \text{ is not a local minimum of } \widetilde{\mathcal{J}}_{s,h}. \tag{7.23}$$

For this purpose, let us observe that the first eigenfunction $e_{1,s}$ of $(-\Delta)^s$ (with homogeneous Dirichlet boundary conditions) is positive in Ω (see Corollary 4.8) and that, by Proposition 3.1,

$$\|e_{1,s}\|^2_{\mathbb{H}^s_0(\Omega)} = \lambda_{1,s} \int_\Omega e_{1,s}(x)^2 dx. \tag{7.24}$$

Thanks to (7.19), there exists $\delta > 0$ such that

$$F(\xi) > \frac{\lambda_{1,s}}{2 \operatorname*{ess\,inf}_{x\in\Omega} h(x)} \xi^2,$$

for every $\xi \in (0,\delta)$. Now, by Theorem 5.2, one has that $e_{1,s} \in C^{0,\alpha}(\overline{\Omega})$. Hence, we can define $\theta_\eta(x) := \eta e_{1,s}(x)$, for every $x \in \Omega$, where

$$\eta \in \Lambda_\delta := \left(0, \frac{\delta}{\max_{x\in\Omega} e_{1,s}(x)}\right).$$

Taking into account (7.24), we easily get that

$$\begin{aligned}\int_\Omega h(x)F(\theta_\eta(x))dx &> \frac{\lambda_{1,s}\displaystyle\int_\Omega h(x)\theta_\eta(x)^2dx}{2\operatorname*{ess\,inf}_{x\in\Omega} h(x)} \\ &\geq \frac{\lambda_{1,s}}{2}\int_\Omega \theta_\eta(x)^2dx \\ &= \frac{1}{2}\|\theta_\eta\|^2_{\mathbb{H}^s_0(\Omega)};\end{aligned}$$

that is,

$$\widetilde{\mathcal{J}}_{s,h}(\theta_\eta) = \frac{1}{2}\|\theta_\eta\|^2_{\mathbb{H}^s_0(\Omega)} - \int_\Omega h(x)F(\theta_\eta(x))dx < 0,$$

for every $\eta \in \Lambda_\delta$. Hence, (7.23) holds and the proof is complete. □

A consequence of Theorem 7.5 is given by the following result:

Corollary 7.6 Let $s \in (0,1)$, $n > 2s$, and let Ω be an open, bounded subset of $\mathbb{R}^n$ with continuous boundary. Moreover, let $f : [0,+\infty) \to [0,+\infty)$ be a continuous function satisfying (7.18) and such that

$$\lim_{\xi\to+\infty}\frac{F(\xi)}{\xi^2} = 0 \quad \textit{and} \quad \liminf_{\xi\to 0^+}\frac{F(\xi)}{\xi^2} > 0.$$

Then, for every

$$\alpha > \frac{\lambda_{1,s}}{2\liminf\limits_{\xi\to 0^+}\dfrac{F(\xi)}{\xi^2}},$$

the nonlocal parametric problem

$$\begin{cases} (-\Delta)^s u = \alpha f(u) & \text{in } \Omega \\ u = 0 & \text{in } \mathbb{R}^n \setminus \Omega, \end{cases}$$

admits at least one nonnegative and nontrivial weak solution $u_\alpha \in \mathbb{H}^s_0(\Omega)$.

A direct application of this result is given in the next example.

Example 7.7 Let us consider the following problem

$$\begin{cases} (-\Delta)^s u = \dfrac{\alpha u}{1+u^2} & \text{in } \Omega \\ u = 0 & \text{in } \mathbb{R}^n \setminus \Omega, \end{cases}$$

By virtue of Theorem 7.5, for every $\alpha > \lambda_{1,s}$, this problem admits at least one nonnegative and nontrivial weak solution $u_\alpha \in \mathbb{H}^s_0(\Omega)$.

The existence of infinitely many weak solutions for problem (7.2) comes up from an adequate iteration of Theorem 7.3; see the recent paper [154] for details. By prescribing, in particular, the behavior of $t \mapsto F(t)/t^2$ at 0^+, in addition to some global properties of F expressed in terms of the auxiliary function ψ, we are able to guarantee that these solutions are contained in balls smaller and smaller. However, assumptions on $t \mapsto F(t)/t^2$ at $+\infty$ will result in the existence of infinitely many solutions diverging to $+\infty$ in the $X^s_0(\Omega)$-norm. Here and in what follows, we focus on the case of infinitely many solutions with the $X^s_0(\Omega)$-norm decaying to zero.

Theorem 7.8 *Let $s \in (0,1)$, $n > 2s$, and Ω be an open, bounded subset of $\mathbb{R}^n$ with continuous boundary. Moreover, let f satisfy* (7.3) *and* (7.4) *and $h \in L^\infty(\Omega) \setminus \{0\}$ with $h \geq 0$ in Ω. Let $x_0 \in \Omega$ and $\rho > 0$ be such that $B(x_0,\rho) \subset \Omega$ and*

$$h(x) > \frac{\|h\|_{L^\infty(\Omega)}}{2} \quad \text{a.e. in } B(x_0,\rho).$$

Furthermore, let $0 \leq a < b \leq +\infty$, and assume that there exists $\psi \in \mathfrak{F}_q$ such that, for each $\lambda \in (a,b)$, the function $\lambda\psi - F$ is coercive and has a unique global minimum in $\mathbb{R}$. Finally, suppose that

$$\alpha(F,\psi,b) \leq 0 < \beta(F,\psi,a),$$

$$\liminf_{r\to 0^+} \frac{\sup_{\psi^{-1}(r)} F}{r^{2/q}} < \frac{1}{2c_q^2\left(\gamma_\psi \|h\|_{L^\infty(\Omega)}\right)^{2/q} \|h\|_{L^1(\Omega)}^{(q-2)/q}},$$

there exists $l > 0$ such that

$$\liminf_{t\to 0^+} \frac{F(t)}{t^2} > -l, \quad \limsup_{t\to 0^+} \frac{F(t)}{t^2} > 2(2^n-1)\frac{l\theta\rho^2 \|h\|_{L^\infty(\Omega)} + 2\kappa_1\kappa_2}{\theta\rho^2 \|h\|_{L^\infty(\Omega)}},$$

where θ is given in (1.56), *κ_1 and κ_2 are as in* (1.39), *and c_q is as in* (6.59). *Then problem* (7.2) *admits a sequence of nontrivial weak solutions $\{u_j\}_{j\in\mathbb{N}}$ with the following properties*:

$$\lim_{j\to+\infty} \|u_j\|_{X_0^s(\Omega)} = 0$$

and

$$\left\{\mathcal{J}_{K,h}(u_j)\right\}_{j\in\mathbb{N}} \quad \text{is increasing and} \quad \mathcal{J}_{K,h}(u_j) < 0 \quad \text{for any } j \in \mathbb{N}.$$

7.2 A doubly parametric problem

In this section, we study the existence of solutions for the following general nonlocal equation depending on two real parameters:

$$\begin{cases} \mathcal{L}_K u + \lambda u + \mu f(x,u) = 0 & \text{in } \Omega \\ u = 0 & \text{in } \mathbb{R}^n \setminus \Omega, \end{cases} \tag{7.25}$$

using again variational techniques. To prove our main result, stated in Theorem 7.10, in the following, we will perform the variational principle contained in [176] in the form given here:

Theorem 7.9 ([176]) *Let $(E, \|\cdot\|)$ be a reflexive real Banach space, and let $\Phi, \Psi : E \to \mathbb{R}$ be two Gâteaux differentiable functionals such that Φ is strongly continuous, sequentially weakly lower semicontinuous and coercive in E, and Ψ is sequentially weakly upper semicontinuous in E. Let J_μ be the functional defined as $J_\mu := \Phi - \mu\Psi$, $\mu \in \mathbb{R}$, and, for any $r > \inf_{u\in E} \Phi(u)$, let φ be the function defined as*

$$\varphi(r) := \inf_{u\in\Phi^{-1}((-\infty,r))} \frac{\sup_{v\in\Phi^{-1}((-\infty,r))} \Psi(v) - \Psi(u)}{r - \Phi(u)}.$$

Then for any $r > \inf_{u\in E}\Phi(u)$ *and any* $\mu \in (0, 1/\varphi(r))$[1], *the restriction of the functional* J_μ *to* $\Phi^{-1}((-\infty, r))$ *admits a global minimum, which is a critical point (precisely a local minimum) of* J_μ *in* E.

In this section, the nonlinearity in (7.25) is a Carathéodory function $f : \Omega \times \mathbb{R} \to \mathbb{R}$ verifying the following growth condition[2]:

$$\begin{gathered}\textit{there exist } a_1, a_2 \geq 0, \textit{ and } q \in (1, 2_s^*), 2_s^* := 2n/(n-2s), \textit{ such that}\\ |f(x,t)| \leq a_1 + a_2|t|^{q-1} \quad \textit{a.e. } x \in \Omega, t \in \mathbb{R}.\end{gathered} \tag{7.26}$$

Moreover, we assume the following condition, which is a sort of subquadratical growth assumption at zero:

$$\begin{gathered}\textit{there exist a nonempty open set } D \subseteq \Omega \textit{ and a set } B \subseteq D\\ \textit{of positive Lebesgue measure such that}\\ \limsup_{t\to 0^+} \frac{\operatorname{essinf}_{x\in B} F(x,t)}{t^2} = +\infty\\ \textit{and } \liminf_{t\to 0^+} \frac{\operatorname{essinf}_{x\in D} F(x,t)}{t^2} > -\infty,\end{gathered} \tag{7.27}$$

where, as usual, F is the function given in (6.8).

7.2.1 Some bifurcation theorems

The main results of this section can be stated as follows:

Theorem 7.10 *Let* $s \in (0,1)$, $n > 2s$, Ω *be an open, bounded subset of* $\mathbb{R}^n$ *with continuous boundary, and let* λ_1 *be the first eigenvalue of the operator* $-\mathscr{L}_K$ *with homogeneous Dirichlet boundary data, where* $K : \mathbb{R}^n \setminus \{0\} \to (0, +\infty)$ *is a function satisfying conditions (1.55) and (1.56). Further, let* $f : \Omega \times \mathbb{R} \to \mathbb{R}$ *be a Carathéodory function verifying (7.26). In addition, if* $f(x,0) = 0$ *for a.e.* $x \in \Omega$, *assume also (7.27). Then for any* $\lambda < \lambda_1$, *there exists* $\mu_\lambda > 0$ *depending on* λ *such that, for any* $\mu \in (0, \mu_\lambda)$, *problem (7.25) admits at least one nontrivial weak solution* $u_\mu \in X_0^s(\Omega)$. *Also,* $\mu_\lambda = +\infty$, *provided that* $q \in (1,2)$.

Moreover,

$$\lim_{\mu\to 0^+} \|u_\mu\|_{X_0^s(\Omega)} = 0,$$

[1] Note that, by definition, $\varphi(r) \geq 0$, for any $r > \inf_{u\in E}\Phi(u)$. Here and in the following, if $\varphi(r) = 0$, by $1/\varphi(r)$, we mean $+\infty$; that is, we set $1/\varphi(r) = +\infty$.

[2] Note that here, differently from (7.26), the exponent q is allowed to take the values in (1, 2].

and, for every $r > 0$, the function

$$\mu \mapsto \frac{1}{2}\int_{\mathbb{R}^n\times\mathbb{R}^n} |u_\mu(x)-u_\mu(y)|^2 K(x-y)\,dx\,dy$$
$$-\frac{\lambda}{2}\int_\Omega |u_\mu(x)|^2\,dx - \mu\int_\Omega F(x,u_\mu(x))\,dx$$

is negative and strictly decreasing in $\big(0,1/\varphi_{K,\lambda}(r)\big)$, where $\varphi_{K,\lambda}(r)$ is defined in (7.32).

Actually, using a truncation argument, we can prove that problem (7.25) admits a nontrivial nonnegative (nonpositive) weak solution, provided that $f(\cdot,0)=0$. Indeed, the following result holds:

Corollary 7.11 Let all the assumptions of Theorem 7.10 be satisfied, and assume that $f(\cdot,0)=0$. Then problem (7.25) admits a nontrivial nonnegative weak solution $u_+ \in X_0^s(\Omega)$.

In general, when $f(\cdot,0)\neq 0$, problem (7.25) admits changing-sign solutions, as happens if we look at the classical case of the Laplacian.

As an application of Theorem 7.10, we can consider the following model problem:

$$\begin{cases} (-\Delta)^s u - \lambda u = \mu(a(x)|u|^{r-2}u + b(x)|u|^{q-2}u + c(x)) & \text{in } \Omega \\ u = 0 & \text{in } \mathbb{R}^n\setminus\Omega. \end{cases} \tag{7.28}$$

In this framework, Theorem 7.10 (here we also take into account the result stated in Corollary 7.11) reduces to the following result:

Theorem 7.12 *Let $s\in(0,1)$, $n>2s$, Ω be an open, bounded subset of $\mathbb{R}^n$ with continuous boundary, and let $\lambda_{1,s}$ be the first eigenvalue of $(-\Delta)^s$ with homogeneous Dirichlet boundary data. Furthermore, assume that $1<r<2\le q<2_s^*$ and $a,b,c:\Omega\to\mathbb{R}$ are functions such that $a,b,c\in L^\infty(\Omega)$. In addition, if $c\equiv 0$ a.e. in Ω, assume that* $\operatorname{essinf}_{x\in\Omega} a(x)>0$. *Then for any $\lambda<\lambda_{1,s}$, there exists $\mu_\lambda>0$ such that, for any $\mu\in(0,\mu_\lambda)$, problem* (7.28) *admits at least one nontrivial weak solution $u_\mu\in\mathbb{H}_0^s(\Omega)$ and*

$$\int_{\mathbb{R}^n\times\mathbb{R}^n}\frac{|u_\mu(x)-u_\mu(y)|^2}{|x-y|^{n+2s}}\,dx\,dy\to 0$$

as $\mu\to 0^+$. Also, $\mu_\lambda=+\infty$, provided that $b\equiv 0$ a.e. in Ω. Furthermore, if $c\equiv 0$ a.e. in Ω, the solution u_μ is nonnegative in $\mathbb{R}^n$.

As a final remark, we give an estimate from below for the parameter μ_λ appearing in Theorem 7.10. Indeed, while when $q\in(1,2)$, Theorem 7.10 assures that $\mu_\lambda=+\infty$, the exact value of μ_λ is not known in the other cases, that is, when $q\in[2,2_s^*)$. Setting

$$m_\lambda := \min\left\{\sqrt{\frac{\lambda_1-\lambda}{\lambda_1}},1\right\}, \tag{7.29}$$

we have that

$$\mu_\lambda := \sup_{r>0} \frac{1}{\varphi_{K,\lambda}(r)} \geq \sup_{r>0} \frac{q m_\lambda^q}{\sqrt{2} a_1 c_1 q m_\lambda^{q-1} r^{-1/2} + 2^{q/2} a_2 c_q^q r^{q/2-1}}.$$

Then

$$\mu_\lambda \geq \begin{cases} \dfrac{m_\lambda^2}{a_2 c_2^2} & \text{if} \quad q = 2 \\ \dfrac{q m_\lambda^q}{\sqrt{2} a_1 c_1 q m_\lambda^{q-1} r_{\max}^{-1/2} + 2^{q/2} a_2 c_q^q r_{\max}^{q/2-1}} & \text{if} \quad q \in (2, 2_s^*), \end{cases}$$

where

$$r_{\max} := \frac{m_\lambda^2}{2} \left(\frac{a_1 c_1 q}{a_2 c_q^q (q-2)} \right)^{2/(q-1)},$$

while a_1 and a_2 are as in (7.26), m_λ is given in (7.29), and c_t is given in (6.59), $t = 1, q$.

7.2.2 Proof of Theorem 7.10

The idea of this proof consists of applying Theorem 7.9 to the functional $\mathcal{J}_{K,\lambda,\mu}$ defined as follows:

$$\mathcal{J}_{K,\lambda,\mu}(u) := \Phi_{K,\lambda}(u) - \mu \Psi(u), \quad u \in X_0^s(\Omega),$$

with

$$\Phi_{K,\lambda}(u) := \frac{1}{2} \int_{\mathbb{R}^n \times \mathbb{R}^n} |u(x) - u(y)|^2 K(x-y)\, dx\, dy - \frac{\lambda}{2} \int_\Omega |u(x)|^2 dx,$$

as well as

$$\Psi(u) := \int_\Omega F(x, u(x))\, dx.$$

First of all, recall that $X_0^s(\Omega)$ is a Hilbert space and that the functionals $\Phi_{K,\lambda}$ and Ψ are Frechét differentiable in $X_0^s(\Omega)$. Also note that the map

$$u \mapsto \|u\|^2_{X_0^s(\Omega),\lambda} := \int_{\mathbb{R}^n \times \mathbb{R}^n} |u(x) - u(y)|^2 K(x-y)\, dx\, dy - \lambda \int_\Omega |u(x)|^2 dx$$

is lower semicontinuous in the weak topology of $X_0(\Omega)$, so the functional $\Phi_{K,\lambda}$ is lower semicontinuous in the weak topology of $X_0^s(\Omega)$.

Moreover, the application

$$u \mapsto \int_\Omega F(x, u(x)) dx$$

is continuous in the weak topology of $X_0^s(\Omega)$. Indeed, if $\{u_j\}_{j\in\mathbb{N}}$ is a sequence in $X_0(\Omega)$ such that $u_j \rightharpoonup u$ weakly in $X_0(\Omega)$, then, by Sobolev embedding and [43, theorem IV.9], up to a subsequence, u_j converges to u strongly in $L^\nu(\Omega)$ and a.e. in Ω as $j \to +\infty$, and it is dominated by some function $\kappa_\nu \in L^\nu(\Omega)$, that is,

$$|u_j(x)| \leq \kappa_\nu(x) \quad \text{a.e. } x \in \Omega, \quad \text{for any } j \in \mathbb{N}, \tag{7.30}$$

for any $\nu \in [1, 2_s^*)$. Then, by the continuity of F and (7.26), it follows that

$$F(x, u_j(x)) \to F(x, u(x)) \quad \text{a.e. } x \in \Omega$$

as $j \to +\infty$ and

$$\left|F(x, u_j(x))\right| \leq a_1 |u_j(x)| + \frac{a_2}{q} |u_j(x)|^q \leq a_1 \kappa_1(x) + \frac{a_2}{q} \left(\kappa_q(x)\right)^q \in L^1(\Omega)$$

a.e. $x \in \Omega$, and for any $j \in \mathbb{N}$. Hence, by applying the Lebesgue dominated convergence theorem in $L^1(\Omega)$, we have that

$$\int_\Omega F(x, u_j(x))dx \to \int_\Omega F(x, u(x))dx$$

as $j \to +\infty$; that is, the map

$$u \mapsto \int_\Omega F(x, u(x))dx$$

is continuous from $X_0^s(\Omega)$ with the weak topology to $\mathbb{R}$. Thus, the functional Ψ is continuous from $X_0^s(\Omega)$ with the weak topology to $\mathbb{R}$.

Hence, we have shown that the functionals $\Phi_{K,\lambda}$ and Ψ have the regularity required by Theorem 7.9. Now, for every $\lambda < \lambda_1$ and for any $u \in X_0^s(\Omega)$, one has

$$\Phi_{K,\lambda}(u) = \frac{1}{2} \|u\|^2_{X_0^s(\Omega),\lambda}, \tag{7.31}$$

so the functional $\Phi_{K,\lambda}$ is coercive in $X_0^s(\Omega)$ and $\inf\limits_{u \in X_0^s(\Omega)} \Phi_{K,\lambda}(u) = 0$.

Let $r > 0$, and let $\varphi_{K,\lambda}$ be the function defined as

$$\varphi_{K,\lambda}(r) := \inf_{u \in \Phi_{K,\lambda}^{-1}((-\infty,r))} \frac{\sup\limits_{v \in \Phi_{K,\lambda}^{-1}((-\infty,r))} \Psi(v) - \Psi(u)}{r - \Phi_{K,\lambda}(u)}. \tag{7.32}$$

It is easy to see that $\varphi_{K,\lambda}(r) \geq 0$, for any $r > 0$. Then, by Theorem 7.9, one has that

$$\begin{gathered}\text{for any } r > 0 \text{ and any } \mu \in \left(0, 1/\varphi_{K,\lambda}(r)\right), \text{ the restriction} \\ \text{of } \mathcal{J}_{K,\lambda,\mu} \text{ to } \Phi_{K,\lambda}^{-1}((-\infty, r)) \text{ admits a global minimum } u_{\mu,r}, \\ \text{which is a critical point (namely, a local minimum) of } \mathcal{J}_{K,\lambda,\mu} \text{ in } X_0^s(\Omega).\end{gathered} \tag{7.33}$$

Remember that when $\varphi_{K,\lambda}(r) = 0$, by $1/\varphi_{K,\lambda}(r)$, we mean $+\infty$.

Let μ_λ be defined as follows:

$$\mu_\lambda := \sup_{r>0} \frac{1}{\varphi_{K,\lambda}(r)}.$$

Note that $\mu_\lambda > 0$ because $\varphi_{K,\lambda}(r) \geq 0$, for any $r > 0$.

Hence, let us fix $\bar{\mu} \in (0, \mu_\lambda)$. First of all, thanks to the definition of μ_λ, it is easy to see that

$$\text{there exists } \bar{r}_{\bar{\mu}} > 0 \text{ such that } \bar{\mu} \leq 1/\varphi_{K,\lambda}(\bar{r}_{\bar{\mu}}). \tag{7.34}$$

Then, by (7.33) applied with $r = \bar{r}_{\bar{\mu}}$, we have that, for any μ such that

$$0 < \mu < \bar{\mu} \leq 1/\varphi_{K,\lambda}(\bar{r}_{\bar{\mu}}),$$

the function

$$u_\mu := u_{\mu, \bar{r}_{\bar{\mu}}}$$

is a global minimum of the functional $\mathcal{J}_{K,\lambda,\mu}$ restricted to $\Phi_{K,\lambda}^{-1}((-\infty, \bar{r}_{\bar{\mu}}))$; that is,

$$\mathcal{J}_{K,\lambda,\mu}(u_\mu) \leq \mathcal{J}_{K,\lambda,\mu}(u) \quad \text{for any } u \in X_0^s(\Omega) \text{ such that } \Phi_{K,\lambda}(u) < \bar{r}_{\bar{\mu}} \tag{7.35}$$

and

$$\Phi_{K,\lambda}(u_\mu) < \bar{r}_{\bar{\mu}}, \tag{7.36}$$

and u_μ is also a critical point of $\mathcal{J}_{K,\lambda,\mu}$ in $X_0^s(\Omega)$, so it is a weak solution of problem (7.25). In this way, we have shown that, for any $\lambda < \lambda_1$ and any $\mu \in (0, \mu_\lambda)$, problem (7.25) admits a weak solution $u_\mu \in X_0^s(\Omega)$.

Now we have to prove that $\mu_\lambda = +\infty$, provided that $q \in (1,2)$. To this end, note that, by (7.26), one has

$$F(x,t) \leq a_1|t| + \frac{a_2}{q}|t|^q, \tag{7.37}$$

for a.e. $x \in \Omega$ and any $t \in \mathbb{R}$. Thus, for any $u \in X_0^s(\Omega)$, we get

$$\Psi(u) = \int_\Omega F(x,u(x))dx \leq a_1 \|u\|_{L^1(\Omega)} + \frac{a_2}{q}\|u\|_{L^q(\Omega)}^q. \tag{7.38}$$

Also, by (7.31), for any $u \in X_0^s(\Omega)$ such that $\Phi_{K,\lambda}(u) < r$, with $r > 0$, we easily get that

$$\|u\|_{X_0^s(\Omega),\lambda} < \sqrt{2r}. \tag{7.39}$$

Hence, by Sobolev embedding, (7.38), and (7.39), we obtain that, for any $u \in X_0^s(\Omega)$ such that $\Phi_{K,\lambda}(u) < r$,

$$\begin{aligned}
\Psi(u) &\leq a_1 \|u\|_{L^1(\Omega)} + \frac{a_2}{q}\|u\|_{L^q(\Omega)}^q \\
&\leq \frac{a_1 c_1}{m_\lambda}\|u\|_{X_0^s(\Omega),\lambda} + \frac{a_2 c_q^q}{q m_\lambda^q}\|u\|_{X_0^s(\Omega),\lambda}^q \\
&< \frac{a_1 c_1}{m_\lambda}(2r)^{1/2} + \frac{a_2 c_q^q}{q m_\lambda^q}(2r)^{q/2},
\end{aligned}$$

so

$$\sup_{u\in\Phi_{K,\lambda}^{-1}((-\infty,r))}\Psi(u)\le\frac{\sqrt{2}a_1c_1}{m_\lambda}r^{1/2}+\frac{2^{q/2}a_2c_q^q}{qm_\lambda^q}r^{q/2},$$

for any $r>0$. Then, denoting by χ the function

$$\chi(r):=\frac{\sup\limits_{u\in\Phi_{K,\lambda}^{-1}((-\infty,r))}\Psi(u)}{r}\qquad r>0,$$

we have

$$\chi(r)\le\frac{\sqrt{2}a_1c_1}{m_\lambda}r^{-1/2}+\frac{2^{q/2}a_2c_q^q}{qm_\lambda^q}r^{q/2-1},\tag{7.40}$$

for every $r>0$.

Now observe that, by definition of $\varphi_{K,\lambda}$ and of χ and by (7.40), for any $r>0$, we have

$$\varphi_{K,\lambda}(r)\le\frac{\sup\limits_{u\in\Phi_{K,\lambda}^{-1}((-\infty,r))}\Psi(u)}{r}=\chi(r)\le\frac{\sqrt{2}a_1c_1}{m_\lambda}r^{-1/2}+\frac{2^{q/2}a_2c_q^q}{qm_\lambda^q}r^{q/2-1},$$

just because $\Phi_{K,\lambda}(0)=\Psi(0)=0$. That is,

$$\frac{1}{\varphi_{K,\lambda}(r)}\ge\frac{qm_\lambda^q}{\sqrt{2}a_1c_1qm_\lambda^{q-1}r^{-1/2}+2^{q/2}a_2c_q^q r^{q/2-1}},$$

so

$$\mu_\lambda=\sup_{r>0}\frac{1}{\varphi_{K,\lambda}(r)}\ge\sup_{r>0}\frac{qm_\lambda^q}{\sqrt{2}a_1c_1qm_\lambda^{q-1}r^{-1/2}+2^{q/2}a_2c_q^q r^{q/2-1}}=+\infty,\tag{7.41}$$

provided that $q\in(1,2)$. Hence, $\mu_\lambda=+\infty$ if $q\in(1,2)$, and the assertion of Theorem 7.10 is proved.

We have to show that, for any $\mu\in(0,\mu_\lambda)$, the solution u_μ found here is not the trivial function. If $f(\cdot,0)\neq 0$, then it easily follows that $u_\mu\not\equiv 0$ in $X_0^s(\Omega)$ because the trivial function does not solve problem (7.25).

Let us consider the case where $f(\cdot,0)=0$, and let us fix $\bar\mu\in(0,\mu_\lambda)$ and $\mu\in(0,\bar\mu)$. Finally, let u_μ be as in (7.35) and (7.36). In this setting, to prove that $u_\mu\not\equiv 0$ in $X_0^s(\Omega)$, first we claim that there exists a sequence $\{w_j\}_{j\in\mathbb{N}}$ in $X_0^s(\Omega)$ such that

$$\limsup_{j\to+\infty}\frac{\Psi(w_j)}{\Phi_{K,\lambda}(w_j)}=+\infty.\tag{7.42}$$

By the assumption on the limsup in (7.27), there exists a sequence $\{\xi_j\}_{j\in\mathbb{N}}$ in $\mathbb{R}^+$ such that $\xi_j \to 0^+$ as $j \to +\infty$ and

$$\lim_{j\to+\infty} \frac{\operatorname{essinf}_{x\in B} F(x,\xi_j)}{\xi_j^2} = +\infty, \tag{7.43}$$

we have that, for any $M > 0$ and j sufficiently large,

$$\operatorname{essinf}_{x\in B} F(x,\xi_j) > M\xi_j^2. \tag{7.44}$$

Now let C be a set of positive Lebesgue measure such that $C \subset B$. Also let $v \in X_0^s(\Omega)$ be a function such that

$$\begin{aligned} &v(x) \in [0,1] \quad \text{for every } x \in \mathbb{R}^n, \\ &v(x) = 1 \quad \text{for every } x \in C, \\ &v(x) = 0 \quad \text{for every } x \in \Omega \setminus D. \end{aligned}$$

Of course, C exists because B has positive Lebesgue measure, while the function v exists thanks to the fact that $C_0^2(\Omega) \subseteq X_0^s(\Omega)$.

Finally, let $w_j := \xi_j v$, for any $j \in \mathbb{N}$. It is easily seen that $w_j \in X_0^s(\Omega)$, for any $j \in \mathbb{N}$ (actually, $w_j \in C_0^2(\Omega)$ if v does). With this choice, direct computations ensure that condition (7.42) holds (we omit the details; see [155]).

Now note that

$$\|w_j\|_{X_0^s(\Omega),\lambda} = |\xi_j|\, \|v\|_{X_0^s(\Omega),\lambda} \to 0$$

as $j \to +\infty$, so, for j large enough,

$$\|w_j\|_{X_0^s(\Omega),\lambda} < \sqrt{2\bar{r}_{\bar{\mu}}},$$

where $\bar{r}_{\bar{\mu}}$ is given in (7.34). As a consequence of this and taking into account (7.31),

$$w_j \in \Phi_{K,\lambda}^{-1}((-\infty,\bar{r}_{\bar{\mu}})), \tag{7.45}$$

provided that j is large enough. Also, by (7.42) and the fact that $\mu > 0$,

$$\mathcal{J}_{K,\lambda,\mu}(w_j) = \Phi_{K,\lambda}(w_j) - \mu\Psi(w_j) < 0, \tag{7.46}$$

for j sufficiently large.

Since u_μ is a global minimum of the restriction of $\mathcal{J}_{K,\lambda,\mu}$ to $\Phi_{K,\lambda}^{-1}((-\infty,\bar{r}_{\bar{\mu}}))$ (see (7.35)), by (7.45) and (7.46), we conclude that

$$\mathcal{J}_{K,\lambda,\mu}(u_\mu) \le \mathcal{J}_{K,\lambda,\mu}(w_j) < 0 = \mathcal{J}_{K,\lambda,\mu}(0),$$

so $u_\mu \not\equiv 0$ in $X_0^s(\Omega)$. Thus, u_μ is a nontrivial weak solution of problem (7.25). The arbitrariness of μ and $\bar{\mu}$ gives that $u_\mu \not\equiv 0$, for any $\mu \in (0,\mu_\lambda)$.

The properties of these solutions stated in Theorem 7.10 can be easily proved by direct computation (see [155]). See also paper [156] for related topics.

8
Resonant fractional equations

Nonlinear elliptic problems modeled by

$$\begin{cases} -\Delta u = \lambda u + f(x,u) & \text{in } \Omega \\ u = 0 & \text{on } \partial\Omega, \end{cases} \tag{8.1}$$

where $\Omega \subset \mathbb{R}^n$, $n > 2$, is an open, bounded set, λ is a positive parameter, and the perturbation f is a function satisfying different growth conditions (asymptotically linear, superlinear, subcritical, or critical, for instance), were widely studied in the literature (see, e.g., [9, 46, 174, 212, 220] and references therein).

In this chapter, we consider a resonance problem driven by the nonlocal integrodifferential operator $\mathcal{L}_K$ with homogeneous Dirichlet boundary conditions. This problem has a variational structure, and we find a solution for it using the saddle point theorem. As a particular case, we derive an existence theorem for the following fractional Laplacian equation:

$$\begin{cases} (-\Delta)^s u = \lambda a(x) u + f(x,u) & \text{in } \Omega \\ u = 0 & \text{in } \mathbb{R}^n \setminus \Omega, \end{cases}$$

where λ is an eigenvalue of the related nonhomogenous linear problem with homogeneous Dirichlet boundary data. This existence theorem extends to the nonlocal setting some results already known in the literature in the case of the Laplace operator $-\Delta$. The results of this chapter are based on the papers [102, 196].

8.1 A saddle point result

In some recent papers, these problems were treated in a nonlocal setting; in this framework, see, for instance, [99] for the asymptotically linear case, [51] for subcritical nonlinearities, and [22, 58, 214] for the critical case.

Our aim is to consider the nonlocal version of problem (8.1) in the case where the perturbation $f : \overline{\Omega} \times \mathbb{R} \to \mathbb{R}$ is a function such that (6.3) holds, and

$$\begin{gathered} \textit{there exists a constant } M > 0 \textit{ such that} \\ |f(x,t)| \leq M \textit{ for any } (x,t) \in \Omega \times \mathbb{R}, \end{gathered} \tag{8.2}$$

$$F(x,t) := \int_0^t f(x,\tau)d\tau \to +\infty \textit{ as } |t| \to \infty \textit{ uniformly for } x \in \Omega. \tag{8.3}$$

To be precise, here we deal with the following problem:

$$\begin{cases} -\mathscr{L}_K u = \lambda a(x)u + f(x,u) & \text{in } \Omega \\ u = 0 & \text{in } \mathbb{R}^n \setminus \Omega, \end{cases} \tag{8.4}$$

where $s \in (0,1)$ is fixed, $n > 2s$, $\Omega \subset \mathbb{R}^n$ is an open, bounded set with continuous boundary, $\mathscr{L}_K$ is the integrodifferential operator given in (1.54), and $a : \overline{\Omega} \to \mathbb{R}$ is such that

$$a \textit{ is a positive Lipschitz continuous function in } \overline{\Omega}. \tag{8.5}$$

One of the motivations for studying (8.4) is an attempt to extend some important results that are well known for the classical case of the Laplacian $-\Delta$ (see, e.g., [174, chapter 4 and theorem 4.12]) to a nonlocal setting. The conditions we consider on a and f are classical in the nonlinear analysis (see, e.g., conditions (p_1), (p_2) and (p_7) in [174, Theorem 4.12]) and roughly speaking, they state that problem (8.4) is a suitable perturbation from the following nonhomogenous eigenvalue problem:

$$\begin{cases} -\mathscr{L}_K u = \lambda a(x)u & \text{in } \Omega \\ u = 0 & \text{in } \mathbb{R}^n \setminus \Omega. \end{cases} \tag{8.6}$$

We recall that there exists a nondecreasing sequence of (positive) eigenvalues $\{\lambda_{k,a}\}_{k\in\mathbb{N}}$ for which (8.6) admits nontrivial solutions (see Proposition 8.2).

Finally, note that, thanks to (8.3), the nonlinearity f cannot be the trivial function. As a model for f, we can take the functions

$$f(x,t) := M > 0 \quad \text{or} \quad f(x,t) := b(x)\arctan t,$$

with $b \in Lip(\overline{\Omega})$ and $b > 0$ in Ω. In the first case, $u \equiv 0$ does not solve (8.4), while in the second cases, the trivial function is a solution of (8.4). In general, the function $u \equiv 0$ in $\mathbb{R}^n$ is a solution of problem (8.4) if and only if $f(\cdot,0) = 0$. This is an important difference with respect to the other result in the same subject, where the trivial function is always a solution.

We find weak solutions for (8.4) via variational methods. For this, first, we need the weak formulation of (8.4), which is given by the following problem:

$$\begin{cases} \displaystyle\int_{\mathbb{R}^n \times \mathbb{R}^n} (u(x) - u(y))(\varphi(x) - \varphi(y))K(x-y)dx\,dy \\ \qquad = \lambda \displaystyle\int_\Omega a(x)u(x)\varphi(x)dx + \int_\Omega f(x,u(x))\varphi(x)dx \quad \forall \varphi \in X_0^s(\Omega) \\ u \in X_0^s(\Omega). \end{cases} \tag{8.7}$$

The main result of this chapter can be stated as follows:

Theorem 8.1 *Let* $s \in (0,1)$, $n > 2s$, *and* Ω *be an open, bounded subset of* $\mathbb{R}^n$ *with continuous boundary. Let* $K : \mathbb{R}^n \setminus \{0\} \to (0, +\infty)$ *be a function satisfying* (1.55) *and* (1.56), *and let* $f : \overline{\Omega} \times \mathbb{R} \to \mathbb{R}$ *and* $a : \overline{\Omega} \to \mathbb{R}$ *be two functions verifying* (6.3), (8.2) *and* (8.3), *and* (8.5), *respectively. Moreover, assume that* λ *is an eigenvalue of the nonhomogeneous linear problem* (8.6). *Then problem* (8.4) *admits a weak solution* $u \in X_0^s(\Omega)$.

In the classical case of the Laplacian $-\Delta$, the counterpart of Theorem 8.1 is given in [174, Theorem 4.12]: in this sense, Theorem 8.1 may be seen as the natural extension of classical results to the nonlocal fractional setting.

The strategy for proving Theorem 8.1 is based on the fact that problem (8.7) can be seen as the Euler–Lagrange equation of a suitable functional. Hence, the solutions of (8.7) can be found as critical points of this functional: for this purpose, we will exploit the saddle point theorem by Rabinowitz (see [173, 174]).

8.2 Eigenvalues for linear problems with weights

Consider the problem

$$\begin{cases} \displaystyle\int_{\mathbb{R}^n \times \mathbb{R}^n} (u(x)-u(y))(\varphi(x)-\varphi(y))K(x-y)dx\,dy \\ \qquad\qquad\qquad = \lambda \displaystyle\int_\Omega a(x)u(x)\varphi(x)dx \quad \forall \varphi \in X_0^s(\Omega) \\ u \in X_0^s(\Omega). \end{cases} \tag{8.8}$$

We recall that $\lambda \in \mathbb{R}$ is an eigenvalue of problem (8.8), provided that there exists a nontrivial solution $u \in X_0^s(\Omega)$ of problem (8.8), and in this case, any solution will be called an eigenfunction corresponding to the eigenvalue λ.

For proof of the next result, we refer to Propositions 3.1 and 3.2, where problem (8.8) with $a \equiv 1$ was considered (the case of $a \not\equiv 1$ can be proved similarly, just by replacing the classical space $L^2(\Omega)$ with $L^2(\Omega, \mu)$).

Proposition 8.2 Let $s \in (0,1)$, $n > 2s$, Ω be an open, bounded subset of $\mathbb{R}^n$, and let $K : \mathbb{R}^n \setminus \{0\} \to (0, +\infty)$ be a function satisfying assumptions (1.55) and (1.56). Moreover, let $a : \overline{\Omega} \to \mathbb{R}$ be a function verifying (8.5). Then

(a) Problem (8.8) admits an eigenvalue $\lambda_{1,a}$ that is positive and that can be characterized as

$$\lambda_{1,a} = \min_{\substack{u \in X_0^s(\Omega) \\ \|u\|_{L^2(\Omega,\mu)}=1}} \int_{\mathbb{R}^n \times \mathbb{R}^n} |u(x)-u(y)|^2 K(x-y)dx\,dy$$

or, equivalently,

$$\lambda_{1,a} = \min_{u \in X_0^s(\Omega)\setminus\{0\}} \frac{\displaystyle\int_{\mathbb{R}^n \times \mathbb{R}^n} |u(x) - u(y)|^2 K(x-y)dx\,dy}{\displaystyle\int_\Omega a(x)\,|u(x)|^2\,dx}, \tag{8.9}$$

where $\|\cdot\|_{L^2(\Omega,\mu)}$ denotes the L^2-norm with respect to the measure $\mu(x) = a(x)dx$.

(b) There exists a nonnegative function $e_{1,a} \in X_0^s(\Omega)$ that is an eigenfunction corresponding to $\lambda_{1,a}$, attaining the minimum in (8.9); that is, $\|e_{1,a}\|_{L^2(\Omega,\mu)} = 1$, and

$$\lambda_{1,a} = \int_{\mathbb{R}^n \times \mathbb{R}^n} \left|e_{1,a}(x) - e_{1,a}(y)\right|^2 K(x-y)dx\,dy.$$

(c) $\lambda_{1,a}$ is simple; that is, if $u \in X_0^s(\Omega)$ is a solution of the equation

$$\int_{\mathbb{R}^n \times \mathbb{R}^n} (u(x)-u(y))(\varphi(x)-\varphi(y))K(x-y)dxdy = \lambda_{1,a}\int_\Omega a(x)u(x)\varphi(x)dx,$$

for any $\varphi \in X_0^s(\Omega)$, then $u = \zeta e_{1,a}$, with $\zeta \in \mathbb{R}$.

(d) The set of the eigenvalues of problem (8.8) consists of a sequence $\{\lambda_{k,a}\}_{k\in\mathbb{N}}$ with

$$0 < \lambda_{1,a} < \lambda_{2,a} \le \cdots \le \lambda_{k,a} \le \lambda_{k+1,a} \le \cdots$$

and

$$\lambda_{k,a} \to +\infty \quad \text{as } k \to +\infty.$$

Moreover, for any $k \in \mathbb{N}$, the eigenvalues can be characterized as

$$\lambda_{k+1,a} = \min_{\substack{u \in \mathbb{P}_{k+1,a} \\ \|u\|_{L^2(\Omega,\mu)}=1}} \int_{\mathbb{R}^n \times \mathbb{R}^n} |u(x) - u(y)|^2 K(x-y)dx\,dy$$

or, equivalently,

$$\lambda_{k+1,a} = \min_{u \in \mathbb{P}_{k+1,a}\setminus\{0\}} \frac{\displaystyle\int_{\mathbb{R}^n \times \mathbb{R}^n} |u(x) - u(y)|^2 K(x-y)dx\,dy}{\displaystyle\int_\Omega a(x)\,|u(x)|^2\,dx}, \tag{8.10}$$

where

$$\mathbb{P}_{k+1,a} := \left\{u \in X_0^s(\Omega) : \langle u, e_{j,a}\rangle_{X_0^s(\Omega)} = 0 \quad \forall j = 1,\ldots,k\right\}. \tag{8.11}$$

(e) For any $k \in \mathbb{N}$, there exists a function $e_{k+1,a} \in \mathbb{P}_{k+1,a}$ that is an eigenfunction corresponding to $\lambda_{k+1,a}$, attaining the minimum in (8.10); that is, $\|e_{k+1,a}\|_{L^2(\Omega,\mu)} = 1$, and

$$\lambda_{k+1,a} = \int_{\mathbb{R}^n \times \mathbb{R}^n} \left|e_{k+1,a}(x) - e_{k+1,a}(y)\right|^2 K(x-y)dx\,dy.$$

(f) The sequence $\{e_{k,a}\}_{k\in\mathbb{N}}$ of eigenfunctions corresponding to $\lambda_{k,a}$ is an orthonormal basis of $L^2(\Omega,\mu)$ and an orthogonal basis of $X_0^s(\Omega)$.

(g) Each eigenvalue $\lambda_{k,a}$ has finite multiplicity; more precisely, if $\lambda_{k,a}$ is such that

$$\lambda_{k-1,a} < \lambda_{k,a} = \cdots = \lambda_{k+h,a} < \lambda_{k+h+1,a},$$

for some $h \in \mathbb{N}_0$, then the set of all the eigenfunctions corresponding to $\lambda_{k,a}$ agrees with

$$\operatorname{span}\{e_{k,a},\ldots,e_{k+h}\}.$$

(h) For any $k \in \mathbb{N}$, the eigenvalues can be characterized as

$$\lambda_{k,a} = \max_{u\in\operatorname{span}\{e_{1,a},\ldots,e_{k,a}\}\setminus\{0\}} \frac{\displaystyle\int_{\mathbb{R}^n\times\mathbb{R}^n} |u(x)-u(y)|^2 K(x-y)\,dx\,dy}{\displaystyle\int_\Omega a(x)|u(x)|^2\,dx}.$$

We conclude this section with some notation. In what follows, without loss of generality, we will fix $\lambda = \lambda_{k,a}$ with $k \in \mathbb{N}$ such that $\lambda_{k,a} < \lambda_{k+1,a}$, and we will denote by $\mathbb{H}_{k,a}$ the linear subspace of $X_0^s(\Omega)$ generated by the first k eigenfunctions of $-\mathscr{L}_K$; that is,

$$\mathbb{H}_{k,a} := \operatorname{span}\{e_{1,a},\ldots,e_{k,a}\},$$

while $\mathbb{P}_{k+1,a}$ will be the space defined in (3.11).

It is immediate to observe that $\mathbb{P}_{k+1,a} = \mathbb{H}_{k,a}^{\perp}$ with respect to the scalar product in $X_0^s(\Omega)$. Thus, because $X_0^s(\Omega)$ is a Hilbert space, we can write it as a direct sum as follows:

$$X_0^s(\Omega) = \mathbb{H}_{k,a} \oplus \mathbb{P}_{k+1,a}.$$

Moreover, because $\{e_{1,a},\ldots,e_{k,a},\ldots\}$ is an orthogonal basis of $X_0^s(\Omega)$, it follows that

$$\mathbb{P}_{k+1,a} = \overline{\operatorname{span}\{e_{j,a} : j \geq k+1\}}.$$

Also, we will set

$$\mathbb{E}_{k,a}^0 := \operatorname{span}\{e_{j,a} : \lambda_{j,a} = \lambda_{k,a}\} \quad \text{and} \quad \mathbb{E}_{k,a}^- := \operatorname{span}\{e_{j,a} : \lambda_{j,a} < \lambda_{k,a}\}. \tag{8.12}$$

Note that with this notation, if $u \in \mathbb{H}_{k,a}$, then we can write it as

$$u = u^0 + u^- \quad \text{with} \quad u^0 \in \mathbb{E}_{k,a}^0 \quad \text{and} \quad u^- \in \mathbb{E}_{k,a}^-.$$

8.3 Some technical lemmas

In this section, we prove some technical lemmas that will be useful in applying the saddle point theorem to problem (8.7).

Lemma 8.3 *Let* $K : \mathbb{R}^n \setminus \{0\} \to (0, +\infty)$ *satisfy assumptions* (1.55) *and* (1.56), *and let* $a : \overline{\Omega} \to \mathbb{R}$ *verify* (8.5). *Then for any* $u \in \mathbb{P}_{k+1,a}$,

$$\int_{\mathbb{R}^n \times \mathbb{R}^n} |u(x) - u(y)|^2 K(x-y)\,dx\,dy - \lambda_{k,a} \int_\Omega a(x)|u(x)|^2\,dx \\ \geq \left(1 - \frac{\lambda_{k,a}}{\lambda_{k+1,a}}\right) \|u\|^2_{X_0^s(\Omega)}.$$

Proof If $u \equiv 0$, then the assertion is trivial. Now let $u \in \mathbb{P}_{k+1,a} \setminus \{0\}$. By the variational characterization of $\lambda_{k+1,a}$ given in (8.10), we get that

$$\|u\|^2_{L^2(\Omega,\mu)} \leq \frac{1}{\lambda_{k+1,a}} \|u\|^2_{X_0^s(\Omega)}.$$

As a consequence of this and taking into account that $\lambda_{k,a}$ is positive (because $\lambda_{k,a} \geq \lambda_{1,a} > 0$), we obtain

$$\int_{\mathbb{R}^n \times \mathbb{R}^n} |u(x) - u(y)|^2 K(x-y)\,dx\,dy - \lambda_{k,a} \int_\Omega a(x)|u(x)|^2\,dx \\ \geq \|u\|^2_{X_0^s(\Omega)} - \frac{\lambda_{k,a}}{\lambda_{k+1,a}} \|u\|^2_{X_0^s(\Omega)} \\ = \left(1 - \frac{\lambda_{k,a}}{\lambda_{k+1,a}}\right) \|u\|^2_{X_0^s(\Omega)},$$

concluding the proof. □

Note that if $\lambda_{k,a} = \lambda_{k+1,a}$, then Lemma 8.3 is trivial. The interesting case is when $\lambda_{k,a} < \lambda_{k+1,a}$.

Lemma 8.4 *Let* $K : \mathbb{R}^n \setminus \{0\} \to (0, +\infty)$ *satisfy assumptions* (1.55) *and* (1.56), *and let* $a : \overline{\Omega} \to \mathbb{R}$ *verify* (8.5). *Then there exists a positive constant* M^*, *depending on k, such that*

$$\int_{\mathbb{R}^n \times \mathbb{R}^n} |u(x) - u(y)|^2 K(x-y)\,dx\,dy - \lambda_{k,a} \int_\Omega a(x)|u(x)|^2\,dx \leq -M^* \|u^-\|^2_{X_0^s(\Omega)},$$

for all $u \in \mathbb{H}_{k,a}$, *where* $u = u^- + u^0$, $u^- \in \mathbb{E}^-_{k,a}$, *and* $u^0 \in \mathbb{E}^0_{k,a}$.

Proof Of course, if $u \equiv 0$, then the assertion is trivial. Hence, assume that $u \in \mathbb{H}_{k,a} \setminus \{0\}$. Let $h \in \mathbb{N}$ be the multiplicity of $\lambda_{k,a}$ (h is finite, thanks to Proposition 8.2(g)); that is, suppose that

$$\lambda_{k-h-1,a} < \lambda_{k-h,a} = \cdots = \lambda_{k,a} < \lambda_{k+1,a}. \tag{8.13}$$

With this notation, u can be written as

$$u = u^- + u^0,$$

with

$$u^- \in \mathbb{E}_{k,a}^- = \operatorname{span}\{e_{1,a}, \ldots, e_{k-h-1,a}\} \quad \text{and} \quad u^0 \in \mathbb{E}_{k,a}^0 = \operatorname{span}\{e_{k-h,a}, \ldots, e_{k,a}\}.$$

Notice that u^0 is a linear combination of eigenfunctions corresponding to the same eigenvalue $\lambda_{k-h,a} = \cdots = \lambda_{k,a}$; hence, it is also an eigenfunction corresponding to $\lambda_{k,a}$. Thus, by (8.8),

$$\|u^0\|^2_{X_0^s(\Omega)} = \lambda_{k,a}\|u^0\|^2_{L^2(\Omega,\mu)}.$$

Also, u^- and u^0 are orthogonal in both $X_0^s(\Omega)$ and $L^2(\Omega, \mu)$; therefore,

$$\begin{aligned}
&\|u\|^2_{X_0^s(\Omega)} - \lambda_{k,a}\|u\|^2_{L^2(\Omega,\mu)} \\
&\quad = \|u^-\|^2_{X_0^s(\Omega)} + \|u^0\|^2_{X_0^s(\Omega)} - \lambda_{k,a}\left(\|u^-\|^2_{L^2(\Omega,\mu)} + \|u^0\|^2_{L^2(\Omega,\mu)}\right) \qquad (8.14) \\
&\quad = \|u^-\|^2_{X_0^s(\Omega)} - \lambda_{k,a}\|u^-\|^2_{L^2(\Omega,\mu)}.
\end{aligned}$$

Now note that $u^- \in \mathbb{E}_{k,a}^- = \operatorname{span}\{e_{1,a}, \ldots, e_{k-h-1,a}\}$. Hence, by this and Proposition 8.2, we get

$$\|u^-\|^2_{X_0^s(\Omega)} \le \lambda_{k-h-1,a}\|u^-\|^2_{L^2(\Omega,\mu)}. \qquad (8.15)$$

Finally, (8.14) and (8.15) yield

$$\begin{aligned}
\|u\|^2_{X_0^s(\Omega)} - \lambda_{k,a}\|u\|^2_{L^2(\Omega,\mu)} &= \|u^-\|^2_{X_0^s(\Omega)} - \lambda_{k,a}\|u^-\|^2_{L^2(\Omega,\mu)} \\
&\le \|u^-\|^2_{X_0^s(\Omega)} - \frac{\lambda_{k,a}}{\lambda_{k-h-1,a}}\|u^-\|^2_{X_0^s(\Omega)} \\
&= \left(1 - \frac{\lambda_{k,a}}{\lambda_{k-h-1,a}}\right)\|u^-\|^2_{X_0^s(\Omega)},
\end{aligned}$$

which gives the desired assertion with $M^* := \lambda_{k,a}/(\lambda_{k-h-1,a}) - 1$. Note that $M^* > 0$, thanks to (8.13). □

In the next two results we discuss some properties of the potential F defined as in (8.3).

Lemma 8.5 *Let $f : \overline{\Omega} \times \mathbb{R} \to \mathbb{R}$ satisfy* (6.3), (8.2), *and* (8.3). *Then there exists a positive constant $\widetilde{M}$, depending on Ω, such that*

$$\left|\int_\Omega F(x, u(x))\,dx\right| \le \widetilde{M}\|u\|_{X_0^s(\Omega)},$$

for all $u \in X_0^s(\Omega)$.

Proof Using the definition of F and (8.2), it is easy to see that

$$\left|\int_\Omega F(x, u(x))\,dx\right| = \left|\int_\Omega \int_0^{u(x)} f(x,t)\,dt\,dx\right| \le M\int_\Omega |u(x)|\,dx,$$

so, by the Hölder inequality and Lemma 1.30, we get

$$\left|\int_\Omega F(x,u(x))dx\right| \le M\,|\Omega|^{\frac{1}{2}}\,\|u\|_{L^2(\Omega)} \le \widetilde{M}\,\|u\|_{X_0^s(\Omega)},$$

for all $u \in X_0^s(\Omega)$, where $\widetilde{M}$ is a positive constant depending on Ω. Hence, the assertion is proved. □

Lemma 8.6 *Let* $f : \overline{\Omega}\times\mathbb{R}\to\mathbb{R}$ *satisfy* (6.3), (8.2), *and* (8.3). *Then*

$$\lim_{\substack{u\in\mathbb{E}^0_{k,a}\\ \|u\|_{X_0^s(\Omega)}\to+\infty}} \int_\Omega F(x,u(x))dx = +\infty.$$

Proof We argue by contradiction and suppose that there exists a positive constant C and a sequence $\{u_j\}_{j\in\mathbb{N}} \in \mathbb{E}^0_{k,a}$ such that

$$t_j := \|u_j\|_{X_0^s(\Omega)} \to +\infty \tag{8.16}$$

and

$$\int_\Omega F(x,u_j(x))dx \le C. \tag{8.17}$$

Let $v_j := (1/\|u_j\|_{X_0^s(\Omega)})u_j$, for every $j \in \mathbb{N}$. Of course, $\{v_j\}_{j\in\mathbb{N}}$ is bounded in $X_0^s(\Omega)$. Hence, since $\mathbb{E}^0_{k,a}$ is finite dimensional, there exists $v \in \mathbb{E}^0_{k,a}$ such that $\{v_j\}_{j\in\mathbb{N}}$ converges to v strongly in $X_0^s(\Omega)$. Note also that $v \not\equiv 0$ because

$$\|v\|_{X_0^s(\Omega)} = \lim_{j\to+\infty} \|v_j\|_{X_0^s(\Omega)} = 1.$$

Furthermore, recalling Lemma 1.30,

$$v_j \to v \quad \text{in } L^q(\mathbb{R}^n) \quad \text{for any } q \in [1, 2^*_s), \tag{8.18}$$

and by applying [43, Theorem IV.9], up to a subsequence (still denoted by v_j),

$$v_j \to v \quad \text{a.e. in } \mathbb{R}^n \text{ as } j \to +\infty. \tag{8.19}$$

Now we define $i(r) := \inf_{x\in\overline{\Omega},\,|t|\ge r} F(x,t)$, for $r > 0$. By (8.3), it follows that

$$\lim_{r\to+\infty} i(r) = +\infty. \tag{8.20}$$

Note that

$$\inf_{x\in\overline{\Omega},\,t\in\mathbb{R}} F(x,t) \quad \text{is finite.} \tag{8.21}$$

Indeed, by (8.3), it follows that, for any $H > 0$, there exists $R > 0$ such that

$$F(x,t) > H \quad \text{for any } |t| > R \text{ and any } x \in \Omega. \tag{8.22}$$

Moreover, if $|t| \le R$, by (8.2), we have

$$|F(x,t)| \le M\,|t| \le MR =: C_R, \tag{8.23}$$

for any $x \in \Omega$. Hence, by (8.22) and (8.23), we can conclude that

$$F(x,t) \geq -C_R \quad \text{for any } (x,t) \in \Omega \times \mathbb{R},$$

which implies (8.21).

As a consequence of (8.21), we may define

$$\omega^* := -\min\left\{-1, \inf_{x\in\overline{\Omega}, t\in\mathbb{R}} F(x,t)\right\}.$$

Notice that $\omega^* \geq 0$ and $F(x,t) \geq -\omega^*$, for any $x \in \overline{\Omega}$ and any $t \in \mathbb{R}$. Now we fix $h > 0$ and set

$$\Omega_{j,h} := \left\{x \in \Omega : \left|t_j v_j(x)\right| \geq h\right\}.$$

Thus, we get

$$\begin{aligned}\int_\Omega F(x,t_j v_j(x))dx &= \int_{\Omega_{j,h}} F(x,t_j v_j(x))dx + \int_{\Omega\setminus\Omega_{j,h}} F(x,t_j v_j(x))dx \\ &\geq \left|\Omega_{j,h}\right| i(h) - \omega^* \left|\Omega\right|.\end{aligned} \tag{8.24}$$

Since $v \not\equiv 0$, there exists a set $\Omega^\sharp$ with $\left|\Omega^\sharp\right| > 0$ and a constant $\delta > 0$ such that $|v(x)| \geq \delta$ a.e. $x \in \Omega^\sharp$. Then, by (8.19) and the Egorov theorem, there exists a measurable set $\Omega^* \subseteq \Omega^\sharp$ such that $|\Omega^*| \geq \frac{1}{2}|\Omega^\sharp| > 0$ and the limit in (8.19) is uniform in Ω^*. In particular, if j is large enough,

$$\sup_{x\in\Omega^*} |v_j(x) - v(x)| \leq \frac{\delta}{4},$$

and therefore, for every j sufficiently large, one has $|v_j(x)| \geq 3\delta/4$ for a.e. $x \in \Omega^*$.

Thus, by (8.16), for h fixed earlier, there exists j_h such that $\left|t_j v_j(x)\right| \geq h$, for any $j \geq j_h$ and a.e. $x \in \Omega^*$. As a consequence of this, we have that $\Omega^* \subseteq \Omega_{j,h}$, for $j \geq j_h$. Finally, by (8.17) and (8.24), we have

$$C \geq \int_\Omega F(x,t_j v_j(x))dx \geq \left|\Omega^*\right| i(h) - \omega^* \left|\Omega\right|,$$

for $j \geq j_h$. Passing to the limit as $h \to +\infty$ and taking into account (8.20), we get a contradiction. This proves the assertion. □

8.4 The main result

This section is devoted to the proof of Theorem 8.1. For this purpose, first of all, we observe that problem (8.7) has a variational structure; indeed, it is the Euler–Lagrange equation of the functional $\mathcal{J}_a : X_0^s(\Omega) \to \mathbb{R}$ defined as follows:

$$\begin{aligned}\mathcal{J}_a(u) := &\frac{1}{2}\int_{\mathbb{R}^n\times\mathbb{R}^n} |u(x)-u(y)|^2 K(x-y)\,dx\,dy \\ &-\frac{\lambda}{2}\int_\Omega a(x)|u(x)|^2 dx - \int_\Omega F(x,u(x))dx,\end{aligned}$$

for every $u \in X_0^s(\Omega)$, where F was introduced in (8.3).

Note that the functional $\mathcal{J}_a$ is Fréchet differentiable in $u \in X_0^s(\Omega)$, and for any $\varphi \in X_0^s(\Omega)$, one has

$$\langle \mathcal{J}_a'(u), \varphi \rangle = \int_{\mathbb{R}^n \times \mathbb{R}^n} (u(x) - u(y))(\varphi(x) - \varphi(y))K(x-y)\,dx\,dy$$
$$- \lambda \int_\Omega a(x)u(x)\varphi(x)\,dx - \int_\Omega f(x,u(x))\varphi(x)\,dx.$$

Thus, critical points of $\mathcal{J}_a$ are weak solutions to problem (8.4). To find these critical points, in what follows we will apply the saddle point theorem by Rabinowitz (see [173, 174]). For this, as usual for minimax theorems, we have to check that the functional $\mathcal{J}_a$ has a particular geometric structure (as stated, in our case, in conditions (I_3) and (I_4) of [174, theorem 4.6]) and that it satisfies the Palais–Smale compactness condition (see, e.g., [174, p. 3]).

8.4.1 Geometry of the functional $\mathcal{J}_a$

In this subsection, we prove that the functional $\mathcal{J}_a$ has the geometric features required by the saddle point theorem.

Proposition 8.7 Let $K : \mathbb{R}^n \setminus \{0\} \to (0, +\infty)$ satisfy assumptions (1.55) and (1.56). Moreover, let $\lambda = \lambda_{k,a} < \lambda_{k+1,a}$, for some $k \in \mathbb{N}$, and let f and a be two functions satisfying (6.3), (8.2) and (8.3), and (8.5), respectively. Then

$$\liminf_{\substack{u \in \mathbb{P}_{k+1,a} \\ \|u\|_{X_0^s(\Omega)} \to +\infty}} \frac{\mathcal{J}_a(u)}{\|u\|^2_{X_0^s(\Omega)}} > 0. \tag{8.25}$$

Proof Because $u \in \mathbb{P}_{k+1,a}$, by Lemmas 8.3 and 8.5, we have

$$\mathcal{J}_a(u) \geq \frac{1}{2}\left(1 - \frac{\lambda_{k,a}}{\lambda_{k+1,a}}\right) \|u\|^2_{X_0^s(\Omega)} - \widetilde{M}\|u\|_{X_0^s(\Omega)}.$$

Hence, dividing both sides of this expression by $\|u\|^2_{X_0^s(\Omega)}$ and passing to the limit as $\|u\|_{X_0^s(\Omega)} \to +\infty$, we get (8.25) because $\lambda_{k,a} < \lambda_{k+1,a}$, by assumption. □

Proposition 8.8 Let $K : \mathbb{R}^n \setminus \{0\} \to (0, +\infty)$ satisfy assumptions (1.55) and (1.56). Moreover, let $\lambda = \lambda_{k,a} < \lambda_{k+1,a}$, for some $k \in \mathbb{N}$, and let f and a be two functions satisfying (6.3), (8.2) and (8.3), and (8.5), respectively. Then

$$\lim_{\substack{u \in \mathbb{H}_{k,a} \\ \|u\|_{X_0^s(\Omega)} \to +\infty}} \mathcal{J}_a(u) = -\infty.$$

Proof Because $u \in \mathbb{H}_{k,a}$, we can write $u = u^- + u^0$, with $u^- \in \mathbb{E}^-_{k,a}$ and $u^0 \in \mathbb{E}^0_{k,a}$. Also, $\mathscr{J}_a(u)$ can be written as

$$\begin{aligned}\mathscr{J}_a(u) = &\frac{1}{2}\int_{\mathbb{R}^n\times\mathbb{R}^n} |u(x)-u(y)|^2 K(x-y)\,dx\,dy\\ &-\frac{\lambda_{k,a}}{2}\int_\Omega a(x)|u(x)|^2 dx - \int_\Omega F(x,u^0(x))\,dx \\ &-\int_\Omega \big(F(x,u^0(x)+u^-(x)) - F(x,u^0(x))\big)\,dx.\end{aligned} \tag{8.26}$$

First of all, note that, by (8.2), the Hölder inequality, and Lemma 1.30, it follows that

$$\begin{aligned}\left|\int_\Omega \big(F(x,u^0(x)+u^-(x)) - F(x,u^0(x))\big)dx\right| &= \left|\int_\Omega\int_{u^0(x)}^{u^0(x)+u^-(x)} f(x,t)\,dt\,dx\right|\\ &\le M\int_\Omega |u^-(x)|\,dx\\ &\le M\,|\Omega|^{\frac{1}{2}}\,\|u^-\|_{L^2(\Omega)}\\ &\le \overline{M}\|u^-\|_{X_0^s(\Omega)},\end{aligned} \tag{8.27}$$

where $\overline{M}$ denotes a positive constant depending on Ω.

Thus, by (8.26), (8.27), and Lemma 8.4, we get

$$\mathscr{J}_a(u) \le -M^*\|u^-\|^2_{X_0^s(\Omega)} + \overline{M}\|u^-\|_{X_0^s(\Omega)} - \int_\Omega F(x,u^0(x))\,dx. \tag{8.28}$$

Beware that the first norm on the right-hand side of (8.28) is squared, while the second one is not. Moreover, by orthogonality, we have

$$\|u\|^2_{X_0^s(\Omega)} = \|u^0\|^2_{X_0^s(\Omega)} + \|u^-\|^2_{X_0^s(\Omega)}. \tag{8.29}$$

Then, as $\|u\|_{X_0^s(\Omega)} \to +\infty$, we have that at least one of the two norms, either $\|u^0\|_{X_0^s(\Omega)}$ or $\|u^-\|_{X_0^s(\Omega)}$, goes to infinity.

Suppose that $\|u^0\|_{X_0^s(\Omega)} \to +\infty$ (in this case, $\|u^-\|_{X_0^s(\Omega)}$ can be finite or not; nevertheless, $\|u\|_{X_0^s(\Omega)}$ diverges, due to (8.29)). Then (8.28), the fact that $u^0 \in \mathbb{E}^0_{k,a}$, and Lemma 8.6 show that $\mathscr{J}_a(u) \to -\infty$, so Proposition 8.8 follows.

Otherwise, assume that $\|u^0\|_{X_0^s(\Omega)}$ is finite. In this setting, the divergence of $\|u\|_{X_0^s(\Omega)}$ and (8.29) imply that

$$\|u^-\|_{X_0^s(\Omega)} \to +\infty, \tag{8.30}$$

and by Lemma 8.5, $\int_\Omega F(x,u^0(x))\,dx$ is also finite.

Moreover, by (8.28) and (8.30), we have that $\mathscr{J}_a(u) \to -\infty$ as $\|u\|_{X_0^s(\Omega)} \to +\infty$. This completes the proof of Proposition 8.8. □

8.4.2 The Palais–Smale condition

In this subsection, we discuss a compactness property for the functional $\mathcal{J}_a$, given by the Palais–Smale condition. First of all, as usual when using variational methods, we prove the boundedness of a Palais–Smale sequence for $\mathcal{J}_a$. We say that $\{u_j\}_{j\in\mathbb{N}}$ is a Palais–Smale sequence for $\mathcal{J}_a$ at level $c \in \mathbb{R}$ if

$$|\mathcal{J}_a(u_j)| \leq c \tag{8.31}$$

and

$$\sup\left\{|\langle \mathcal{J}_a'(u_j), \varphi\rangle| : \varphi \in X_0^s(\Omega),\ \|\varphi\|_{X_0^s(\Omega)} = 1\right\} \to 0 \tag{8.32}$$

as $j \to +\infty$ hold true.

Proposition 8.9 Let $K : \mathbb{R}^n \setminus \{0\} \to (0, +\infty)$ satisfy assumptions (1.55) and (1.56). Moreover, assume that $\lambda = \lambda_{k,a} < \lambda_{k+1,a}$, for some $k \in \mathbb{N}$, and let f and a be two functions satisfying (6.3), (8.2) and (8.3), and (8.5), respectively. Finally, let $c \in \mathbb{R}$, and let $\{u_j\}_{j\in\mathbb{N}}$ be a Palais–Smale sequence for $\mathcal{J}_a$ at level c. Then $\{u_j\}_{j\in\mathbb{N}}$ is bounded in $X_0^s(\Omega)$.

Proof Let $u_j = u_j^0 + u_j^- + u_j^+$, where $u_j^0 \in \mathbb{E}_{k,a}^0$, $u_j^- \in \mathbb{E}_{k,a}^-$ and $u_j^+ \in \mathbb{P}_{k+1,a}$. To prove Proposition 8.9, we will show that the sequences $\{u_j^0\}_{j\in\mathbb{N}}$, $\{u_j^-\}_{j\in\mathbb{N}}$ and $\{u_j^+\}_{j\in\mathbb{N}}$ are bounded in $X_0^s(\Omega)$.

First of all, by (8.32), for large j, we get

$$\begin{aligned}
\|u_j^\pm\|_{X_0^s(\Omega)} &\geq \left|\langle \mathcal{J}_a'(u_j), u_j^\pm\rangle\right| \\
&= \left|\int_{\mathbb{R}^n\times\mathbb{R}^n} \big(u_j(x) - u_j(y)\big)\big(u_j^\pm(x) - u_j^\pm(y)\big) K(x-y)\,dx\,dy \right. \\
&\quad \left. - \lambda_{k,a}\int_\Omega a(x)|u_j^\pm(x)|^2\,dx - \int_\Omega f(x, u_j(x))u_j^\pm(x)\,dx\right|,
\end{aligned} \tag{8.33}$$

while, by (8.2), the Hölder inequality, and Lemma 1.30,

$$\left|\int_\Omega f(x, u_j(x))u_j^\pm(x)\,dx\right| \leq \tilde{M}\,\|u_j^\pm\|_{X_0^s(\Omega)}, \tag{8.34}$$

with $\tilde{M}$ positive constant.

Finally, taking into account that $\{e_{1,a}, \ldots, e_{k,a}, \ldots\}$ is a orthogonal basis of $X_0^s(\Omega)$ and of $L^2(\Omega, d\mu)$, $d\mu = a(\cdot)dx$, we have that the scalar product (both in $X_0^s(\Omega)$ and in $L^2(\Omega, d\mu)$) between $u_j = u_j^0 + u_j^- + u_j^+$ and $u_j^\pm$ coincides with the scalar product of $u_j^\pm$ with itself. As a consequence,

$$\begin{aligned}
\langle \mathcal{J}_a'(u_j), u_j^\pm\rangle &= \int_{\mathbb{R}^n\times\mathbb{R}^n} |u_j^\pm(x) - u_j^\pm(y)|^2 K(x-y)\,dx\,dy \\
&\quad - \lambda_{k,a}\int_\Omega a(x)|u_j^\pm(x)|^2\,dx - \int_\Omega f(x, u_j(x))u_j^\pm(x)\,dx.
\end{aligned} \tag{8.35}$$

Now, by Lemma 8.3 (applied with $u = u_j^+ \in \mathbb{P}_{k+1,a}$) and (8.33)–(8.35), we get

$$\left(1 - \frac{\lambda_{k,a}}{\lambda_{k+1,a}}\right) \|u_j^+\|^2_{X_0^s(\Omega)} - \tilde{M}\|u_j^+\|_{X_0^s(\Omega)} \le \|u_j^+\|_{X_0^s(\Omega)},$$

which shows that the sequence $\{u_j^+\}_{j\in\mathbb{N}}$ is bounded in $X_0^s(\Omega)$.

Moreover, again by (8.33)–(8.35) and Lemma 8.4 (applied to $u_j^- \in \mathbb{E}_{k,a}^- \subset \mathbb{H}_{k,a}$), it follows that

$$\|u_j^-\|_{X_0^s(\Omega)} \ge -\langle \mathscr{J}_a'(u_j), u_j^-\rangle \ge M^*\|u_j^-\|^2_{X_0^s(\Omega)} - \tilde{M}\|u_j^-\|_{X_0^s(\Omega)},$$

so $\{u_j^-\}_{j\in\mathbb{N}}$ is also bounded in $X_0^s(\Omega)$.

It remains to show that the sequence $\{u_j^0\}_{j\in\mathbb{N}}$ is bounded in $X_0^s(\Omega)$. For this purpose, we point out that $u_j^0 \in \mathbb{E}_{k,a}^0$, and so, by (8.12), u_j^0 is an eigenfunction corresponding to $\lambda_{k,a}$. Accordingly, by (8.8),

$$\frac{1}{2}\int_{\mathbb{R}^n\times\mathbb{R}^n} |u_j^0(x) - u_j^0(y)|^2 K(x-y)\,dx\,dy = \frac{\lambda_{k,a}}{2}\int_\Omega a(x)|u_j^0(x)|^2\,dx. \tag{8.36}$$

Therefore, by (8.31), (8.36), and orthogonality, we see that

$$\begin{aligned}
c &\ge \left|\mathscr{J}_a(u_j)\right| \\
&= \bigg|\frac{1}{2}\int_{\mathbb{R}^n\times\mathbb{R}^n} \left|u_j^0(x) + u_j^-(x) + u_j^+(x) - u_j^0(y) - u_j^-(y) - u_j^+(y)\right|^2 K(x-y)\,dx\,dy \\
&\quad - \frac{\lambda_{k,a}}{2}\int_\Omega a(x)\left|u_j^0(x) + u_j^-(x) + u_j^+(x)\right|^2 dx - \int_\Omega F(x, u_j(x))\,dx\bigg| \\
&= \bigg|\frac{1}{2}\int_{\mathbb{R}^n\times\mathbb{R}^n} \left(|u_j^0(x) - u_j^0(y)|^2 + |u_j^-(x) - u_j^-(y)|^2 + |u_j^+(x) - u_j^+(y)|^2\right) \\
&\quad \times K(x-y)\,dx\,dy \\
&\quad - \frac{\lambda_{k,a}}{2}\int_\Omega a(x)\left(|u_j^0(x)|^2 + |u_j^-(x)|^2 + |u_j^+(x)|^2\right) dx - \int_\Omega F(x, u_j(x))\,dx\bigg| \\
&= \bigg|\frac{1}{2}\int_{\mathbb{R}^n\times\mathbb{R}^n} \left(|u_j^+(x) - u_j^+(y)|^2 + |u_j^-(x) - u_j^-(y)|^2\right) K(x-y)\,dx\,dy \\
&\quad - \frac{\lambda_{k,a}}{2}\int_\Omega a(x)\left(|u_j^+(x)|^2 + |u_j^-(x)|^2\right) dx \\
&\quad - \int_\Omega \left(F(x, u_j(x)) - F(x, u_j^0(x))\right) dx - \int_\Omega F(x, u_j^0(x))\,dx\bigg|.
\end{aligned} \tag{8.37}$$

By Lemma 1.30 and the Hölder inequality, we get that there exists a positive constant C, possibly depending on Ω, such that

$$\begin{aligned}\left|\lambda_{k,a}\int_\Omega a(x)\left(|u_j^+(x)|^2+|u_j^-(x)|^2\right)dx\right| &\le \lambda_{k,a}\|a\|_{L^\infty(\Omega)}\left(\|u_j^+\|^2_{X_0^s(\Omega)}+\|u_j^-\|^2_{X_0^s(\Omega)}\right)\\ &\le 2C \end{aligned} \tag{8.38}$$

and

$$\begin{aligned}\left|\int_\Omega \left(F(x,u_j(x))-F(x,u_j^0(x))\right)dx\right| &\le \int_\Omega\left|\int_{u_j^0(x)}^{u_j^0(x)+u_j^-(x)+u_j^+(x)} f(x,t)dt\right|dx\\ &\le M\int_\Omega \left(|u_j^-(x)|+|u_j^+(x)|\right)dx \\ &\le M_*\left(\|u_j^-\|_{X_0^s(\Omega)}+\|u_j^+\|_{X_0^s(\Omega)}\right)\\ &\le C\end{aligned} \tag{8.39}$$

because the sequences $\{u_j^-\}_{j\in\mathbb{N}}$ and $\{u_j^+\}_{j\in\mathbb{N}}$ are bounded in $X_0^s(\Omega)$, and (8.2) holds true. Here M_* is a positive constant. Hence, by (8.37)–(8.39), it is easy to see that

$$\begin{aligned}&\left|\int_\Omega F(x,u_j^0(x))dx\right|\\ &\le |\mathcal{J}_a(u_j)|+\left|\frac{1}{2}\int_{\mathbb{R}^n\times\mathbb{R}^n}\left(|u_j^+(x)-u_j^+(y)|^2+|u_j^-(x)-u_j^-(y)|^2\right)K(x-y)dx\,dy\right.\\ &\qquad\left.-\frac{\lambda_{k,a}}{2}\int_\Omega a(x)(|u_j^+(x)|^2+|u_j^-(x)|^2)\,dx-\int_\Omega \left(F(x,u_j(x))-F(x,u_j^0(x))\right)dx\right|\\ &\le c+\frac{1}{2}\left(\|u^+\|^2_{X_0^s(\Omega)}+\|u^-\|^2_{X_0^s(\Omega)}\right)+2C\\ &\le \tilde{C},\end{aligned}$$

where $\tilde{C}$ is a positive constant independent of j. Here we have again used the fact that the sequences $\{u_j^-\}_{j\in\mathbb{N}}$ and $\{u_j^+\}_{j\in\mathbb{N}}$ are bounded in $X_0^s(\Omega)$.

Hence, the integral $\int_\Omega F(x,u_j^0(x))dx$ is bounded. As a consequence, because $u^0\in\mathbb{E}^0_{k,a}$, by Lemma 8.6, it follows that the sequence $\{u_j^0\}_{j\in\mathbb{N}}$ is also bounded in $X_0^s(\Omega)$, concluding the proof of Proposition 8.9. □

Now it remains to check the validity of the Palais–Smale condition; that is, we have to show that every Palais–Smale sequence $\{u_j\}_{j\in\mathbb{N}}$ for $\mathcal{J}_a$ at level $c\in\mathbb{R}$

strongly converges in $X_0^s(\Omega)$, up to a subsequence. This will be done in the next result.

Proposition 8.10 Let $K : \mathbb{R}^n \setminus \{0\} \to (0, +\infty)$ satisfy assumptions (1.55) and (1.56). Moreover, assume that $\lambda = \lambda_{k,a} < \lambda_{k+1,a}$, for some $k \in \mathbb{N}$, and let f and a be two functions satisfying (6.3), (8.2) and (8.3), and (8.5), respectively. Let $\{u_j\}_{j\in\mathbb{N}}$ be a Palais–Smale sequence for $\mathcal{J}_a$ at level c. Then there exists $u_\infty \in X_0^s(\Omega)$ such that $\{u_j\}_{j\in\mathbb{N}}$ strongly converges to u_∞ in $X_0^s(\Omega)$.

Proof Since, by Proposition 8.9, $\{u_j\}_{j\in\mathbb{N}}$ is bounded in $X_0^s(\Omega)$ and $X_0^s(\Omega)$ is a reflexive space (being a Hilbert space, by Lemma 1.29), up to a subsequence, there exists $u_\infty \in X_0^s(\Omega)$ such that $\{u_j\}_{j\in\mathbb{N}}$ converges to u_∞ weakly in $X_0^s(\Omega)$; that is,

$$\begin{aligned} &\int_{\mathbb{R}^n\times\mathbb{R}^n} (u_j(x)-u_j(y))(\varphi(x)-\varphi(y))K(x-y)\,dx\,dy \\ &\quad \to \int_{\mathbb{R}^n\times\mathbb{R}^n} (u_\infty(x)-u_\infty(y))(\varphi(x)-\varphi(y))K(x-y)\,dx\,dy, \end{aligned} \tag{8.40}$$

for any $\varphi \in X_0^s(\Omega)$, as $j \to +\infty$. Moreover, by applying Lemma 1.30 and [43, theorem IV.9], up to a subsequence,

$$\begin{aligned} &u_j \to u_\infty \quad \text{in } L^q(\mathbb{R}^n) \quad \text{for any } q \in [1, 2_s^*) \\ &u_j \to u_\infty \quad \text{a.e. in } \mathbb{R}^n \quad \text{as } j \to +\infty. \end{aligned} \tag{8.41}$$

By (8.32), we have

$$\begin{aligned} 0 \leftarrow \langle \mathcal{J}_a'(u_j), u_j - u_\infty \rangle = &\int_{\mathbb{R}^n\times\mathbb{R}^n} |u_j(x)-u_j(y)|^2 K(x-y)dxdy \\ &- \int_{\mathbb{R}^n\times\mathbb{R}^n} (u_j(x)-u_j(y))(u_\infty(x)-u_\infty(y))K(x-y)dxdy \\ &- \lambda_{k,a}\int_\Omega a(x)u_j(x)(u_j(x)-u_\infty(x))dx \\ &- \int_\Omega f(x,u_j(x))(u_j(x)-u_\infty(x))dx. \end{aligned} \tag{8.42}$$

Now, by using the Hölder inequality, (8.2), and (8.41), we get

$$\begin{aligned} &\left|\lambda_{k,a}\int_\Omega a(x)u_j(x)(u_j(x)-u_\infty(x))dx + \int_\Omega f(x,u_j(x))(u_j(x)-u_\infty(x))dx\right| \\ &\quad \le \left(\lambda_{k,a}\|a\|_{L^\infty(\Omega)}\|u_j\|_{L^2(\Omega)} + M\,|\Omega|^{\frac{1}{2}}\right)\|u_j - u_\infty\|_{L^2(\Omega)} \to 0 \end{aligned} \tag{8.43}$$

as $j \to +\infty$. We observe that the preceding computation also takes into account a term that involves the nonlinearity f.

Hence, passing to the limit in (8.42) and taking into account (8.40) and (8.43), it follows that

$$\int_{\mathbb{R}^n\times\mathbb{R}^n} \left|u_j(x)-u_j(y)\right|^2 K(x-y)\,dx\,dy \to \int_{\mathbb{R}^n\times\mathbb{R}^n} |u_\infty(x)-u_\infty(y)|^2 K(x-y)\,dx\,dy;$$

that is,

$$\|u_j\|_{X_0^s(\Omega)} \to \|u_\infty\|_{X_0^s(\Omega)}. \tag{8.44}$$

Finally, we have that

$$\begin{aligned}\|u_j-u_\infty\|^2_{X_0^s(\Omega)} &= \|u_j\|^2_{X_0^s(\Omega)} + \|u_\infty\|^2_{X_0^s(\Omega)} - 2\int_{\mathbb{R}^n\times\mathbb{R}^n} (u_j(x)-u_j(y))(u_\infty(x) \\ &\quad - u_\infty(y))K(x-y)\,dx\,dy \\ &\to 2\|u_\infty\|^2_{X_0^s(\Omega)} - 2\|u_\infty\|^2_{X_0^s(\Omega)} = 0 \quad \text{as } j\to+\infty,\end{aligned}$$

thanks to (8.40) and (8.44). Hence, $u_j \to u_\infty$ strongly in $X_0^s(\Omega)$ as $j\to+\infty$, and this completes the proof. □

8.4.3 Proof of Theorem 8.1

In this section, we will prove Theorem 8.1 as an application of the saddle point theorem [174, theorem 4.6]. At first, we prove that $\mathcal{J}_a$ satisfies the geometric structure required by the saddle point theorem. For this, note that by Proposition 8.7, for any $H>0$, there exists $R>0$ such that if $u\in\mathbb{P}_{k+1,a}$ and $\|u\|_{X_0^s(\Omega)}\geq R$, then

$$\mathcal{J}_a(u)\geq H, \tag{8.45}$$

while, if $u\in\mathbb{P}_{k+1,a}$ with $\|u\|_{X_0^s(\Omega)}\leq R$, by applying (8.2), the Hölder inequality, and Lemma 1.30, we have

$$\begin{aligned}\mathcal{J}_a(u) &\geq -\frac{\lambda_{k,a}}{2}\int_\Omega a(x)|u(x)|^2dx - \int_\Omega F(x,u(x))dx \\ &\geq -\frac{\lambda_{k,a}}{2}\|a\|_{L^\infty(\Omega)}\|u\|^2_{L^2(\Omega)} - M\int_\Omega |u(x)|\,dx \\ &\geq -\frac{\lambda_{k,a}}{2}\|a\|_{L^\infty(\Omega)}\|u\|^2_{X_0^s(\Omega)} - M_*\|u\|_{X_0^s(\Omega)} \\ &\geq -\frac{\lambda_{k,a}}{2}\|a\|_{L^\infty(\Omega)}R^2 - M_*R =: -C_R.\end{aligned} \tag{8.46}$$

Here M_* is a positive constant. Hence, by (8.45) and (8.46), we get

$$\mathcal{J}_a(u)\geq -C_R \quad \text{for any } u\in\mathbb{P}_{k+1,a}. \tag{8.47}$$

Moreover, by Proposition 8.8, there exists $T>0$ such that, for any $u\in\mathbb{H}_{k,a}$ with $\|u\|_{X_0^s(\Omega)}=T$, we have

$$\mathcal{J}_a(u) < -C_R. \tag{8.48}$$

Thus, by (8.47) and (8.48), it easily follows that

$$\sup_{\substack{u\in\mathbb{H}_{k,a},\\ \|u\|_{X_0^s(\Omega)}=T}} \mathscr{J}_a(u) < -C_R \le \inf_{u\in\mathbb{P}_{k+1,a}} \mathscr{J}_a(u),$$

so the functional $\mathscr{J}_a$ has the geometric structure of the saddle point theorem (see assumptions (I_3) and (I_4) of [174, theorem 4.6]).

Since $\mathscr{J}_a$ also satisfies the Palais–Smale condition, by Proposition 8.10, the saddle point theorem provides the existence of a critical point $u \in X_0^s(\Omega)$ for the functional $\mathscr{J}_a$. This concludes the proof of Theorem 8.1.

9

A pseudoindex approach to nonlocal problems

The aim of this chapter is to investigate the existence and multiplicity of weak solutions to nonlocal equations involving a general integrodifferential operator of fractional type when the nonlinearity is odd symmetric and asymptotically linear at infinity. In particular, we consider a nonresonant setting, as will be explained in an exhaustive way in Section 9.1.

These results, gotten by using a pseudoindex theory related to the genus, extend to the nonlocal setting some theorems, already known in the literature, in the case of the classical Laplace operator. This chapter is based on the paper [28].

9.1 A multiplicity result

In recent years, there has been a considerable amount of research related to nonresonant elliptic equations. For instance, the semilinear problem

$$\begin{cases} -\Delta u = g(x,u) & \text{in } \Omega \\ u = 0 & \text{on } \partial\Omega, \end{cases} \tag{9.1}$$

where Ω is an open, bounded domain of $\mathbb{R}^n$, with continuous boundary $\partial\Omega$, and g is a given real function on $\Omega \times \mathbb{R}$, asymptotically linear and possibly odd, has been widely investigated in the literature (see [7, 26, 174] and references therein, as well as [43] for the linear case).

In this section, we provide a multiplicity result for the nonlocal counterpart of (9.1), that is, for the following problem:

$$\begin{cases} \mathscr{L}_K u + g(x,u) = 0 & \text{in } \Omega \\ u = 0 & \text{in } \mathbb{R}^n \setminus \Omega. \end{cases} \tag{9.2}$$

Here $s \in (0,1)$ is fixed, $n > 2s$, $\Omega \subset \mathbb{R}^n$ is an open, bounded set with continuous boundary, and $\mathscr{L}_K$ is the integrodifferential operator given in (1.54). Moreover, we suppose that there exist $\lambda_\infty \in \mathbb{R}$ and $f : \Omega \times \mathbb{R} \to \mathbb{R}$ such that

$$g(x,t) := \lambda_\infty t + f(x,t) \quad \text{for a.e. } x \in \Omega,\ t \in \mathbb{R}. \tag{9.3}$$

Hence, by (9.3), problem (9.2) assumes the form

$$\begin{cases} \mathscr{L}_K u + \lambda_\infty u + f(x,u) = 0 & \text{in } \Omega \\ u = 0 & \text{in } \mathbb{R}^n \setminus \Omega. \end{cases} \tag{9.4}$$

In what follows, we work under these assumptions on the nonlinear term f:

$$f(x,\cdot) \text{ is odd a.e. } x \in \Omega, \tag{9.5}$$

$$f \text{ is a Carathéodory function and } \sup_{|t|\le a} |f(\cdot,t)| \in L^\infty(\Omega) \text{ for any } a > 0, \tag{9.6}$$

$$\lim_{|t|\to\infty} \frac{f(x,t)}{t} = 0 \text{ uniformly with respect to a.e. } x \in \Omega, \tag{9.7}$$

$$\lim_{t\to 0} \frac{f(x,t)}{t} = \lambda_0 \in \mathbb{R} \text{ uniformly with respect to a.e. } x \in \Omega. \tag{9.8}$$

The weak solutions of (9.4) can be found as critical points of the energy functional naturally associated with it. When looking for multiple solutions of (9.4), we will exploit a pseudoindex theory related to the Krasnoselskii genus due to Benci (see [33] and Section 9.2) and multiplicity results for critical points of even functionals, as stated in [26].

The application of these kinds of abstract tools to our setting leads to the following multiplicity result for (9.4) when λ_∞ is not an eigenvalue of $-\mathscr{L}_K$ with homogeneous Dirichlet boundary data (here and in what follows, we use the notation of Proposition 3.1, namely, λ_k denotes the kth eigenvalue of $-\mathscr{L}_K$):

Theorem 9.1 *Let $s \in (0,1)$, $n > 2s$, and Ω be an open, bounded subset of $\mathbb{R}^n$ with continuous boundary. Let $K : \mathbb{R}^n \setminus \{0\} \to (0,+\infty)$ be a function satisfying and let $f : \Omega \times \mathbb{R} \to \mathbb{R}$ verify (9.5)–(9.8). Assume that*

$$\begin{gathered} \text{there exist } h,k \in \mathbb{N}, \text{ with } k \ge h, \text{ such that} \\ \lambda_0 + \lambda_\infty < \lambda_h \le \lambda_k < \lambda_\infty. \end{gathered} \tag{9.9}$$

Then problem (9.4) has at least $k-h+1$ distinct pairs of nontrivial weak solutions in $X_0^s(\Omega)$, provided that λ_∞ is not an eigenvalue of $-\mathscr{L}_K$ with homogeneous Dirichlet boundary data.

Remark 9.2 We emphasize that Theorem 9.1 still holds if we require that

$$\lambda_\infty < \lambda_h \le \lambda_k < \lambda_0 + \lambda_\infty$$

instead of condition (9.9). This can be proved by combining the proof of Theorem 9.1 with that of [27, theorem 3.1].

When dealing with the classical case of the Laplacian (namely, problem (9.1)), similar results can be found in the celebrated paper [26, theorems 0.1 and 0.3] and in [27, theorem 3.1]. Moreover, we point out that our result suggests that this line of research could be successfully applied to similar contexts.

9.1.1 Variational formulation

The weak solutions of problem (9.4) are the critical points of the C^1-functional $\mathscr{J}_{K,\lambda_\infty} : X_0^s(\Omega) \to \mathbb{R}$ defined as

$$\mathscr{J}_{K,\lambda_\infty}(u) := \frac{1}{2}\int_{\mathbb{R}^n\times\mathbb{R}^n} |u(x)-u(y)|^2 K(x-y)\,dx\,dy - \frac{\lambda_\infty}{2}\int_\Omega |u(x)|^2\,dx \\ - \int_\Omega F(x,u(x))\,dx, \tag{9.10}$$

where, as usual, F is the function defined in (6.8).

Indeed, as is easy to see, by (9.6) and (9.7), for all $\varepsilon > 0$, there exists $a_\varepsilon > 0$ such that

$$|f(x,t)| \le \varepsilon|t| + a_\varepsilon \quad \text{for a.e. } x \in \Omega,\ t \in \mathbb{R}. \tag{9.11}$$

Hence, the functional $\mathscr{J}_{K,\lambda_\infty}$ is Fréchet differentiable at any $u \in X_0^s(\Omega)$, and

$$\langle \mathscr{J}'_{K,\lambda_\infty}(u),\varphi\rangle = \int_{\mathbb{R}^n\times\mathbb{R}^n} (u(x)-u(y))(\varphi(x)-\varphi(y))K(x-y)\,dx\,dy \\ - \lambda_\infty\int_\Omega u(x)\varphi(x)\,dx - \int_\Omega f(x,u(x))\varphi(x)\,dx, \tag{9.12}$$

for every $\varphi \in X_0^s(\Omega)$.

9.2 A pseudoindex theorem

In this section, we present the main classical tools that we use for studying problem (9.4). In particular, we recall some basic notions of pseudoindex theory for an even functional with symmetry group $\mathbb{Z}_2 = \{\mathrm{id}, -\mathrm{id}\}$.

Throughout this section, we denote by $(E, \|\cdot\|)$ a Banach space, Φ a C^1-functional on E, $\Phi^b := \{e \in E : \Phi(e) \le b\}$ the sublevel of Φ corresponding to $b \in \overline{\mathbb{R}} := \mathbb{R} \cup \{\pm\infty\}$, and by

$$K_c := \{e \in E :\ \Phi(e) = c,\ \Phi'(e) = 0\}$$

the set of critical points of Φ in E at the critical level $c \in \mathbb{R}$.

Let us set

$$\Sigma := \big\{A \subseteq E : A \text{ is closed and symmetric w.r.t. the origin, i.e. } -e \in A \text{ if } e \in A\big\}$$

and

$$\mathscr{H} := \{h \in C(E,E) : h \text{ odd}\}.$$

For $A \in \Sigma$, $A \neq \emptyset$, *the genus of* A is denoted by $\gamma(A)$ and is defined as follows:

$$\gamma(A) := \inf\big\{\ell \in \mathbb{N} :\ \exists\psi \in C(A,\mathbb{R}^\ell\setminus\{0\}) \text{ s.t. } \psi(-e) = -\psi(e) \text{ for all } e \in A\big\},$$

if such an infimum exists; otherwise, $\gamma(A) = +\infty$. We also assume that $\gamma(\emptyset) = 0$.

The index theory $(\Sigma, \mathscr{H}, \gamma)$ related to $\mathbb{Z}_2$ is also called the *genus* (for more details, we refer to [212, section II.5]). The pseudoindex related to the genus and $S \in \Sigma$ is the triplet $(S, \mathscr{H}^*, \gamma^*)$ such that $\mathscr{H}^*$ is a group of odd homeomorphisms and $\gamma^* : \Sigma \to \mathbb{N} \cup \{+\infty\}$ is the map defined by

$$\gamma^*(A) := \min_{h \in \mathscr{H}^*} \gamma(h(A) \cap S), \quad \forall A \in \Sigma$$

(see [33] for more details).

The proof of our main result, stated in Theorem 9.1, is based on the following abstract result:

Theorem 9.3 ([26]) *Let $a, b, c_0, c_\infty \in \overline{\mathbb{R}}$, $-\infty \le a < c_0 < c_\infty < b \le +\infty$, Φ be an even functional, $(\Sigma, \mathscr{H}, \gamma)$ the genus theory on E, $S \in \Sigma$, and $(S, \mathscr{H}^*, \gamma^*)$ the pseudoindex theory related to the genus and S, with*

$$\mathscr{H}^* := \{h \in \mathscr{H} : h \text{ bounded homeomorphism s.t. } h(e) = e \text{ if } e \notin \Phi^{-1}((a,b))\}.$$

Assume that

(*a*) *The functional Φ satisfies* (PS) *in (a, b),*
(*b*) *$S \subseteq \Phi^{-1}([c_0, +\infty[)$, and*
(*c*) *There exist $\tilde{k} \in \mathbb{N}$ and $\tilde{A} \in \Sigma$ such that $\tilde{A} \subseteq \Phi^{c_\infty}$ and $\gamma^*(\tilde{A}) \ge \tilde{k}$.*

Then, setting

$$\Sigma_i^* := \left\{A \in \Sigma : \gamma^*(A) \ge i\right\},$$

the numbers

$$c_i := \inf_{A \in \Sigma_i^*} \sup_{e \in A} \Phi(e) \quad i = 1, \ldots, \tilde{k}, \tag{9.13}$$

are critical values for Φ, and

$$c_0 \le c_1 \le \cdots \le c_{\tilde{k}} \le c_\infty.$$

Furthermore, if $c = c_i = \cdots = c_{i+r}$, with $i \ge 1$ and $i + r \le \tilde{k}$, then $\gamma(K_c) \ge r + 1$.

Remark 9.4 In the applications, a lower bound for the pseudoindex of a suitable $\tilde{A}$ as in (c) of Theorem 9.3 is needed. For this purpose, we would recall that considering the genus theory $(\Sigma, \mathscr{H}, \gamma)$ on E and two closed subspaces V, W of E, the following relation holds:

$$\text{if } \dim V < +\infty \ \text{ and } \operatorname{codim} W < +\infty,$$
$$\text{then } \gamma(V \cap h(\partial B \cap W)) \ge \dim V - \operatorname{codim} W,$$

for every bounded $h \in \mathscr{H}$ and every open bounded symmetric neighborhood B of 0 in E (see [26, theorem A.2]). In this setting, codim W denotes the *codimension* of W.

9.3 The Palais–Smale condition

In this section, we prove that in the nonresonant case the functional $\mathcal{J}_{K,\lambda_\infty}$ satisfies the Palais–Smale condition at any level $c \in \mathbb{R}$.

Proposition 9.5 Assume that (9.6) and (9.7) hold and that λ_∞ is not an eigenvalue of $-\mathcal{L}_K$ with homogeneous Dirichlet boundary data. Then the functional $\mathcal{J}_{K,\lambda_\infty}$ satisfies the Palais–Smale condition in $\mathbb{R}$.

Proof Let $c \in \mathbb{R}$ and $\{u_j\}_{j\in\mathbb{N}}$ be a sequence in $X_0^s(\Omega)$ such that

$$\begin{aligned} &\mathcal{J}_{K,\lambda_\infty}(u_j) \to c \quad \text{and} \\ &\sup\left\{|\langle \mathcal{J}'_{K,\lambda_\infty}(u_j),\varphi\rangle| : \varphi \in X_0^s(\Omega),\ \|\varphi\|_{X_0^s(\Omega)} = 1\right\} \to 0 \end{aligned} \tag{9.14}$$

as $j \to +\infty$. Then in particular,

$$\langle u_j,\varphi\rangle_{X_0^s(\Omega)} - \lambda_\infty \int_\Omega u_j(x)\varphi(x)\,dx - \int_\Omega f(x,u_j(x))\varphi(x)\,dx = o(1), \tag{9.15}$$

for all $\varphi \in X_0^s(\Omega)$, as $j \to +\infty$.

First, we prove the following claim:

$$\left\{\|u_j\|_{X_0^s(\Omega)}\right\}_{j\in\mathbb{N}} \quad \text{is bounded in } \mathbb{R}. \tag{9.16}$$

To this end, arguing by contradiction, let us assume that

$$\|u_j\|_{X_0^s(\Omega)} \to +\infty \quad \text{as } j \to +\infty. \tag{9.17}$$

Then, setting $w_j := u_j/\|u_j\|_{X_0^s(\Omega)}$, for any $j \in \mathbb{N}$, the sequence $\{w_j\}_{j\in\mathbb{N}}$ is bounded in $X_0^s(\Omega)$, and there exists $w_\infty \in X_0^s(\Omega)$ such that, up to subsequences, it results in

$$w_j \rightharpoonup w_\infty \quad \text{weakly in } X_0^s(\Omega)$$

and

$$w_j \to w_\infty \quad \text{strongly in } L^2(\Omega) \tag{9.18}$$

as $j \to +\infty$ (see Lemma 1.30).

Evaluating (9.15) in $w_j - w_\infty$ and dividing by $\|u_j\|_{X_0^s(\Omega)}$, we get

$$\begin{aligned} \langle w_j, w_j - w_\infty\rangle_{X_0^s(\Omega)} &= \lambda_\infty \int_\Omega w_j(x)(w_j - w_\infty)(x)\,dx \\ &\quad - \int_\Omega \frac{f(x,u_j(x))}{\|u_j\|_{X_0^s(\Omega)}}(w_j - w_\infty)(x)\,dx + o(1) \end{aligned} \tag{9.19}$$

as $j \to +\infty$.

Now, by (9.18), it follows that

$$\left|\int_\Omega w_j(x)(w_j - w_\infty)(x)\,dx\right| \le \|w_j\|_{L^2(\Omega)}\|w_j - w_\infty\|_{L^2(\Omega)} = o(1),$$

while (9.11), (9.17), and (9.18) imply that

$$\left| \int_\Omega \frac{f(x,u_j(x))}{\|u_j\|_{X_0^s(\Omega)}} (w_j - w_\infty)(x)\,dx \right|$$

$$\leq \varepsilon \|w_j\|_{L^2(\Omega)} \|w_j - w_\infty\|_{L^2(\Omega)} + \frac{a_\varepsilon \|w_j - w_\infty\|_{L^1(\Omega)}}{\|u_j\|_{X_0^s(\Omega)}} = o(1)$$

as $j \to +\infty$. Then, as a consequence of this and by (9.19), it follows that

$$\langle w_j, w_j - w_\infty \rangle_{X_0^s(\Omega)} = o(1);$$

hence,

$$w_j \to w_\infty \quad \text{strongly in } X_0^s(\Omega) \tag{9.20}$$

as $j \to +\infty$. Thus, by the definition of w_j, it follows that $w_\infty \not\equiv 0$.

Now, dividing (9.15) by $\|u_j\|_{X_0^s(\Omega)}$, we have

$$\langle w_j, \varphi \rangle_{X_0^s(\Omega)} = \lambda_\infty \int_\Omega w_j(x)\varphi(x)\,dx - \int_\Omega \frac{f(x,u_j(x))}{\|u_j\|_{X_0^s(\Omega)}} \varphi(x)\,dx + o(1) \tag{9.21}$$

as $j \to +\infty$. Again, (9.11), (9.17), and (9.18) give

$$\int_\Omega \frac{f(x,u_j(x))}{\|u_j\|_{X_0^s(\Omega)}} \varphi(x)\,dx \to 0 \quad \forall \varphi \in X_0^s(\Omega) \tag{9.22}$$

as $j \to +\infty$. Therefore, by (9.20) and (9.22), passing to the limit in (9.21), we get that

$$\langle w, \varphi \rangle_{X_0^s(\Omega)} = \lambda_\infty \int_\Omega w(x)\varphi(x)\,dx \quad \forall \varphi \in X_0^s(\Omega).$$

This means that λ_∞ is an eigenvalue of $-\mathscr{L}_K$, against our assumption. Thus (9.16) is proved.

Since $X_0^s(\Omega)$ is a reflexive space, up to subsequence, there exists $u_\infty \in X_0^s(\Omega)$ such that

$$u_j \rightharpoonup u_\infty \quad \text{weakly in } X_0^s(\Omega) \tag{9.23}$$

and

$$u_j \to u_\infty \quad \text{strongly in } L^2(\Omega) \tag{9.24}$$

as $j \to +\infty$. Furthermore, by (9.14) and (9.23), we get

$$\langle \mathscr{J}'_{K,\lambda_\infty}(u_j), u_j - u_\infty \rangle \to 0, \tag{9.25}$$

and by (9.11), we also have

$$\int_\Omega f(x,u_j(x))(u_j(x) - u_\infty(x))\,dx \to 0$$

as $j \to +\infty$.

Hence, by (9.12) and (9.25), it follows that

$$\begin{aligned}&\int_{\mathbb{R}^n\times\mathbb{R}^n}\left(u_j(x)-u_j(y)\right)^2K(x-y)\,dx\,dy\\&\quad-\int_{\mathbb{R}^n\times\mathbb{R}^n}(u_j(x)-u_j(y))(u_\infty(x)-u_\infty(y))K(x-y)\,dx\,dy\to 0\end{aligned}\tag{9.26}$$

as $j\to+\infty$. Then, by (9.23) and (9.26), we have

$$u_j\to u_\infty \quad \text{strongly in } X_0^s(\Omega)$$

as $j\to+\infty$, and the proof is complete. □

9.4 Some preparatory lemmas

In order to prove Theorem 9.1, we need the following lemmas:

Lemma 9.6 *Assume that conditions* (9.6)–(9.8) *hold. Let* λ_h *be as in* (9.9) *and* $\mathbb{P}_h$ *as in* (3.11). *Then there exist* $\rho>0$ *and* $c_0>0$ *such that*

$$\mathcal{J}_{K,\lambda_\infty}(u)\geq c_0 \quad \forall u\in S_\rho\cap\mathbb{P}_h,\tag{9.27}$$

where

$$S_\rho:=\{u\in X_0^s(\Omega):\|u\|_{X_0^s(\Omega)}=\rho\}.$$

Proof By (9.7) and (9.8), it follows that

$$\lim_{|t|\to\infty}\frac{F(x,t)}{t^2}=0$$

and

$$\lim_{t\to0}\frac{F(x,t)}{t^2}=\frac{\lambda_0}{2},$$

uniformly with respect to a.e. $x\in\Omega$. Moreover, condition (9.9) implies that $\lambda_0<0$. Therefore, for every $\varepsilon>0$, there exist $r_\varepsilon(\geq1)$ and $\delta_\varepsilon>0$ such that

$$|F(x,t)|\leq\frac{\varepsilon}{2}|t|^2 \quad \text{if } |t|>r_\varepsilon\tag{9.28}$$

and

$$\left|F(x,t)-\frac{\lambda_0}{2}t^2\right|\leq\frac{\varepsilon}{2}|t|^2 \quad \text{if } |t|<\delta_\varepsilon,\tag{9.29}$$

for a.e. $x\in\Omega$. However, by (9.6), taking any constant

$$q\in\left[0,\frac{4s}{n-2s}\right),$$

there exists $k_{r_\varepsilon}>0$ such that

$$|F(x,t)|\leq k_{r_\varepsilon}|t|^{q+2} \quad \text{if } \delta_\varepsilon\leq|t|\leq r_\varepsilon, \text{ a.e. } x\in\Omega.\tag{9.30}$$

Further, inequalities (9.28)–(9.30) imply that, for any $\varepsilon > 0$, there exists $k_\varepsilon > 0$ such that

$$F(x,t) \le \frac{\lambda_0 + \varepsilon}{2}|t|^2 + k_\varepsilon |t|^{q+2},$$

for a.e. $x \in \Omega$ and for all $t \in \mathbb{R}$. Thus, we infer that, for any $u \in X_0^s(\Omega)$,

$$\int_\Omega F(x,u(x))\,dx \le \frac{\lambda_0 + \varepsilon}{2}\|u\|^2_{L^2(\Omega)} + k_\varepsilon \|u\|^{q+2}_{L^{q+2}(\Omega)}.$$

Plainly, by the Sobolev inequalities, for a suitable $k'_\varepsilon > 0$, we have

$$\mathcal{J}_{K,\lambda_\infty}(u) \ge \frac{1}{2}\|u\|^2_{X_0^s(\Omega)} - \frac{\lambda_\infty + \lambda_0 + \varepsilon}{2}\|u\|^2_{L^2(\Omega)} - k'_\varepsilon \|u\|^{q+2}_{X_0^s(\Omega)}, \tag{9.31}$$

for every $u \in X_0^s(\Omega)$. Thus, by (3.10) and (9.31), it follows that

$$\mathcal{J}_{K,\lambda_\infty}(u) \ge \frac{1}{2}\left(1 - \frac{\lambda_\infty + \lambda_0 + \varepsilon}{\lambda_h}\right)\|u\|^2_{X_0^s(\Omega)} - k'_\varepsilon \|u\|^{q+2}_{X_0^s(\Omega)},$$

for every $u \in \mathbb{P}_h$. Hence, by (9.9), for a suitable ε, there exists $k''_\varepsilon > 0$ such that

$$\mathcal{J}_{K,\lambda_\infty}(u) \ge k''_\varepsilon \|u\|^2_{X_0^s(\Omega)} - k'_\varepsilon \|u\|^{q+2}_{X_0^s(\Omega)} \quad \forall\, u \in \mathbb{P}_h.$$

Then we conclude that if ρ is small enough, there exists $c_0 > 0$ such that inequality (9.27) holds. □

Lemma 9.7 *Assume that* (9.6) *and* (9.7) *hold. Let* λ_k *be as in* (9.9), $\mathbb{H}_k :=$ span$\{e_1,\dots,e_k\}$, *and* c_0 *as in Lemma 9.6. Then there exists* $c_\infty > c_0$ *such that*

$$\mathcal{J}_{K,\lambda_\infty}(u) \le c_\infty \quad \forall\, u \in \mathbb{H}_k. \tag{9.32}$$

Proof By (9.11), fixing any $\varepsilon > 0$, there exists $C_\varepsilon > 0$ such that

$$\mathcal{J}_{K,\lambda_\infty}(u) \le \frac{1}{2}\|u\|^2_{X_0^s(\Omega)} - \frac{\lambda_\infty}{2}|u|_2^2 + \frac{\varepsilon}{2}\|u\|^2_{L^2(\Omega)} + C_\varepsilon \|u\|_{L^2(\Omega)}, \tag{9.33}$$

for every $u \in X_0^s(\Omega)$. Hence, by (9.33), taking λ_k as in (9.9) and $\varepsilon > 0$ such that $\lambda_k + \varepsilon < \lambda_\infty$, it results that

$$\mathcal{J}_{K,\lambda_\infty}(u) \le \frac{1}{2}(\lambda_k + \varepsilon - \lambda_\infty)\|u\|^2_{L^2(\Omega)} + C_\varepsilon \|u\|_{L^2(\Omega)},$$

for every $u \in \mathbb{H}_k$, and because $\mathbb{H}_k$ is a k-dimensional subspace, the functional $\mathcal{J}_{K,\lambda_\infty}$ tends to $-\infty$ as $\|u\|_{X_0^s(\Omega)}$ diverges in $\mathbb{H}_k$. Thus, there exists $c_\infty = c_\infty(\varepsilon)$ (with $c_\infty > c_0$) such that inequality (9.32) holds. □

9.5 $k-h+1$ distinct pairs of solutions

Finally, in this section, we can prove the main result of this chapter.

Proof of Theorem 9.1 First, by Proposition 9.5, the functional $\mathscr{J}_{K,\lambda_\infty}$ satisfies the Palais–Smale condition at any level $c \in \mathbb{R}$, and by assumption, it is even.

Let us take $\lambda_h, \mathbb{P}_h, \rho$, and c_0 as in Lemma 9.6 and $\lambda_k, \mathbb{H}_k$, and c_∞ as in Lemma 9.7. Then we consider the pseudoindex theory $(S_\rho \cap \mathbb{P}_h, \mathscr{H}^*, \gamma^*)$ related to the genus, $S_\rho \cap \mathbb{P}_h$, and $\mathscr{J}_{K,\lambda_\infty}$. By Remark 9.4 applied to $V := \mathbb{H}_k$, $\partial B := S_\rho$, and $W := \mathbb{P}_h$, we get

$$\gamma\left(\mathbb{H}_k \cap h\left(S_\rho \cap \mathbb{P}_h\right)\right) \geq \dim \mathbb{H}_k - \operatorname{codim} \mathbb{P}_h, \quad \forall h \in \mathscr{H}^*,$$

which implies

$$\gamma^*(\mathbb{H}_k) \geq (k-h+1).$$

Then Theorem 9.3 applies with $\tilde{A} := \mathbb{H}_k$ and $S := S_\rho \cap \mathbb{P}_h$, so $\mathscr{J}_{K,\lambda_\infty}$ has at least $k-h+1$ distinct pairs of critical points corresponding to at most $k-h+1$ distinct critical values c_i, where c_i is as in (9.13). This ends the proof of Theorem 9.1. □

10

Multiple solutions for parametric equations

In this chapter, we are interested in nonlocal equations depending on parameters. In many mathematical problems deriving from applications, the presence of one parameter (or more) is a relevant feature, and the study of how solutions depend on parameters is an important topic.

Most of the results in this direction were obtained through bifurcation theory (for an extensive treatment of this matters, we refer to [9, 10] and their bibliographies). However, some interesting results also can be obtained by means of variational techniques (for this, see [18] and the references therein) or as a combination of the two methods.

In particular, here we study a class of nonlocal fractional Laplacian problems depending on some real parameters. By using the analytical context of fractional Sobolev spaces developed in Chapter 1, we discuss the existence of multiple weak solutions for different classes of these equations via a recent abstract critical point result for differentiable and parametric functionals proved by Ricceri in [182, theorem 3]. The results of this chapter are based on the papers [146, 147].

10.1 Two abstract critical points results

In this section, we recall the abstract results we will use in this chapter. Let E be a nonempty set, and let $I, \Psi, \Phi : E \to \mathbb{R}$ be three given functions. If $\mu > 0$ and $r \in \left(\inf_{u\in E} \Phi(u), \sup_{u\in E} \Phi(u)\right)$, we put

$$\beta(\mu I + \Psi, \Phi, r) := \sup_{u\in\Phi^{-1}((r,+\infty))} \frac{\mu I(u) + \Psi(u) - \inf_{u\in\Phi^{-1}((-\infty,r))} (\mu I(u) + \Psi(u))}{r - \Phi(u)}.$$

Also, when the map $\Psi+\Phi$ is bounded from below, for any $r \in \left(\inf_{u\in E}\Phi(u), \sup_{u\in E}\Phi(u)\right)$ such that

$$\inf_{u\in\Phi^{-1}((-\infty,r])} I(u) \le \inf_{u\in\Phi^{-1}(r)} I(u),$$

we put

$$\mu^\star(I,\Psi,\Phi,r) := \inf\left\{\frac{\Psi(u)-\gamma+r}{\eta_r - I(u)} : u \in E, \Phi(u) < r, I(u) < \eta_r\right\},$$

where $\gamma := \inf_{u\in E}\big(\Psi(u)+\Phi(u)\big)$ and $\eta_r := \inf_{u\in\Phi^{-1}(r)} I(u)$.

With this notation, for the reader's convenience, here we can recall one of the abstract tools we consider in this chapter (see [182, theorem 3]). For completeness, we mention that other related results are contained in [178, 179, 180, 181].

Theorem 10.1 ([182]) *Let $(E, \|\cdot\|)$ be a reflexive Banach space, let $I : E \to \mathbb{R}$ be a sequentially weakly lower semicontinuous, bounded on each bounded subset of E, C^1-functional whose derivative admits a continuous inverse on the topological dual E' of E, and let $\Phi, \Psi : E \to \mathbb{R}$ be two C^1-functionals with compact derivative. Assume also that the functional $\Psi + \lambda\Phi$ is bounded from below for all $\lambda > 0$ and that*

$$\liminf_{\|u\|\to+\infty} \frac{\Psi(u)}{I(u)} = -\infty. \tag{10.1}$$

Then, for any $r > \sup_{u\in S}\Phi(u)$, where S is the set of all global minima of I, for each

$$\mu > \max\{0, \mu^\star(I,\Psi,\Phi,r)\}$$

and each compact interval $[a,b] \subset \big(0, \beta(\mu I+\Psi,\Psi,r)\big)$, there exists a number $\rho > 0$ with the following property: for every $\lambda \in [a,b]$ and every C^1-functional $\Gamma : E \to \mathbb{R}$ with compact derivative, there exists $\delta > 0$ such that, for each $\nu \in [0,\delta]$, the equation

$$\mu I'(u) + \Psi'(u) + \lambda\Phi'(u) + \nu\Gamma'(u) = 0$$

has at least three solutions in E whose norms are less than ρ.

We recall that if I is a C^1-functional, its derivative $I' : E \to E'$ admits a continuous inverse on E', provided that there exists a continuous operator $J : E' \to E$ such that

$$J(I'(u)) = u\,,$$

for any $u \in E$.

In the next result, we denote by $\mathscr{W}_E$ the class of all functionals $I : E \to \mathbb{R}$ such that

$$\text{if } u_j \rightharpoonup u \text{ weakly in } E \text{ as } j \to +\infty \text{ and } \liminf_{j\to+\infty} I(u_j) \le I(u),$$

$$\text{then, up to a subsequence, } u_j \to u \text{ strongly as } j \to +\infty.$$

Theorem 10.2 ([179]) *Let $(E, \|\cdot\|)$ be a separable and reflexive real Banach space, let $\Phi : E \to \mathbb{R}$ be a coercive, sequentially weakly lower semicontinuous C^1-functional belonging to $\mathscr{W}_E$, bounded on each bounded subset of E and whose derivative admits a continuous inverse on E', and let $J : E \to \mathbb{R}$ be a C^1-functional with compact derivative. Assume that Φ has a strict local minimum z_0 with $\Phi(z_0) = J(z_0) = 0$. Finally, assume that*

$$\max\left\{\limsup_{\|z\|\to+\infty}\frac{J(z)}{\Phi(z)}, \limsup_{z\to z_0}\frac{J(z)}{\Phi(z)}\right\} \leq 0 \tag{10.2}$$

and

$$\sup_{z\in E}\min\{\Phi(z), J(z)\} > 0. \tag{10.3}$$

Set

$$\Theta := \inf\left\{\frac{\Phi(z)}{J(z)} : z \in E,\ \min\{\Phi(z), J(z)\} > 0\right\}. \tag{10.4}$$

Then, for any compact interval $[a,b] \subset (\Theta, +\infty)$, there exists a number $\rho > 0$ with the following property: for every $\lambda \in [a,b]$ and every C^1-functional $\Psi : E \to \mathbb{R}$ with compact derivative, there exists $\tilde{\mu} > 0$ such that, for each $\mu \in [0, \tilde{\mu}]$, the equation

$$\Phi'(z) - \lambda J'(z) - \mu\Psi'(z) = 0 \tag{10.5}$$

has at least three solutions whose norms are less than ρ.

10.2 Three weak solutions

In this section, we consider the following problem

$$\begin{cases} -\mathscr{L}_K u = \varepsilon f(x,u) - \lambda g(x,u) + \nu h(x,u) & \text{in } \Omega \\ u = 0 & \text{in } \mathbb{R}^n \setminus \Omega, \end{cases} \tag{10.6}$$

where $s \in (0,1)$ is fixed, $n > 2s$, $\Omega \subset \mathbb{R}^n$ is an open, bounded set with continuous boundary, and $\mathscr{L}_K$ is the integrodifferential operator given in (1.54).

In what follows, we denote by $\mathscr{A}$ the class of all continuous (or, more generally, Carathéodory) functions $p : \Omega \times \mathbb{R} \to \mathbb{R}$ such that, for some $q \in [1, 2_s^*)$,

$$\sup_{(x,t)\in\Omega\times\mathbb{R}}\frac{|p(x,t)|}{1+|t|^{q-1}} < +\infty \tag{10.7}$$

by P its primitive with respect to the second variable (see the function given in (6.8)) and by

$$J_p(u) := \int_\Omega P(x, u(x))dx. \tag{10.8}$$

Note that J_p is C^1 with derivatives given by

$$\langle J_p'(u), \varphi\rangle = \int_\Omega p(x, u(x))\varphi(x)dx,$$

for any $\varphi \in X_0^s(\Omega)$.

Let $f,g:\Omega\times\mathbb{R}\to\mathbb{R}$ be two functions belonging to $\mathscr{A}$ such that $G-F$ is bounded from below. For any $r>0$, set

$$\widetilde{\mu}(f,g,r):=2\inf\left\{\frac{r-\widetilde{\gamma}-J_f(u)}{\widetilde{\eta}_r-\|u\|^2_{X_0^s(\Omega)}}:u\in X_0^s(\Omega),J_g(u)<r,\|u\|^2_{X_0^s(\Omega)}<\widetilde{\eta}_r\right\},$$

where

$$\widetilde{\gamma}:=\int_\Omega\inf_{\xi\in\mathbb{R}}\big(G(x,\xi)-F(x,\xi)\big)dx$$

and

$$\widetilde{\eta}_r:=\inf_{u\in J_g^{-1}(r)}\|u\|^2_{X_0^s(\Omega)}.$$

Finally, for any $\varepsilon\in(0,1/\max\{0,\widetilde{\mu}(f,g,r)\})$, we denote by $\widetilde{\beta}(\varepsilon,f,g,r)$ the quantity

$$\widetilde{\beta}(\varepsilon,f,g,r):=\sup_{u\in J_g^{-1}((r,+\infty))}\frac{\|u\|^2_{X_0^s(\Omega)}-2\varepsilon J_f(u)-\inf\limits_{u\in J_g^{-1}((-\infty,r))}(\|u\|^2_{X_0^s(\Omega)}-2\varepsilon J_f(u))}{2(r-J_g(u))}.$$

With this notation, the main result of this section reads as follows:

Theorem 10.3 *Let $s\in(0,1)$, $n>2s$, and Ω be an open, bounded subset of $\mathbb{R}^n$, with continuous boundary, and let $K:\mathbb{R}^n\setminus\{0\}\to(0,+\infty)$ be a function satisfying* (1.55) *and* (1.56). *Further, let $\sigma\in(2,2_s^*)$ and $f,g:\Omega\times\mathbb{R}\to\mathbb{R}$ be two functions belonging to $\mathscr{A}$ such that*

$$\lim_{\xi\to+\infty}\frac{\inf_{x\in\Omega}F(x,\xi)}{\xi^2}=+\infty,\tag{10.9}$$

$$\limsup_{|\xi|\to+\infty}\frac{\sup_{x\in\Omega}F(x,\xi)}{|\xi|^\sigma}<+\infty,\tag{10.10}$$

and

$$\lim_{|\xi|\to+\infty}\frac{\inf_{x\in\Omega}G(x,\xi)}{|\xi|^\sigma}=+\infty.\tag{10.11}$$

Then for any $r>0$, any $\varepsilon\in\left]0,1/\max\{0,\widetilde{\mu}(f,g,r)\}\right[$, and any compact interval $[a,b]\subset(0,\widetilde{\beta}(\varepsilon,f,g,r))$, there exists a number $\rho>0$ with the following property: for every $\lambda\in[a,b]$ and every function $h\in\mathscr{A}$, there exists $\delta>0$ such that, for each $\nu\in[0,\delta]$, problem (10.6) *admits at least three weak solutions whose norms in $X_0^s(\Omega)$ are less than ρ.*

Proof The idea consists of applying Theorem 10.1, taking for any $u\in E:=X_0^s(\Omega)$

$$I(u):=\frac{1}{2}\|u\|^2_{X_0^s(\Omega)}\quad\text{and}\quad\Psi(u):=-J_f(u),$$

as well as

$$\Phi(u):=J_g(u)\quad\text{and}\quad\Gamma(u):=J_h(u).$$

First, let us prove that

$$\limsup_{\|u\|_{X_0^s(\Omega)}\to+\infty} \frac{J_f(u)}{\|u\|^2_{X_0^s(\Omega)}} = +\infty. \tag{10.12}$$

For this purpose, it is enough to show that

$$\lim_{k\to+\infty} \frac{J_f(ke_1)}{\|ke_1\|^2_{X_0^s(\Omega)}} = +\infty, \tag{10.13}$$

where e_1 is the first eigenfunction of $-\mathscr{L}_K$ with homogeneous Dirichlet boundary conditions (see Chapter 3). To this end, fix two positive numbers M and M_1 such that $M < M_1/2$. Since (10.9) holds, there exists $\eta > 0$ such that

$$F(x,\xi) \geq \lambda_1 M_1 \xi^2,$$

for all $(x,\xi) \in \Omega \times (\eta, +\infty)$.

Now, for any $k \in \mathbb{N}$, set

$$A_k := \left\{ x \in \Omega : e_1(x) \geq \frac{\eta}{k} \right\}.$$

Note that $A_k \subseteq A_{k+1}$, for any $k \in \mathbb{N}$, so the sequence $\left\{ \int_{A_k} |e_1(x)|^2 dx \right\}_{k\in\mathbb{N}}$ is nondecreasing; that is,

$$\int_{A_k} |e_1(x)|^2 dx \leq \int_{A_{k+1}} |e_1(x)|^2 dx,$$

for every $k \in \mathbb{N}$, and

$$\int_{A_k} |e_1(x)|^2 dx \to \int_\Omega |e_1(x)^2| dx$$

as $k \to +\infty$. At this point, fix $\tilde{k} \in \mathbb{N}$ so that

$$\int_{A_{\tilde{k}}} |e_1(x)|^2 dx > \frac{2M}{M_1} \int_\Omega |e_1(x)|^2 dx.$$

Moreover, since $f \in \mathscr{A}$, there exists a constant $c > 0$ such that

$$|F(x,\xi)| \leq c(|\xi| + |\xi|^q),$$

for every $(x,\xi) \in \Omega \times \mathbb{R}$. Thus,

$$\sup_{(x,\xi)\in\Omega\times[0,\eta]} |F(x,\xi)| \leq c(\eta + \eta^q) < +\infty.$$

Then, for any $k \in \mathbb{N}$ satisfying

$$k > \max\left\{\tilde{k}, \left(\frac{|\Omega| \sup\limits_{(x,\xi)\in\Omega\times[0,\eta]} |F(x,\xi)|}{M\|e_1\|^2_{X_0^s(\Omega)}}\right)^{1/2}\right\},$$

we have

$$\begin{aligned}\frac{J_f(ke_1)}{\|ke_1\|^2_{X_0^s(\Omega)}} &= \frac{\int_{A_k} F(x,ke_1(x))dx}{k^2\|e_1\|^2_{X_0^s(\Omega)}} + \frac{\int_{\Omega\setminus A_k} F(x,ke_1(x))dx}{k^2\|e_1\|^2_{X_0^s(\Omega)}} \\ &\geq \frac{\lambda_1 k^2 M_1 \int_{A_k} e_1(x)^2 dx}{k^2\|e_1\|^2_{X_0^s(\Omega)}} + \frac{\int_{\Omega\setminus A_k} F(x,ke_1(x))dx}{k^2\|e_1\|^2_{X_0^s(\Omega)}} \\ &> \frac{2\lambda_1 M \int_{\Omega} e_1(x)^2 dx}{\|e_1\|^2_{X_0^s(\Omega)}} - \frac{|\Omega| \sup\limits_{(x,\xi)\in\Omega\times[0,\eta]} |F(x,\xi)|}{k^2\|e_1\|^2_{X_0^s(\Omega)}} \\ &> 2M - M = M,\end{aligned}$$

which shows (10.13) due to the arbitrariness of M.

Now observe that, by (10.10), there exists $a > 0$ such that

$$F(x,\xi) \leq a(|\xi|^\sigma + 1), \tag{10.14}$$

while, by (10.11) for any $b > 0$, there exists a constant $c_b > 0$ such that

$$G(x,\eta) \geq b|\xi|^\sigma - c_b, \tag{10.15}$$

for any $(x,\xi) \in \Omega \times \mathbb{R}$.

Let $\lambda > 0$ be fixed, and choose b in (10.15) such that $b \geq \lambda a$. Hence, by (10.14) and (10.15), it follows that

$$\begin{aligned}(G - \lambda F)(x,\xi) &\geq b|\xi|^\sigma - c_b - \lambda a(|\xi|^\sigma + 1) \\ &= (b - \lambda a)|\xi|^\sigma - (c_b + a\lambda) \\ &\geq -(c_b + a\lambda),\end{aligned}$$

for any $(x,\xi) \in \Omega \times \mathbb{R}$. As a consequence, we get

$$\int_\Omega (G - \lambda F)(x,u(x))dx \geq -(c_b + a\lambda)|\Omega|.$$

Then the functional $J_g - \lambda J_f$ is bounded from below in $X_0^s(\Omega)$.

Hence, all the assumptions of Theorem 10.1 are satisfied, so the assertion is proved. □

A model for problem (10.6) is given by

$$\begin{cases} (-\Delta)^s u = \varepsilon|u|^{p-1}u - \lambda|u|^{q-1}u + \nu h(u) & \text{in } \Omega \\ u = 0 & \text{in } \mathbb{R}^n \setminus \Omega. \end{cases} \tag{10.16}$$

In this setting, Theorem 10.3 reads as follows:

Theorem 10.4 *Let $s \in (0,1)$, $n > 2s$, and Ω be an open, bounded subset of $\mathbb{R}^n$ with continuous boundary. Moreover, let*

$$1 < p < q < \frac{n+2s}{n-2s}.$$

Then, for any $\varepsilon > 0$ small enough, there exists $\lambda_\varepsilon > 0$ such that, for every compact interval $[a,b] \subset (0,\lambda_\varepsilon)$, there exists $\rho > 0$ with the following property: for every $\lambda \in [a,b]$ and every continuous function $h : \mathbb{R} \to \mathbb{R}$ satisfying

$$\limsup_{|\xi|\to\infty} \frac{|h(\xi)|}{|\xi|^\alpha} < +\infty,$$

for some $\alpha \in (0,(n+2s)/(n-2s))$, there exists $\delta > 0$ such that, for every $\nu \in [0,\delta]$, problem (10.16) has at least three distinct weak solutions $\{u_j\}_{j=1}^3$ in $\mathbb{H}_0^s(\Omega)$ such that $\|u_j\|_{\mathbb{H}_0^s(\Omega)} < \rho$, for $j = 1,2,3$.

It is worth pointing out that the variational approach used to attack such problems is not often easy to perform; indeed, due to the presence of the nonlocal term, variational methods do not to work when applied to these classes of equations.

Finally, we observe that contrary to most of the known applications in the classical Laplacian case, no condition on the behavior of the nonlinearities near zero is assumed in the results stated here.

10.3 Two weak solutions

In this section, we study the existence and nonexistence of weak solutions of the following fractional problem:

$$\begin{cases} -\mathcal{L}_K u = \lambda\beta(x)f(u) + \mu g(x,u) & \text{in } \Omega \\ u = 0 & \text{in } \mathbb{R}^n \setminus \Omega, \end{cases} \tag{10.17}$$

where, as usual, $s \in (0,1)$ is fixed, $n > 2s$, $\Omega \subset \mathbb{R}^n$ is an open, bounded set with continuous boundary, and $\mathcal{L}_K$ is the integrodifferential operator given in (1.54). Moreover, λ and μ are real parameters,$\beta \in L^\infty(\Omega)$ satisfies

$$\operatorname*{ess\,inf}_{x\in\Omega} \beta(x) > 0, \tag{10.18}$$

while the nonlinearity $g : \Omega \times \mathbb{R} \to \mathbb{R}$ is only assumed to be a Carathéodory function verifying (7.26). Finally, we assume that $f : \mathbb{R} \to \mathbb{R}$ is continuous, superlinear at zero, and sublinear at infinity; that is,

$$\lim_{t\to 0} \frac{f(t)}{t} = 0, \tag{10.19}$$

$$\lim_{|t|\to\infty} \frac{f(t)}{t} = 0, \tag{10.20}$$

and

$$\sup_{t\in\mathbb{R}} F(t) > 0, \tag{10.21}$$

where F is as in (7.5). Note that conditions (10.19) and (10.20) are standard assumptions to be satisfied in the presence of subcritical terms (see, e.g., [125] and [126]).

The main multiplicity result of this section can be stated as follows:

Theorem 10.5 *Let $s \in (0,1)$, $n > 2s$, and Ω be an open, bounded subset of $\mathbb{R}^n$ with continuous boundary. Let $f : \mathbb{R} \to \mathbb{R}$ and $g : \Omega \times \mathbb{R} \to \mathbb{R}$ be two functions that satisfy* (10.19)–(10.21) *and* (7.26)*, respectively, and let $\beta \in L^\infty(\Omega)$ such that condition* (10.18) *holds. Then there exists $\lambda^\star$ such that, for every $\lambda > \lambda^\star$, there is $\widetilde{\mu} > 0$ with the property that, for every $\mu \in [-\widetilde{\mu}, \widetilde{\mu}]$, problem* (10.17) *has at least two distinct nontrivial weak solutions in $\mathbb{H}_0^s(\Omega)$.*

In the proof of Theorem 10.5, we also have that, for every compact interval $[a,b] \subset (\lambda^\star, +\infty)$, there exists a number $\rho > 0$ such that, for every $\lambda \in [a,b]$, the weak solutions u_λ^i, $i = 1,2$, of problem (10.17) verify $\|u_\lambda^i\|_{\mathbb{H}_0^s(\Omega)} \leq \rho$. The key tool for the proof of Theorem 10.5 (that we omit here) is an abstract critical point theorem for differentiable functionals proved in [179, theorem 2].

When $\mu = 0$ in (10.17), the problem becomes

$$\begin{cases} -\mathscr{L}_K u = \lambda\beta(x)f(u) & \text{in } \Omega \\ u = 0 & \text{in } \mathbb{R}^n \setminus \Omega. \end{cases} \tag{10.22}$$

When the nonlinear term satisfies the assumptions

$$\lim_{t\to 0^+} \frac{f(t)}{t} = 0, \quad \lim_{t\to +\infty} \frac{f(t)}{t} = 0, \tag{10.23}$$

and

$$\sup_{t\in[0,+\infty)} F(t) > 0, \tag{10.24}$$

with respect to (10.22), the following result holds true:

Theorem 10.6 *Let $s \in (0,1)$, $n > 2s$, let Ω be an open, bounded subset of $\mathbb{R}^n$ with continuous boundary $\partial\Omega$, $\beta \in L^\infty(\Omega)$ satisfying* (10.18)*, and let $f : [0,+\infty) \to \mathbb{R}$ be a continuous function verifying* (10.23) *and* (10.24)*. Then*

(*a*) *Problem* (10.22) *admits only the trivial solution whenever*

$$0 \leq \lambda < \lambda_1 \|\beta\|_{L^\infty(\Omega)}^{-1} \left(\max_{t>0} \frac{f(t)}{t} \right)^{-1}. \tag{10.25}$$

(*b*) *There exists $\lambda^\star > 0$ such that problem* (10.22) *admits at least two distinct nontrivial and nonnegative weak solutions whenever $\lambda > \lambda^\star$.*

Proof Due to (10.23), it follows that $f(0) = 0$, and one can extend the function f to $\mathbb{R}$ by $f(t) = 0$, for $t \leq 0$. Moreover, under our assumptions, standard computations ensure that every weak solution of problem (10.22) is nonnegative.

Let us prove part (a). Arguing by contradiction, we assume that there exists a weak solution $u_0 \in X_0^s(\Omega) \setminus \{0\}$ of the problem (10.22); that is,

$$\begin{cases} \displaystyle\int_{\mathbb{R}^n \times \mathbb{R}^n} (u_0(x) - u_0(y))(\varphi(x) - \varphi(y))K(x-y)\,dx\,dy \\ \qquad = \lambda \displaystyle\int_\Omega \beta(x) f(u_0(x))\varphi(x)\,dx \quad \forall\, \varphi \in X_0^s(\Omega). \end{cases} \tag{10.26}$$

In particular, testing (10.26) with $\varphi = u_0$, we have

$$\|u_0\|^2_{X_0^s(\Omega)} = \lambda \int_\Omega \beta(x) f(u_0(x)) u_0(x)\,dx. \tag{10.27}$$

However, since (10.23) holds, one has that the quantity

$$c_f := \max_{t>0} \frac{f(t)}{t}$$

is well defined and positive. Hence, because u_0 is nonnegative, it follows that

$$\begin{aligned} \int_\Omega \beta(x) f(u_0(x)) u_0(x)\,dx &\leq \|\beta\|_{L^\infty(\Omega)} \int_\Omega f(u_0(x)) u_0(x)\,dx \\ &\leq c_f \|\beta\|_{L^\infty(\Omega)} \|u_0\|^2_{L^2(\Omega)} \\ &\leq \frac{c_f}{\lambda_1} \|\beta\|_{L^\infty(\Omega)} \|u_0\|^2_{X_0^s(\Omega)}, \end{aligned} \tag{10.28}$$

thanks to Proposition 3.1.

Thus, by (10.27), (10.28), and (10.25), we get

$$\|u_0\|^2_{X_0^s(\Omega)} \leq \lambda \frac{c_f}{\lambda_1} \|\beta\|_{L^\infty(\Omega)} \|u_0\|^2_{X_0^s(\Omega)} < \|u_0\|^2_{X_0^s(\Omega)},$$

which is a contradiction.

Part (b) of Theorem 10.6 can be proved via direct minimization and the mountain pass theorem (as well as exploiting Theorem 10.2 again). □

The expression $\lambda^\star$ in Theorems 10.5 and 10.6 has the form

$$\lambda^\star := \inf_{\substack{u \in X_0^s(\Omega) \\ \Psi(u) > 0}} \frac{\dfrac{1}{2} \displaystyle\int_{\mathbb{R}^n \times \mathbb{R}^n} |u(x) - u(y)|^2 K(x-y)\,dx\,dy}{\displaystyle\int_\Omega \beta(x) F(u(x))\,dx}. \tag{10.29}$$

This definition is quite involved. For this, we give an upper estimate that can be easily computed. Let us consider the truncated function defined in (1.35) and assume that the kernel K, instead of (1.55) and (1.56), satisfies the following more restrictive hypothesis:

$$\text{there exists } \theta \in (0,1] \text{ such that, for any } x \in \mathbb{R}^n \setminus \{0\},$$
$$\theta|x|^{-(n+2s)} \leq K(x) \leq \theta^{-1}|x|^{-(n+2s)}.$$

First of all, we need the following lemma:

Lemma 10.7 *Let $f : \mathbb{R} \to \mathbb{R}$ be a continuous function satisfying* (10.21). *Then there exist $t_0 \in \mathbb{R}$ and $\sigma_0 \in (0,1)$ such that*

$$\int_\Omega F(u^{t_0}_{\sigma_0}(x))dx \geq \operatorname*{ess\,inf}_{x\in\Omega}\beta(x)\left(F(t_0)\sigma_0^n - (1-\sigma_0^n)\max_{|t|\leq|t_0|}|F(t)|\right)\omega_n\tau^n > 0,$$

where ω_n denotes the volume of the unit ball in $\mathbb{R}^n$.

Proof By hypothesis (10.21), there exists $t_0 \in \mathbb{R}$ such that $F(t_0) > 0$. Pick σ_0 in $(0,1)$ so that

$$F(t_0)\sigma_0^n - (1-\sigma_0^n)\max_{|t|\leq|t_0|}|F(t)| > 0.$$

Now let $u^{t_0}_{\sigma_0}$ be the truncated function defined in (1.35). Because

$$\max_{x\in\overline{\Omega}}|u^{t_0}_{\sigma_0}(x)| \leq |t_0|,$$

one has

$$\int_{B(x_0,\tau)\setminus B(x_0,\sigma_0\tau)} F(u^{t_0}_{\sigma_0}(x))dx \geq -(1-\sigma_0^n)\max_{|t|\leq|t_0|}|F(t)|\tau^n\omega_n,$$

and therefore,

$$\begin{aligned}
\Psi(u^{t_0}_{\sigma_0}) &:= \int_\Omega \beta(x)F(u^{t_0}_{\sigma_0}(x))dx \\
&= \int_{B(x_0,\sigma_0\tau)} \beta(x)F(u^{t_0}_{\sigma_0}(x))dx + \int_{B(x_0,\tau)\setminus B(x_0,\sigma_0\tau)} \beta(x)F(u^{t_0}_{\sigma_0}(x))dx \\
&\geq \operatorname*{ess\,inf}_{x\in\Omega}\beta(x)\left(F(t_0)\sigma_0^n\tau^n\omega_n + \int_{B(x_0,\tau)\setminus B(x_0,\sigma_0\tau)} F(u^{t_0}_{\sigma_0}(x))dx\right) \\
&\geq \operatorname*{ess\,inf}_{x\in\Omega}\beta(x)\left(F(t_0)\sigma_0^n - (1-\sigma_0^n)\max_{|t|\leq|t_0|}|F(t)|\right)\omega_n\tau^n.
\end{aligned}$$

This concludes the proof. □

Hence, by using Proposition 1.16 and Lemma 10.7, we have that $\lambda^\star < \lambda_0$, where

$$\lambda_0 := \frac{t_0^2\pi^{\frac{n}{2}}(1-\sigma_0^n)\kappa_1\kappa_2}{2\theta(1-\sigma_0)^2\operatorname*{ess\,inf}_{x\in\Omega}\beta(x)\,\omega_n\tau^2\Gamma\left(1+\frac{n}{2}\right)\left(F(t_0)\sigma_0^n - (1-\sigma_0^n)\max_{|t|\leq|t_0|}|F(t)|\right)}.$$

Finally, for completeness, we mention that in [147], by using related variational arguments, the following multiplicity result has been proved:

Theorem 10.8 *Let $s \in (0,1)$, $n > 2s$, and Ω be an open, bounded subset of $\mathbb{R}^n$ with continuous boundary. Let $f : \mathbb{R} \to \mathbb{R}$ be a continuous function verifying* (10.19)–(10.21). *Then there exist an open interval $\Lambda \subset (0,+\infty)$ and a real positive number κ such that, for every $\lambda \in \Lambda$, the problem*

$$\begin{cases} (-\Delta)^s u = \lambda f(u) & \text{in } \Omega \\ u = 0 & \text{in } \mathbb{R}^n \setminus \Omega \end{cases} \tag{10.30}$$

has at least two distinct nontrivial weak solutions whose $\mathbb{H}_0^s(\Omega)$-norms are less than κ.

Furthermore, we give additional information as far as the localization of interval Λ is concerned. More precisely, we have that

$$\Lambda \subset \left[\frac{\lambda_{1,s}}{\max\limits_{t \in \mathbb{R}} \dfrac{f(t)}{t}}, \frac{2}{(1-\sigma_0)^2} \frac{(1-\sigma_0^n)\kappa_1\kappa_2}{\left(F(t_0)\sigma_0^n - (1-\sigma_0^n) \max\limits_{|t| \le |t_0|} |F(t)| \right)} \frac{t_0^2}{\tau^2} \right].$$

We also cite, for completeness, the papers [97, 157], where the existence of multiple weak solutions for nonlocal problems has been studied by using different linking methods.

11
Infinitely many solutions

In Chapter 6, we studied the existence of nontrivial solutions for a nonlocal fractional problem whose prototype was given as

$$\begin{cases} (-\Delta)^s u - \lambda u = |u|^{q-2}u & \text{in } \Omega \\ u = 0 & \text{in } \mathbb{R}^n \setminus \Omega. \end{cases}$$

In this chapter, we consider a perturbation of this problem, namely, problems modeled as follows:

$$\begin{cases} (-\Delta)^s u - \lambda u = |u|^{q-2}u + h & \text{in } \Omega \\ u = 0 & \text{in } \mathbb{R}^n \setminus \Omega, \end{cases} \tag{11.1}$$

where $s \in (0,1)$ is a fixed parameter, $(-\Delta)^s$ is the fractional Laplace operator, while λ is a real parameter, the exponent $q \in (2, 2_s^*)$, with $2_s^* := 2n/(n-2s)$, $n > 2s$, the function h belongs to the space $L^2(\Omega)$, and finally, the set Ω is an open, bounded subset of $\mathbb{R}^n$ with continuous boundary.

Adapting the classical variational techniques used to study the standard Laplace equation with subcritical growth nonlinearities to the nonlocal framework, we prove in this chapter that the preceding problem admits infinitely many weak solutions $\{u_k\}_{k\in\mathbb{N}}$ with the property that their Sobolev norm goes to infinity as $k \to +\infty$, provided that the exponent

$$q < 2_s^* - \frac{2s}{n-2s}.$$

In this sense, the results presented here may be seen as the extension of some classical nonlinear analysis theorems to the case of fractional operators.

The results of this chapter are based on the paper [195]. We point out that under different superlinear growth assumptions on the nonlinear term, in [36], three different theorems about the existence of infinitely many weak solutions for nonlocal problems have been proved using the classical fountain theorem, extending some classical results for semilinear Laplacian equations to the nonlocal fractional setting.

11.1 The main results

In Chapter 6, the case $h \equiv 0$ in (11.1) was considered, and a nontrivial weak solution was obtained using classical minimax theorems, namely, the mountain pass and the linking theorems, based on different values of the parameter λ. A natural question is whether or not classical existence results for Laplacian equations still hold in the nonlocal framework of (11.1). The aim of this chapter is to answer this question with respect to the existence of infinitely many weak solutions for the subcritical equation (11.1).

In [19, 20, 211], the classical Laplacian case was studied using different methods, and the existence of infinitely many weak solutions (with the property that the L^2-norm of their gradient goes to infinity) was proved (see also [120, 212]). We will show that that result still holds true in the nonlocal setting. Precisely, our existence result with respect to problem (11.1) can be stated as follows:

Theorem 11.1 *Let $s \in (0,1)$, $n > 2s$, and Ω be an open, bounded subset of $\mathbb{R}^n$ with continuous boundary. Assume also that $h \in L^2(\Omega)$ and $q \in (2, 2_s^* - 2s/(n-2s))$, with $2_s^* := 2n/(n-2s)$. Then, for any $\lambda \in \mathbb{R}$, problem* (11.1) *admits infinitely many weak solutions $\{u_k\}_{k\in\mathbb{N}}$ in $\mathbb{H}_0^s(\Omega)$ and*

$$\int_{\mathbb{R}^n \times \mathbb{R}^n} \frac{|u_k(x) - u_k(y)|^2}{|x-y|^{n+2s}}\, dx\, dy \to +\infty$$

as $k \to +\infty$.

Equation (11.1) represents a model only for the general problem studied here. Indeed, in what follows, we consider the following equation:

$$\begin{cases} \mathscr{L}_K u + \lambda u + f(x,u) + h(x) = 0 & \text{in } \Omega \\ u = 0 & \text{in } \mathbb{R}^n \setminus \Omega, \end{cases} \tag{11.2}$$

where the set $\Omega \subset \mathbb{R}^n$, $n > 2s$, is open, bounded, and with continuous boundary, and $s \in (0,1)$ is fixed, while $\mathscr{L}_K$ is defined in (1.54). Moreover, the nonlinear term in equation (11.2) is a function $f : \overline{\Omega} \times \mathbb{R} \to \mathbb{R}$, verifying (6.3)–(6.5) and the following conditions:

$$f(x,\cdot) \textit{ is a } C^1(\mathbb{R}) \textit{ continuous function, for any } x \in \overline{\Omega}; \tag{11.3}$$

there exist $\beta \in (0,1)$ and $r > 0$ such that, for any $x \in \overline{\Omega}$, $t \in \mathbb{R}$, $|t| \geq r$,

$$0 < \frac{f(x,t)}{t} \leq \beta f_t(x,t), \tag{11.4}$$

where f_t denotes the derivative of f with respect to the second variable; that is,

$$f_t(x,t) = \frac{\partial f}{\partial t}(x,t).$$

Note that these assumptions are the standard ones when dealing with partial differential equations driven by the Laplace operator (or, more generally, by uniformly elliptic operators) with homogeneous Dirichlet boundary conditions. For this, we refer to [19, theorem 6.1], [120, p. 233] and [212, chapter II, theorem 7.2] (see also [20, 211]). As a model for f, we can take the function given in (6.9). Finally, the perturbation $h : \Omega \to \mathbb{R}$ is a function such that

$$h \in L^2(\Omega). \tag{11.5}$$

Clearly, the weak formulation of (11.2) is given by the following problem:

$$\begin{cases} \displaystyle\int_{\mathbb{R}^n \times \mathbb{R}^n} (u(x) - u(y))(\varphi(x) - \varphi(y))K(x-y)\,dx\,dy - \lambda \int_\Omega u(x)\varphi(x)\,dx \\ \displaystyle\quad = \int_\Omega f(x, u(x))\varphi(x)\,dx + \int_\Omega h(x)\varphi(x)\,dx \quad \forall\, \varphi \in X_0^s(\Omega) \\ u \in X_0^s(\Omega). \end{cases} \tag{11.6}$$

The main result of this chapter is given as follows:

Theorem 11.2 *Let* $s \in (0,1)$, $n > 2s$, *and* Ω *be an open, bounded subset of* $\mathbb{R}^n$ *with continuous boundary. Let* $K : \mathbb{R}^n \setminus \{0\} \to (0, +\infty)$ *be a function satisfying conditions* (1.55) *and* (1.56), *let* $f : \overline{\Omega} \times \mathbb{R} \to \mathbb{R}$ *verify* (6.3)–(6.5) *with* $q \in (2, 2_s^* - 2s/(n-2s))$, $2_s^* := 2n/(n-2s)$, *and* (11.3) *and* (11.4), *and let* $h : \Omega \to \mathbb{R}$ *satisfy* (11.5). *Then, for any* $\lambda \in \mathbb{R}$, *problem* (11.2) *admits infinitely many weak solutions* $\{u_k\}_{k\in\mathbb{N}}$ *in* $X_0^s(\Omega)$ *such that*

$$\int_{\mathbb{R}^n \times \mathbb{R}^n} |u_k(x) - u_k(y)|^2 K(x-y)\,dx\,dy \to +\infty$$

as $k \to +\infty$.

Of course, when K and f are exactly as in the model, that is, $K(x) = |x|^{-(n+2s)}$ and $f(x,t) = |t|^{q-2}t$, $t \in (2, 2_s^*)$, then Theorem 11.2 reduces to Theorem 11.1. When $s = 1$, our problem reduces to the standard semilinear Laplace partial differential equation. In this framework, the bound from above on the exponent q becomes $q < 2(n-1)/(n-2)$, which is the usual one when dealing with the Laplace equation (see [120, chapter 5, theorem 4.6 and remark 4.7]). For all these reasons, Theorems 11.1 and 11.2 may be seen as the fractional version of the classical existence results given in [120, chapter 5, theorem 4.6] in the model case $f(x,t) := |t|^{q-2}t$ and in [19, theorem 6.1] for more general nonlinearities (see also [212, chapter II, theorem 7.2] and [20, 211]). Also, when $h \equiv 0$ (i.e., when there is no perturbation), Theorem 11.1 represents the nonlocal counterpart of the classical results obtained, for instance, in [11] (see also [212] and references therein).

The classical results stated in [120, chapter 5, theorem 4.6] and in [19, theorem 6.1] are proved by means of variational and topological methods. We will adapt these techniques here to the nonlocal framework to get Theorem 11.2. The main difficulty, as in the standard case of the Laplacian, will be to prove the Palais–Smale condition for the energy functional associated with the problem.

11.2 Abstract approach

In this section, for the reader's convenience, we illustrate our strategy to prove Theorem 11.2. First of all, we observe that problem (11.6) is the Euler–Lagrange equation of the functional $\mathcal{J}_{K,\lambda,h} : X_0^s(\Omega) \to \mathbb{R}$ defined as follows:

$$\mathcal{J}_{K,\lambda,h}(u) := \frac{1}{2}\int_{\mathbb{R}^n\times\mathbb{R}^n} |u(x)-u(y)|^2 K(x-y)\,dx\,dy - \frac{\lambda}{2}\int_\Omega |u(x)|^2\,dx$$
$$- \int_\Omega F(x,u(x))\,dx - \int_\Omega h(x)u(x)\,dx,$$

for every $u \in X_0^s(\Omega)$. Here the function F is the one given in (6.8).

Notice that the functional $\mathcal{J}_{K,\lambda,h}$ is well defined, thanks to the assumptions on f, to Lemma 1.31(a), and to condition (11.5). Moreover, $\mathcal{J}_{K,\lambda,h}$ is Fréchet differentiable at $u \in X_0^s(\Omega)$, and

$$\langle \mathcal{J}'_{K,\lambda,h}(u), \varphi\rangle = \int_{\mathbb{R}^n\times\mathbb{R}^n} (u(x)-u(y))(\varphi(x)-\varphi(y))K(x-y)\,dx\,dy$$
$$-\lambda\int_\Omega u(x)\varphi(x)\,dx - \int_\Omega f(x,u(x))\varphi(x)\,dx - \int_\Omega h(x)\varphi(x)\,dx,$$

for any $\varphi \in X_0^s(\Omega)$.

Hence, to prove Theorem 11.2, we will look for critical points of the functional $\mathcal{J}_{K,\lambda,h}$. To this end, let us introduce the functional $\mathcal{I}_{K,\lambda,h} : X_0^s(\Omega) \to \mathbb{R}$, defined as follows:

$$\mathcal{I}_{K,\lambda,h}(u) := \max_{\tau>0} \mathcal{J}_{K,\lambda,h}(\tau u),$$

for every $u \in X_0^s(\Omega)$.

As we will show later, the critical points of functionals $\mathcal{J}_{K,\lambda,h}$ and $\mathcal{I}_{K,\lambda,h}$ are strictly related to each other; indeed, the following result holds true:

Proposition 11.3 Let $\lambda \in \mathbb{R}$, $s \in (0,1)$, $n > 2s$, and Ω be an open, bounded subset of $\mathbb{R}^n$ with continuous boundary. Let $K : \mathbb{R}^n \setminus \{0\} \to (0,+\infty)$ be a function satisfying conditions (1.55) and (1.56), let $f : \overline{\Omega}\times\mathbb{R} \to \mathbb{R}$ verify (6.3)–(6.5) with $q \in (2, 2_s^* - 2s/(n-2s))$, $2_s^* := 2n/(n-2s)$, and (11.3) and (11.4), and let $h : \Omega \to \mathbb{R}$ satisfy (11.5). Then there exists a positive constant α such that

(a) The functional $\mathcal{I}_{K,\lambda,h}$ is of class C^1 in $\{\mathcal{I}_{K,\lambda,h} > \alpha\}$, and
(b) For any $v \in \{\mathcal{I}_{K,\lambda,h} > \alpha\}$, there exists a unique $\sigma(v) > 0$ such that

$$\mathcal{I}_{K,\lambda,h}(v) = \mathcal{J}_{K,\lambda,h}(\sigma(v)v).$$

As a consequence, for any $v \in \{\mathcal{I}_{K,\lambda,h} > \alpha\}$,

$$\mathcal{I}'_{K,\lambda}(v) = \sigma(v)\mathcal{J}'_{K,\lambda}(\sigma(v)v).$$

Proof Let us consider $v \in X_0^s(\Omega)$. First of all, note that, by definition of $\mathcal{I}_{K,\lambda,h}$, we have

$$\mathcal{I}_{K,\lambda,h}(tv) = \mathcal{I}_{K,\lambda,h}(v) \quad \text{for any } t > 0.$$

Hence, without loss of generality, we can assume that $\|v\|_{L^q(\Omega)} = 1$.

From now on, thanks in particular to assumptions (11.3) and (11.4), we can argue exactly as in [19, propositions 6.2 and 6.3], where the classical case of the Laplacian (in general of second-order elliptic operators) was considered: we have just to replace the L^2-norm of the gradient with the $X_0^s(\Omega)$-norm. □

As a consequence of Proposition 11.3, we have the next result, which relates the critical points of $\mathscr{I}_{K,\lambda,h}$ to those of $\mathscr{J}_{K,\lambda,h}$. In the following, M will denote the set

$$M := \{u \in X_0^s(\Omega) : \|u\|_{L^q(\Omega)} = 1\},$$

where $q \in (2, 2_s^*)$. Note that M is well defined because the space $X_0^s(\Omega)$ is compactly embedded into $L^q(\Omega)$.

Corollary 11.4 Let all the assumptions of Proposition 11.3 be satisfied, and let α and $\sigma(v)$ be as in Proposition 11.3. Then the following assertions hold true:

(a) If $v \in \{\mathscr{I}_{K,\lambda,h} > \alpha\}$ is a critical point of $\mathscr{I}_{K,\lambda,h}$, then $\sigma(v)v$ is a critical point of $\mathscr{J}_{K,\lambda,h}$, and

(b) If $v \in \{\mathscr{I}_{K,\lambda,h} > \alpha\}$ is a critical point of $\mathscr{I}_{K,\lambda,h}$ on M, then v is a critical point of $\mathscr{I}_{K,\lambda,h}$ on $X_0^s(\Omega)$.

Proof Assertion (a) is a consequence of the fact that $\sigma(v) > 0$ and of Proposition 11.3(b). For part (b), note that if $v \in \{\mathscr{I}_{K,\lambda,h} > \alpha\}$ is a critical point of $\mathscr{I}_{K,\lambda,h}$ on M, then there exists $\rho \in \mathbb{R}$ such that

$$\langle \mathscr{I}'_{K,\lambda}(v), \varphi\rangle = \rho \int_\Omega |v(x)|^{q-2} v(x)\varphi(x)\,dx, \tag{11.7}$$

for any $\varphi \in X_0^s(\Omega)$. In particular, taking $\varphi = v$ because $v \in M$, we get

$$\rho = \langle \mathscr{I}'_{K,\lambda}(v), v\rangle.$$

Hence, by Proposition 11.3(b) and by part (a), we have

$$\rho = \langle \mathscr{I}'_{K,\lambda}(v), v\rangle = \sigma(v)\langle \mathscr{J}'_{K,\lambda}(\sigma(v)v), v\rangle = 0.$$

As a consequence of this and of (11.7), we deduce that $\langle \mathscr{I}'_{K,\lambda}(v), \varphi\rangle = 0$ for any $\varphi \in X_0^s(\Omega)$, which proves assertion (b). □

Remark 11.5 We would like to note that in the model case $f(x,t) = |t|^{q-2}t$, $q \in (2, 2_s^*)$ and in the assertions of Proposition 11.3 and Corollary 11.4, the open set $\{\mathscr{I}_{K,\lambda,h} > \alpha\}$ can be replaced with $\{\mathscr{I}_{K,\lambda,h} > 0\}$.

Thanks to Corollary 11.4, to find the critical points of $\mathscr{J}_{K,\lambda,h}$, it is enough to look for critical points $u \in \{\mathscr{I}_{K,\lambda,h} > \alpha\}$ of the functional $\mathscr{I}_{K,\lambda,h}$ on M. Hence, to prove Theorem 11.2, this will be our strategy; namely, we will look for critical points $u \in \{\mathscr{I}_{K,\lambda,h} > \alpha\}$ of $\mathscr{I}_{K,\lambda,h}$ on M. To this end, we will apply [120, chapter 5, theorem 2.2] to the functional $\mathscr{I}_{K,\lambda,h}$. For this, we have to prove that

1. The functional $\mathscr{I}_{K,\lambda,h}$ satisfies the *Palais–Smale compactness condition* on M, and
2. For $k \in \mathbb{N}$ large enough, it hold true that $a_k \in \mathbb{R}$ and $a_k < b_k$, where a_k and b_k are defined as follows:

$$a_k := \inf_{A \in \mathscr{A}_k} \max_{v \in A} \mathscr{I}_{K,\lambda,h}(v) \tag{11.8}$$

and

$$b_k := \inf_{B \in \mathscr{B}_k} \max_{v \in B} \mathscr{I}_{K,\lambda,h}(v), \tag{11.9}$$

with $\mathscr{A}_k$ and $\mathscr{B}_k$ given by

$$\mathscr{A}_k := \{g(S^{k-1}) \text{ such that } g : S^{k-1} \to M \text{ is continuous and odd}\}$$

and

$$\mathscr{B}_k := \{g(S^k_+) \text{ such that } g : S^k_+ \to M \text{ is continuous and } g_{|S^k_+} \text{ is odd}\}.$$

Here S^k is the unit ball in $\mathbb{R}^{k+1}$, while S^k_+ is the northern hemisphere of S^k that is the set

$$S^k_+ := \{(u_1, \dots, u_{k+1}) \in S^k \text{ such that } u_{k+1} \geq 0\}.$$

To prove these properties with respect to the functional $\mathscr{I}_{K,\lambda,h}$, we need some preliminary results.

11.2.1 Some preliminary lemmas

In this subsection, we prove some auxiliary results on $\mathscr{I}_{K,\lambda,h}$ that will be useful in what follows. Here assumptions (11.3) and (11.4) will be crucial.

First, we need some notation. In what follows, $\mathscr{I}_{K,\lambda}$ will be the even functional $\mathscr{I}_{K,\lambda,h}$ obtained in the particular case when $h \equiv 0$; that is,

$$\mathscr{I}_{K,\lambda}(u) = \max_{\tau > 0} \mathscr{J}_{K,\lambda}(\tau u),$$

where $\mathscr{J}_{K,\lambda}$ is the functional defined in (6.20). Moreover, in this case, we also will denote by $\sigma_0(v)$ the element given in Proposition 11.3(b); that is, $\sigma_0(v)$ will be such that

$$\mathscr{I}_{K,\lambda}(v) = \mathscr{J}_{K,\lambda}(\sigma_0(v)v).$$

Based on this notation, we have the following result:

Lemma 11.6 *Let $\lambda \in \mathbb{R}$, $s \in (0,1)$, $n > 2s$, and Ω be an open, bounded subset of $\mathbb{R}^n$ with continuous boundary. Let $K : \mathbb{R}^n \setminus \{0\} \to (0, +\infty)$ be a function satisfying conditions* (1.55) *and* (1.56), *let $f : \overline{\Omega} \times \mathbb{R} \to \mathbb{R}$ verify* (6.3)–(6.5) *with $q \in (2, 2^*_s - 2s/(n-2s))$, $2^*_s := 2n/(n-2s)$, and* (11.3) *and* (11.4), *and let $h : \Omega \to \mathbb{R}$ satisfy* (11.5). *Finally, let α be as in Proposition 11.3. Then there exist two positive constants*

$\bar{R}$ and $\bar{c}$, depending only on α, λ, q, $|\Omega|$, and $\|h\|_{L^2(\Omega)}$, such that, for any $v \in M$ with $\|v\|_{X_0^s(\Omega)} \geq \bar{R}$, the following assertions hold true:

(a) $\mathcal{I}_{K,\lambda,h}(v) > \alpha$,

(b) $|\mathcal{I}_{K,\lambda,h}(v) - \mathcal{I}_{K,\lambda}(v)| \leq \bar{c}\left(1 + \mathcal{I}_{K,\lambda,h}(v)^{1/q}\right)$, *and*

(c) $|\mathcal{I}_{K,\lambda,h}(v) - \mathcal{I}_{K,\lambda}(v)| \leq \bar{c}\left(1 + \mathcal{I}_{K,\lambda}(v)^{1/q}\right)$.

Proof Let us start by proving assertion (a). Thanks to the definition of $\mathcal{I}_{K,\lambda,h}$, to obtain our goal, it is enough to show that

$$\mathcal{I}_{K,\lambda,h}(v) > \alpha \tag{11.10}$$

when $v \in M$ is such that $\|v\|_{X_0^s(\Omega)}$ is large enough.

To this end, note that (6.4), (11.5), and the Hölder inequality yield

$$\begin{aligned}
\mathcal{I}_{K,\lambda,h}(v) &= \frac{1}{2}\|v\|^2_{X_0^s(\Omega)} - \frac{\lambda}{2}\|v\|^2_{L^2(\Omega)} - \int_\Omega F(x,v(x))\,dx - \int_\Omega h(x)v(x)\,dx \\
&\geq \frac{1}{2}\|v\|^2_{X_0^s(\Omega)} - \frac{\lambda}{2}\|v\|^2_{L^2(\Omega)} - a_1\|v\|_{L^1(\Omega)} - \frac{a_2}{q}\|v\|^q_{L^q(\Omega)} - \|h\|_{L^2(\Omega)}\|v\|_{L^2(\Omega)} \\
&\geq \frac{1}{2}\|v\|^2_{X_0^s(\Omega)} - \kappa - a_1|\Omega|^{(q-1)/q}\|v\|_{L^q(\Omega)} - \frac{a_2}{q}\|v\|^q_{L^q(\Omega)} \\
&= \frac{1}{2}\|v\|^2_{X_0^s(\Omega)} - \kappa - a_1|\Omega|^{(q-1)/q} - \frac{a_2}{q}
\end{aligned}$$

when $v \in M$. Here κ is a positive constant given by

$$\kappa := \max\left\{(\lambda|\Omega|^{(q-2)/q})/2 + |\Omega|^{(q-2)/(2q)}\|h\|_{L^2(\Omega)}, |\Omega|^{(q-2)/(2q)}\|h\|_{L^2(\Omega)}\right\}.$$

Hence, if $\|v\|_{X_0^s(\Omega)}$ is sufficiently large, we get (11.10), and this ends the proof of (a).

Taking into account that part (a) holds true, assertions (b) and (c) can be proved as in [19, lemma 6.8]. □

Also, in what follows, we denote by a_k^0 and b_k^0 the quantities

$$a_k^0 := \inf_{A \in \mathcal{A}_k} \max_{v \in A} \mathcal{I}_{K,\lambda}(v) \tag{11.11}$$

and

$$b_k^0 := \inf_{B \in \mathcal{B}_k} \max_{v \in B} \mathcal{I}_{K,\lambda}(v), \tag{11.12}$$

with $\mathcal{A}_k$ and $\mathcal{B}_k$ given as earlier, for any $k \in \mathbb{N}$.

About the sequence a_k defined in (11.8), we have the following result:

Lemma 11.7 *Let $\lambda \in \mathbb{R}$, $s \in (0,1)$, $n > 2s$, and Ω be an open, bounded subset of $\mathbb{R}^n$ with continuous boundary. Let $K : \mathbb{R}^n \setminus \{0\} \to (0,+\infty)$ be a function satisfying conditions* (1.55) *and* (1.56)*, let $f : \overline{\Omega} \times \mathbb{R} \to \mathbb{R}$ verify* (6.3)–(6.5) *with*

$q \in (2, 2_s^* - 2s/(n-2s))$, $2_s^* := 2n/(n-2s)$, *and* (11.3) *and* (11.4), *and let* $h : \Omega \to \mathbb{R}$ *satisfy* (11.5). *Then there exist* $\tilde{c} > 0$ *and* $\tilde{\kappa} \in \mathbb{N}$ *such that, for any* $k \geq \tilde{\kappa}$,

$$a_k \geq \tilde{c}\, k^{2sq/(n(q-2))}.$$

In particular, $a_k \to +\infty$ *as* $k \to +\infty$.

Proof Here we can argue exactly as in [19, proposition 6.10]. To this end, we also can use the properties of the eigenvalues of $(-\Delta)^s$ given in Chapter 3. □

In what follows, we also need the following result, which is a sort of Gronwall lemma in a discrete framework; its proof can be found in [120, chapter 5, lemma 4.5].

Lemma 11.8 *Let* $p > 1$ *and* α_k *be a sequence such that* $\alpha_k > 0$, *for any* $k \in \mathbb{N}$. *Assume that there exist a positive constant* c_o *and* $\kappa_o \in \mathbb{N}$ *such that*

$$\alpha_{k+1} \leq \alpha_k + c_o(1 + \alpha_k^{1/(p+1)}) \quad \textit{for any } k \geq \kappa_o.$$

Then there exist a positive constant c^* *and* $\kappa^* \in \mathbb{N}$ *such that*

$$\alpha_k \leq c^* k^{(p+1)/p} \quad \textit{for any } k \geq \kappa^*. \tag{11.13}$$

Now we are ready to prove the properties stated earlier in properties (1) and (2); this will be done in the following sections.

11.3 Some compactness conditions

In this section, we prove that the functional $\mathcal{I}_{K,\lambda,h}$ satisfies property (1); that is, the Palais–Smale compactness condition on M. To this end, first, we need to show that the energy functional $\mathcal{J}_{K,\lambda,h}$ associated with problem (1.54) satisfies the Palais–Smale condition in $X_0^s(\Omega)$ at any level $c \in \mathbb{R}$ (see (6.21)). This will be accomplished in the next subsection.

11.3.1 The Palais–Smale condition for $\mathcal{J}_{K,\lambda,h}$

This subsection is devoted to proof of the Palais–Smale condition for the functional $\mathcal{J}_{K,\lambda,h}$. The proof can be performed as in Propositions 6.8, 6.9, and 6.15, taking also into account the presence of the extra term $\int_\Omega h(x)u(x)dx$ in the functional $\mathcal{J}_{K,\lambda,h}$.

For this purpose, we need the estimates on the nonlinearity f and its primitive F given in Lemma 6.3. Further, with the next result, we show that the function F (i.e., the primitive of f, see (18.6)) satisfies a superquadratic growth condition and, moreover, that F and f verify a sort of Ambrosetti–Rabinowitz condition.

Lemma 11.9 *Assume that* $f : \overline{\Omega} \times \mathbb{R} \to \mathbb{R}$ *is a function satisfying conditions* (6.3), (11.3), *and* (11.4). *Let* F *be as in* (6.8) *and* $\mu = (\beta + 1)/\beta > 2$. *Then* (*a*) *There exist two positive constant a_3 and a_4 such that, for any $x \in \overline{\Omega}$ and $t \in \mathbb{R}$,*

$$|F(x,t)| \geq a_3|t|^\mu - a_4, \quad \textit{and}$$

(*b*) *There exists a positive constant a_5 such that, for any $x \in \overline{\Omega}$ and $t \in \mathbb{R}$, $|t| \geq r$,*

$$0 < \mu F(x,t) \leq t f(x,t) + a_5.$$

Proof First, let us prove assertion (a). Let $r > 0$ be as in (11.4); then, for any $x \in \overline{\Omega}$ and $t \in \mathbb{R}$ with $|t| \geq r > 0$,

$$\frac{f_t(x,t)}{f(x,t)} \geq \frac{1}{\beta t}.$$

Here we also use the fact that $tf(x,t) > 0$, for $|t| \geq r$.

Now suppose that $t > r$. Integrating both terms in $[r,t]$, we obtain

$$|f(x,t)| = f(x,t) \geq \frac{f(x,r)}{r^{1/\beta}} t^\beta.$$

With the same arguments and again using the fact that $tf(x,t) > 0$ when $|t| \geq r$, it is easy to prove that if $x \in \overline{\Omega}$ and $t < -r$, then it holds that

$$|f(x,t)| \geq \frac{|f(x,-r)|}{r^{1/\beta}} |t|^{1/\beta},$$

so, for any $x \in \overline{\Omega}$ and $t \in \mathbb{R}$ with $|t| \geq r$, we get

$$|f(x,t)| \geq \tilde{m}(x)|t|^{1/\beta}, \tag{11.14}$$

where $\tilde{m}(x) := r^{-1/\beta} \min\{f(x,r), |f(x,-r)|\}$.

Note that $\tilde{m}$ is a continuous functions because $x \mapsto f(x,\cdot)$ is. Moreover, $\tilde{m}$ is positive, being $f(x,t) > 0$, for any $x \in \overline{\Omega}$ and $t \in \mathbb{R}$, such that $|t| \geq r$ (see (11.4)). Hence, for any $x \in \overline{\Omega}$,

$$\tilde{m}(x) \geq \min_{x \in \overline{\Omega}} \tilde{m}(x) =: \tilde{m} > 0.$$

This and (11.14) give

$$|f(x,t)| \geq \tilde{m}\,|t|^{1/\beta}, \tag{11.15}$$

for any $x \in \overline{\Omega}$ and $t \in \mathbb{R}$ such that $|t| \geq r$.

Now let $t > r$. Integrating both terms in (11.15) in $[r,t]$, we get

$$F(x,t) \geq a_3\,|t|^\mu + F(x,r) - a_3 r^\mu, \tag{11.16}$$

where a_3 is a positive constant and $\mu = (\beta+1)/\beta$. Note that $\mu > 2$ because $\beta \in (0,1)$ by assumption.

Arguing in the same way, it is easily seen that if $t < -r$, then

$$F(x,t) \geq a_3\,|t|^\mu + F(x,-r) - a_3 r^\mu,$$

so, by this and (11.16), we obtain that, for any $x \in \overline{\Omega}$ and $t \in \mathbb{R}$ with $|t| \geq r$,

$$F(x,t) \geq a_3 |t|^\mu + \hat{m} - a_3 r^\mu \geq a_3 |t|^\mu - a_3 r^\mu, \tag{11.17}$$

where $\hat{m}(x) := \min\{F(x,r),\, F(x,-r)\} > 0$ (note that $F(x,r) > 0$ because $tf(x,t) > 0$, for any $x \in \overline{\Omega}$ and $t \in \mathbb{R}$, with $|t| \geq r$).

Also, the function F is continuous for any $x \in \overline{\Omega}$ and $t \in \mathbb{R}$ such that $|t| \leq r$, so, by the Weierstrass theorem, it is bounded, say, by

$$|F(x,t)| \leq \widetilde{M} \quad \text{in } \overline{\Omega} \times \{|t| \leq r\}, \tag{11.18}$$

where $\widetilde{M}$ is a positive constant.

Taking $a_4 := \widetilde{M} + a_3 r^\mu > 0$ from (11.17) and (11.18), it follows that

$$F(x,t) \geq a_3 |t|^\mu - a_4, \tag{11.19}$$

for any $x \in \overline{\Omega}$ and $t \in \mathbb{R}$. This concludes the proof of part (a).

Now let us show assertion (b). For this, note that because $tf(x,t) > 0$ when $x \in \overline{\Omega}$ and $t \in \mathbb{R}$ with $|t| \geq r$, then $F(x,t) > 0$ in the same set.

Let $r > 0$ be as in (11.4); then, for any $x \in \overline{\Omega}$ and $t \in \mathbb{R}$ with $|t| \geq r > 0$,

$$f(x,t) \leq \beta t f_t(x,t).$$

Suppose that $t > r$. Integrating both terms in $[r,t]$, we obtain

$$0 < \mu F(x,t) \leq tf(x,t) + \mu F(x,r) - rf(x,r),$$

so, taking into account that $rf(x,r) > 0$, we have

$$0 < \mu F(x,t) \leq tf(x,t) + \mu F(x,r), \tag{11.20}$$

for any $x \in \overline{\Omega}$ and $t \in \mathbb{R}$, $|t| \geq r$.

With the same arguments, it is easy to prove that if $t < -r$, then it holds that

$$0 < \mu F(x,t) \leq tf(x,t) + \mu F(x,-r) + rf(x,-r),$$

so

$$0 < \mu F(x,t) \leq tf(x,t) + \mu F(x,-r)$$

because $rf(x,-r) < 0$.

Hence, from this and (11.20), for any $x \in \overline{\Omega}$ and $t \in \mathbb{R}$ with $|t| \geq r$, we get

$$0 < \mu F(x,t) \leq tf(x,t) + M(x), \tag{11.21}$$

where

$$M(x) := \mu \max\{F(x,r), F(x,-r)\} > 0,$$

being $F(x,t) > 0$. Taking $a_5 := \max_{x\in\overline{\Omega}} M(x)$, we get assertion (b). Note that a_5 exists by assumption (6.3) and the Weierstrass theorem. □

Assertion (b) in Lemma 11.9 is classical, and it is a sort of Ambrosetti–Rabinowitz condition that usually is assumed to prove that the energy functional satisfies the Palais–Smale compactness property. Of course, our condition (11.4) is stronger than assertion (b), but in a sense, they have the same nature.

Thanks to Lemma 11.9 and arguing as in Chapter 6, we get the following compactness result:

Proposition 11.10 Let $\lambda \in \mathbb{R}$, $s \in (0,1)$, $n > 2s$, and Ω be an open, bounded set of $\mathbb{R}^n$ with continuous boundary. Let $K : \mathbb{R}^n \setminus \{0\} \to (0,+\infty)$ be a function satisfying conditions (1.55) and (1.56), let $f : \overline{\Omega} \times \mathbb{R} \to \mathbb{R}$ verify (6.3)–(6.5) with $q \in (2, 2^*_s - 2s/(n-2s))$, $2^*_s := 2n/(n-2s)$, and (11.3) and (11.4), and let $h : \Omega \to \mathbb{R}$ satisfy (11.5). Then $\mathscr{I}_{k,\lambda,h}$ satisfies the Palais–Smale condition at any level $c \in \mathbb{R}$.

As a consequence of Proposition 11.10, we can prove that the functional $\mathscr{I}_{K,\lambda,h}$ satisfies the Palais–Smale condition on M for any value of the parameter $\lambda \in \mathbb{R}$. This will be done in the next subsection.

11.3.2 The Palais–Smale condition on M for $\mathscr{I}_{K,\lambda,h}$

In this subsection, we show that the functional $\mathscr{I}_{K,\lambda,h}$ satisfies the Palais–Smale condition on M; that is, we show that the following result holds true:

Proposition 11.11 Let $\lambda \in \mathbb{R}$, $s \in (0,1)$, $n > 2s$, and Ω be an open, bounded set of $\mathbb{R}^n$ with continuous boundary. Let $K : \mathbb{R}^n \setminus \{0\} \to (0,+\infty)$ be a function satisfying conditions (1.55) and (1.56), let $f : \overline{\Omega} \times \mathbb{R} \to \mathbb{R}$ verify (6.3)–(6.5) with $q \in (2, 2^*_s - 2s/(n-2s))$, $2^*_s := 2n/(n-2s)$, and (11.3) and (11.4), and let $h : \Omega \to \mathbb{R}$ satisfy (11.5). Finally, let $\bar{R} > 0$ be the constant given in Lemma 11.6. Let $c \in \mathbb{R}$, and let $\{(u_j, \rho_j)\}_{j\in\mathbb{N}}$ be a sequence in $M \times \mathbb{R}$ such that $\|u_j\|_{X^s_0(\Omega)} \geq \bar{R}$, for $j \in \mathbb{N}$ sufficiently large,

$$\mathscr{I}_{K,\lambda,h}(u_j) \to c \tag{11.22}$$

and

$$\sup\left\{|\langle \mathscr{I}'_{K,\lambda,h}(u_j) + \rho_j |u_j|^{q-2} u_j, \varphi\rangle| : \varphi \in X^s_0(\Omega), \|\varphi\|_{X^s_0(\Omega)} = 1\right\} \to 0 \tag{11.23}$$

as $j \to +\infty$. Then there exists $u_\infty \in X^s_0(\Omega)$ such that, up to a subsequence,

$$\|u_j - u_\infty\|_{X^s_0(\Omega)} \to 0$$

and

$$\rho_j \to 0$$

as $j \to +\infty$.

Proof First, we want to show that the sequence $\{u_j\}_{j\in\mathbb{N}}$ is bounded in $X_0^s(\Omega)$. By (11.22), there exists $\kappa > 0$ such that, for any $j \in \mathbb{N}$,

$$|\mathscr{I}_{K,\lambda,h}(u_j)| \le \kappa. \tag{11.24}$$

Also, by definition of $\mathscr{I}_{K,\lambda,h}$, it is easily seen that

$$\mathscr{J}_{K,\lambda,h}(u_j) \le \mathscr{I}_{K,\lambda,h}(u_j),$$

so, by this and (11.24),

$$\mathscr{J}_{K,\lambda,h}(u_j) \le \kappa,$$

for any $j \in \mathbb{N}$.

Hence,

$$\begin{aligned}
\|u_j\|^2_{X_0^s(\Omega)} &= 2\,\mathscr{J}_{K,\lambda,h}(u_j) + \lambda\,\|u_j\|^2_{L^2(\Omega)} + 2\int_\Omega F(x,u_j(x))\,dx + 2\int_\Omega h(x)u_j(x)\,dx \\
&\le 2\kappa + \lambda\,\|u_j\|^2_{L^2(\Omega)} + 2a_1\,\|u_j\|_{L^1(\Omega)} + \frac{2a_2}{q}\,\|u_j\|^q_{L^q(\Omega)} + 2\|h\|_{L^2(\Omega)}\,\|u_j\|_{L^2(\Omega)} \\
&\le 2\kappa + \kappa^* + 2a_1\,|\Omega|^{(q-1)/q}\,\|u_j\|_{L^q(\Omega)} + \frac{2a_2}{q}\,\|u_j\|^q_{L^q(\Omega)} \\
&= 2\kappa + \kappa^* + 2a_1\,|\Omega|^{(q-1)/q} + \frac{2a_2}{q},
\end{aligned}$$

thanks to condition (6.4), the Hölder inequality (here used with $q/2 > 1$ and its conjugate as well as with q and its conjugate), and because $u_j \in M$. Here

$$\kappa^* = \max\{(\lambda\,|\Omega|^{(q-2)/q}) + 2|\Omega|^{(q-2)/2q}\,\|h\|_{L^2(\Omega)}, 2|\Omega|^{(q-2)/2q}\,\|h\|_{L^2(\Omega)}\} > 0.$$

Hence,

$$\text{the sequence } \{u_j\}_{j\in\mathbb{N}} \text{ is bounded in } X_0^s(\Omega). \tag{11.25}$$

We would like to note that because $\|u_j\|_{X_0^s(\Omega)} \ge \bar{R}$, for j sufficiently large, say, $j \ge j^*$, with $j^* \in \mathbb{N}$, then $u_j \not\equiv 0$, for any $j \ge j^*$. Moreover, by Lemma 11.6,

$$\mathscr{I}_{K,\lambda,h}(u_j) > \alpha \quad \text{for any } j \ge j^*. \tag{11.26}$$

Hence, the functional $\mathscr{I}_{K,\lambda,h}$ is differentiable in u_j, for $j \ge j^*$, by Proposition 11.3.

Now we claim that

$$\langle \mathscr{I}'_{K,\lambda}(u_j), u_j\rangle = 0 \quad \text{for } j \ge j^*. \tag{11.27}$$

To this end, fix $j \ge j^*$, and let $g(t) := \mathscr{I}_{K,\lambda,h}(tu_j)$, for any $t > 0$. Then, by the definition of $\mathscr{I}_{K,\lambda,h}$, the function g turns out to be constant, so

$$0 = g'(t) = \langle \mathscr{I}'_{K,\lambda}(tu_j), u_j\rangle,$$

for any $t > 0$. Taking $t = 1$, we get (11.27).

By (11.23)–(11.27) and the fact that $u_j \in M$, it is easily seen that

$$
\begin{aligned}
|\rho_j| &= \rho_j \|u_j\|^q_{L^q(\Omega)} \\
&= |\langle \mathscr{I}'_{K,\lambda}(u_j) + \rho_j |u_j|^{q-2} u_j, u_j \rangle| \\
&\le \sup\left\{ |\langle \mathscr{I}'_{K,\lambda,h}(u_j) + \rho_j |u_j|^{q-2} u_j, \varphi \rangle| \,:\, \varphi \in X^s_0(\Omega), \|\varphi\|_{X^s_0(\Omega)} = 1 \right\} \|u_j\|_{X^s_0(\Omega)} \\
&\to 0
\end{aligned}
$$

as $j \to +\infty$. Thus, one of the assertion of Proposition 11.11 is proved.

Now it remains to show that, up to a subsequence, $\{u_j\}_{j\in\mathbb{N}}$ converges strongly to some u_∞ in $X^s_0(\Omega)$. To this end, for any $j \ge j^*$, let us denote by v_j the element

$$v_j := \sigma(u_j) u_j, \tag{11.28}$$

where $\sigma(u_j) > 0$ is the unique constant (see Proposition 11.3(b)) such that

$$\mathscr{I}_{K,\lambda,h}(u_j) = \mathscr{J}_{K,\lambda,h}(v_j). \tag{11.29}$$

Note that we can apply Proposition 11.3 because (11.26) holds true.

By (11.22) and (11.29), we get

$$\mathscr{J}_{K,\lambda,h}(v_j) \to c \quad \text{as } j \to +\infty. \tag{11.30}$$

Moreover, again by Proposition 11.3(b), (11.23), (11.25), and the fact that $\rho_j \to 0$, we have that, for any $\varphi \in X^s_0(\Omega)$,

$$\sigma(u_j) \langle \mathscr{J}'_{K,\lambda}(v_j), \varphi \rangle = \langle \mathscr{I}'_{K,\lambda}(u_j), \varphi \rangle \to 0 \tag{11.31}$$

as $j \to +\infty$.

We claim that

$$\text{there exists } \delta > 0 \text{ such that } \sigma(u_j) \ge \delta, \text{ for } j \text{ large enough.} \tag{11.32}$$

For this, we argue by contradiction, and we suppose that

$$\sigma(u_j) \to 0 \tag{11.33}$$

as $j \to +\infty$. Then, by this, (11.25), and (11.28), we deduce that $v_j \rightharpoonup 0$ weakly in $X^s_0(\Omega)$, so, by Lemma 1.30, up to a subsequence,

$$
\begin{aligned}
&v_j \to 0 \quad \text{in } L^\nu(\mathbb{R}^n) \\
&\text{and} \\
&v_j \to 0 \quad \text{a.e. in } \mathbb{R}^n
\end{aligned}
\tag{11.34}
$$

as $j \to +\infty$, and by [43, theorem IV.9], there exists $\ell_\nu \in L^\nu(\mathbb{R}^n)$ such that

$$|u_j(x)| \le \ell_\nu(x) \quad \text{a.e. in } \mathbb{R}^n, \text{ for any } j \in \mathbb{N}, \tag{11.35}$$

for any $\nu \in [1, 2^*_s)$.

By Lemma 6.3, (11.34), and the dominated convergence theorem, we have that

$$\int_\Omega F(x,v_j(x))\,dx \to \int_\Omega F(x,0)\,dx = 0 \tag{11.36}$$

as $j\to+\infty$ because $F(\cdot,0)=0$. Thus, by (11.30), (11.34), and (11.36), we get

$$\begin{aligned}
c &= \lim_{j\to+\infty} \mathscr{J}_{K,\lambda,h}(v_j)\\
&= \lim_{j\to+\infty}\left(\frac{1}{2}\int_{\mathbb{R}^n\times\mathbb{R}^n}|v_j(x)-v_j(y)|^2K(x-y)\,dx\,dy\right.\\
&\qquad\left.-\frac{\lambda}{2}\int_\Omega |v_j(x)|^2\,dx-\int_\Omega F(x,v_j(x))\,dx-\int_\Omega h(x)v_j(x)\,dx\right)\\
&=\frac{1}{2}\lim_{j\to+\infty}\|v_j\|^2_{X_0^s(\Omega)};
\end{aligned}$$

that is,

$$\|v_j\|^2_{X_0^s(\Omega)} \to 2c \geq 2\alpha > 0 \tag{11.37}$$

as $j\to+\infty$. For the last inequality, we use the fact that $c=\lim_{j\to+\infty}\mathscr{J}_{K,\lambda,h}(u_j)\geq\alpha$, thanks to (11.22) and (11.26).

By (11.28) and the fact that $\sigma(u_j)>0$ by construction, we deduce that

$$\|u_j\|_{X_0^s(\Omega)} = (\sigma(u_j))^{-1}\|v_j\|_{X_0^s(\Omega)},$$

so, by (11.33) and (11.37), we conclude that

$$\|u_j\|_{X_0^s(\Omega)} \to +\infty$$

as $j\to+\infty$. This contradicts (11.25), so assertion (11.32) is proved.

As a consequence of (11.31) and (11.32), we deduce that, for any $\varphi\in X_0^s(\Omega)$,

$$\langle \mathscr{J}'_{K,\lambda}(v_j),\varphi\rangle \to 0 \tag{11.38}$$

as $j\to+\infty$. Since (11.30) and (11.38) hold true, by Proposition 11.10, we get that the sequence $\{v_j\}_{j\in\mathbb{N}}$ is bounded in $X_0^s(\Omega)$. Hence, the sequence $\{v_j\}_{j\in\mathbb{N}}$ is bounded in $L^q(\Omega)$ by the embedding properties of $X_0^s(\Omega)$ into the classical Lebesgue spaces. As a consequence of this, of (11.28), and of the fact that $u_j\in M$, we get

$$\sigma(u_j)=\sigma(u_j)\|u_j\|_{L^q(\Omega)}=\|v_j\|_{L^q(\Omega)}<\kappa,$$

for any $j\geq j^*$, for a suitable positive constant κ, that is, the sequence

$$\{\sigma(u_j)\}_{j\in\mathbb{N}} \text{ is bounded in } [\delta,+\infty). \tag{11.39}$$

Now we can apply Proposition 11.10 to the sequence $\{v_j\}_{j\in\mathbb{N}}$ (remember that that the sequence $\{v_j\}_{j\in\mathbb{N}}$ is bounded in $X_0^s(\Omega)$, and again, (11.38) holds true). Then this and (11.39) yield that there exist $v_\infty \in X_0^s(\Omega)$ and $\sigma^* \in \mathbb{R}$ such that, up to subsequences (still denoted by v_j and $\sigma(u_j)$),

$$v_j \to v_\infty \quad \text{strongly in } X_0^s(\Omega) \tag{11.40}$$

and

$$\sigma(u_j) \to \sigma^* \quad \text{in } [\delta, +\infty) \tag{11.41}$$

as $j \to +\infty$.

Note that, by (11.32),

$$\sigma^* \geq \delta > 0. \tag{11.42}$$

Finally, as a consequence of (11.28) (remember that $\sigma(u_j) > 0$, for any $j \geq j^*$, by Proposition 11.3(b)) and (11.40)–(11.42), we have

$$u_j = (\sigma(u_j))^{-1} v_j \to (\sigma^*)^{-1} v_\infty =: u_\infty \quad \text{strongly in } X_0^s(\Omega)$$

as $j \to +\infty$. This concludes the proof. □

11.4 Existence of infinitely many solutions

This section is devoted to the proof of Theorem 11.2. The arguments we use here are variational ones. To perform our proof, we consider the preliminary results stated in Subsection 11.2.1, and we apply [120, chapter 5, theorem 2.2].

In the proof of Theorem 11.2, the assumptions on q will be crucial for our argument. In particular, we use the fact that

$$q < 2_s^* - \frac{2s}{n-2s}. \tag{11.43}$$

We would like to note that, until now, we never used this condition, and all the results proved in preceding sections are valid for any $q \in (2, 2_s^*)$. Assumption (11.43) is fundamental for the proof of the validity of property (2) (see Section 11.2).

If we consider the classical Laplace setting, the existence of infinitely many solutions for this problem was proved under some restriction on q, namely, for any $q > 2$ such that

$$q < \frac{2(n-1)}{n-2}.$$

For more details on this, we refer to [120, chapter 5, remark 4.7]. Note that this bound from above on q corresponds to (11.43) when $s = 1$ (which gives the classical Laplacian case).

11.4.1 Proof of Theorem 11.2

As we said in Section 11.2, because problem (11.2) has a variational nature, the proof of Theorem 11.2 reduces to finding critical points of the functional $\mathcal{J}_{K,\lambda,h}$. To this end, by Corollary 11.4, it is enough to look for critical points $u \in \{\mathcal{J}_{K,\lambda,h} > \alpha\}$ of the functional $\mathcal{I}_{K,\lambda,h}$ on M (here $\alpha > 0$ is the constant given in Proposition 11.3). For this, we will apply [120, chapter 5, theorem 2.2].

First, note that the functional $\mathcal{I}_{K,\lambda,h} \in C^1(\{\mathcal{I}_{K,\lambda,h} > \alpha\})$ by Proposition 11.3(a), and it verifies the Palais–Smale condition on M, thanks to Proposition 11.11. Hence, by [120, chapter 5, theorem 2.2], we get that, for any $k \in \mathbb{N}$,

$$a_k \leq b_k,$$

where a_k and b_k are as in (11.8) and (11.9), respectively.

Now, in order to achieve our goal, we need to prove that $a_k \in \mathbb{R}$ and that $a_k < b_k$, for $k \in \mathbb{N}$ sufficiently large. To this end, note that, by Lemma 11.7 and the choice of $q > 2$, we have that $a_k \to +\infty$ as $k \to +\infty$, so $a_k > \alpha$, for $k \geq \tilde{\kappa}$, with $\tilde{\kappa} \in \mathbb{N}$. This means that $a_k \in \mathbb{R}$, for any $k \geq \tilde{\kappa}$.

Now it remains to show that $a_k < b_k$, for any $k \in \mathbb{N}$ large enough. To this end, note that, for any $k \in \mathbb{N}$ and any $A \in \mathcal{A}_k$, there exists $u_A \in A \cap \mathbb{P}_k$ (for this, see [120, chapter 2, theorem 4.4 and the proof of chapter 5, lemma 3.3]), where

$$\mathbb{P}_k := \{u \in X_0^s(\Omega) \colon \langle u, e_j \rangle_{X_0^s(\Omega)} = 0 \quad \forall j = 1, \dots, k-1\}.$$

Let us fix $k \in \mathbb{N}$. By the variational characterization of λ_k (see Proposition 3.1), we have that

$$\|u_A\|_{X_0^s(\Omega)} \geq \sqrt{\lambda_k}\, \|u_A\|_{L^2(\Omega)}. \tag{11.44}$$

Also, by the Hölder inequality applied in $L^2(\Omega)$ (for this, we need that $2 < q < 2_s^* - 2s/(n-2s)$, which is our assumption), and the embedding properties of the space $X_0^s(\Omega)$, we get

$$\begin{aligned}
\|u_A\|_{L^q(\Omega)}^q &= \int_\Omega |u_A(x)|^{q-1} |u_A(x)|\, dx \\
&\leq \|u_A\|_{L^2(\Omega)} \|u_A\|_{L^{2(q-1)}(\Omega)}^{q-1} \\
&\leq C \|u_A\|_{L^2(\Omega)} \|u_A\|_{X_0^s(\Omega)}^{q-1},
\end{aligned}$$

for a suitable $C > 0$. As a consequence of this and taking into account that $u_A \in A \subseteq M$, we get

$$\|u_A\|_{L^2(\Omega)} \geq C^{-1} \|u_A\|_{L^q(\Omega)}^q \|u_A\|_{X_0^s(\Omega)}^{1-q} = C^{-1} \|u_A\|_{X_0^s(\Omega)}^{1-q},$$

so this and (11.44) yield

$$\|u_A\|_{X_0^s(\Omega)} \geq C^{-1} \sqrt{\lambda_k}\, \|u_A\|_{X_0^s(\Omega)}^{1-q};$$

that is,

$$\|u_A\|_{X_0^s(\Omega)} \geq C^{-1/q} \lambda_k^{1/(2q)}.$$

Because $\lambda_k \to +\infty$ as $k \to +\infty$ (see Proposition 3.1), for k large enough, we get that

$$\|u_A\|_{X_0^s(\Omega)} \geq \bar{R},$$

where $\bar{R}$ is the positive constant given in Lemma 11.6. This proves that, for $k \in \mathbb{N}$ sufficiently large, say, $k \geq \breve{\kappa}$, with $\breve{\kappa} \in \mathbb{N}$, and for any $A \in \mathscr{A}_k$,

$$\text{the set } A \cap \{\|v\|_{X_0^s(\Omega)} \geq \bar{R}\} \text{ is not empty.}$$

With this assertion and using the fact that $\mathscr{I}_{K,\lambda,h}(v) = \mathscr{I}_{K,\lambda,h}(tv)$, for any $t > 0$, it is easy to see that, for $k \geq \breve{\kappa}$,

$$a_k = \inf_{A \in \mathscr{A}_k} \max_{\substack{v \in A \\ \|v\|_{X_0^s(\Omega)} \geq \bar{R}}} \mathscr{I}_{K,\lambda,h}(v) \tag{11.45}$$

and

$$b_k = \inf_{B \in \mathscr{B}_k} \max_{\substack{v \in B \\ \|v\|_{X_0^s(\Omega)} \geq \bar{R}}} \mathscr{I}_{K,\lambda,h}(v). \tag{11.46}$$

Also, the same holds true for a_k^0 and b_k^0 (see formulas (11.11) and (11.12)).

Finally, we can show that

$$a_k < b_k \quad \text{for any } k \in \mathbb{N} \text{ sufficiently large.} \tag{11.47}$$

For this, we argue by contradiction, and we suppose that there exists $\bar{\kappa} \in \mathbb{N}$ such that $a_k = b_k$, for any $k \geq \bar{\kappa}$ (note that we know that $a_k \leq b_k$, for any $k \in \mathbb{N}$). Let $\hat{\kappa} = \max\{\bar{\kappa}, \tilde{\kappa}, \breve{\kappa}\}$.

By (11.45) and (11.46) and the analogous formulas for a_k^0 and b_k^0 and Lemma 11.6, we get that for any $k \in \mathbb{N}$ with $k \geq \breve{\kappa}$

$$a_{k+1}^0 = b_k^0 \leq b_k + \bar{c}(1 + b_k^{1/q})$$

and

$$a_{k+1} \leq a_{k+1}^0 + \bar{c}(1 + (a_{k+1}^0)^{1/q}),$$

so

$$a_{k+1} \leq b_k + \bar{c}(1 + b_k^{1/q}) + \bar{c}[1 + (b_k + \bar{c}(1 + b_k^{1/q}))^{1/q}].$$

As a consequence of this and using the assumption that $a_k = b_k$, for any $k \geq \hat{\kappa}$, we deduce that

$$a_{k+1} \leq a_k + \bar{c}(1 + a_k^{1/q}) + \bar{c}[1 + (a_k + \bar{c}(1 + a_k^{1/q}))^{1/q}],$$

which gives

$$a_{k+1} \leq a_k + \hat{c}(1 + a_k^{1/q}), \tag{11.48}$$

for a suitable positive constant $\hat{c}$ and for any $k \geq \hat{\kappa}$. Here we also use the fact that a_k is large enough (since $a_k \to +\infty$ as $k \to +\infty$).

By (11.48) and Lemma 11.8 (applied here with $p = q - 1 > 1$), there exist $c^* > 0$ and $\kappa^* \in \mathbb{N}$ such that

$$a_k \le c^* k^{q/(q-1)} \quad \text{for } k \ge \kappa^*. \tag{11.49}$$

Moreover, combining (11.49) and Lemma 11.7, we deduce that, for k sufficiently large,

$$\tilde{c}\, k^{2sq/n(q-2)} \le a_k \le c^* k^{q/(q-1)},$$

which implies that

$$\frac{2sq}{n(q-2)} \le \frac{q}{q-1};$$

that is,

$$q \ge 1 + n/(n-2s) = 2_s^* - 2s/(n-2s).$$

Clearly, this contradicts the assumption on q (see (11.43)). Thus, (11.47) is proved.

Hence, by [120, chapter 5, theorem 2.2], the functional $\mathscr{I}_{K,\lambda,h}$ admits a sequence of critical points $\{v_k\}_{k\in\mathbb{N}}$ on M with critical value $c_k = \mathscr{I}_{K,\lambda,h}(v_k) \ge b_k$. Because

$$\mathscr{I}_{K,\lambda,h}(v_k) \ge b_k > a_k > \alpha,$$

for k large enough (as $a_k \to +\infty$ as $k \to +\infty$), then by Corollary 11.4(a), the sequence $\{u_k\}_{k\in\mathbb{N}}$, defined as

$$u_k := \sigma(v_k) v_k,$$

where $\{\sigma(v_k)\}_{k\in\mathbb{N}}$ is given as in Proposition 11.3(b), is a sequence of critical points of $\mathscr{J}_{K,\lambda,h}$ on $X_0^s(\Omega)$. In addition, again by Proposition 11.3(b),

$$\mathscr{J}_{K,\lambda,h}(u_k) = \mathscr{J}_{K,\lambda,h}(\sigma(v_k)v_k) = \mathscr{I}_{K,\lambda,h}(v_k) \ge b_k > a_k \to +\infty \tag{11.50}$$

as $k \to +\infty$.

Finally, we have to show that $\|u_k\|_{X_0^s(\Omega)} \to +\infty$ as $k \to +\infty$. For this, we can argue again by contradiction. Indeed, if the sequence $\{u_k\}_{k\in\mathbb{N}}$ were bounded in $X_0^s(\Omega)$, then, by the embeddings properties of $X_0^s(\Omega)$ into the usual Lebesgue spaces and assumption (6.4), we would get that the sequence $\{\mathscr{J}_{K,\lambda,h}(u_k)\}_{k\in\mathbb{N}}$ is bounded in $\mathbb{R}$, but this contradicts (11.50). Thus,

$$\|u_k\|_{X_0^s(\Omega)} \to +\infty$$

as $k \to +\infty$, and this concludes the proof of Theorem 11.2.

For completeness, we also point out that in [36, 153], requiring different superlinear growth assumptions on the nonlinearity term f and exploiting the classical fountain theorem, three different results about the existence of infinitely many weak solutions for nonlocal equations have been obtained. All these theorems extend some classical results for semilinear Laplacian equations to the nonlocal fractional setting. See also the recent paper [145] in which the existence of infinitely many weak solutions for fractional problems has been proved using a $\mathbb{Z}_2$-symmetric version of the mountain pass theorem.

12
Fractional Kirchhoff-type problems

In this chapter, we are interested in the existence of weak solutions for the following fractional Kirchhoff-type problem:

$$\begin{cases} -M(\|u\|^2_{X^s_0(\Omega)})\mathscr{L}_K u = \upsilon(\lambda,\mu,x,u) & \text{in } \Omega \\ u = 0 & \text{in } \mathbb{R}^n \setminus \Omega, \end{cases} \tag{12.1}$$

where $\Omega \subset \mathbb{R}^n$ is a bounded domain with continuous boundary, $n > 2s$ with $s \in (0,1)$, $M : [0,+\infty) \to [0,+\infty)$ is the Kirchhoff term, and $\upsilon : \mathbb{R} \times \mathbb{R} \times \Omega \times \mathbb{R} \to \mathbb{R}$ is a suitable function. We consider both the degenerate and the nondegenerate settings.

The analogous and classical counterpart of our problem models several interesting phenomena studied in mathematical physics, even in the one-dimensional case. Its origins, as is well known, date back to 1883, when Kirchhoff proposed his celebrated equation

$$\rho\partial^2_{tt}u - \left(\frac{P_0}{h} + \frac{E}{2L}\int_0^L |\partial_x u(x)|^2\,dx\right)\partial^2_{xx}u = 0 \tag{12.2}$$

as a nonlinear extension of D'Alembert's wave equation for free vibrations of elastic strings. Here $u = u(x,t)$ is the transverse string displacement at the space coordinate x and time t, L is the length of the string, h is the area of the cross section, E is Young's modulus of the material, ρ is the mass density, and P_0 is the initial tension (see the classical manuscript [122]).

In a very recent paper [105], the authors have proposed an interesting and fascinating physical interpretation of Kirchhoff's equation in the fractional scenario. In their correction of the early (one-dimensional) model, the tension on the string, which classically has a "nonlocal" nature arising from the average of the kinetic energy $|\partial_x u|^2/2$ on $[0,L]$, possesses a further nonlocal behavior provided by the H^s-norm (or other more general fractional norms) of the function u. See also the paper [16], where a fractional Kirchhoff equation involving a critical term is investigated using suitable variational methods.

From a purely mathematical point of view, it is worth mentioning that the early classical studies dedicated to equation (12.2) were given by Bernstein [34] and Pohozaev [166]. An important motivation for this study was provided by Lions' work [134], where a functional analysis approach was proposed to attack it. The results presented in this chapter are based on the papers [98, 144, 150, 152, 158, 159].

12.1 Nondegenerate Kirchhoff equations

The typical Kirchhoff function is as follows:

$$M(t) := m_0 + m_1 t \quad \text{where } m_0 \geq 0 \text{ and } m_1 > 0. \tag{12.3}$$

In the presence of the term M given in (12.3), problem (12.1) is said to be nondegenerate when $m_0 > 0$, while it is degenerate if $m_0 = 0$.

In this section, we focus on nondegenerate Kirchhoff equations, giving both existence and multiplicity results for some special classes of problems.

12.1.1 Mountain pass solution

First of all, we consider the following problem:

$$\begin{cases} M\left(\displaystyle\int_{\mathbb{R}^n\times\mathbb{R}^n} \frac{|u(x)-u(y)|^2}{|x-y|^{n+2s}}\,dxdy\right)(-\Delta)^s u = f(x,u) & \text{in } \Omega \\ u = 0 & \text{in } \mathbb{R}^n \setminus \Omega, \end{cases} \tag{12.4}$$

where $s \in (0,1)$, Ω is an open, bounded subset of $\mathbb{R}^n$, $n > 2s$, with continuous boundary, and $(-\Delta)^s$ is the fractional Laplace operator given in (1.20).

On the Kirchhoff term $M : [0,+\infty) \to [0,+\infty)$, we assume the following conditions:

$$M \textit{ is continuous in } [0,+\infty); \tag{12.5}$$

$$\begin{aligned} &\textit{there exists a constant } m_0 \textit{ such that} \\ &0 < m_0 \leq M(t), \textit{ for any } t \in [0,+\infty); \textit{ and} \end{aligned} \tag{12.6}$$

$$\textit{there exists } t_0 \geq 0 \textit{ such that } \widehat{M}(t) \geq t M(t), \textit{ for any } t \in [t_0,+\infty), \tag{12.7}$$

where

$$\widehat{M}(t) := \int_0^t M(\tau)\,d\tau \quad t \in [0,+\infty). \tag{12.8}$$

Note that conditions (12.6) and (12.7) ensure that the potential $\widehat{M}$ has a sublinear growth (see [152] for the details).

Moreover, the function $f : \overline{\Omega} \times \mathbb{R} \to \mathbb{R}$ satisfies (6.3), (6.4), (6.7), and the following condition:

$$\begin{aligned} &\textit{there exists } \gamma < m_0\lambda_{1,s} \textit{ such that} \\ &\limsup_{t\to 0} \frac{f(x,t)}{t} \leq \gamma \textit{ uniformly in } x \in \overline{\Omega}, \end{aligned} \tag{12.9}$$

where $\lambda_{1,s}$ is the first eigenvalue of $(-\Delta)^s$ with homogeneous Dirichlet boundary data (see Proposition 3.1).

As a model for M, we can take the function

$$M(t) := 2 + \frac{\sin t}{1+t^2} \quad t \in [0,+\infty),$$

while a prototype for f is given by

$$f(t) := t^3 + t^4 \quad t \in \mathbb{R}.$$

Under these assumptions, the existence of at least one nontrivial weak solution for (12.4) can be proved. More precisely, the following result holds:

Theorem 12.1 *Let $s \in (0,1)$, $n > 2s$, and Ω be an open, bounded subset of $\mathbb{R}^n$ with continuous boundary. Let $M : [0,+\infty) \to [0,+\infty)$ be a function satisfying* (12.5)–(12.7), *and let $f : \overline{\Omega} \times \mathbb{R} \to \mathbb{R}$, verifying* (6.3), (6.4), (6.7), *and* (12.9). *Then problem* (12.4) *has at least one nontrivial weak solution in $\mathbb{H}^s_0(\Omega)$.*

A detailed proof of Theorem 12.1, which is based on the mountain pass theorem, can be found in the paper [152].

Another class of Kirchhoff-type problems we are interested in is given by the following problem:

$$\begin{cases} -(m_0 + m_1 \|u\|^2_{X^s_0(\Omega)})\mathscr{L}_K u = f(x,u) & \text{in } \Omega \\ u = 0 & \text{in } \mathbb{R}^n \setminus \Omega. \end{cases} \tag{12.10}$$

Here $f : \Omega \times \mathbb{R} \to \mathbb{R}$ satisfies (6.3), (6.6), and the following assumptions:

$$\begin{gathered} \textit{there exist } a_1, a_2 > 0 \textit{ and } q \in (4, 2^*_s), 2^*_s := 2n/(n-2s), \textit{ such that} \\ |f(x,t)| \le a_1 + a_2|t|^{q-1}, \textit{ for any } x \in \Omega, t \in \mathbb{R}; \end{gathered} \tag{12.11}$$

there exists $\sigma \ge 1$ such that

$$tf(x,t) - 4F(x,t) \ge \frac{1}{\sigma}\big(\zeta t f(x,\zeta t) - 4F(x,\zeta t)\big), \tag{12.12}$$

for any $(x,t) \in \Omega \times \mathbb{R}$ and any $\zeta \in [0,1]$;

$$\begin{gathered} \textit{there exists } \delta > 0 \textit{ such that} \\ F(x,t) \le m_0 \frac{\lambda_1}{2} t^2, \textit{ for any } x \in \Omega \textit{ and } t \in (-\delta,\delta); \textit{ and} \end{gathered} \tag{12.13}$$

$$\limsup_{|t|\to\infty} \frac{f(x,t)}{t^3} = +\infty \textit{ uniformly in } x \in \Omega. \tag{12.14}$$

Here λ_1 is the first eigenvalue of $-\mathscr{L}_K$ with homogeneous Dirichlet boundary data (see Proposition 3.1). A model for f is given by

$$f(t) := t^3 \log(1+|t|) \quad t \in \mathbb{R}.$$

In this setting, we have the following existence result:

Theorem 12.2 *Let $s \in (0,1)$, $2s < n < 4s$, and Ω be an open, bounded subset of $\mathbb{R}^n$ with continuous boundary. Let $K : \mathbb{R}^n \setminus \{0\} \to (0,+\infty)$ be a function satisfying* (1.55) *and* (1.56), *and let $f : \Omega \times \mathbb{R} \to \mathbb{R}$, verifying* (6.3), (6.6), *and* (12.11)–(12.14). *Then problem* (12.10) *has at least one nontrivial weak solution in $X_0^s(\Omega)$.*

The abstract tool used for proving Theorem 12.2 is a suitable form of the mountain pass theorem obtained in [190]. As usual, the Ambrosetti–Rabinowitz condition (12.12) plays a crucial role in proving that every Palais–Smale sequence is bounded, as well as in showing that the so-called mountain pass geometry is satisfied. However, even dealing with different problems than the ones treated here, several authors studied more general or different assumptions that still allow us to apply minmax methods to get the existence of critical points. The proof of Theorém 12.2 moves along this direction; indeed, a monotonicity trick that allows us to prove the compactness Cerami condition is exploited. For more details, we refer to the paper [158].

12.1.2 Multiple solutions

In this section, we discuss the existence of multiple weak solutions for Kirchhoff-type problems.

12.1.3 Kirchhoff equations with bounded primitive

Here we consider the following nonlocal problem:

$$\begin{cases} -M(\|u\|^2_{X_0^s(\Omega)})\mathscr{L}_K u = \mu h\left(\int_\Omega \left(\int_0^{u(x)} f(x,t)dt\right)dx - \lambda\right) f(x,u) & \text{in } \Omega \\ u = 0 & \text{in } \mathbb{R}^n \setminus \Omega, \end{cases} \tag{12.15}$$

where $s \in (0,1)$, Ω is an open, bounded subset of $\mathbb{R}^n$, $n > 2s$, with continuous boundary, and $\mathscr{L}_K$ is the integrodifferential operator given in (1.54), while M and h are two suitable continuous functions, λ and μ are real parameters, and $f : \Omega \times \mathbb{R} \to \mathbb{R}$ is a Carathéodory function with subcritical growth. More precisely, we assume that the Kirchhoff term $M : [0,+\infty) \to \mathbb{R}$ verifies (12.5) and the following conditions:

$$\inf_{t \geq 0} M(t) > 0, \textit{ and} \tag{12.16}$$

$$\begin{gathered} \textit{there exists a continuous function } v_M : [0,+\infty) \to \mathbb{R} \textit{ such that} \\ v_M(tM(t^2)) = t, \textit{ for any } t \in [0,+\infty). \end{gathered} \tag{12.17}$$

Moreover, the nonlinearity $f : \Omega \times \mathbb{R} \to \mathbb{R}$ is a Carathéodory function satisfying (10.7) and the following assumptions:

$$\sup_{(x,\xi) \in \Omega \times \mathbb{R}} |F(x,\xi)| < +\infty, \textit{ and} \tag{12.18}$$

$$\sup_{u\in X_0^s(\Omega)} \Big|\int_\Omega F(x,u(x))dx\Big| > 0. \tag{12.19}$$

We recall that a *weak solution* of problem (12.15) is a function $u \in X_0^s(\Omega)$ such that

$$M(\|u\|^2_{X_0^s(\Omega)})\int_{\mathbb{R}^n\times\mathbb{R}^n}(u(x)-u(y))(\varphi(x)-\varphi(y))K(x-y)dxdy$$
$$= \mu h\left(\int_\Omega F(x,u(x))dx-\lambda\right)\int_\Omega f(x,u(x))\varphi(x)dx,$$

for any $\varphi \in X_0^s(\Omega)$. This is the Euler–Lagrange equation of the functional $\mathcal{J}_{K,M}$: $X_0^s(\Omega) \to \mathbb{R}$ defined as follows:

$$\mathcal{J}_{K,M}(u) := \frac{1}{2}\widehat{M}(\|u\|^2_{X_0^s(\Omega)}) - \mu H\left(\int_\Omega F(x,u(x))dx-\lambda\right),$$

where, as usual,

$$H(t) := \int_0^t h(\tau)d\tau \quad t\in\mathbb{R}.$$

Thus, critical points of $\mathcal{J}_{K,M}$ are solutions of problem (12.15).

The abstract tool for proving the existence of these critical points (namely, the main multiplicity result for problem (12.15)) is [181, theorem 1.6]: we recall it here for the reader's convenience.

Theorem 12.3 ([181]) *Let $(E,\|\cdot\|)$ be a separable and reflexive real Banach space, and let $\eta, J : E\to\mathbb{R}$ be two C^1-functionals with compact derivative and $J(0) = \eta(0) = 0$. Assume also that J is bounded and nonconstant and that η is bounded above. Then, for any sequentially weakly lower semicontinuous and coercive C^1-functional $\Phi : E\to\mathbb{R}$ whose derivative admits a continuous inverse on E' and with $\Phi(0) = 0$, for every convex C^1-function*

$$\varphi : \left(-\operatorname*{osc}_{u\in E} J(u), \operatorname*{osc}_{u\in E} J(u)\right) \to [0,+\infty),$$

with $\varphi^{-1}(0) = \{0\}$, for which the number

$$\theta^\star := \inf_{u\in J^{-1}\left(\left(\inf_{u\in E} J(u),\sup_{u\in E} J(u)\right)\setminus\{0\}\right)} \frac{\Phi(u)-\eta(u)}{\varphi(J(u))}$$

is nonnegative, and for every $\mu > \theta^\star$, there exists an open interval

$$\Lambda \subseteq \left(\inf_{u\in E} J(u), \sup_{u\in E} J(u)\right)$$

and a number $\rho > 0$ such that, for each $\lambda\in\Lambda$,

$$\Phi'(u) = \mu\varphi'(J(u)-\lambda)J'(u)+\eta'(u)$$

has at least three distinct solutions whose norms are less than ρ.

For a generic function $\Phi : E \to \mathbb{R}$, the symbol $\operatorname*{osc}_{u\in E} \Phi(u)$ denotes the oscillation, possibly infinite, given by

$$\operatorname*{osc}_{u\in E} \Phi(u) := \sup_{u\in E} \Phi(u) - \inf_{u\in E} \Phi(u).$$

We emphasize that in [181, theorem 1.6] the author established a theorem tailor-made for a class of nonlocal problems involving nonlinearities with bounded primitive. This result follows from [183, theorem 3], and the main novelty obtained in the most recent paper [181] is that, in contrast with a large part of the existing literature, the abstract energy functional does not depend on the parameter λ in an affine way. See also the related paper [151].

In what follows, let

$$\alpha_f := \inf_{u\in X_0^s(\Omega)} \int_\Omega F(x,u(x))dx, \quad \beta_f := \sup_{u\in X_0^s(\Omega)} \int_\Omega F(x,u(x))dx,$$

$$\omega_f := \beta_f - \alpha_f,$$

and

$$\mathscr{R}_f := \left\{ u \in X_0^s(\Omega) : \int_\Omega F(x,u(x))dx \in (\alpha_f, \beta_f) \setminus \{0\} \right\}.$$

With this notations, the multiplicity result for (12.15) reads as follows:

Theorem 12.4 *Let $s \in (0,1)$, $n > 2s$, and let Ω be an open, bounded subset of $\mathbb{R}^n$ with continuous boundary. Let $K : \mathbb{R}^n \setminus \{0\} \to (0,+\infty)$ be a function satisfying* (1.55) *and* (1.56), *and let $f : \Omega \times \mathbb{R} \to \mathbb{R}$ satisfy* (10.7), (12.18), *and* (12.19). *Then, for any $M : [0,+\infty) \to \mathbb{R}$ satisfying* (12.5), (12.16), *and* (12.17), *any continuous and nondecreasing function $h : (-\omega_f, \omega_f) \to \mathbb{R}$ with $h^{-1}(0) = \{0\}$, and any*

$$\mu > \inf_{u\in\mathscr{R}_f} \left\{ \frac{\widehat{M}(\|u\|^2_{X_0^s(\Omega)})}{2H\left(\int_\Omega F(x,u(x))dx\right)} \right\},$$

there exists an open interval $\Lambda \subseteq (\alpha_f, \beta_f)$ and a number $\rho > 0$ such that, for each $\lambda \in \Lambda$, problem (12.15) *has at least three distinct weak solutions whose norms in $X_0^s(\Omega)$ are less than ρ.*

Proof Let us apply Theorem 12.3 by choosing $E := X_0^s(\Omega)$, $\eta = 0$,

$$J(u) := \int_\Omega F(x,u(x))dx, \quad \Phi(u) := \frac{1}{2}\widehat{M}(\|u\|^2_{X_0^s(\Omega)}) \quad u \in E,$$

and

$$\varphi(t) := H(t) \quad t \in \mathbb{R}.$$

Due to (10.7), J is a C^1-functional with compact derivative (see Lemma 1.30) and by (12.18), J is clearly bounded in E. Moreover, it is easy to see that Φ is a

C^1-functional and, because $\widehat{M}$ is increasing, Φ is also sequentially weakly lower semicontinuous in E.

Let us prove that the derivative $\Phi' : E \to E'$ admits a continuous inverse. Since E is reflexive, we identify E with its topological dual E'. For our goal, let $T : E \to E$ be the operator defined by

$$T(v) := \begin{cases} \dfrac{v_M(\|v\|_{X_0^s(\Omega)})}{\|v\|_{X_0^s(\Omega)}} v & \text{if } v \neq 0 \\ 0 & \text{if } v = 0, \end{cases}$$

where v_M is the function in (12.16).

Thanks to the continuity of v_M and the fact that $v_M(0) = 0$, the operator T is continuous in E. Moreover, for any $u \in E \setminus \{0\}$, since $M(\|u\|^2_{X_0^s(\Omega)}) > 0$ (thanks to (12.16)), we have

$$\begin{aligned} T(\Phi'(u)) &= T(M(\|u\|^2_{X_0^s(\Omega)})u) \\ &= \frac{v_M(M(\|u\|^2_{X_0^s(\Omega)})\|u\|_{X_0^s(\Omega)})}{M(\|u\|^2_{X_0^s(\Omega)})\|u\|_{X_0^s(\Omega)}} M(\|u\|^2_{X_0^s(\Omega)})u \\ &= u, \end{aligned}$$

as desired. Now put

$$\gamma := \inf_{t \geq 0} M(t).$$

By (12.16), it is easy to see that $\gamma > 0$, and

$$\widehat{M}(t) \geq \gamma t, \quad \text{for any } t \in [0, +\infty).$$

In particular, this implies that Φ is coercive in E.

Finally, due to the hypotheses on h, it follows that the function φ is nonnegative, convex and $\varphi^{-1}(0) = \{0\}$. Then all the assumptions of Theorem 12.3 are verified, so the existence of three distinct weak solutions for problem (12.15) is established. □

In the model case of the fractional Laplacian, Theorem 12.4 reads as follows:

Theorem 12.5 *Let $s \in (0,1)$, $n > 2s$, and let Ω be an open, bounded subset of $\mathbb{R}^n$ with continuous boundary. Moreover, let $f : \mathbb{R} \to \mathbb{R}$ be a nontrivial continuous function such that*

$$\sup_{\xi \in \mathbb{R}} |F(\xi)| < +\infty,$$

and let $h : \left(-|\Omega| \operatorname{osc}_{\xi \in \mathbb{R}} F(\xi), |\Omega| \operatorname{osc}_{\xi \in \mathbb{R}} F(\xi) \right) \to \mathbb{R}$ be a continuous and nondecreasing function such that $h^{-1}(0) = \{0\}$. Then, for $a, b > 0$ fixed, for any μ sufficient large, there exist an open interval

$$\Lambda \subseteq \left(|\Omega| \inf_{\xi \in \mathbb{R}} F(\xi), |\Omega| \sup_{\xi \in \mathbb{R}} F(\xi) \right)$$

and a number $\rho > 0$ such that, for every $\lambda \in \Lambda$,

$$\begin{cases} \left(a+b\|u\|^2_{X^s_0(\Omega)}\right)(-\Delta)^s u = \mu h\left(\int_\Omega F(x,u(x))dx - \lambda\right) f(x,u) & \text{in } \Omega \\ u = 0 & \text{in } \mathbb{R}^n \setminus \Omega \end{cases}$$

has at least three distinct weak solutions $\{u_j\}_{j=1}^3$ in $\mathbb{H}^s_0(\Omega)$ and $\|u_j\|_{\mathbb{H}^s_0(\Omega)} < \rho$, for $j = 1,2,3$.

12.2 Degenerate Kirchhoff equations

In this section, we treat degenerate Kirchhoff problems; that is, we allow the Kirchhoff function to take the zero value. This case is quite delicate and not very covered in literature, even more in the fractional setting, as explicitly pointed out in [105]. We also mention, for the sake of completeness, that degenerate Kirchhoff-type problems driven by nonhomogeneous elliptic operators have been recently taken into account, for instance, in [72, 167].

12.2.1 Some multiplicity results

Here we are interested in the existence of weak solutions for the following fractional Kirchhoff-type problem:

$$\begin{cases} -M(\|u\|^2_{X^s_0(\Omega)})\mathscr{L}_K u = \lambda f(x,u) + \mu g(x,u) & \text{in } \Omega \\ u = 0 & \text{in } \mathbb{R}^n \setminus \Omega, \end{cases} \tag{12.20}$$

where $s \in (0,1)$, $2s < n < 4s$, $\Omega \subset \mathbb{R}^n$ is a bounded domain with continuous boundary, $f,g : \Omega \times \mathbb{R} \to \mathbb{R}$ are two suitable Carathéodory functions with subcritical growth, and λ and μ are two positive real parameters.

In our setting, the Kirchhoff function $M : [0,+\infty) \to [0,+\infty)$ is assumed to satisfy (12.5) and the following structural assumptions:

$$M \textit{ is nondecreasing in } [0,+\infty); \tag{12.21}$$

$$\textit{there exists } m_1 > 0 \textit{ such that } M(t) \geq m_1 t, \textit{ for any } t \in [0,+\infty); \textit{ and} \tag{12.22}$$

$$\textit{there exists } \sigma > 1 \textit{ such that}$$
$$m_\sigma := \liminf_{t\to+\infty} \frac{\widehat{M}(t)}{t^\sigma} > 0, \tag{12.23}$$

where $\widehat{M}$ is defined in (12.8). Of course, a model for M is given by

$$M(t) := m_1 t \quad m_1 > 0. \tag{12.24}$$

Furthermore, on the nonlinearity $f : \Omega \times \mathbb{R} \to \mathbb{R}$, we assume that it is a Caratéodory function verifying (10.7) and that

$$\sup_{u\in X^s_0(\Omega)} \int_\Omega F(x,u(x))dx > 0, \tag{12.25}$$

$$\limsup_{t\to 0} \frac{\sup_{x\in\Omega} F(x,t)}{|t|^4} \leq 0, \tag{12.26}$$

and

$$\limsup_{|t|\to\infty} \frac{\sup_{x\in\Omega} F(x,t)}{|t|^{2\sigma}} \leq 0, \tag{12.27}$$

with F defined in (6.8).

The main result of this section reads as follows:

Theorem 12.6 *Let $s \in (0,1)$, $2s < n < 4s$, and let $\Omega \subset \mathbb{R}^n$ be an open, bounded subset with continuous boundary. Let $K : \mathbb{R}^n \setminus \{0\} \to (0,+\infty)$ be a function verifying* (1.55) *and* (1.56), *let $M : [0,+\infty) \to [0,+\infty)$ satisfy* (12.5) *and* (12.21)–(12.23), *and let $f : \Omega \times \mathbb{R} \to \mathbb{R}$ satisfy* (10.7) *and* (12.25)–(12.27). *Finally, let Θ be defines as follows:*

$$\Theta := \frac{1}{2} \inf \left\{ \frac{\widehat{M}\left(\|u\|^2_{X_0^s(\Omega)}\right)}{\displaystyle\int_\Omega F(x,u(x))dx} : u \in X_0^s(\Omega), \int_\Omega F(x,u(x))dx > 0 \right\}.$$

Then, for each compact interval $[a,b] \subset (\Theta,+\infty)$, there exists a number $\rho > 0$ with the following property: for every $\lambda \in [a,b]$ and every g satisfying (10.7), *there exists $\tilde{\mu} > 0$ such that, for each $\mu \in [0,\tilde{\mu}]$, problem* (12.20) *has at least three distinct weak solutions whose norms in $X_0^s(\Omega)$ are less than ρ.*

Proof Let us invoke Theorem 10.2 with the choices

$$E := X_0^s(\Omega), \quad \Phi(u) := \frac{1}{2}\widehat{M}(\|u\|^2_{X_0^s(\Omega)}),$$

and $J := J_f$ (see (10.8)). In view of assumption (12.22), the functional Φ is coercive in $X_0^s(\Omega)$, as

$$\Phi(u) \geq \frac{m_1}{4} \|u\|^4_{X_0^s(\Omega)},$$

for every $u \in X_0^s(\Omega)$. Moreover, since $\widehat{M}$ is increasing, Φ is bounded on any bounded subset of $X_0^s(\Omega)$. Also, Φ satisfies all the other assumptions of Theorem 10.2, thanks to (12.21) and (12.22) (for the details, see [159]). It is straightforward to realize that $u_0 = 0$ is the only global minimum of Φ and that $\Phi(u_0) = J(u_0) = 0$.

Now (10.7) implies that

$$|f(x,t)| \leq c\left(1+|t|^{q-1}\right), \tag{12.28}$$

for all $(x,t) \in \Omega \times \mathbb{R}$, for some $c > 0$, and for some $q \in [1,2_s^*)$. Since $2s < n < 4s$, $4 < 2_s^*$, and it is possible to pick $r \in \mathbb{R}$ such that

$$\max\{4,q\} < r < 2_s^*.$$

Now fix $\varepsilon > 0$. Due to (12.26), there exists $T > 0$ such that

$$F(x,t) \leq \varepsilon |t|^4,$$

for any $(x,t) \in \Omega \times [-T,T]$; however,

$$|F(x,t)| \le c(1+|t|^r),$$

for any $(x,t) \in \Omega \times \mathbb{R}$, and the function $\mathbb{R} \setminus [-T,T] \ni t \mapsto |t|^r$ attains its minimum at $t = T$, so

$$|F(x,t)| \le c|t|^r,$$

for any $(x,t) \in \Omega \times (\mathbb{R} \setminus [-T,T])$. As a result, we get that

$$|F(x,t)| \le \varepsilon |t|^4 + c|t|^r,$$

for any $(x,t) \in \Omega \times \mathbb{R}$.

Thus, by the embeddings $X_0^s(\Omega) \hookrightarrow L^4(\Omega)$ and $X_0^s(\Omega) \hookrightarrow L^r(\Omega)$ and by (12.22), we obtain

$$\begin{aligned} J(u) &\le \varepsilon \|u\|_{L^4(\Omega)}^4 + c\|u\|_{L^r(\Omega)}^r \\ &\le \varepsilon c_1 \|u\|_{X_0^s(\Omega)}^4 + c_2 \|u\|_{X_0^s(\Omega)}^r \\ &\le \frac{2\varepsilon c_1}{m_1} \widehat{M}\left(\|u\|_{X_0^s(\Omega)}^2\right) + c_2 \left(\frac{2}{m_1} \widehat{M}\left(\|u\|_{X_0^s(\Omega)}^2\right)\right)^{r/4} \\ &\le \frac{4\varepsilon c_1}{m_1} \Phi(u) + c_2 \left(\frac{4}{m_1} \Phi(u)\right)^{r/4}, \end{aligned}$$

for any $u \in X_0^s(\Omega)$ and for suitable positive constants c_1 and c_2. Hence, due to the choice of r, we have

$$\limsup_{u \to 0} \frac{J(u)}{\Phi(u)} \le \frac{4\varepsilon c_1}{m_1}. \tag{12.29}$$

When $\sigma < n/(n-2s)$, thanks to (12.27), it is easy to obtain the estimate

$$|F(x,t)| \le \varepsilon |t|^{2\sigma} + c_3,$$

for all $(x,t) \in \Omega \times \mathbb{R}$, for some $c_3 > 0$; however, when $\sigma > n/(n-2s)$, owing to the fact that $|t|^{r-2_s^*} \to 0$ as $|t| \to \infty$, we get

$$|F(x,t)| \le \varepsilon |t|^{2_s^*} + c_4,$$

for all $(x,t) \in \Omega \times \mathbb{R}$ and some $c_4 > 0$. Then we obtain

$$|F(x,t)| \le \varepsilon |t|^{\beta} + c_5,$$

for all $(x,t) \in \Omega \times \mathbb{R}$ and some $c_5 > 0$, where

$$\beta := 2\min\left\{\sigma, \frac{n}{n-2s}\right\},$$

so

$$J(u) \leq c_6(1 + \varepsilon \|u\|^{\beta}_{X_0^s(\Omega)}), \tag{12.30}$$

for every $u \in X_0^s(\Omega)$ and some $c_6 > 0$.

Furthermore, by (12.23), we have

$$\widehat{M}(t) \geq m_\sigma |t|^\sigma - c_7,$$

for every $t \in [0, +\infty)$ and for some $c_7 > 0$. So, taking into account (12.30), we get

$$\frac{J(u)}{\Phi(u)} \leq \frac{c_6(\varepsilon \|u\|^{\beta}_{X_0^s(\Omega)} + 1)}{m_\sigma \|u\|^{2\sigma}_{X_0^s(\Omega)} - c_8},$$

for every $u \neq 0$ and some $c_8 > 0$. Due to the definition of β,

$$\limsup_{\|u\| \to +\infty} \frac{J(u)}{\Phi(u)} \leq \frac{c_6 \varepsilon}{m_\sigma}. \tag{12.31}$$

Conditions (12.29) and (12.31), together with the arbitrariness of ε, yield

$$\max\left\{\limsup_{u \to 0} \frac{J(u)}{\Phi(u)}, \limsup_{\|u\|_{X_0^s(\Omega)} \to +\infty} \frac{J(u)}{\Phi(u)}, \right\} \leq 0,$$

and all the assumptions of Theorem 10.2 are satisfied. Hence, for any compact interval $[a,b] \subset (\Theta, +\infty)$, there exists a number $\rho > 0$ with the property described in the conclusion of Theorem 10.2.

Fix $\lambda \in [a,b]$, a function g satisfying (10.7), and set

$$\Psi(u) := \int_\Omega G(x, u(x))dx,$$

for all $u \in X_0^s(\Omega)$; clearly, $\Psi \in C^1(X_0^s(\Omega), \mathbb{R})$ and Ψ' is compact, so there exists $\tilde{\mu} > 0$ such that, for each $\mu \in [0, \tilde{\mu}]$, the equation

$$\mathscr{E}'_{\lambda,\mu}(u) = \Phi'(u) - \lambda J'(u) - \mu \Psi'(u) = 0,$$

has at least three solutions whose $X_0^s(\Omega)$-norms are less than ρ. This ends the proof.

□

In the model case (12.24), Theorem 12.6 becomes

Corollary 12.7 Let s, n, Ω, K be as in Theorem 12.6, and let f satisfy (10.7) and

$$\sup_{u \in X_0^s(\Omega)} \int_\Omega F(x, u(x))dx > 0 \tag{12.32}$$

and

$$\max\left\{\limsup_{t\to 0}\frac{\sup_{x\in\Omega}F(x,t)}{|t|^4},\limsup_{|t|\to+\infty}\frac{\sup_{x\in\Omega}F(x,t)}{|t|^4}\right\}\leq 0. \tag{12.33}$$

Then, if we pick $m_1 > 0$ and set

$$\Theta := \frac{m_1}{4}\inf\left\{\frac{\|u\|^4_{X^s_0(\Omega)}}{\displaystyle\int_\Omega F(x,u(x))dx} : u\in X^s_0(\Omega), \int_\Omega F(x,u(x))dx > 0\right\},$$

for each compact interval $[a,b]\subset(\Theta,+\infty)$, there exists a number $\rho > 0$ with the following property: for every $\lambda\in[a,b]$ and every g satisfying (10.7), there exists $\tilde{\mu} > 0$ such that, for each $\mu\in[0,\tilde{\mu}]$, the problem

$$\begin{cases} -m_1\|u\|^2_{X^s_0(\Omega)}\mathscr{L}_K u = \lambda f(x,u)+\mu g(x,u) & \text{in } \Omega\\ u = 0 & \text{in } \mathbb{R}^n\setminus\Omega \end{cases} \tag{12.34}$$

admits at least three weak solutions whose norms in $X^s_0(\Omega)$ are less than ρ.

Proof Fix $m_1 > 0$, and apply Theorem 12.6 with $M(t) := m_1 t$, for all $t\geq 0$. Condition (12.23) follows at once with the choice $\sigma = 2$; this choice itself implies (12.27) via assumption (12.33). □

Finally, if we drop the dependence of f from x, we obtain the following meaningful result:

Corollary 12.8 Let s,n,Ω,K be as in Theorem 12.6, and let $f:\mathbb{R}\to\mathbb{R}$ be a continuous function such that

$$\sup_{t\in\mathbb{R}}F(t) > 0,$$

$$\limsup_{t\to 0}\frac{F(t)}{|t|^4}\leq 0,$$

and

$$\limsup_{|t|\to\infty}\frac{|f(t)|}{|t|^q} < +\infty, \quad \text{for some } q\in[0,3).$$

Then, if we pick $m_1 > 0$ and set

$$\Theta := \frac{m_1}{4}\inf\left\{\frac{\|u\|^4_{X^s_0(\Omega)}}{\displaystyle\int_\Omega F(u(x))dx} : u\in X^s_0(\Omega), \int_\Omega F(u(x))dx > 0\right\},$$

for each compact interval $[a,b]\subset(\Theta,+\infty)$, there exists a number $\rho > 0$ with the following property: for every $\lambda\in[a,b]$ and every g satisfying (10.7), there exists

$\tilde{\mu} > 0$ such that, for each $\mu \in [0, \tilde{\mu}]$, the problem

$$\begin{cases} -m_1 \|u\|^2_{X_0^s(\Omega)} \mathscr{L}_K u = \lambda f(u) + \mu g(x,u) & \text{in } \Omega \\ u = 0 & \text{in } \mathbb{R}^n \setminus \Omega \end{cases} \tag{12.35}$$

admits at least three weak solutions whose norms in $X_0^s(\Omega)$ are less than ρ.

Proof The continuity of f and its behavior at infinity imply that f satisfies (10.7) and, in turn, that

$$\frac{F(t)}{t^4} \leq c_9|t|^{-3} + c_{10}|t|^{q-3},$$

for all $t \in \mathbb{R} \setminus \{0\}$ and for some $c_9, c_{10} > 0$. Passing to the lim sup for $|t| \to \infty$ in the last inequality, we obtain that (12.33) is completely satisfied, and hence, we get the assertion. □

12.2.2 A Clark-type result for a Kirchhoff model

To end this chapter, we study the following nonlocal equation:

$$\begin{cases} -M\left(\|u\|^2_{X_0^s(\Omega)}\right) \mathscr{L}_K u = \left(\int_\Omega F(x,u(x))dx\right)^r f(x,u) & \text{in } \Omega \\ u = 0 & \text{in } \mathbb{R}^n \setminus \Omega, \end{cases} \tag{12.36}$$

where $r \geq 0$, $s \in (0,1)$, Ω is an open, bounded subset of $\mathbb{R}^n$, $n > 2s$, with continuous boundary, $f : \overline{\Omega} \times \mathbb{R} \to \mathbb{R}$ and $M : [0,+\infty) \to [0,+\infty)$ are suitable functions verifying the conditions stated in the sequel, and F is, as usual, as in (6.8).

Our aim is to get the existence of multiple weak solutions for problem (12.36). More precisely, by a weak solution for (12.36), we mean a function $u : \mathbb{R}^n \to \mathbb{R}$ such that

$$\begin{cases} M(\|u\|^2_{X_0^s(\Omega)}) \displaystyle\int_{\mathbb{R}^n \times \mathbb{R}^n} (u(x) - u(y))(\varphi(x) - \varphi(y))K(x-y)dx\,dy \\ \quad = \left(\displaystyle\int_\Omega F(x,u(x))dx\right)^r \displaystyle\int_\Omega f(x,u(x))\varphi(x)dx \quad \forall \varphi \in X_0^s(\Omega) \\ u \in X_0^s(\Omega). \end{cases} \tag{12.37}$$

We assume that the function $M : [0,+\infty) \to [0,+\infty)$ verifies (12.5) and is such that

$$\begin{gathered} \textit{there are positive constants } m_0, m_1, m_2, m_3, \alpha, \beta,\ \alpha \leq \beta, \textit{such that} \\ m_0 + m_1 t^\alpha \leq M(t) \leq m_2 + m_3 t^\beta, \quad \textit{for any } t \in [0,+\infty), \end{gathered} \tag{12.38}$$

while $f : \overline{\Omega} \times \mathbb{R} \to \mathbb{R}$ satisfies (6.3) and the following assumptions:

$$f(x,-t) = -f(x,t), \quad \textit{for any } (x,t) \in \overline{\Omega} \times \mathbb{R}, \textit{ and} \tag{12.39}$$

$$\begin{gathered} \textit{there are constants } k_1, k_2 > 0, \textit{ and } 1 < \gamma_1 \leq \gamma_2 < 2/(r+1), r < 1, \textit{ such that} \\ k_1 t^{\gamma_1 - 1} \leq f(x,t) \leq k_2 t^{\gamma_2 - 1}, \quad \textit{for every } x \in \overline{\Omega} \textit{and any } t \in [0,+\infty). \end{gathered} \tag{12.40}$$

Remark 12.9 Of course, conditions (6.3), (12.39), and (12.40) immediately yield

$$|f(x,t)| \le k_2|t|^{\gamma_2-1}, \quad \text{for every } x \in \overline{\Omega} \text{ and any } t \in \mathbb{R}. \tag{12.41}$$

Furthermore, note that the assumption on γ_2 stated in (12.40) implies that

$$\gamma_2 < 2_s^* := \frac{2n}{n-2s} \tag{12.42}$$

owing to the fact that $2/(r+1) < 2_s^*$, for any $r \ge 0$.

A simple model case for problem (12.36) can be exhibited considering the Kirchhoff function $M : [0,+\infty) \to [0,+\infty)$ given by

$$M(t) = m_0 + m_1 t^{\alpha} \tag{12.43}$$

and the nonlinearity $f : \overline{\Omega} \times \mathbb{R} \to \mathbb{R}$ defined by

$$f(t) = a(x)|t|^{\sigma-2}t,$$

where $a : \overline{\Omega} \to \mathbb{R}$ is a nonnegative continuous map, and

$$1 < \sigma < \frac{2}{r+1}.$$

Note that because $f(x,0) = 0$, for any $x \in \overline{\Omega}$, by (12.40), the function $u \equiv 0$ is a solution of problem (12.36). In the following result we show that (12.36) admits infinitely many weak solutions.

Theorem 12.10 *Let $s \in (0,1)$, $n > 2s$, Ω be an open, bounded subset of $\mathbb{R}^n$ with continuous boundary and $r \ge 0$. Let $K : \mathbb{R}^n \setminus \{0\} \to (0,+\infty)$ be a function satisfying* (1.55) *and* (1.56), *and let $M : [0,+\infty) \to [0,+\infty)$ satisfy* (12.5) *and* (12.38) *and $f : \overline{\Omega} \times \mathbb{R} \to \mathbb{R}$ verify* (6.3) *and* (12.39) *and* (12.40). *Then problem* (12.36) *admits infinitely many weak solutions.*

As a consequence of (12.38), it is easy to see that $M(0) \neq 0$. One of the novelty of this section is that we manage to treat the degenerate case; that is, we allow the Kirchhoff function M to take the zero value at 0. Indeed, our result stated in Theorem 12.10 continues to hold if, instead of condition (12.38), we require the following weaker hypothesis:

$$\begin{gathered} \textit{there are } \alpha,\beta > 0 \textit{ and nonnegative constants } m_0, m_1, m_2, m_3 \\ \textit{with } m_0 + m_1 > 0 \textit{ and } m_2 + m_3 > 0 \textit{ such that} \\ m_0 + m_1 t^{\alpha} \le M(t) \le m_2 + m_3 t^{\beta}, \quad \textit{for any } t \in [0,+\infty). \end{gathered} \tag{12.44}$$

We would point out that condition (12.44) allows us to consider both the degenerate setting for the Kirchhoff function (i.e., the case when $m_0 = m_2 = 0$) and the nondegenerate one (i.e., the case where $m_1 = 0$). In this framework, our multiplicity result can be stated as follows:

Theorem 12.11 *Let $s \in (0,1)$, $n > 2s$, Ω be an open, bounded subset of $\mathbb{R}^n$ with continuous boundary, and $r \geq 0$. Let $K : \mathbb{R}^n \setminus \{0\} \to (0,+\infty)$ be a function satisfying* (1.55) *and* (1.56), *and let $M : [0,+\infty) \to [0,+\infty)$ satisfy* (12.5) *and* (12.38) *and $f : \overline{\Omega} \times \mathbb{R} \to \mathbb{R}$ verify* (6.3), (12.39), *and* (12.40). *Then problem* (12.36) *admits infinitely many weak solutions.*

A special case of Theorem 12.11, whose classical and anisotropic counterpart was proved in [76, theorem 1.1], can be stated as follows:

Theorem 12.12 *Let $s \in (0,1)$, $n > 2s$, Ω be an open, bounded subset of $\mathbb{R}^n$ with continuous boundary, and $r \geq 0$. Let $K : \mathbb{R}^n \setminus \{0\} \to (0,+\infty)$ be a function satisfying* (1.55) *and* (1.56), *and let $f : \overline{\Omega} \times \mathbb{R} \to \mathbb{R}$ verify* (6.3), (12.39), *and* (12.40). *Then, for any nonnegative constants m_0, m_1 such that $m_0 + m_1 > 0$, the problem*

$$\begin{cases} -\left(m_0 + m_1 \|u\|^2_{X_0^s(\Omega)}\right) \mathscr{L}_K u = \left(\int_\Omega F(x,u(x))\,dx\right)^r f(x,u) & \text{in } \Omega \\ u = 0 & \text{in } \mathbb{R}^n \setminus \Omega \end{cases} \tag{12.45}$$

admits infinitely many weak solutions.

Note that problem (12.45) is degenerate if $a = 0$ and nondegenerate if $b = 0$. In the degenerate context, the simplest model we can deal with is given by

$$\begin{cases} \left(\int_{\mathbb{R}^n \times \mathbb{R}^n} \frac{|u(x) - u(y)|^2}{|x-y|^{n+2s}}\,dx\,dy\right)(-\Delta)^s u = f(x,u) & \text{in } \Omega \\ u = 0 & \text{in } \mathbb{R}^n \setminus \Omega, \end{cases} \tag{12.46}$$

and the corresponding multiplicity result is as follows:

Theorem 12.13 *Let $s \in (0,1)$, $n > 2s$, Ω be an open, bounded subset of $\mathbb{R}^n$ with continuous boundary. Further, let f be a function verifying* (6.3), (12.39), *and* (12.40). *Then problem* (12.46) *admits infinitely many weak solutions.*

If we consider the prototype when $-\mathscr{L}_K = (-\Delta)^s$, in the limit case when $s = 1$, we get the Laplace operator $-\Delta$. Hence, the classical counterpart of problem (12.36) is given by

$$\begin{cases} -M\left(\int_\Omega |\nabla u(x)|^2 dx\right) \Delta u = \left(\int_\Omega F(x,u(x))\,dx\right)^r f(x,u) & \text{in } \Omega \\ u = 0 & \text{on } \partial\Omega. \end{cases} \tag{12.47}$$

A more general problem involving the p-Laplacian operator was first studied in [77] in the case where $r = 0$, while in [76] the authors considered a $p(x)$-Kirchhoff equation with $r > 0$. In [77], the authors got a similar version of Theorem 12.13 for (12.47), thanks to a variational approach developed in the paper. Adapting, in a suitable way, this type of argument to the nonlocal setting, we can prove Theorems 12.10 and 12.11.

The proofs of Theorems 12.10 and 12.11, as well as of their consequences, rely on a classical Clark's result (see [71]) based on the genus theory introduced by Krasnoselskii (see [9, 61, 123, 174]). The idea consists of looking for critical points of the Euler–Lagrange functionals associated with the problems under consideration, showing that their geometric structure and their compactness properties fit with the requirements of the Clark's result. Among some natural analytic properties, the condition required in the abstract result by Clark is that the functional has a suitable behavior on a symmetric compact set of assigned Krasnoselskii's genus (for full details, we refer to [98]).

13

On fractional Schrödinger equations

In this chapter, we focus our attention on the so-called fractional Schrödinger equation of the form

$$i\frac{\partial \psi}{\partial t} = (-\Delta)^s \psi + V(x)\psi - |\psi|^{p-1}\psi, \tag{13.1}$$

where $(x,t) \in \mathbb{R}^n \times (0,+\infty)$, and $V : \mathbb{R}^n \to \mathbb{R}$ is a suitable potential. Equation (13.1) comes from an expansion of the Feynman path integral from Brownian-like to Lévy-like quantum mechanical paths and was considered for the first time in the literature by Laskin in [130, 131].

When $s = 1$, equation (13.1) gives back the classical Schrödinger equation. In this case, standing-wave solutions are of the form

$$\psi(x,t) = e^{-i\omega t}u(x),$$

where ω is a suitable constant, and u solves the elliptic equation

$$-\Delta u + V(x)u - |u|^{p-1}u = 0. \tag{13.2}$$

We do not even try to review the huge bibliography of equations like (13.2), we just emphasize that the potential $V : \mathbb{R}^n \to \mathbb{R}$ has a crucial role concerning the existence and behavior of solutions. For instance, when V is a positive constant or V is radially symmetric, it is natural to look for radially symmetric solutions (see e.g., [210, 220]). However, after the seminal paper of Rabinowitz [175], where the potential V is assumed to be coercive, several different assumptions were adopted to obtain existence and multiplicity results (see [29, 30, 31, 109, 113]).

While the classical Laplacian case has been widely investigated in the literature, in the nonlocal fractional setting, the situation seems to be under development, and an always increasing interest for this topic is shown by the mathematical community (see, e.g., among others, [68, 84, 95, 108]).

In this spirit, in [192], the author looked for standing-wave solutions of a more general equation than (13.1). More precisely, in that paper, the nonlocal fractional equation

$$(-\Delta)^s u + V(x)u = f(x,u) \quad x \in \mathbb{R}^n \tag{13.3}$$

is studied under certain hypotheses on the potential V and the nonlinearity f (see also [193, 194]). Moreover, fractional Schrödinger-type problems have been considered in some recent papers, such as [17, 65, 81, 170, 171], just to name a few. The results of this chapter are based on the paper [149].

13.1 The main problem

Motivated by an increasing interest in the current literature, under suitable conditions on the potential V and exploiting variational methods, here we study the existence of multiple solutions for the following fractional parametric problem:

$$(-\Delta)^s u + V(x)u = \lambda(f(x,u) + \mu g(x,u)) \quad x \in \mathbb{R}^n, \tag{13.4}$$

where $n > 2$, $V : \mathbb{R}^n \to \mathbb{R}$, and $f, g : \mathbb{R}^n \times \mathbb{R} \to \mathbb{R}$ are continuous functions verifying suitable growth conditions, while λ, μ are real parameters.

13.1.1 Assumptions on the potential and the nonlinearity

In this chapter, we assume that the potential V satisfies the following conditions:

$$V \in C(\mathbb{R}^n) \textit{ with } \inf_{x\in\mathbb{R}^n} V(x) > 0, \textit{ and} \tag{13.5}$$

$$\begin{gathered} \textit{for any } M > 0, \textit{ there exists } r_0 > 0 \textit{ such that} \\ \lim_{|y|\to+\infty} \left|\{x \in B(y,r_0) : V(x) \le M\}\right| = 0, \end{gathered} \tag{13.6}$$

where $B(y,r)$ denotes the ball in $\mathbb{R}^n$ with center y and radius $r > 0$, and $|\cdot|$ denotes the standard Lebesgue measure in $\mathbb{R}^n$. Note that conditions (13.5) and (13.6) are not new in the literature; see, for instance, the papers [65, 215].

We also require that $f, g : \mathbb{R}^n \times \mathbb{R} \to \mathbb{R}$ are two continuous functions verifying the following standard assumptions:

$$\begin{gathered} \textit{there exist } W \in L^1(\mathbb{R}^n) \cap L^\infty(\mathbb{R}^n), W \not\equiv 0, \textit{ and } q \in (0,1) \textit{ such that} \\ \max\left\{|f(x,s)|, |g(x,s)|\right\} \le W(x)|s|^q \textit{ for any } (x,t) \in \mathbb{R}^n \times \mathbb{R}, \end{gathered} \tag{13.7}$$

$$\lim_{s\to 0} \frac{f(x,s)}{s} = \lim_{s\to 0} \frac{g(x,s)}{s} = 0 \textit{ uniformly for any } x \in \mathbb{R}^n, \textit{ and} \tag{13.8}$$

$$\textit{there exists } t_0 \in \mathbb{R} \textit{ such that } \sup_{\sigma>0} \left(\min_{|x|\le\sigma} \int_0^{t_0} f(x,\tau)\,d\tau \right) > 0. \tag{13.9}$$

13.1.2 The abstract framework

Let us consider the Hilbert space

$$E_s^n(V) := \left\{ u \in H^s(\mathbb{R}^n) : \int_{\mathbb{R}^n} |\xi|^{2s} |\mathcal{F}u(\xi)|^2 d\xi + \int_{\mathbb{R}^n} V(x)|u(x)|^2 dx < \infty \right\},$$

with the norm

$$\|u\|_{E_s^n(V)} := \left(\int_{\mathbb{R}^n} |\xi|^{2s} |\mathcal{F}u(\xi)|^2 d\xi + \int_{\mathbb{R}^n} V(x)|u(x)|^2 dx \right)^{1/2},$$

where $H^s(\mathbb{R}^n)$ denotes the fractional Sobolev space defined in Subsection 1.2.2. The nonlocal analysis we perform in this chapter is quite general and is successfully exploited for other goals in recent contributions (see [65, 68, 215] and references therein).

By a *weak solution* of (13.4), we mean a function $u \in E_s^n(V)$ such that

$$\begin{aligned} &\int_{\mathbb{R}^n} |\xi|^{2s} \mathcal{F}u(\xi)\mathcal{F}v(\xi) d\xi + \int_{\mathbb{R}^n} V(x)u(x)v(x) dx \\ &\quad = \lambda \int_{\mathbb{R}^n} f(x,u(x))v(x) dx + \lambda\mu \int_{\mathbb{R}^n} g(x,u(x))v(x) dx, \end{aligned} \tag{13.10}$$

for any $v \in E_s^n(V)$.

As direct computations prove, problem (13.10) represents the Euler–Lagrange equation of the C^1-functional $\mathcal{J}_{s,V} : E_s^n(V) \to \mathbb{R}$ defined as

$$\begin{aligned} \mathcal{J}_{s,V}(u) := &\frac{1}{2} \left(\int_{\mathbb{R}^n} |\xi|^{2s} |\mathcal{F}u(\xi)|^2 d\xi + \int_{\mathbb{R}^n} V(x)|u(x)|^2 dx \right) \\ &- \lambda \int_{\mathbb{R}^n} F(x,u(x)) dx - \lambda\mu \int_{\mathbb{R}^n} G(x,u(x)) dx, \end{aligned}$$

where F and G are as in (6.8).

In what follows, we need some embedding properties of $E_s^n(V)$ into the classical Lebesgue spaces. For this purpose, first, we prove the following fractional Gagliardo–Nirenberg inequality:

Lemma 13.1 *Let $s \in (0,1)$, $p \in [1,+\infty)$, and $n > 2s$. Then, for any $u \in H^s(\mathbb{R}^n)$, the following inequality holds:*

$$\|u\|_{L^p(\mathbb{R}^n)} \le C(n,s,\alpha)[u]_{H^s(\mathbb{R}^n)}^{\alpha} \|u\|_{L^r(\mathbb{R}^n)}^{1-\alpha}, \tag{13.11}$$

with

$$\frac{n}{p} = \alpha\frac{n-2s}{2} + (1-\alpha)\frac{n}{r},$$

where $r \ge 1$, $\alpha \in [0,1]$, and $C(n,s,\alpha)$ is a positive constant.

Proof The conclusion is trivial if $p = 1$. Indeed, in such a case, it is suffices to take $r = 1$ and $\alpha = 0$. Hence, let us suppose that $p > 1$. Since

$$\frac{1}{p} = \frac{\alpha}{2_s^*} + \frac{1-\alpha}{r},$$

by the Hölder inequality and Theorem 1.4, it follows that

$$\|u\|_{L^p(\mathbb{R}^n)} \le \|u\|^{\alpha}_{L^{2_s^*}(\mathbb{R}^n)} \|u\|^{1-\alpha}_{L^r(\mathbb{R}^n)} \le C(n,s,\alpha)[u]^{\alpha}_{H^s(\mathbb{R}^n)} \|u\|^{1-\alpha}_{L^r(\mathbb{R}^n)}, \tag{13.12}$$

where we set

$$C(n,s,\alpha) := K(n,s)^{\alpha/2}.$$

The proof is complete. □

As a by-product of Lemma 13.1, the next embedding result, which is crucial in our approach, can be easily proved.

Lemma 13.2 *Let $V : \mathbb{R}^n \to \mathbb{R}$ be a potential satisfying* (13.5) *and* (13.6). *Then the embedding*

$$E_s^n(V) \hookrightarrow L^p(\mathbb{R}^n)$$

is compact for any $p \in [2, 2_s^*)$.

Proof By [65], we know that the Hilbert space $E_s^n(V)$ is compactly embedded into $L^2(\mathbb{R}^n)$. Therefore, we only consider the case $p \in (2, 2_s^*)$. To do this, we use the fractional Gagliardo–Nirenberg inequality proved in Lemma 13.1.

Hence, let $\{u_j\}_{j\in\mathbb{N}} \subset E_s^n(V)$ be a sequence such that $u_j \rightharpoonup u_0$ in $E_s^n(V)$; that is, $\{u_j\}_{j\in\mathbb{N}}$ weakly converges to u_0 in $E_s^n(V)$ as $j \to +\infty$. Then $\{u_j\}_{j\in\mathbb{N}}$ is bounded in $E_s^n(V)$, and by Lemma 13.1, for $r = 2$ and

$$\alpha := \frac{(p-2)n}{2sp} \in (0,1),$$

there exists a constant $C_1(n,s,\alpha) > 0$ such that

$$\|u_j - u_0\|_{L^p(\mathbb{R}^n)} \le C_1(n,s,\alpha)\|u_j - u_0\|^{\alpha}_{E_s^n(V)} \|u_j - u_0\|^{1-\alpha}_{L^2(\mathbb{R}^n)}, \tag{13.13}$$

for any $p \in (2, 2_s^*)$. Thus, since $\{u_j\}_{j\in\mathbb{N}}$ is bounded in $E_s^n(V)$, by (13.13), we have

$$\|u_j - u_0\|_{L^p(\mathbb{R}^n)} \le C_1(n,s,\alpha)(M + \|u_0\|^{\alpha}_{E_s^n(V)}) \|u_j - u_0\|^{1-\alpha}_{L^2(\mathbb{R}^n)} \to 0;$$

that is, $u_j \to u_0$ in $L^p(\mathbb{R}^n)$ as $j \to +\infty$. This completes the proof. □

Finally, we mention the following result, which is a consequence of (13.9) and will be crucial in Section 13.2. For a proof we refer to Lemma 10.7.

Lemma 13.3 *Let $f : \mathbb{R}^n \times \mathbb{R} \to \mathbb{R}$ be a continuous function satisfying* (13.9). *Then there exists $u_0 \in E_s^n(V)$ such that*

$$\int_{\mathbb{R}^n} F(x, u_0(x))\,dx > 0.$$

13.2 Multiple solutions

In this section, we state some multiplicity results for problem (13.4). Their proofs are based on variational techniques. In particular, we perform a direct consequence of some general results given in [177, 183], which ensures the existence of multiple critical points for a functional under suitable regularity assumptions on it.

The first multiplicity result of this section is the following:

Theorem 13.4 *Let* $V : \mathbb{R}^n \to \mathbb{R}$ *be a potential satisfying* (13.5) *and* (13.6), *and let* $f, g : \mathbb{R}^n \times \mathbb{R} \to \mathbb{R}$ *be two continuous functions verifying* (13.7)–(13.9). *Then there exists* $\mu_0 > 0$ *such that to every* $\mu \in [-\mu_0, \mu_0]$ *it corresponds a nonempty open interval* $\Sigma_\mu \subset (0, +\infty)$ *and a number* $\kappa_\mu > 0$ *for which* (13.4) *has at least two distinct, nontrivial weak solutions* $v_{\lambda\mu}$ *and* $w_{\lambda\mu}$ *with the property that*

$$\max\{\|v_{\lambda\mu}\|_{E^n_s(V)}, \|w_{\lambda\mu}\|_{E^n_s(V)}\} \le \kappa_\mu$$

whenever $\lambda \in \Sigma_\mu$.

The constant μ_0 given in Theorem 13.4 can be computed as follows:

$$\mu_0 := \frac{\displaystyle\int_{\mathbb{R}^n} F(x, u_0(x))\,dx}{\left|\displaystyle\int_{\mathbb{R}^n} G(x, u_0(x))\,dx\right| + 1},$$

where $u_0 \in E^n_s(V)$ is the function given in Lemma 13.3. Also, in this case, the Σ_μ is such that

$$\Sigma_\mu \subset \left[0, \frac{2\|u_0\|^2_{E^n_s(V)}}{\displaystyle\int_{\mathbb{R}^n} F(x, u_0(x))\,dx}\left(1 + \left|\int_{\mathbb{R}^n} G(x, u_0(x))\,dx\right|\right)\right].$$

From the point of view of the parameter λ, the counterpart of Theorem 13.4 is the following:

Theorem 13.5 *Let* $V : \mathbb{R}^n \to \mathbb{R}$ *be a potential satisfying* (13.5) *and* (13.6), *and let* $f, g : \mathbb{R}^n \times \mathbb{R} \to \mathbb{R}$ *be two continuous functions verifying* (13.7)–(13.9). *Then there exists* $\lambda_0 > 0$ *such that to every* $\lambda \in (\lambda_0, +\infty)$ *it corresponds a nonempty open interval* $\Lambda_\lambda \subset \mathbb{R}$ *and a number* $\kappa_\lambda > 0$ *for which* (13.4) *has at least two distinct, nontrivial weak solutions* $v_{\lambda\mu}$ *and* $w_{\lambda\mu}$ *with*

$$\mathcal{J}_{s,V}(v_{\lambda\mu}) < 0 < \mathcal{J}_{s,V}(w_{\lambda\mu})$$

and

$$\max\{\|v_{\lambda\mu}\|_{E^n_s(V)}, \|w_{\lambda\mu}\|_{E^n_s(V)}\} \le \kappa_\lambda$$

whenever $\mu \in \Lambda_\lambda$.

The proof of Theorem 13.5 is based on the validity of Lemma 13.2 and on the classical mountain pass theorem. In such a case,

$$\lambda_0 := \frac{\|u_0\|^2_{E_s^n(V)}}{2\displaystyle\int_{\mathbb{R}^n} F(x,u_0(x))dx},$$

where, again, u_0 is given in Lemma 13.3, and for any $\lambda > \lambda_0$, Λ_λ is as follows:

$$\Lambda_\lambda \equiv (-\mu_\lambda^*, \mu_\lambda^*),$$

where

$$\mu_\lambda^* := \frac{1}{1+\displaystyle\int_{\mathbb{R}^n} |G(x,u_0(x))|\,dx}\left(1-\frac{\lambda_0}{\lambda}\right)\int_{\mathbb{R}^n} F(x,u_0(x))dx. \tag{13.14}$$

Although Theorems 13.4 and 13.5 are completely independent, as a simple by-product of Theorem 13.5, we obtain the following result, whose conclusion partially goes back to Theorem 13.4:

Theorem 13.6 *Let* $V : \mathbb{R}^n \to \mathbb{R}$ *be a potential satisfying* (13.5) *and* (13.6), *and let* $f, g : \mathbb{R}^n \times \mathbb{R} \to \mathbb{R}$ *be two continuous functions verifying* (13.7)–(13.9). *Then there exists* $\overline{\mu} > 0$ *such that, for every* $\mu \in [-\overline{\mu}, \overline{\mu}]$, *the set*

$$\Sigma := \{\lambda > 0 : (13.4) \textit{ has at least two distinct, nontrivial weak solutions}\}$$

contains an interval.

The abstract approach adopted here is employed for fractional Scrödinger equations and is patterned after [124, problem (1.1)]. Let us note that in our setting, the situation is more delicate with respect to that treated in the cited paper. Indeed, the fractional framework produces some technical difficulties that we overcome by using an appropriate variational formulation, as well as the embedding properties proved in Lemmas 13.1 and 13.2. The multiplicity results given here represent a nontrivial fractional counterpart of [124, theorems 2.1–2.3].

13.3 Nonexistence results

The theorems stated in Section 13.2 do not work in general for any $\lambda \in \mathbb{R}$. For instance, consider

$$f(x,t) := \frac{(\sin t)^2}{(1+|x|^\alpha)^\beta} \quad \forall (x,t) \in \mathbb{R}^n \times \mathbb{R}$$

and

$$g(x,t) := \frac{(\arctan t)^2}{(1+|x|^\alpha)^\beta} \quad \forall (x,t) \in \mathbb{R}^n \times \mathbb{R},$$

with $\alpha,\beta > 0$ such that $\alpha\beta > n \geq 3$. In such a case, an easy computation shows that the problem

$$(-\Delta)^s u + V(x)u = \frac{\lambda}{(1+|x|^{\alpha})^{\beta}}((\sin u)^2 + \mu(\arctan u)^2) \quad x \in \mathbb{R}^n, \tag{13.15}$$

possesses only the trivial solution whenever

$$0 < \lambda < \frac{1}{(1+|\mu|\pi)S_2^2},$$

and μ is arbitrary. Here S_2 denotes the best Sobolev embedding constant of the injection $E_s^n(V) \hookrightarrow L^2(\mathbb{R}^n)$.

More generally, the following nonexistence result holds:

Theorem 13.7 *Let $V : \mathbb{R}^n \to \mathbb{R}$ be a potential satisfying* (13.5) *and* (13.6), *and let $f,g : \mathbb{R}^n \times \mathbb{R} \to \mathbb{R}$ be two continuous functions verifying* (13.8). *Assume that*

$$f(x,t) := W(x)h(t),$$

and

$$g(x,s) := W(x)k(t)$$

for any $(x,t) \in \mathbb{R}^n \times \mathbb{R}$, where h and k are Lipschitz continuous functions with Lipschitz constants $L_h > 0$ and $L_k > 0$, respectively, and $W \in L^1(\mathbb{R}^n) \cap L^{\infty}(\mathbb{R}^n)$ satisfies (13.7). *Then for any parameter $\mu \in \mathbb{R}$ and any λ such that*

$$0 < \lambda < \frac{1}{\|W\|_{L^{\infty}(\mathbb{R}^n)}(L_h + |\mu|L_k)S_2^2},$$

the problem

$$(-\Delta)^s u + V(x)u = \lambda W(x)(h(u) + \mu k(u)) \quad x \in \mathbb{R}^n \tag{13.16}$$

admits only the trivial solution.

Proof Arguing by contradiction, let us assume that there exists a weak solution $\bar{u} \in E_s^n(V) \setminus \{0\}$ of the problem (13.16); that is,

$$\begin{aligned} &\int_{\mathbb{R}^n} |\xi|^{2s} \mathscr{F}\bar{u}(\xi)\mathscr{F}v(\xi)\,d\xi + \int_{\mathbb{R}^n} V(x)\bar{u}(x)v(x)\,dx \\ &= \lambda \int_{\mathbb{R}^n} W(x)h(\bar{u}(x))v(x)\,dx + \lambda\mu \int_{\mathbb{R}^n} W(x)k(\bar{u}(x))v(x)\,dx, \end{aligned} \tag{13.17}$$

for any $v \in E_s^n(V)$.

In particular, testing (13.17) with $v = \bar{u}$, we have

$$\|\bar{u}\|^2_{E_s^n(V)} = \lambda \int_{\mathbb{R}^n} W(x)h(\bar{u}(x))\bar{u}(x)\,dx + \lambda\mu \int_{\Omega} W(x)k(\bar{u}(x))\bar{u}(x)\,dx. \tag{13.18}$$

Since h and k are Lipschitz continuous functions and (13.8) holds, then $h(0)=k(0)=0$. Hence, by (13.18), it follows that

$$\begin{aligned}\|\bar{u}\|^2_{E^n_s(V)} &\leq \lambda\|W\|_{L^\infty(\mathbb{R}^n)}\left(\int_{\mathbb{R}^n}|h(\bar{u}(x))\bar{u}(x)|\,dx+\mu\int_{\mathbb{R}^n}|k(\bar{u}(x))\bar{u}(x)|\,dx\right)\\ &= \lambda\|W\|_{L^\infty(\mathbb{R}^n)}\left(\int_{\mathbb{R}^n}|h(\bar{u}(x)-h(0))||\bar{u}(x)|\,dx\right.\\ &\qquad\left.+|\mu|\int_{\mathbb{R}^n}|k(\bar{u}(x))-k(0)||\bar{u}(x)|\,dx\right)\\ &\leq \lambda\|W\|_{L^\infty(\mathbb{R}^n)}(L_h+|\mu|L_k)\|\bar{u}\|^2_{L^2(\mathbb{R}^n)}.\end{aligned} \tag{13.19}$$

Then, by the choice of λ, (13.19), and Lemma 13.2, we easily get

$$\|\bar{u}\|^2_{E^n_s(V)} \leq \lambda\|W\|_{L^\infty(\mathbb{R}^n)}(L_h+|\mu|L_k)S_2^2\|\bar{u}\|^2_{E^n_s(V)} < \|\bar{u}\|^2_{E^n_s(V)},$$

which is a contradiction. In conclusion, problem (13.16) admits only the trivial solution. □

13.4 Perturbed Schrödinger equations

The last part of this chapter is dedicated to the existence of multiple weak solutions for the following nonlocal Schrödinger equation:

$$(-\Delta)^s u + V(x)u = \lambda W(x)f(u)+\mu g(x,u), \quad x\in\mathbb{R}^n \tag{13.20}$$

where $V:\mathbb{R}^n\to\mathbb{R}$ is a function satisfying (13.5) and(13.6), λ and μ are real parameters, and W is such that

$$W\in L^1(\mathbb{R}^n)\cap L^\infty_+(\mathbb{R}^n) \quad\text{and}\quad \sup_{\sigma>0}\left(\inf_{|x|\leq\sigma} W(x)\right)>0. \tag{13.21}$$

Also, we assume that the nonlinear term $f:\mathbb{R}\to\mathbb{R}$ is a continuous function such that

$$\begin{gathered}\textit{there exists } q\in(0,1) \textit{ such that}\\ |f(t)|\leq|t|^q \textit{ for any } t\in\mathbb{R}, \textit{ and}\end{gathered} \tag{13.22}$$

$$\lim_{t\to 0}\frac{f(t)}{t}=0. \tag{13.23}$$

Finally, we require that the perturbation term $g:\mathbb{R}^n\times\mathbb{R}\to\mathbb{R}$ is a continuous function such that

$$\begin{gathered}\textit{there exist } Z\in L^1(\mathbb{R}^n)\cap L^\infty(\mathbb{R}^n), Z\not\equiv 0, \textit{ and } r\in(0,1) \textit{ such that}\\ |g(x,t)|\leq Z(x)|t|^r \textit{ for any } (x,t)\in\mathbb{R}^n\times\mathbb{R}.\end{gathered} \tag{13.24}$$

The main result of this section ensures that, for $\lambda>0$ large enough, problem (13.20) admits at least two nontrivial weak solutions, as well as the stability of this problem with respect to an arbitrary subcritical perturbation of the Schrödinger equation. More precisely, our result reads as follows:

Theorem 13.8 *Let $f : \mathbb{R} \to \mathbb{R}$ be a continuous function satisfying* (13.22) *and* (13.23). *Furthermore, assume that condition* (13.21) *holds, and set*

$$\Theta := \frac{1}{2} \inf \left\{ \frac{\|u\|^2_{E^n_s(V)}}{\displaystyle\int_{\mathbb{R}^n} W(x)F(u(x))dx} : u \in E^n_s(V), \int_{\mathbb{R}^n} W(x)F(u(x))dx > 0 \right\}.$$

Then, for any compact interval $[a,b] \subset (\Theta, +\infty)$, there exists a number $\rho > 0$ with the following property: for every $\lambda \in [a,b]$ and every g satisfying (13.24), *there exists $\tilde{\mu} > 0$ such that, for each $\mu \in [0, \tilde{\mu}]$, problem* (13.20) *has at least three weak solutions whose norms in $E^n_s(V)$ are less than ρ.*

As a model for problem (13.20), we can consider the following problem:

$$(-\Delta)^s u + V(x)u = \frac{\lambda |u|^q \sin u + \mu |\sin u|^r}{(1 + |x|^n)^2} \qquad x \in \mathbb{R}^n$$

with $r, q \in (0, 1)$.

Part III

Nonlocal critical problems

14

The Brezis–Nirenberg result for the fractional Laplacian

In the literature, there are many papers related to the study of the following critical elliptic equation:

$$\begin{cases} -\Delta u - \lambda u = |u|^{2_*-2}u & \text{in } \Omega \\ u = 0 & \text{on } \partial\Omega \quad 2_* := 2n/(n-2), \end{cases} \tag{14.1}$$

where $\Omega \subset \mathbb{R}^n$ (with $n > 2$) is an open, bounded set, and λ is a positive parameter (see, e.g., [9, 212, 220] and references therein).

Equation (14.1) is particularly relevant for its relation to problems arising in differential geometry and physics, where a lack of compactness occurs. One of the most important problems that give rise to equation (14.1) is the Yamabe problem: given an n-dimensional compact Riemannian manifold $(\mathcal{M}, g)$, $n > 2$, with scalar curvature $k = k(x)$, find a metric $\tilde{g}$ conformal to g with constant scalar curvature $\tilde{k}$. Putting

$$\tilde{g} := u^{4/(n-2)}g,$$

where $u > 0$ is the conformal factor, then the Yamabe problem can be formulated as follows: find $u > 0$ satisfying the equation

$$-4\frac{n-1}{n-2}\Delta_{\mathcal{M}}u = \tilde{k}u^{2_s^*-1} - k(x)u \quad \text{in } \mathcal{M}.$$

Here $\Delta_{\mathcal{M}}$ is the Laplace–Beltrami operator on $\mathcal{M}$ with respect to the metric g.

Motivated by the interest shown in the literature for nonlocal operators of elliptic type, in this chapter we study the nonlocal counterpart of problem (14.1); that is, we consider the following critical nonlocal problem:

$$\begin{cases} (-\Delta)^s u - \lambda u = |u|^{2_s^*-2}u & \text{in } \Omega \\ u = 0 & \text{in } \mathbb{R}^n \setminus \Omega, \end{cases} \tag{14.2}$$

where $s \in (0,1)$ is fixed, and $(-\Delta)^s$ is the fractional Laplace operator defined, up to normalization factors, as in (1.20), while $\Omega \subset \mathbb{R}^n$, $n > 2s$, is open, bounded and with continuous boundary, $\lambda > 0$, and $2_s^* := 2n/(n-2s)$ is the fractional critical Sobolev exponent (see Lemma 1.31).

It would be interesting to understand if it is possible to set a fractional Yamabe problem, as just described, directly in the nonlocal framework and if such a problem coincides with (14.2). Answering this question goes beyond the scope of this book. We would just recall that in a nonlocal setting, one also could define a notion of nonlocal scalar curvature (see, e.g., [1]). The results of this chapter have been obtained from [204].

14.1 A critical fractional Laplace equation

In the classical critical setting (see problem (14.1)) in their famous paper [46], Brezis and Nirenberg proved that

1. If $n \geq 4$, then, for any $\lambda \in (0, \widetilde{\lambda}_1)$, problem (14.1) has a positive solution;
2. If $n = 3$, then there exists a constant $\lambda_* \in (0, \widetilde{\lambda}_1)$ such that, for any $\lambda \in (\lambda_*, \widetilde{\lambda}_1)$, problem (14.1) has a positive solution.

In the case where Ω is a ball in (2), the authors calculated explicitly λ_* and proved that the following result (stronger than (2)) holds true:

3. When $n = 3$ and Ω is a ball, problem (14.1) has a positive solution if and only if $\lambda \in (\widetilde{\lambda}_1/4, \widetilde{\lambda}_1)$.

Here $\widetilde{\lambda}_1$ is the first eigenvalue of the Laplacian with homogeneous Dirichlet boundary conditions. In [60], Capozzi, Fortunato, and Palmieri extended the existence result of [46], proving that if $n \geq 4$, then problem (14.1) has a solution for any $\lambda > 0$.

Our aim is to show that all these results can be extended to problem (14.2). To do this, as in the classical case, we will use a variational technique: we can approach the problem at least in two different ways. The first consists of noting that the solutions of (14.2) correspond to critical points of the "free" functional defined as follows:

$$u \mapsto \frac{1}{2}\int_{\mathbb{R}^n\times\mathbb{R}^n} \frac{|u(x)-u(y)|^2}{|x-y|^{n+2s}}\,dx\,dy - \frac{\lambda}{2}\int_\Omega |u(x)|^2\,dx - \frac{1}{2_s^*}\int_\Omega |u(x)|^{2_s^*}\,dx.$$

The second variational approach consists of looking for critical points of the functional

$$u \mapsto \int_{\mathbb{R}^n\times\mathbb{R}^n} \frac{|u(x)-u(y)|^2}{|x-y|^{n+2s}}\,dx\,dy - \lambda\int_\Omega |u(x)|^2\,dx$$

on the sphere $\{u : \|u\|_{L^{2_s^*}(\Omega)} = 1\}$. This second approach is related to the existence of the best fractional critical Sobolev constant.

In the model case (14.2), these two approaches are equivalent. However, for more general nonlinearities with critical growth, this is not true because it is not always possible to reduce a boundary value problem to a constrained minimization problem: in this case, use of the "free" functional therefore becomes somewhat unavoidable.

The weak formulation of (14.2) is given by

$$\begin{cases} \displaystyle\int_{\mathbb{R}^n\times\mathbb{R}^n} \frac{(u(x)-u(y))(\varphi(x)-\varphi(y))}{|x-y|^{n+2s}}\,dx\,dy - \lambda\int_\Omega u(x)\varphi(x)dx \\[2ex] \displaystyle\quad = \int_\Omega |u(x)|^{2_s^*-2}u(x)\varphi(x)dx \quad \forall\,\varphi\in\mathbb{H}_0^s(\Omega) \\[2ex] u\in\mathbb{H}_0^s(\Omega), \end{cases} \tag{14.3}$$

which is the Euler–Lagrange equation of the functional $\mathscr{J}_{s,\lambda}:\mathbb{H}_0^s(\Omega)\to\mathbb{R}$ defined as

$$\begin{aligned} \mathscr{J}_{s,\lambda}(u) := &\frac{1}{2}\int_{\mathbb{R}^n\times\mathbb{R}^n} \frac{|u(x)-u(y)|^2}{|x-y|^{n+2s}}\,dx\,dy \\ &-\frac{\lambda}{2}\int_\Omega |u(x)|^2dx - \frac{1}{2_s^*}\int_\Omega |u(x)|^{2_s^*}dx. \end{aligned} \tag{14.4}$$

As in the classical case of the Laplacian, in the nonlocal setting, the issue that needs to be overcome is that the functional $\mathscr{J}_{s,\lambda}$ does not satisfy the Palais–Smale condition. This lack of compactness is due to the fact that the embedding $H^s(\mathbb{R}^n)\hookrightarrow L^{2_s^*}(\mathbb{R}^n)$ is not compact. For this, the functional $\mathscr{J}_{s,\lambda}$ does not verify the Palais–Smale condition globally, but – as we will see in what follows – only in a suitable range related to the best fractional critical Sobolev constant S_s in the embedding $H^s(\mathbb{R}^n)\hookrightarrow L^{2_s^*}(\mathbb{R}^n)$ (see Lemma 1.32). As a consequence, an estimate of the critical level of $\mathscr{J}_{s,\lambda}$ is necessary. For this, it is convenient to define the constant

$$S_{s,\lambda} := \inf_{v\in H^s(\mathbb{R}^n)\setminus\{0\}} S_{s,\lambda}(v), \tag{14.5}$$

where, for any $v\in H^s(\mathbb{R}^n)\setminus\{0\}$,

$$S_{s,\lambda}(v) := \frac{\displaystyle\int_{\mathbb{R}^n\times\mathbb{R}^n} \frac{|v(x)-v(y)|^2}{|x-y|^{n+2s}}\,dx\,dy - \lambda\int_{\mathbb{R}^n} |v(x)|^2dx}{\displaystyle\left(\int_{\mathbb{R}^n}|v(x)|^{2_s^*}dx\right)^{2/2_s^*}}. \tag{14.6}$$

The consistency of all these definitions relies on Lemma 1.32. We stress that $S_{s,\lambda}$ and S_s (see formula (1.92)) do not depend on Ω because the minimization occurs on the whole of $H^s(\mathbb{R}^n)$. Notice that $S_{s,\lambda}(v) < S_s(v)$, for any $v \in H^s(\mathbb{R}^n) \setminus \{0\}$ and any $\lambda > 0$, so

$$S_{s,\lambda} \leq S_s. \tag{14.7}$$

In the forthcoming theorems we will consider the case in which the *strict* inequality occurs. The main result of this chapter can be stated as follows:

Theorem 14.1 *Let $s \in (0,1)$, $n \geq 4s$, and Ω be an open, bounded subset of $\mathbb{R}^n$ with continuous boundary.*

Then, denoting by $\lambda_{1,s}$ the first eigenvalue of the nonlocal operator $(-\Delta)^s$ with homogeneous Dirichlet boundary datum, for any $\lambda \in (0, \lambda_{1,s})$, problem (14.2) *admits a nontrivial weak solution $u \in \mathbb{H}^s_0(\Omega)$.*

For the proof of Theorem 14.1, we will use the fact that the infimum in (1.92) is attained by a suitable function $\widetilde{u}$, as is proved in [82, theorem 1.1]. Then, starting from $\widetilde{u}$, we will define a family u_ε, and we will estimate $S_{s,\lambda}(u_\varepsilon)$ (see Sections 14.3 and 14.4 for the details). With respect to the classical case of the Laplacian, here this estimate is more delicate because of the nonlocal nature of the operator $(-\Delta)^s$. In particular, in the nonlocal framework, in order to control the Gagliardo seminorm of u_ε, pointwise estimates on u_ε are not enough. Indeed, we also need some integral estimates that allow us to control the interactions between B_δ and $\mathbb{R}^n \setminus B_\delta$ (for more details, see Proposition 14.11).

As we said earlier, in this chapter we will adapt the variational approach used in [46] to the nonlocal framework. For this, we will work in a functional analytic setting that is inspired by (but not equivalent to) the fractional Sobolev spaces in order to correctly encode the Dirichlet boundary datum in the variational formulation. Also, the functional setting we will use allows us to overcome the problems related to the lack of compactness and to show that the Palais–Smale condition holds true in a suitable range related to the best fractional Sobolev constant S_s (see Subsection 14.3.1 for more details). In fact, this range of validity of the Palais–Smale condition may be explicitly related to the strict sign of the inequality (14.7), as we will discuss in Proposition 14.13.

The proofs are different in several ways from those of the classical case of the Laplacian: these additional complications are not only technical, but sometimes they reflect the different distribution of the energy density due to the nonlocal interactions. For instance, the far-away part of the energy in our case, though it is small, gives a contribution to the functional that cannot be neglected because it is of order comparable with the perturbation.

Finally, note that when $s = 1$, Theorem 14.1 reads as part (a) of the results obtained in [46] for the Laplacian. For all these reasons, we think that Theorem 14.1 may be seen as the natural extension of the results proved in [46] (see also [212, 220]) to the nonlocal fractional framework. In this respect, we recall a recent paper [214] where a nonlocal version of the Brezis–Nirenberg result also was considered (in the case $s = 1/2$ and for an operator conceptually different from the one treated here). See also the recent paper [101], where nonlocal Hardy–Sobolev critical elliptic Dirichlet problems have been investigated.

14.2 Geometry of the functional $\mathcal{J}_{s,\lambda}$

In this section, we study problem (14.3). First of all, we note that the energy functional $\mathcal{J}_{s,\lambda}$ associated with (14.3) is well defined thanks to Lemma 1.31(b). Moreover, $\mathcal{J}_{s,\lambda}$ is Fréchet differentiable in $u \in \mathbb{H}_0^s(\Omega)$, and, for any $\varphi \in \mathbb{H}_0^s(\Omega)$,

$$\langle \mathcal{J}'_{s,\lambda}(u), \varphi \rangle = \int_{\mathbb{R}^n \times \mathbb{R}^n} \frac{(u(x)-u(y))(\varphi(x)-\varphi(y))}{|x-y|^{n+2s}} \, dx\, dy - \lambda \int_\Omega u(x)\varphi(x)\, dx$$
$$- \int_\Omega |u(x)|^{2_s^*-2} u(x)\varphi(x)\, dx.$$

Thus, critical points of $\mathcal{J}_{s,\lambda}$ are solutions to problem (14.3). To find these critical points, we will apply a variant of the mountain pass theorem without the Palais–Smale condition, as given in [11] (see also [46, theorem 2.2]).

Indeed, here we cannot apply the classical mountain pass theorem because, owing to the lack of compactness in the embedding $H^s(\mathbb{R}^n) \hookrightarrow L^{2_s^*}(\mathbb{R}^n)$ (see Lemma 1.31(b)), the functional $\mathcal{J}_{s,\lambda}$ does not verify the Palais–Smale condition globally, but only in an energy range determined by the best fractional critical Sobolev constant S_s given in formula (1.92) (see also Lemma 1.32).

This variant of the mountain pass theorem requires that the functional $\mathcal{J}_{s,\lambda}$ have a suitable *geometric structure*, as stated, for eample, in conditions (2.9) and (2.10) of [46, theorem 2.2].

Hence, we start proving that $\mathcal{J}_{s,\lambda}$ has this geometric feature.

Proposition 14.2 Let $\lambda \in (0, \lambda_{1,s})$. Then there exist $\rho > 0$ and $\beta > 0$ such that, for any $u \in \mathbb{H}_0^s(\Omega)$ with $\|u\|_{\mathbb{H}_0^s(\Omega)} = \rho$, it results that $\mathcal{J}_{s,\lambda}(u) \geq \beta$.

Proof Let u be a function in $\mathbb{H}_0^s(\Omega)$. By the variational characterization of $\lambda_{1,s}$ (see Proposition 3.1), we get

$$\begin{aligned}
\mathcal{J}_{s,\lambda}(u) &= \frac{1}{2}\int_{\mathbb{R}^n \times \mathbb{R}^n} \frac{|u(x)-u(y)|^2}{|x-y|^{n+2s}} \, dx\, dy - \frac{\lambda}{2}\int_\Omega |u(x)|^2\, dx - \frac{1}{2_s^*}\int_\Omega |u(x)|^{2_s^*} dx \\
&\geq \frac{1}{2}\left(1 - \frac{\lambda}{\lambda_{1,s}}\right) \|u\|^2_{\mathbb{H}_0^s(\Omega)} - \frac{1}{2_s^*}\|u\|^{2_s^*}_{L^{2_s^*}(\Omega)} \\
&\geq \frac{1}{2}\left(1 - \frac{\lambda}{\lambda_{1,s}}\right) \|u\|^2_{\mathbb{H}_0^s(\Omega)} - \frac{1}{2_s^* S_s^{2_s^*/2}}\|u\|^{2_s^*}_{L^{2_s^*}(\Omega)},
\end{aligned}$$

thanks to Lemma 6.5 (here we need $0 < \lambda < \lambda_{1,s}$) and to Lemma 1.32. Hence, it easily follows that

$$\mathcal{J}_{s,\lambda}(u) \geq \alpha \|u\|^2_{\mathbb{H}_0^s(\Omega)} \left(1 - \kappa \|u\|^{2_s^*-2}_{\mathbb{H}_0^s(\Omega)}\right),$$

for suitable positive constants α and κ.

Now let $u \in \mathbb{H}_0^s(\Omega)$ be such that $\|u\|_{\mathbb{H}_0^s(\Omega)} = \rho > 0$. Since $2_s^* > 2$, we can choose ρ sufficiently small (i.e., ρ such that $1 - \kappa\rho^{2_s^*-2} > 0$) that

$$\inf_{\substack{u \in \mathbb{H}_0^s(\Omega)\\ \|u\|_{\mathbb{H}_0^s(\Omega)}=\rho}} \mathcal{J}_{s,\lambda}(u) \geq \alpha\rho^2(1 - \kappa\rho^{2_s^*-2}) =: \beta > 0.$$

Hence, Proposition 14.2 is proved. □

Proposition 14.3 Let $\lambda \in (0, \lambda_{1,s})$. Then there exists $e \in \mathbb{H}_0^s(\Omega)$ such that $e \geq 0$ a.e. in $\mathbb{R}^n$, $\|e\|_{\mathbb{H}_0^s(\Omega)} > \rho$, and $\mathcal{J}_{s,\lambda}(e) < 0$, where ρ is given in Proposition 14.2.

Proof We fix $u \in \mathbb{H}_0^s(\Omega)$ such that $\|u\|_{\mathbb{H}_0^s(\Omega)} = 1$ and $u \geq 0$ a.e. in $\mathbb{R}^n$; we remark that this choice is possible thanks to Lemma 1.20 (alternatively, one can replace any $u \in \mathbb{H}_0^s(\Omega)$ with its positive part, which belongs to $\mathbb{H}_0^s(\Omega)$ too, thanks to Lemma 1.22). Also, let $\zeta > 0$. By the fact that λ is positive, we have

$$\begin{aligned}\mathcal{J}_{s,\lambda}(\zeta u) &= \frac{\zeta^2}{2}\int_{\mathbb{R}^n\times\mathbb{R}^n} \frac{|u(x)-u(y)|^2}{|x-y|^{n+2s}}\,dx\,dy - \frac{\lambda}{2}\zeta^2\int_\Omega |u(x)|^2\,dx - \frac{\zeta^{2_s^*}}{2_s^*}\int_\Omega |u(x)|^{2_s^*}\,dx\\ &\leq \frac{\zeta^2}{2}\|u\|^2_{\mathbb{H}_0^s(\Omega)} - \frac{\zeta^{2_s^*}}{2_s^*}\int_\Omega |u(x)|^{2_s^*}\,dx\\ &= \frac{\zeta^2}{2} - \frac{\zeta^{2_s^*}}{2_s^*}\int_\Omega |u(x)|^{2_s^*}\,dx.\end{aligned}$$

Since $2_s^* > 2 > 1$, passing to the limit as $\zeta \to +\infty$ (for u fixed), we get that

$$\mathcal{J}_{s,\lambda}(\zeta u) \to -\infty,$$

so the assertion follows, taking $e = \zeta u$, with ζ sufficiently large. □

14.3 Some crucial estimates

This section is devoted to the proof of some estimates, that will be crucial in proving that the critical level of the functional $\mathcal{J}_{s,\lambda}$ stays below the threshold where the Palais–Smale condition is satisfied.

First of all, we need the following result:

Proposition 14.4 For any $u_0 \in \mathbb{H}_0^s(\Omega)\setminus\{0\}$, the following relation holds true:

$$\sup_{\zeta\geq 0} \mathcal{J}_{s,\lambda}(\zeta u_0) = \frac{s}{n}\, S_{s,\lambda}^{n/(2s)}(u_0),$$

where the function $H^s(\mathbb{R}^n)\setminus\{0\} \ni v \mapsto S_{s,\lambda}(v)$ is defined in (14.6).

Proof Fix $u_0 \in \mathbb{H}_0^s(\Omega)$, and let $g : [0,+\infty) \to \mathbb{R}$ be the function

$$g(\zeta) = \frac{\zeta^2}{2} \int_{\mathbb{R}^n \times \mathbb{R}^n} \frac{|u_0(x) - u_0(y)|^2}{|x-y|^{n+2s}} \, dx\, dy - \frac{\zeta^2 \lambda}{2} \int_\Omega |u_0(x)|^2 \, dx - \frac{\zeta^{2_s^*}}{2_s^*} \int_\Omega |u_0(x)|^{2_s^*} \, dx.$$

Note that g is differentiable in $(0,+\infty)$, and

$$g'(\zeta) = \zeta \int_{\mathbb{R}^n \times \mathbb{R}^n} \frac{|u_0(x) - u_0(y)|^2}{|x-y|^{n+2s}} \, dx\, dy - \zeta \lambda \int_\Omega |u_0(x)|^2 \, dx - \zeta^{2_s^*-1} \int_\Omega |u_0(x)|^{2_s^*} \, dx,$$

so $g'(\zeta) \geq 0$ if and only if

$$\zeta \leq \bar{\zeta} = \left(\frac{\displaystyle\int_{\mathbb{R}^n \times \mathbb{R}^n} \frac{|u_0(x) - u_0(y)|^2}{|x-y|^{n+2s}} \, dx\, dy - \lambda \int_\Omega |u_0(x)|^2 \, dx}{\displaystyle\int_\Omega |u_0(x)|^{2_s^*} \, dx} \right)^{1/(2_s^*-2)}.$$

Therefore, $\bar{\zeta}$ is a maximum for g, and

$$\sup_{\zeta \geq 0} g(\zeta) = \max_{\zeta \geq 0} g(\zeta) = g(\bar{\zeta}) = \frac{s}{n} S_{s,\lambda}^{n/(2s)}(u_0).$$

This concludes the proof. □

Now we proceed by proving some lemmas.

Lemma 14.5 *The infimum in formula* (1.92) *is attained; that is,*

$$S_s = S_s(\widetilde{u}),$$

where

$$\widetilde{u}(x) := \kappa \left(\mu^2 + |x - x_0|^2 \right)^{-(n-2s)/2} \quad x \in \mathbb{R}^n, \tag{14.8}$$

with $\kappa \in \mathbb{R} \setminus \{0\}$, $\mu > 0$ and $x_0 \in \mathbb{R}^n$ fixed constants.

Equivalently, the function $\bar{u}$ defined as

$$\bar{u}(x) := \frac{\widetilde{u}(x)}{\|\widetilde{u}\|_{L^{2_s^*}(\mathbb{R}^n)}} \tag{14.9}$$

is such that

$$S_s = \inf_{\substack{v \in H^s(\mathbb{R}^n) \\ \|v\|_{L^{2_s^*}(\mathbb{R}^n)} = 1}} \int_{\mathbb{R}^n \times \mathbb{R}^n} \frac{|v(x) - v(y)|^2}{|x-y|^{n+2s}} \, dx\, dy = \int_{\mathbb{R}^n \times \mathbb{R}^n} \frac{|\bar{u}(x) - \bar{u}(y)|^2}{|x-y|^{n+2s}} \, dx\, dy.$$

Proof This assertion comes from [82, theorem 1.1]. □

As in the classical case of the Laplacian, starting from $\bar{u}$, we can construct an explicit solution of the following limiting problem

$$(-\Delta)^s u = |u|^{2^*_s - 2} u \quad \text{in } \mathbb{R}^n. \tag{14.10}$$

This is shown in the following result:

Lemma 14.6 *The function*

$$u^*(x) := \bar{u}\left(\frac{x}{S_s^{1/(2s)}}\right) \quad x \in \mathbb{R}^n \tag{14.11}$$

is a solution of problem (14.10) *satisfying the property*

$$\|u^*\|^{2^*_s}_{L^{2^*_s}(\mathbb{R}^n)} = S_s^{n/(2s)}. \tag{14.12}$$

Proof By Lemma 14.5, there exists a Lagrange multiplier $\gamma \in \mathbb{R}$ such that

$$\begin{aligned} &\int_{\mathbb{R}^n \times \mathbb{R}^n} \frac{(\bar{u}(x) - \bar{u}(y))(\varphi(x) - \varphi(y))}{|x-y|^{n+2s}}\, dx\, dy \\ &\qquad = \gamma \int_{\mathbb{R}^n} |\bar{u}(x)|^{2^*_s - 2} \bar{u}(x) \varphi(x) dx, \end{aligned} \tag{14.13}$$

for any $\varphi \in H^s(\mathbb{R}^n)$. Taking $\varphi = \bar{u}$ as a test function in (14.13) and using again Lemma 14.5, we get

$$S_s = \int_{\mathbb{R}^n \times \mathbb{R}^n} \frac{|\bar{u}(x) - \bar{u}(y)|^2}{|x-y|^{n+2s}}\, dx\, dy = \gamma \tag{14.14}$$

because $\|\bar{u}\|_{L^{2^*_s}(\mathbb{R}^n)} = 1$ by construction (see (14.9)).

Let u^* be the function defined as in (14.11). Note that $u^* \in H^s(\mathbb{R}^n) \setminus \{0\}$. Moreover, for any $\varphi \in H^s(\mathbb{R}^n)$, by (14.13) and (14.14), we have

$$\begin{aligned} &\int_{\mathbb{R}^n \times \mathbb{R}^n} \frac{(u^*(x) - u^*(y))(\varphi(x) - \varphi(y))}{|x-y|^{n+2s}}\, dx\, dy \\ &\quad = S_s^{(n-2s)/(2s)} \int_{\mathbb{R}^n \times \mathbb{R}^n} \frac{(\bar{u}(x) - \bar{u}(y))(\varphi(S_s^{1/(2s)} x) - \varphi(S_s^{1/(2s)} y))}{|x-y|^{n+2s}}\, dx\, dy \\ &\quad = S_s^{n/(2s)} \int_{\mathbb{R}^n} |\bar{u}(x)|^{2^*_s - 2} \bar{u}(x) \varphi(S_s^{1/(2s)} x) dx \\ &\quad = \int_{\mathbb{R}^n} |u^*(x)|^{2^*_s - 2} u^*(x) \varphi(x) dx; \end{aligned} \tag{14.15}$$

that is, u^* is a solution of problem (14.10).

Finally, taking $\varphi = u^*$ as a test function in (14.15) and using (14.9) and (14.11), we get

$$\begin{aligned}\int_{\mathbb{R}^n\times\mathbb{R}^n} \frac{|u^*(x)-u^*(y)|^2}{|x-y|^{n+2s}}\,dx\,dy &= \int_{\mathbb{R}^n} |u^*(x)|^{2_s^*}\,dx \\ &= S_s^{n/(2s)} \int_{\mathbb{R}^n} |\bar{u}(x)|^{2_s^*}\,dx \\ &= S_s^{n/(2s)},\end{aligned}$$

so the proof is complete. □

Now we consider the family of functions U_ε defined as

$$U_\varepsilon(x) := \varepsilon^{-(n-2s)/2}\,u^*(x/\varepsilon) \quad x \in \mathbb{R}^n, \tag{14.16}$$

for any $\varepsilon > 0$. The scaling invariance of the problem is pointed out in the following observation:

Lemma 14.7 *The function U_ε is a solution of problem* (14.10) *and verifies the following equality:*

$$\int_{\mathbb{R}^n\times\mathbb{R}^n} \frac{|U_\varepsilon(x)-U_\varepsilon(y)|^2}{|x-y|^{n+2s}}\,dx\,dy = \int_{\mathbb{R}^n} |U_\varepsilon(x)|^{2_s^*}\,dx = S_s^{n/(2s)},$$

for any $\varepsilon > 0$.

Proof By (14.16) and Lemma 14.6, it easily follows that $U_\varepsilon \in H^s(\mathbb{R}^n)$, and

$$\begin{aligned}&\int_{\mathbb{R}^n\times\mathbb{R}^n} \frac{(U_\varepsilon(x)-U_\varepsilon(y))(\varphi(x)-\varphi(y))}{|x-y|^{n+2s}}\,dx\,dy \\ &\quad = \varepsilon^{(n-2s)/2} \int_{\mathbb{R}^n\times\mathbb{R}^n} \frac{(u^*(x)-u^*(y))(\varphi(\varepsilon x)-\varphi(\varepsilon y))}{|x-y|^{n+2s}}\,dx\,dy \\ &\quad = \varepsilon^{(n-2s)/2} \int_{\mathbb{R}^n} |u^*(x)|^{2_s^*-2}u^*(x)\varphi(\varepsilon x)dx \\ &\quad = \int_{\mathbb{R}^n} |U_\varepsilon(x)|^{2_s^*-2}U_\varepsilon(x)\varphi(x)dx,\end{aligned} \tag{14.17}$$

for any $\varphi \in H^s(\mathbb{R}^n)$ and $\varepsilon > 0$. Hence, U_ε is a solution of problem (14.10) for any $\varepsilon > 0$. Taking $\varphi = U_\varepsilon$ in (14.17) (this choice is admissible, because $U_\varepsilon \in H^s(\mathbb{R}^n)$), we get

$$\int_{\mathbb{R}^n\times\mathbb{R}^n} \frac{|U_\varepsilon(x)-U_\varepsilon(y)|^2}{|x-y|^{n+2s}}\,dx\,dy = \int_{\mathbb{R}^n} |U_\varepsilon(x)|^{2_s^*}\,dx,$$

for any $\varepsilon > 0$. The scale invariance of the norm in $L^{2_s^*}(\mathbb{R}^n)$ under the scaling $u^* \mapsto U_\varepsilon$ and (14.12) gives

$$\int_{\mathbb{R}^n} |U_\varepsilon(x)|^{2_s^*}\,dx = \int_{\mathbb{R}^n} |u^*(x)|^{2_s^*}\,dx = S_s^{n/(2s)},$$

for any $\varepsilon > 0$, so the assertion is proved. □

Now it is necessary to appropriately put U_ε to zero outside Ω, according to the following procedure. Let us fix $\delta > 0$ such that

$$B_{4\delta} \subset \Omega, \tag{14.18}$$

and let $\eta \in C^\infty(\mathbb{R}^n)$ be such that $0 \le \eta \le 1$ in $\mathbb{R}^n$, $\eta \equiv 1$ in B_δ, and $\eta \equiv 0$ in $\mathscr{C} B_{2\delta}$, where $B_\delta = B(0,\delta)$ and $\mathscr{C} B_\delta = \mathbb{R}^n \setminus B_\delta$. For every $\varepsilon > 0$, we denote by u_ε the following function:

$$u_\varepsilon(x) := \eta(x) U_\varepsilon(x) \quad x \in \mathbb{R}^n, \tag{14.19}$$

where U_ε is given in (14.16). Note that $u_\varepsilon \in \mathbb{H}^s_0(\Omega)$ and $u_\varepsilon = 0$ a.e. in $\mathbb{R}^n \setminus \Omega$.

Now we need some preliminary lemmas. In what follows, we suppose that, up to a translation, $x_0 = 0$ in (14.8).

Lemma 14.8 *Let $\rho > 0$ and μ be as in* (14.8). *If $x \in \mathscr{C} B_\rho$, then*

$$|u_\varepsilon(x)| \le |U_\varepsilon(x)| \le C\varepsilon^{(n-2s)/2},$$

for any $\varepsilon > 0$ and for some positive constant C, possibly depending on μ, ρ, s, and n.

Proof Using (14.8) (with $x_0 = 0$), (14.9), (14.11), and (14.16), we have that

$$U_\varepsilon(x) = \widetilde{\kappa}\varepsilon^{-(n-2s)/2} \left(\mu^2 + \left| \frac{x}{\varepsilon S_s^{1/(2s)}} \right|^2 \right)^{-(n-2s)/2}, \tag{14.20}$$

for a suitable $\widetilde{\kappa} \in \mathbb{R} \setminus \{0\}$. If $x \in \mathscr{C} B_\rho$, we deduce that

$$|U_\varepsilon(x)| \le C\varepsilon^{-(n-2s)/2} \left(\mu^2 + \left(\frac{\rho}{\varepsilon S_s^{1/(2s)}} \right)^2 \right)^{-(n-2s)/2} \le C\varepsilon^{(n-2s)/2},$$

so the assertion follows because $\eta \le 1$ in $\mathbb{R}^n$. □

Lemma 14.9 *Let $\rho > 0$ and μ be as in* (14.8). *If $x \in \mathscr{C} B_\rho$, then*

$$|\nabla u_\varepsilon(x)| \le C\varepsilon^{(n-2s)/2},$$

for any $\varepsilon > 0$ and for some positive constant C, possibly depending on μ, ρ, s, and n.

Proof First of all, we observe that, for any $|\xi| \ge \rho$, we have that

$$\left(1 + \left| \frac{\xi}{\varepsilon} \right|^2 \right)^{-(n-2s)/2} + \frac{|\xi|}{\varepsilon^2} \left(1 + \left| \frac{\xi}{\varepsilon} \right|^2 \right)^{-1-(n-2s)/2}$$

$$
\begin{aligned}
&\le \left(1+\left|\frac{\xi}{\varepsilon}\right|^2\right)^{-(n-2s)/2} + \frac{1}{\rho}\left|\frac{\xi}{\varepsilon}\right|^2\left(1+\left|\frac{\xi}{\varepsilon}\right|^2\right)^{-1-(n-2s)/2}\\
&\le \left(1+\frac{1}{\rho}\right)\left(1+\left|\frac{\xi}{\varepsilon}\right|^2\right)^{-(n-2s)/2}\\
&\le \left(1+\frac{1}{\rho}\right)\left(1+\left|\frac{\rho}{\varepsilon}\right|^2\right)^{-(n-2s)/2}\\
&\le \left(1+\frac{1}{\rho}\right)\left(\frac{\varepsilon}{\rho}\right)^{n-2s}.
\end{aligned}
\tag{14.21}
$$

As a consequence of (14.19) and (14.21), we have that, for any $x \in \mathscr{C} B_\rho$,

$$
\begin{aligned}
|\nabla u_\varepsilon(x)| &\le C\varepsilon^{-(n-2s)/2}\left[\left(\mu^2+\left|\frac{x}{\varepsilon S_s^{1/(2s)}}\right|^2\right)^{-(n-2s)/2}\right.\\
&\qquad\left. + \frac{1}{\varepsilon S_s^{1/(2s)}}\left|\frac{x}{\varepsilon S_s^{1/(2s)}}\right|\left(\mu^2+\left|\frac{x}{\varepsilon S_s^{1/(2s)}}\right|^2\right)^{-1-(n-2s)/2}\right]\\
&\le C\varepsilon^{-(n-2s)/2}\cdot\varepsilon^{n-2s} = C\varepsilon^{(n-2s)/2},
\end{aligned}
$$

which proves the assertion. □

Lemma 14.10 *Let δ be as in* (14.18) *and μ be as in* (14.8). *Then the following assertions hold true:*

(*a*) *For any $x \in \mathbb{R}^n$ and $y \in \mathscr{C} B_\delta$, with $|x-y| \le \delta/2$,*

$$
|u_\varepsilon(x) - u_\varepsilon(y)| \le C\varepsilon^{(n-2s)/2}|x-y|, \quad \textit{and} \tag{14.22}
$$

(*b*) *For any $x, y \in \mathscr{C} B_\delta$,*

$$
|u_\varepsilon(x) - u_\varepsilon(y)| \le C\varepsilon^{(n-2s)/2}\min\{1, |x-y|\}, \tag{14.23}
$$

for any $\varepsilon > 0$ and for some positive constant C, possibly depending on μ, δ, s, and n.

Proof Let us start by proving assertion (a). For this, let $x \in \mathbb{R}^n$ and $y \in \mathscr{C} B_\delta$ with $|x-y| \le \delta/2$, and let ξ be any point on the segment joining x and y. Then

$$
\xi = tx + (1-t)y, \quad \text{for some } t \in [0,1],
$$

so

$$
|\xi| = |y + t(x-y)| \ge |y| - t|x-y| \ge \delta - t(\delta/2) \ge \delta/2.
$$

This and Lemma 14.9 (here used with $\rho := \delta/2$) imply that $|\nabla u_\varepsilon(\xi)| \le C\varepsilon^{(n-2s)/2}$, so, by a first-order Taylor expansion,

$$
|u_\varepsilon(x) - u_\varepsilon(y)| \le C\varepsilon^{(n-2s)/2}|x-y|,
$$

which proves (14.22).

Now we show (b). For this, let $x, y \in \mathscr{C} B_\delta$. If $|x-y| \le \delta/2$, then (b) follows from (a), so we may suppose $|x-y| > \delta/2$. Then

$$|u_\varepsilon(x) - u_\varepsilon(y)| \le |u_\varepsilon(x)| + |u_\varepsilon(y)| \le C\varepsilon^{(n-2s)/2},$$

thanks to Lemma 14.8 (here used with $\rho := \delta$), and this completes the proof of (14.23). □

Now we can estimate the Gagliardo seminorm of u_ε according to the following result:

Proposition 14.11 Let $s \in (0,1)$ and $n > 2s$. Then the following estimate holds true:

$$\int_{\mathbb{R}^n \times \mathbb{R}^n} \frac{|u_\varepsilon(x) - u_\varepsilon(y)|^2}{|x-y|^{n+2s}} \, dx \, dy \le S_s^{n/(2s)} + O(\varepsilon^{n-2s})$$

as $\varepsilon \to 0$.

Proof The proof makes use of the previous estimates and is a bit complicated – definitely more difficult than the one for similar results in the case of the Laplacian. The additional complications arise not only from technical obstacles (differentiating is, of course, simpler than integrating) but also from a different distribution of the energy density. In particular, in the fractional case, the interaction between the interior and the exterior of the cutoff in (14.19) is small but not negligible (indeed, it will contribute to the total energy according to estimates (14.28), (14.29), and (14.32)). To keep track of these nonlocal interactions, we introduce the notations

$$\mathbb{D} := \{(x,y) \in \mathbb{R}^n \times \mathbb{R}^n : x \in B_\delta, \ y \in \mathscr{C} B_\delta \text{ and } |x-y| > \delta/2\}$$

and

$$\mathbb{E} := \{(x,y) \in \mathbb{R}^n \times \mathbb{R}^n : x \in B_\delta, \ y \in \mathscr{C} B_\delta \text{ and } |x-y| \le \delta/2\},$$

where δ is as in (14.18).

By (14.19), we have that

$$\begin{aligned}\int_{\mathbb{R}^n \times \mathbb{R}^n} \frac{|u_\varepsilon(x) - u_\varepsilon(y)|^2}{|x-y|^{n+2s}} \, dx \, dy = &\int_{B_\delta \times B_\delta} \frac{|U_\varepsilon(x) - U_\varepsilon(y)|^2}{|x-y|^{n+2s}} \, dx \, dy \\ &+ 2 \int_{\mathbb{D}} \frac{|u_\varepsilon(x) - u_\varepsilon(y)|^2}{|x-y|^{n+2s}} \, dx \, dy \\ &+ 2 \int_{\mathbb{E}} \frac{|u_\varepsilon(x) - u_\varepsilon(y)|^2}{|x-y|^{n+2s}} \, dx \, dy \\ &+ \int_{(\mathscr{C} B_\delta) \times (\mathscr{C} B_\delta)} \frac{|u_\varepsilon(x) - u_\varepsilon(y)|^2}{|x-y|^{n+2s}} \, dx \, dy.\end{aligned} \tag{14.24}$$

By (14.19) and Lemma 14.10 (here, in particular, we use (14.23)), we have

$$\int_{(\mathscr{C}B_\delta)\times(\mathscr{C}B_\delta)} \frac{|u_\varepsilon(x)-u_\varepsilon(y)|^2}{|x-y|^{n+2s}}\,dx\,dy \le C\varepsilon^{n-2s}\int_{B_{2\delta}\times\mathbb{R}^n} \frac{\min\{1,|x-y|^2\}}{|x-y|^{n+2s}}\,dx\,dy$$
$$= O(\varepsilon^{n-2s}), \tag{14.25}$$

while, by (14.22), one has

$$\begin{aligned}\int_{\mathbb{E}} \frac{|u_\varepsilon(x)-u_\varepsilon(y)|^2}{|x-y|^{n+2s}}\,dx\,dy &\le C\varepsilon^{n-2s}\int_{\substack{x\in B_\delta,\,y\in\mathscr{C}B_\delta\\|x-y|\le\delta/2}} \frac{|x-y|^2}{|x-y|^{n+2s}}\,dx\,dy\\ &\le C\varepsilon^{n-2s}\int_{|x|\le\delta} dx\int_{|\xi|\le\delta/2}\frac{1}{|\xi|^{n+2s-2}}\,d\xi\\ &= O(\varepsilon^{n-2s})\end{aligned} \tag{14.26}$$

as $\varepsilon\to 0$. In both these estimates, we use that $s\in(0,1)$.

Now, in (14.24), it remains to estimate the integral on $\mathbb{D}$; that is,

$$\int_{\mathbb{D}} \frac{|u_\varepsilon(x)-u_\varepsilon(y)|^2}{|x-y|^{n+2s}}\,dx\,dy. \tag{14.27}$$

For this, recalling that $u_\varepsilon(x)=U_\varepsilon(x)$, for any $x\in B_\delta$, thanks to (14.19), we note that, for any $(x,y)\in\mathbb{D}$,

$$\begin{aligned}|u_\varepsilon(x)-u_\varepsilon(y)|^2 &= |U_\varepsilon(x)-u_\varepsilon(y)|^2\\ &= |(U_\varepsilon(x)-U_\varepsilon(y))+(U_\varepsilon(y)-u_\varepsilon(y))|^2\\ &\le |U_\varepsilon(x)-U_\varepsilon(y)|^2+|U_\varepsilon(y)-u_\varepsilon(y)|^2\\ &\quad +2|U_\varepsilon(x)-U_\varepsilon(y)|\,|U_\varepsilon(y)-u_\varepsilon(y)|,\end{aligned}$$

so

$$\begin{aligned}\int_{\mathbb{D}} \frac{|u_\varepsilon(x)-u_\varepsilon(y)|^2}{|x-y|^{n+2s}}\,dx\,dy &\le \int_{\mathbb{D}} \frac{|U_\varepsilon(x)-U_\varepsilon(y)|^2}{|x-y|^{n+2s}}\,dx\,dy\\ &\quad+\int_{\mathbb{D}} \frac{|U_\varepsilon(y)-u_\varepsilon(y)|^2}{|x-y|^{n+2s}}\,dx\,dy\\ &\quad+2\int_{\mathbb{D}} \frac{|U_\varepsilon(x)-U_\varepsilon(y)|\,|U_\varepsilon(y)-u_\varepsilon(y)|}{|x-y|^{n+2s}}\,dx\,dy.\end{aligned} \tag{14.28}$$

Hence, to estimate (14.27), we bound the last two terms on the right-hand side of (14.28).

By exploiting Lemma 14.8 (here used with $\rho = \delta$), we obtain

$$\begin{aligned}
\int_{\mathbb{D}} \frac{|U_\varepsilon(y) - u_\varepsilon(y)|^2}{|x-y|^{n+2s}}\,dx\,dy &\le \int_{\mathbb{D}} \frac{(|U_\varepsilon(y)| + |u_\varepsilon(y)|)^2}{|x-y|^{n+2s}}\,dx\,dy \\
&\le 4\int_{\mathbb{D}} \frac{|U_\varepsilon(y)|^2}{|x-y|^{n+2s}}\,dx\,dy \\
&\le C\varepsilon^{n-2s} \int_{\substack{x\in B_\delta,\, y\in \mathscr{C}B_\delta \\ |x-y|>\delta/2}} \frac{1}{|x-y|^{n+2s}}\,dx\,dy \qquad (14.29)\\
&\le C\varepsilon^{n-2s} \int_{|\zeta|\le\delta} d\zeta \int_{|\xi|>\delta/2} \frac{1}{|\xi|^{n+2s}}\,d\xi \\
&= O(\varepsilon^{n-2s})
\end{aligned}$$

as $\varepsilon \to 0$.

In order to estimate the last term on the right-hand side of (14.28), first, we note that, once more by (14.20) (which is valid for any $x \in \mathbb{R}^n$) and Lemma 14.8,

$$|U_\varepsilon(x)|\,|U_\varepsilon(y)| \le C\left(\mu^2 + \left|\frac{x}{\varepsilon S_s^{1/(2s)}}\right|^2\right)^{-(n-2s)/2},$$

for any $(x, y) \in \mathbb{D}$. Therefore, by using the short-hand notation

$$\delta_\varepsilon := \delta/(\varepsilon S_s^{1/(2s)})$$

and the change of variable $\zeta := x/(\varepsilon S_s^{1/(2s)})$ and $\xi := x - y$ (and up to renaming C), we have that

$$\begin{aligned}
\int_{\mathbb{D}} \frac{|U_\varepsilon(x)|\,|U_\varepsilon(y)|}{|x-y|^{n+2s}}\,dx\,dy &\le C\int_{\mathbb{D}} \left(\mu^2 + \left|\frac{x}{\varepsilon S_s^{1/(2s)}}\right|^2\right)^{-(n-2s)/2} |x-y|^{-(n+2s)}\,dx\,dy \\
&\le C\varepsilon^n \int_{\substack{\zeta\in B_{\delta_\varepsilon} \\ |\xi|>\delta/2}} (\mu^2 + |\zeta|^2)^{-(n-2s)/2} |\xi|^{-(n+2s)}\,d\zeta\,d\xi \\
&\le C\varepsilon^n \int_{\zeta\in B_{\delta_\varepsilon}} (\mu^2 + |\zeta|^2)^{-(n-2s)/2}\,d\zeta \\
&\le C\varepsilon^n \left[1 + \int_{\zeta\in B_{\delta_\varepsilon}\setminus B_1} (\mu^2 + |\zeta|^2)^{-(n-2s)/2}\,d\zeta\right] \\
&\le C\varepsilon^n \left[1 + \int_{\zeta\in B_{\delta_\varepsilon}\setminus B_1} |\zeta|^{-(n-2s)}\,d\zeta\right]
\end{aligned}$$

$$= C\varepsilon^n \left[1 + \int_1^{\delta/(\varepsilon S_s^{1/(2s)})} \rho^{-(n-2s)+(n-1)}\, d\rho \right]$$
$$= C\varepsilon^n (1 + \varepsilon^{-2s})$$
$$= O(\varepsilon^{n-2s})$$

as $\varepsilon \to 0$. Here we use again that $s \in (0,1)$.

As a consequence, recalling (14.19) and Lemma 14.8, we get

$$\begin{aligned} \int_{\mathbb{D}} \frac{|U_\varepsilon(x)|\,|U_\varepsilon(y) - u_\varepsilon(y)|}{|x-y|^{n+2s}}\, dx\, dy &\le \int_{\mathbb{D}} \frac{|U_\varepsilon(x)|\,(|U_\varepsilon(y)| + |u_\varepsilon(y)|)}{|x-y|^{n+2s}}\, dx\, dy \\ &\le 2 \int_{\mathbb{D}} \frac{|U_\varepsilon(x)|\,|U_\varepsilon(y)|}{|x-y|^{n+2s}}\, dx\, dy \\ &= O(\varepsilon^{n-2s}) \end{aligned} \tag{14.30}$$

as $\varepsilon \to 0$.

However, again by (14.19) and Lemma 14.8,

$$\begin{aligned} \int_{\mathbb{D}} \frac{|U_\varepsilon(y)|\,|U_\varepsilon(y) - u_\varepsilon(y)|}{|x-y|^{n+2s}}\, dx\, dy &\le 2 \int_{\mathbb{D}} \frac{|U_\varepsilon(y)|^2}{|x-y|^{n+2s}}\, dx\, dy \\ &\le C\varepsilon^{n-2s} \int_{\substack{x \in B_\delta,\, y \in \mathscr{C} B_\delta \\ |x-y| > \delta/2}} \frac{1}{|x-y|^{n+2s}}\, dx\, dy \\ &= O(\varepsilon^{n-2s}) \end{aligned} \tag{14.31}$$

as $\varepsilon \to 0$ (here we argue as in (14.29)). Putting together (14.30) with (14.31), we infer that

$$\begin{aligned} \int_{\mathbb{D}} \frac{|U_\varepsilon(x) - U_\varepsilon(y)|\,|U_\varepsilon(y) - u_\varepsilon(y)|}{|x-y|^{n+2s}}\, dx\, dy &\le \int_{\mathbb{D}} \frac{|U_\varepsilon(x)|\,|U_\varepsilon(y) - u_\varepsilon(y)|}{|x-y|^{n+2s}}\, dx\, dy \\ &\quad + \int_{\mathbb{D}} \frac{|U_\varepsilon(y)|\,|U_\varepsilon(y) - u_\varepsilon(y)|}{|x-y|^{n+2s}}\, dx\, dy \\ &= O(\varepsilon^{n-2s}) \end{aligned} \tag{14.32}$$

as $\varepsilon \to 0$.

Finally, by (14.24), (14.26), (14.28), (14.29), and (14.32), we get[1]

$$\begin{aligned}\int_{\mathbb{R}^n\times\mathbb{R}^n}\frac{|u_\varepsilon(x)-u_\varepsilon(y)|^2}{|x-y|^{n+2s}}\,dx\,dy &= \int_{B_\delta\times B_\delta}\frac{|U_\varepsilon(x)-U_\varepsilon(y)|^2}{|x-y|^{n+2s}}\,dx\,dy\\ &\quad+2\int_{\mathbb{D}}\frac{|U_\varepsilon(x)-U_\varepsilon(y)|^2}{|x-y|^{n+2s}}\,dx\,dy+O(\varepsilon^{n-2s})\\ &\le \int_{\mathbb{R}^n\times\mathbb{R}^n}\frac{|U_\varepsilon(x)-U_\varepsilon(y)|^2}{|x-y|^{n+2s}}\,dx\,dy+O(\varepsilon^{n-2s})\end{aligned}$$

as $\varepsilon\to 0$.

Then the desired result now follows from Lemma 14.7. □

Now we consider the L^2 and the $L^{2^*_s}$-norms of the function u_ε; in this case, our estimates can be proved exactly as in the case of the Laplacian, but we prefer to repeat them for the reader's convenience.

Proposition 14.12 Let $s\in(0,1)$ and $n>2s$.

Then the following estimates hold true:

$$\int_{\mathbb{R}^n}|u_\varepsilon(x)|^2\,dx \ge \begin{cases} C_s\varepsilon^{2s}+O(\varepsilon^{n-2s}) & \text{if}\quad n>4s\\ C_s\varepsilon^{2s}|\log\varepsilon|+O(\varepsilon^{2s}) & \text{if}\quad n=4s\\ C_s\varepsilon^{n-2s}+O(\varepsilon^{2s}) & \text{if}\quad n<4s\end{cases} \tag{14.33}$$

and

$$\int_{\mathbb{R}^n}|u_\varepsilon(x)|^{2^*_s}\,dx = S_s^{n/(2s)}+O(\varepsilon^n) \tag{14.34}$$

as $\varepsilon\to 0$, for some positive constant C_s depending on s.

Proof By definition of u_ε and (14.16), we get

$$\begin{aligned}\int_{\mathbb{R}^n}|u_\varepsilon(x)|^2\,dx &= \int_{B_\delta}|U_\varepsilon(x)|^2\,dx+\int_{B_{2\delta}\setminus B_\delta}|\eta(x)U_\varepsilon(x)|^2\,dx\\ &\ge \varepsilon^{-(n-2s)}\int_{B_\delta}|u^*(x/\varepsilon)|^2\,dx\\ &\ge \varepsilon^{2s}\int_R^{\delta/\varepsilon}|u^*(r)|^2r^{n-1}\,dr,\end{aligned}$$

[1] It is interesting to observe that the energy interaction outside $B_\delta\times B_\delta$ is not negligible; indeed, while the contributions in (14.25), (14.26), (14.29), and (14.32) are all $O(\varepsilon^{n-2s})$, the integral in $\mathbb{D}$ provides a relevant part that needs to be taken into account in the full energy.

for any $0 < R < \delta/\varepsilon$. Now, by (14.11) and Lemma 14.5, we get

$$\begin{aligned}\int_{\mathbb{R}^n} |u_\varepsilon(x)|^2 dx &\geq \varepsilon^{2s} \int_R^{\delta/\varepsilon} |u^*(r)|^2 r^{n-1} dr \\ &\cong \varepsilon^{2s} \int_R^{\delta/\varepsilon} r^{-n+4s-1} dr \\ &= \begin{cases} -C_{1,s}\varepsilon^{n-2s} + C_{2,s}\varepsilon^{2s} & \text{if} \quad n > 4s \\ C_{1,s}\varepsilon^{2s}|\log\varepsilon| + C_{2,s}\varepsilon^{2s} & \text{if} \quad n = 4s \\ C_{1,s}\varepsilon^{n-2s} - C_{2,s}\varepsilon^{2s} & \text{if} \quad n < 4s, \end{cases} \\ &= \begin{cases} C_s\varepsilon^{2s} + O(\varepsilon^{n-2s}) & \text{if} \quad n > 4s \\ C_s\varepsilon^{2s}|\log\varepsilon| + O(\varepsilon^{2s}) & \text{if} \quad n = 4s \\ C_s\varepsilon^{n-2s} + O(\varepsilon^{2s}) & \text{if} \quad n < 4s \end{cases}\end{aligned}$$

as $\varepsilon \to 0$, for some positive constant $C_{1,s}$, $C_{2,s}$, and C_s depending on s and for ε sufficiently small. This gives (14.33), and we now prove (14.34). Arguing in the same way and using Lemma 14.7, we have

$$\begin{aligned}\int_{\mathbb{R}^n} |u_\varepsilon(x)|^{2_s^*} dx &= \int_{\mathbb{R}^n} |U_\varepsilon(x)|^{2_s^*} dx + \int_{\mathbb{R}^n} \left(|\eta(x)|^{2_s^*} - 1\right) |U_\varepsilon(x)|^{2_s^*} dx \\ &= S_s^{n/(2s)} + \int_{\mathbb{R}^n \setminus B_\delta} \left(|\eta(x)|^{2_s^*} - 1\right) |U_\varepsilon(x)|^{2_s^*} dx \\ &= S_s^{n/(2s)} + \varepsilon^{-n} \int_{\mathbb{R}^n \setminus B_\delta} \left(|\eta(x)|^{2_s^*} - 1\right) |u^*(x/\varepsilon)|^{2_s^*} dx \\ &\cong S_s^{n/(2s)} + C\varepsilon^n \int_{\mathbb{R}^n \setminus B_\delta} |x|^{-2n} dx \\ &= S_s^{n/(2s)} + O(\varepsilon^n)\end{aligned}$$

as $\varepsilon \to 0$. This ends the proof. □

14.3.1 Some remarks on condition $S_{s,\lambda} < S_s$

As in the classical case (see the discussion in [46, pp. 458–9]), as well as in the nonlocal setting, the Palais–Smale condition may globally fail for the functional $\mathcal{J}_{s,\lambda}$. Therefore, it is necessary to restrict our attention only to a suitable range related to the best fractional critical Sobolev constant S_s of the embedding $H^s(\mathbb{R}^n) \hookrightarrow L^{2_s^*}(\mathbb{R}^n)$. Indeed, $\mathcal{J}_{s,\lambda}$ satisfies the Palais–Smale condition at any level c, with c satisfying

$$c < \frac{s}{n}\, S_s^{n/(2s)}. \tag{14.35}$$

As discussed in Section 14.1, another natural condition to look at is given by

$$S_{s,\lambda} < S_s. \tag{14.36}$$

It turns out that conditions (14.35) and (14.36) are related as follows:

Proposition 14.13 Let $\lambda \in (0, \lambda_{1,s})$, and let

$$c := \inf_{P\in\mathcal{P}} \sup_{v\in P([0,1])} \mathcal{J}_{s,\lambda}(v), \tag{14.37}$$

where

$$\mathcal{P} := \left\{P \in C([0,1]; \mathbb{H}_0^s(\Omega)) : P(0) = 0,\ P(1) = e\right\},$$

with e given in Proposition 14.3. Then conditions (14.35) and (14.36) are equivalent.

Proof It is enough to show that

$$c = \frac{s}{n}\, S_{s,\lambda}^{n/(2s)}. \tag{14.38}$$

For this, let $u \in H^s(\mathbb{R}^n) \setminus \{0\}$, and let $P(\zeta) = \zeta u$, with $\zeta \in [0,1]$. By (14.37) and Proposition 14.4, it is easily seen that

$$c \le \sup_{\zeta\ge 0} \mathcal{J}_{s,\lambda}(\zeta u) = \frac{s}{n}\, S_{s,\lambda}^{n/(2s)}(u),$$

so, taking the infimum on $u \in H^s(\mathbb{R}^n) \setminus \{0\}$ and using the definition (14.5), we obtain

$$c \le \frac{s}{n}\, S_{s,\lambda}^{n/(2s)}. \tag{14.39}$$

Now let us prove the other inequality. To do this, let us consider the continuous function

$$[0,1] \ni \zeta \mapsto \langle \mathcal{J}'_{s,\lambda}(P(\zeta)), P(\zeta)\rangle,$$

for any fixed $P \in \mathcal{P}$. Let

$$\tilde{\zeta} := \sup\{\zeta \in [0,1] \ :\ P(\zeta) = 0\}. \tag{14.40}$$

Since $P(0) = 0$, $P(1) = e \neq 0$, and P is continuous, it follows that

$$\tilde{\zeta} \in [0,1). \tag{14.41}$$

Since $2 < 2_s^*$ and $\mathcal{J}_{s,\lambda}(e) < 0$ by assumption, we have that

$$\begin{aligned}\langle \mathcal{J}'_{s,\lambda}(e), e\rangle &= \int_{\mathbb{R}^n\times\mathbb{R}^n} \frac{|e(x)-e(y)|^2}{|x-y|^{n+2s}}\,dx\,dy - \lambda\int_\Omega |e(x)|^2\,dx - \int_\Omega |e(x)|^{2_s^*}\,dx\\ &= 2\mathcal{J}_{s,\lambda}(e) + \left(\frac{2}{2_s^*}-1\right)\int_\Omega |e(x)|^{2_s^*}\,dx \qquad (14.42)\\ &< 2\mathcal{J}_{s,\lambda}(e) < 0.\end{aligned}$$

However, since $0 < \lambda < \lambda_{1,s}$, we get

$$\begin{aligned}\langle \mathcal{J}'_{s,\lambda}(P(\zeta)), P(\zeta)\rangle &= \int_{\mathbb{R}^n\times\mathbb{R}^n} \frac{|P(\zeta)(x)-P(\zeta)(y)|^2}{|x-y|^{n+2s}}\,dx\,dy\\ &\quad - \lambda\int_\Omega |P(\zeta)(x)|^2\,dx - \int_\Omega |P(\zeta)(x)|^{2_s^*}\,dx\\ &\geq \left(1-\frac{\lambda}{\lambda_{1,s}}\right)\int_{\mathbb{R}^n\times\mathbb{R}^n} \frac{|P(\zeta)(x)-P(\zeta)(y)|^2}{|x-y|^{n+2s}}\,dx\,dy\\ &\quad - \|P(\zeta)\|_{L^{2_s^*}(\Omega)}^{2_s^*}.\end{aligned}$$

Thus, by Lemma 1.28(a) and the fact that $2 < 2_s^*$,

$$\langle \mathcal{J}'_{s,\lambda}(P(\zeta)), P(\zeta)\rangle \geq \frac{1}{c}\left(1-\frac{\lambda}{\lambda_{1,s}}\right)\|P(\zeta)\|^2_{L^{2_s^*}(\Omega)} - \|P(\zeta)\|^{2_s^*}_{L^{2_s^*}(\Omega)} > 0,$$

provided that

$$\|P(\zeta)\|_{L^{2_s^*}(\Omega)} \in \left(0, \left[\frac{1}{c}\left(1-\frac{\lambda}{\lambda_{1,s}}\right)\right]^{1/(2_s^*-2)}\right).$$

In particular, since P is continuous, recalling (14.40) and (14.41), we obtain that there exists $\widehat{\zeta} \in (\widetilde{\zeta}, 1)$ such that

$$\langle \mathcal{J}'_{s,\lambda}(P(\zeta)), P(\zeta)\rangle > 0 \quad \text{for any } \zeta \in (\widetilde{\zeta}, \widehat{\zeta}). \qquad (14.43)$$

As a consequence of (14.42) and (14.43), we get that there exists $\bar{\zeta} \in (\widehat{\zeta}, 1)$ such that $\bar{u} := P(\bar{\zeta})$ satisfies

$$\begin{aligned}\langle \mathcal{J}'_{s,\lambda}(\bar{u}), \bar{u}\rangle &= \int_{\mathbb{R}^n\times\mathbb{R}^n} \frac{|\bar{u}(x)-\bar{u}(y)|^2}{|x-y|^{n+2s}}\,dx\,dy - \lambda\int_\Omega |\bar{u}(x)|^2\,dx - \int_\Omega |\bar{u}(x)|^{2_s^*}\,dx\\ &= 0. \qquad (14.44)\end{aligned}$$

We stress that

$$\bar{u} \not\equiv 0 \tag{14.45}$$

because $\bar{\zeta} > \widehat{\zeta} > \widetilde{\zeta}$ by construction, so $P(\bar{\zeta}) \neq 0$ by (14.40).

Also, using the definition of $\mathcal{J}_{s,\lambda}$ and (14.44), we easily get

$$\mathcal{J}_{s,\lambda}(\bar{u}) = \left(\frac{1}{2} - \frac{1}{2_s^*}\right) \|\bar{u}\|_{L^{2_s^*}(\Omega)}^{2_s^*} = \frac{s}{n} \|\bar{u}\|_{L^{2_s^*}(\Omega)}^{2_s^*}. \tag{14.46}$$

Hence, by (14.5), it holds true that

$$\begin{aligned} S_{s,\lambda} &\leq S_{s,\lambda}(\bar{u}) \\ &= \frac{\displaystyle\int_{\mathbb{R}^n \times \mathbb{R}^n} \frac{|\bar{u}(x) - \bar{u}(y)|^2}{|x-y|^{n+2s}}\, dx\, dy - \lambda \int_{\mathbb{R}^n} |\bar{u}(x)|^2\, dx}{\left(\displaystyle\int_{\mathbb{R}^n} |\bar{u}(x)|^{2_s^*}\, dx\right)^{2/2_s^*}} \\ &\leq \frac{\displaystyle\int_{\mathbb{R}^n \times \mathbb{R}^n} \frac{|\bar{u}(x) - \bar{u}(y)|^2}{|x-y|^{n+2s}}\, dx\, dy - \lambda \int_{\Omega} |\bar{u}(x)|^2\, dx}{\left(\displaystyle\int_{\Omega} |\bar{u}(x)|^{2_s^*}\, dx\right)^{2/2_s^*}} \\ &= \left(\int_{\Omega} |\bar{u}(x)|^{2_s^*}\, dx\right)^{(2_s^*-2)/2_s^*} \\ &= \left(\frac{n}{s} \mathcal{J}_{s,\lambda}(\bar{u})\right)^{2s/n} \\ &\leq \left(\frac{n}{s} \sup_{v \in P([0,1])} \mathcal{J}_{s,\lambda}(v)\right)^{2s/n}, \end{aligned}$$

thanks to (14.44)–(14.46). Taking the infimum in $P \in \mathscr{P}$, we conclude that

$$S_{s,\lambda} \leq \left(\frac{n}{s} c\right)^{2s/n};$$

that is,

$$c \geq \frac{s}{n} S_{s,\lambda}^{n/(2s)}. \tag{14.47}$$

By (14.39) and (14.47), we deduce (14.38), and this ends the proof. □

14.4 End of the proof of Theorem 14.1

Propositions 14.2 and 14.3 give that the geometry of the variant of the mountain pass theorem stated in [46, theorem 2.2] is fulfilled by $\mathcal{J}_{s,\lambda}$. Moreover, since $F(0) = 0$, we easily get that $\mathcal{J}_{s,\lambda}(0) = 0 < \beta$, with β given in Proposition 14.2. Now let us consider the constant c defined in (14.37).

Claim 14.1 *The constant c given in* (14.37) *is such that*

$$\beta \leq c < \frac{s}{n} S_s^{n/(2s)},$$

where β is given in Proposition 14.2 and S_s is defined in formula (1.92).

Proof First of all, let us prove that $c \geq \beta$. For this, we first observe that, for any $P \in \mathscr{P}$, the function $t \mapsto \|P(t)\|_{\mathbb{H}_0^s(\Omega)}$ is continuous in $[0,1]$, that $\|P(0)\|_{\mathbb{H}_0^s(\Omega)} = \|0\|_{\mathbb{H}_0^s(\Omega)} = 0 < \rho$, and that $\|P(1)\|_{\mathbb{H}_0^s(\Omega)} = \|e\|_{\mathbb{H}_0^s(\Omega)} > \rho$, with ρ as in Proposition 14.2. Accordingly, there exists $\bar{t} \in (0,1)$ such that $\|P(\bar{t})\|_{\mathbb{H}_0^s(\Omega)} = \rho$. Thus,

$$\max_{v \in P([0,1])} \mathscr{J}_{s,\lambda}(v) \geq \mathscr{J}_{s,\lambda}(P(\bar{t})) \geq \inf_{\substack{v \in \mathbb{H}_0^s(\Omega) \\ \|v\|_{\mathbb{H}_0^s(\Omega)} = \rho}} \mathscr{J}_{s,\lambda}(v).$$

As a consequence, we get that $c \geq \beta$, as desired.

Now let us show that (14.35) is verified. For this, as a consequence of Proposition 14.13, it is enough to show that (14.36) holds true. For this, let us consider two different cases.

Case 1: $n > 4s$. By the definition of the function $S_{s,\lambda}(\cdot)$ given in formula (14.6) and by Propositions 14.11 and 14.12, we get

$$\begin{aligned} S_{s,\lambda}(u_\varepsilon) &\leq \frac{S_s^{n/(2s)} + O(\varepsilon^{n-2s}) - \lambda C_s \varepsilon^{2s}}{\left(S_s^{n/(2s)} + O(\varepsilon^n)\right)^{2/2_s^*}} \\ &\leq S_s + O(\varepsilon^{n-2s}) - \lambda \widetilde{C}_s \varepsilon^{2s} \\ &= S_s + \varepsilon^{2s}\left(O(\varepsilon^{n-4s}) - \lambda \widetilde{C}_s\right) \\ &< S_s, \end{aligned}$$

for any $\lambda > 0$, if $\varepsilon > 0$ is sufficiently small, for a suitable positive constant $\widetilde{C}_s$. Hence, condition (14.36) is satisfied for any $\lambda > 0$ when $n > 4s$.

Case 2: $n = 4s$. Arguing as in Case 1 and taking into account that $n = 4s$ for some positive constant $\widetilde{C}_s$ we get

$$\begin{aligned} S_{s,\lambda}(u_\varepsilon) &\leq \frac{S_s^{n/(2s)} + O(\varepsilon^{n-2s}) - \lambda C_s \varepsilon^{2s}|\log\varepsilon| + O(\varepsilon^{2s})}{\left(S_s^{n/(2s)} + O(\varepsilon^n)\right)^{2/2_s^*}} \\ &\leq S_s + O(\varepsilon^{2s}) - \lambda \widetilde{C}_s \varepsilon^{2s}|\log\varepsilon| \\ &= S_s + \varepsilon^{2s}\left(O(1) - \lambda \widetilde{C}_s|\log\varepsilon|\right) \\ &< S_s, \end{aligned}$$

for any $\lambda > 0$, if $\varepsilon > 0$ is sufficiently small. Hence, condition (14.36) is satisfied for any $\lambda > 0$ when $n = 4s$, Claim 14.1 is proved. □

By [46, theorem 2.2], there exists a sequence $\{u_j\}_{j\in\mathbb{N}}$ in $\mathbb{H}_0^s(\Omega)$ such that

$$\mathcal{J}_{s,\lambda}(u_j) \to c \tag{14.48}$$

and

$$\sup\left\{|\langle \mathcal{J}'_{s,\lambda}(u_j),\varphi\rangle| : \varphi \in \mathbb{H}_0^s(\Omega), \|\varphi\|_{\mathbb{H}_0^s(\Omega)} = 1\right\} \to 0 \tag{14.49}$$

as $j \to +\infty$. Now we prove some properties of the sequence $\{u_j\}_{j\in\mathbb{N}}$.

Claim 14.2 *The sequence* $\{u_j\}_{j\in\mathbb{N}}$ *is bounded in* $\mathbb{H}_0^s(\Omega)$.

Proof For any $j \in \mathbb{N}$, by (14.48) and (14.49), it easily follows that there exists $\kappa > 0$ such that

$$|\mathcal{J}_{s,\lambda}(u_j)| \le \kappa \tag{14.50}$$

and

$$\left|\langle \mathcal{J}'_{s,\lambda}(u_j), \frac{u_j}{\|u_j\|_{\mathbb{H}_0^s(\Omega)}}\rangle\right| \le \kappa. \tag{14.51}$$

As a consequence of (14.50) and (14.51), we have

$$\mathcal{J}_{s,\lambda}(u_j) - \frac{1}{2}\langle \mathcal{J}'_{s,\lambda}(u_j), u_j\rangle \le \kappa\left(1 + \|u_j\|_{\mathbb{H}_0^s(\Omega)}\right). \tag{14.52}$$

Moreover, we have

$$\mathcal{J}_{s,\lambda}(u_j) - \frac{1}{2}\langle \mathcal{J}'_{s,\lambda}(u_j), u_j\rangle = -\left(\frac{1}{2_s^*} - \frac{1}{2}\right)\|u_j\|_{L^{2_s^*}(\Omega)}^{2_s^*} = \frac{s}{n}\|u_j\|_{L^{2_s^*}(\Omega)}^{2_s^*},$$

so, by this and (14.52), we get that, for any $j \in \mathbb{N}$,

$$\|u_j\|_{L^{2_s^*}(\Omega)}^{2_s^*} \le \kappa_*\left(1 + \|u_j\|_{\mathbb{H}_0^s(\Omega)}\right) \tag{14.53}$$

for a suitable positive constant κ_*.

Finally, by (14.50) and Lemma 6.5 (which holds true because $0 < \lambda < \lambda_{1,s}$), it follows that

$$\kappa \ge \mathcal{J}_{s,\lambda}(u_j) \ge \frac{1}{2}\left(1 - \frac{\lambda}{\lambda_{1,s}}\right)\|u_j\|_{\mathbb{H}_0^s(\Omega)}^2 - \frac{1}{2_s^*}\|u_j\|_{L^{2_s^*}(\Omega)}^{2_s^*},$$

so, recalling (14.53), we see that, for any $j \in \mathbb{N}$,

$$\|u_j\|_{\mathbb{H}_0^s(\Omega)}^2 \le \kappa_{**}\left(1 + \|u_j\|_{\mathbb{H}_0^s(\Omega)}\right)$$

for a suitable positive constant κ_{**}. Hence, the proof of Claim 14.2 is complete. □

Claim 14.3 *Problem* (14.2) *admits a solution* $u_\infty \in \mathbb{H}_0^s(\Omega)$.

Proof Since $\{u_j\}_{j\in\mathbb{N}}$ is bounded in $\mathbb{H}_0^s(\Omega)$ (thanks to Claim 14.2) and $\mathbb{H}_0^s(\Omega)$ is a reflexive space (being a Hilbert space, by Lemma 1.28(c)), up to a subsequence, still denoted by u_j, there exists $u_\infty \in \mathbb{H}_0^s(\Omega)$ such that $u_j \to u_\infty$ weakly in $\mathbb{H}_0^s(\Omega)$; that is,

$$\begin{aligned}&\int_{\mathbb{R}^n\times\mathbb{R}^n} \frac{(u_j(x)-u_j(y))(\varphi(x)-\varphi(y))}{|x-y|^{n+2s}}\,dx\,dy \\ &\to \int_{\mathbb{R}^n\times\mathbb{R}^n} \frac{(u_\infty(x)-u_\infty(y))(\varphi(x)-\varphi(y))}{|x-y|^{n+2s}}\,dx\,dy \quad \forall\,\varphi\in\mathbb{H}_0^s(\Omega)\end{aligned} \tag{14.54}$$

as $j\to+\infty$. Moreover, since the sequence $\{u_j\}_{j\in\mathbb{N}}$ is bounded in $\mathbb{H}_0^s(\Omega)$, we deduce from (14.53) that

$$\{u_j\}_{j\in\mathbb{N}} \text{ is bounded in } L^{2_s^*}(\Omega).$$

Consequently, by Lemma 1.31(b) and the fact that $L^{2_s^*}(\mathbb{R}^n)$ is a reflexive space, we have that, up to a subsequence,

$$u_j \to u_\infty \quad \text{weakly in } L^{2_s^*}(\mathbb{R}^n) \tag{14.55}$$

as $j\to+\infty$, while, by Lemma 1.31(a), up to a subsequence,

$$u_j \to u_\infty \quad \text{in } L^{\nu}(\mathbb{R}^n) \tag{14.56}$$

$$u_j \to u_\infty \quad \text{a.e. in } \mathbb{R}^n \tag{14.57}$$

as $j\to+\infty$, for any $\nu\in[1,2_s^*)$ (see, e.g., [43, theorem IV.9]).

By (14.55) and the fact that $\{|u_j|^{2_s^*-2}u_j\}_{j\in\mathbb{N}}$ is bounded in $L^{2_s^*/(2_s^*-1)}(\Omega)$, we see that

$$|u_j|^{2_s^*-2}u_j \to |u_\infty|^{2_s^*-2}u_\infty \quad \text{weakly in } L^{2_s^*/(2_s^*-1)}(\Omega) \tag{14.58}$$

as $j\to+\infty$.

Since (14.49) holds true, for any $\varphi\in\mathbb{H}_0^s(\Omega)$,

$$\begin{aligned}0\leftarrow\langle \mathscr{J}'_{s,\lambda}(u_j),\varphi\rangle &= \int_{\mathbb{R}^n\times\mathbb{R}^n} \frac{(u_j(x)-u_j(y))(\varphi(x)-\varphi(y))}{|x-y|^{n+2s}}\,dx\,dy \\ &\quad -\lambda\int_\Omega u_j(x)\varphi(x)\,dx - \int_\Omega |u_j(x)|^{2_s^*-2}u_j(x)\varphi(x)\,dx,\end{aligned}$$

so, passing to the limit in this expression as $j\to+\infty$ and taking into account (14.54), (14.56), and (14.58), we get

$$\begin{aligned}&\int_{\mathbb{R}^n\times\mathbb{R}^n} \frac{(u_\infty(x)-u_\infty(y))(\varphi(x)-\varphi(y))}{|x-y|^{n+2s}}\,dx\,dy - \lambda\int_\Omega u_\infty(x)\varphi(x)\,dx \\ &\quad - \int_\Omega |u_\infty(x)|^{2_s^*-2}u_\infty(x)\varphi(x)\,dx = 0,\end{aligned}$$

for any $\varphi\in\mathbb{H}_0^s(\Omega)$; that is, u_∞ is a solution of problem (14.2), and Claim 14.3 follows. □

Claim 14.4 *The solution u_∞ is nontrivial; that is, $u_\infty \not\equiv 0$ in Ω.*

Proof Suppose, by contradiction, that $u_\infty \equiv 0$ in Ω (and so $u_\infty \equiv 0$ in $\mathbb{R}^n$ because $u_\infty \in \mathbb{H}^s_0(\Omega)$). Since $\{u_j\}_{j\in\mathbb{N}}$ bounded in $\mathbb{H}^s_0(\Omega)$ by Claim 14.2, from (14.49), it follows that

$$0 \leftarrow \langle \mathscr{J}'_{K,\lambda}(u_j), u_j\rangle = \int_{\mathbb{R}^n\times\mathbb{R}^n} |u_j(x)-u_j(y)|^2 K(x-y)\,dx\,dy \\ -\lambda\int_\Omega |u_j(x)|^2\,dx - \int_\Omega |u_j(x)|^{2^*_s}\,dx,$$

which, thanks to (14.56), gives

$$\int_{\mathbb{R}^n\times\mathbb{R}^n} |u_j(x)-u_j(y)|^2 K(x-y)\,dx\,dy - \int_\Omega |u_j(x)|^{2^*_s}\,dx \to 0 \tag{14.59}$$

as $j\to+\infty$.

Now, by Claim 14.2, the sequence $\{\|u_j\|_{\mathbb{H}^s_0(\Omega)}\}_{j\in\mathbb{N}}$ is bounded in $\mathbb{R}$. Hence, up to a subsequence, if necessary, we can assume that

$$\|u_j\|^2_{\mathbb{H}^s_0(\Omega)} = \int_{\mathbb{R}^n\times\mathbb{R}^n} |u_j(x)-u_j(y)|^2 K(x-y)\,dx\,dy \to L \tag{14.60}$$

and, as a consequence of (14.59)

$$\int_\Omega |u_j(x)|^{2^*_s}\,dx \to L \tag{14.61}$$

as $j\to+\infty$. Of course, $L\in[0,+\infty)$. Furthermore, by (14.48), we have that

$$\frac{1}{2}\int_{\mathbb{R}^n\times\mathbb{R}^n} |u_j(x)-u_j(y)|^2 K(x-y)\,dx\,dy - \frac{\lambda}{2}\int_\Omega |u_j(x)|^2\,dx \\ -\frac{1}{2^*_s}\int_\Omega |u_j(x)|^{2^*_s}\,dx \to c$$

as $j\to+\infty$, so, using (14.56) with $u_\infty\equiv 0$ in $\mathbb{R}^n$, (14.60), and (14.61), it follows that

$$c = \left(\frac{1}{2}-\frac{1}{2^*_s}\right)L = \frac{s}{n}L. \tag{14.62}$$

Since $c\ge\beta>0$ by Claim 14.1, it is easily seen that $L>0$. Moreover, by Lemma 1.31(b) and (1.92),

$$\int_{\mathbb{R}^n\times\mathbb{R}^n} |u_j(x)-u_j(y)|^2 K(x-y)\,dx\,dy \ge S_s\|u_j\|^2_{L^{2^*_s}(\Omega)},$$

so, passing to the limit as $j\to+\infty$ and taking into account (14.60) and (14.61), we get

$$L \ge S_s L^{2/2^*_s},$$

which, combined with (14.62), gives

$$c \geq \frac{s}{n} S_s^{2_s^*/(2_s^*-2)} = \frac{s}{n} S_s^{n/(2s)}.$$

This contradicts Claim 14.1. Hence, $u_\infty \not\equiv 0$ in Ω, and this ends the proof of Claim 14.4. Hence, the proof of Theorem 14.1 is complete. □

15
Generalization of the Brezis–Nirenberg result

In Chapter 6 we studied problems of the type

$$\begin{cases} (-\Delta)^s u - \lambda u = |u|^{q-2}u & \text{in } \Omega \\ u = 0 & \text{in } \mathbb{R}^n \setminus \Omega, \end{cases}$$

where $q \in (2, 2_s^*)$; that is, we considered nonlocal equations with subcritical growth, and in this setting, we have extended the validity of some existence results known in the classical subcritical case of the Laplacian to the nonlocal framework.

In addition, in Chapter 14 we considered the critical equation (14.2) in the case where the parameter $\lambda \in (0, \lambda_{1,s})$, where $\lambda_{1,s}$ denoted the first eigenvalue of the fractional Laplace operator $(-\Delta)^s$ with homogeneous Dirichlet boundary conditions. In this framework, we proved a nonlocal Brezis–Nirenberg-type result; that is, we showed that problem (14.2) admits a nontrivial solution for any $\lambda \in (0, \lambda_{1,s})$, provided that $n \geq 4s$.

The aim of this chapter is to complete the study of problem (14.2) by showing that the existence result obtained in Theorem 14.1 holds true for any $\lambda > 0$ different from the eigenvalues of $(-\Delta)^s$. This chapter is based on the paper [196].

15.1 Main results

In the setting of critical fractional Laplace equations, the main result of this chapter is as follows:

Theorem 15.1 *Let $s \in (0, 1)$, $n \geq 4s$, Ω be an open, bounded subset of $\mathbb{R}^n$ with continuous boundary, and λ be a positive parameter. Then problem* (14.2) *admits a nontrivial weak solution $u \in \mathbb{H}_0^s(\Omega)$, provided that λ is not an eigenvalue of $(-\Delta)^s$ with homogeneous Dirichlet boundary data.*

This existence theorem may be seen as the natural extension to the nonlocal framework of a well-known result obtained by Capozzi, Fortunato, and Palmieri in [60], where the classical critical Laplace equation (14.1) was studied (for more general critical nonlinearities, see [114]). Indeed, when $s = 1$, Theorem 15.1 reads as [60, theorem 0.1] (see also [114, corollary 1] and [220, theorem 2.24]).

The proof of Theorem 15.1 is based on the fact that in the fractional Laplace setting, a natural condition to look at is the strict inequality $S_{s,\lambda} < S_s$, which holds true in any dimension n provided that $n \geq 4s$. For more details, see Subsection 14.3.1.

When dealing with a nonlocal setting, in the pure power case, all the results presented in this chapter extend the existence theorems obtained in Chapter 6 to the critical case. Furthermore, the results given here extend some classical theorems known for the Laplacian to the case of nonlocal fractional operators (see, e.g., [12, 46, 60, 114]).

15.1.1 Strategy for proving Theorem 15.1

To prove the existence of solutions for problem (14.2) for any $\lambda > 0$, we will apply some variants of the mountain pass theorem and the linking theorem (see, e.g., [11, 44, 172, 174]) that take into account the fact that in the critical setting there is a lack of compactness. As a consequence, as it happens in the classical case, as well as in the nonlocal critical setting the functional $\mathcal{J}_{s,\lambda}$ does not verify the Palais–Smale condition globally, but rather, only in an energy range determined by the best fractional critical Sobolev constant S_s given in formula (1.92).

In the case where $\lambda \in (0, \lambda_{1,s})$, problem (14.2) was studied in Chapter 14. Hence, here we consider only the case where $\lambda \geq \lambda_{1,s}$, and in order to achieve our goal, we will use appropriately the linking theorem of Rabinowitz (see [174, theorem 5.3]).

Since $\lambda \geq \lambda_{1,s}$, we can suppose that

$$\lambda \in [\lambda_{k,s}, \lambda_{k+1,s}), \quad \text{for some } k \in \mathbb{N},$$

where $\lambda_{k,s}$ is the kth eigenvalue of the operator $(-\Delta)^s$ with homogeneous Dirichlet boundary conditions. We recall that in what follows, $e_{k,s}$ will be the kth eigenfunction corresponding to the eigenvalue $\lambda_{k,s}$ of $(-\Delta)^s$, and

$$\mathbb{P}_{k+1,s} := \left\{u \in \mathbb{H}^s_0(\Omega) : \langle u, e_{j,s}\rangle_{\mathbb{H}^s_0(\Omega)} = 0 \ \forall j = 1, \ldots, k\right\}, \tag{15.1}$$

while $\operatorname{span}\{e_{1,s}, \ldots, e_{k,s}\}$ will denote the linear subspace generated by the first k eigenfunctions of $(-\Delta)^s$ for any $k \in \mathbb{N}$.

To prove Theorem 15.1, our strategy consists of adapting the techniques used in the classical case of the Laplacian in the critical setting (see [9, 212, 220] and references therein) to the nonlocal critical framework. Roughly, to apply [174, theorem 5.3], we will prove these three facts:

- *Compactness conditions for* $\mathcal{J}_{s,\lambda}$. The functional $\mathcal{J}_{s,\lambda}$ satisfies the Palais–Smale condition at any level c smaller than a certain threshold related to the best fractional critical Sobolev constant, that is, to be precise, for any c such that $c < (s/n)S_s^{n/(2s)}$.
- *Geometric structure of* $\mathcal{J}_{s,\lambda}$. The functional $\mathcal{J}_{s,\lambda}$ has the geometry required by the linking theorem.

- *Estimate of the critical level of $\mathcal{J}_{s,\lambda}$.* The linking critical level of $\mathcal{J}_{s,\lambda}$ lies below the threshold where the Palais–Smale condition holds true.

15.2 A local Palais–Smale condition for the functional $\mathcal{J}_{s,\lambda}$

We start by proving that the functional $\mathcal{J}_{s,\lambda}$ satisfies the Palais–Smale condition in a suitable energy range involving the best fractional critical Sobolev constant S_s given in (1.92).

Proposition 15.2 Let $\lambda > 0$, and let $c \in \mathbb{R}$ be such that

$$c < \frac{s}{n} S_s^{n/(2s)}. \tag{15.2}$$

Then the functional $\mathcal{J}_{s,\lambda}$ satisfies the Palais–Smale condition at level c.

Proof Let $\{u_j\}_{j\in\mathbb{N}}$ be a Palais–Smale sequence for the functional $\mathcal{J}_{s,\lambda}$ in $\mathbb{H}_0^s(\Omega)$ at level c; that is, let $\{u_j\}_{j\in\mathbb{N}}$ be such that

$$\mathcal{J}_{s,\lambda}(u_j) \to c \tag{15.3}$$

and

$$\sup\left\{ |\langle \mathcal{J}'_{s,\lambda}(u_j), \varphi\rangle| \,:\, \varphi \in \mathbb{H}_0^s(\Omega), \|\varphi\|_{\mathbb{H}_0^s(\Omega)} = 1 \right\} \to 0 \tag{15.4}$$

as $j \to +\infty$. To prove Proposition 15.2, we proceed by steps. Arguing as in Claims 14.2 and 14.3, we get that

Step 15.1 *The sequence $\{u_j\}_{j\in\mathbb{N}}$ is bounded in $\mathbb{H}_0^s(\Omega)$.*

Step 15.2 *Problem* (14.3) *admits a solution $u_\infty \in \mathbb{H}_0^s(\Omega)$.*

Moreover, the next fact is valid.

Step 15.3 *The following relation holds true:*

$$\mathcal{J}_{s,\lambda}(u_\infty) = \frac{s}{n} \int_\Omega |u_\infty(x)|^{2_s^*}\, dx \geq 0.$$

Proof By Step 15.2, taking $\varphi = u_\infty \in \mathbb{H}_0^s(\Omega)$ as a test function in (14.2), we get

$$\int_{\mathbb{R}^n\times\mathbb{R}^n} \frac{|u_\infty(x) - u_\infty(y)|^2}{|x-y|^{n+2s}}\, dx\, dy - \lambda \int_\Omega |u_\infty(x)|^2\, dx = \int_\Omega |u_\infty(x)|^{2_s^*}\, dx,$$

so

$$\mathcal{J}_{s,\lambda}(u_\infty) = \left(\frac{1}{2} - \frac{1}{2_s^*}\right) \int_\Omega |u_\infty(x)|^{2_s^*}\, dx = \frac{s}{n} \int_\Omega |u_\infty(x)|^{2_s^*}\, dx \geq 0.$$

Hence, Step 15.3 is proved. □

Step 15.4 *The following equality holds true:*

$$\mathscr{J}_{s,\lambda}(u_j) = \mathscr{J}_{s,\lambda}(u_\infty) + \frac{1}{2}\int_{\mathbb{R}^n\times\mathbb{R}^n} \frac{|u_j(x) - u_\infty(x) - u_j(y) + u_\infty(y)|^2}{|x-y|^{n+2s}}\,dx\,dy$$
$$- \frac{1}{2_s^*}\int_\Omega |u_j(x) - u_\infty(x)|^{2_s^*}\,dx + o(1)$$

as $j \to +\infty$.

Proof First of all, we observe that, by Step 15.1 and Lemma 1.31(b), the sequence $\{u_j\}_{j\in\mathbb{N}}$ is bounded in $\mathbb{H}_0^s(\Omega)$ and in $L^{2_s^*}(\Omega)$. Hence, by the Brezis–Lieb lemma (see [45, theorem 1]), we get

$$\int_{\mathbb{R}^n\times\mathbb{R}^n} \frac{|u_j(x) - u_j(y)|^2}{|x-y|^{n+2s}}\,dx\,dy$$
$$= \int_{\mathbb{R}^n\times\mathbb{R}^n} \frac{|u_j(x) - u_\infty(x) - u_j(y) + u_\infty(y)|^2}{|x-y|^{n+2s}}\,dx\,dy \tag{15.5}$$
$$+ \int_{\mathbb{R}^n\times\mathbb{R}^n} \frac{|u_\infty(x) - u_\infty(y)|^2}{|x-y|^{n+2s}}\,dx\,dy + o(1)$$

and

$$\int_\Omega |u_j(x)|^{2_s^*}\,dx = \int_\Omega |u_j(x) - u_\infty(x)|^{2_s^*}\,dx + \int_\Omega |u_\infty(x)|^{2_s^*}\,dx + o(1) \tag{15.6}$$

as $j \to +\infty$. Therefore, using also the definition of $\mathscr{J}_{s,\lambda}$, by (14.56), (15.5), and (15.6), we deduce that

$$\mathscr{J}_{s,\lambda}(u_j) = \frac{1}{2}\int_{\mathbb{R}^n\times\mathbb{R}^n} \frac{|u_j(x) - u_\infty(x) - u_j(y) + u_\infty(y)|^2}{|x-y|^{n+2s}}\,dx\,dy$$
$$+ \frac{1}{2}\int_{\mathbb{R}^n\times\mathbb{R}^n} \frac{|u_\infty(x) - u_\infty(y)|^2}{|x-y|^{n+2s}}\,dx\,dy - \frac{\lambda}{2}\int_\Omega |u_\infty(x)|^2 dx$$
$$- \frac{1}{2_s^*}\int_\Omega |u_j(x) - u_\infty(x)|^{2_s^*}\,dx - \frac{1}{2_s^*}\int_\Omega |u_\infty(x)|^{2_s^*}\,dx + o(1)$$
$$= \mathscr{J}_{s,\lambda}(u_\infty) + \frac{1}{2}\int_{\mathbb{R}^n\times\mathbb{R}^n} \frac{|u_j(x) - u_\infty(x) - u_j(y) + u_\infty(y)|^2}{|x-y|^{n+2s}}\,dx\,dy$$
$$- \frac{1}{2_s^*}\int_\Omega |u_j(x) - u_\infty(x)|^{2_s^*}\,dx + o(1)$$

as $j \to +\infty$, which gives the desired assertion. □

Step 15.5 *The following equality holds true:*

$$\int_{\mathbb{R}^n\times\mathbb{R}^n} \frac{|u_j(x)-u_\infty(x)-u_j(y)+u_\infty(y)|^2}{|x-y|^{n+2s}}\,dx\,dy$$
$$=\int_\Omega |u_j(x)-u_\infty(x)|^{2_s^*}dx+o(1)$$

as $j\to+\infty$.

Proof First of all, note that, as a consequence of (14.55), (14.58), and (15.6), we get

$$\begin{aligned}
&\int_\Omega \left(|u_j(x)\,|^{2_s^*-2}u_j(x)-|u_\infty(x)|^{2_s^*-2}u_\infty(x)\right)\left(u_j(x)-u_\infty(x)\right)dx\\
&=\int_\Omega |u_j(x)|^{2_s^*}\,dx-\int_\Omega |u_\infty(x)|^{2_s^*-1}u_\infty(x)u_j(x)\,dx\\
&\quad-\int_\Omega |u_j(x)|^{2_s^*-2}u_j(x)u_\infty(x)\,dx+\int_\Omega |u_\infty(x)|^{2_s^*}\,dx \qquad (15.7)\\
&=\int_\Omega |u_j(x)|^{2_s^*}\,dx-\int_\Omega |u_\infty(x)|^{2_s^*}\,dx+o(1)\\
&=\int_\Omega |u_j(x)-u_\infty(x)|^{2_s^*}\,dx+o(1)
\end{aligned}$$

as $j\to+\infty$. By (15.4) and Steps 15.1 and 15.2, it is easily seen that

$$o(1)=\langle \mathcal{J}'_{s,\lambda}(u_j),u_j-u_\infty\rangle=\langle \mathcal{J}'_{s,\lambda}(u_j)-\mathcal{J}'_{s,\lambda}(u_\infty),u_j-u_\infty\rangle \tag{15.8}$$

as $j\to+\infty$. Moreover, by (14.56) and (15.7),

$$\begin{aligned}
&\langle \mathcal{J}'_{s,\lambda}(u_j)-\mathcal{J}'_{s,\lambda}(u_\infty),u_j-u_\infty\rangle\\
&=\int_{\mathbb{R}^n\times\mathbb{R}^n} \frac{|u_j(x)-u_\infty(x)-u_j(y)+u_\infty(y)|^2}{|x-y|^{n+2s}}\,dx\,dy\\
&\quad-\lambda\int_\Omega |u_j(x)-u_\infty(x)|^2\,dx \qquad (15.9)\\
&\quad-\int_\Omega \left(|u_j(x)|^{2_s^*-2}u_j(x)-|u_\infty(x)|^{2_s^*-2}u_\infty(x)\right)(u_j(x)-u_\infty(x))\,dx\\
&=\int_{\mathbb{R}^n\times\mathbb{R}^n} \frac{|u_j(x)-u_\infty(x)-u_j(y)+u_\infty(y)|^2}{|x-y|^{n+2s}}\,dx\,dy\\
&\quad-\int_\Omega |u_j(x)-u_\infty(x)|^{2_s^*}\,dx+o(1)
\end{aligned}$$

as $j\to+\infty$. Hence, the assertion of Step 15.5 comes from (15.8) and (15.9). □

Now we can conclude the proof of Proposition 15.2. By Step 15.5, it easily follows that

$$\begin{aligned}
&\frac{1}{2}\int_{\mathbb{R}^n\times\mathbb{R}^n}\frac{|u_j(x)-u_\infty(x)-u_j(y)+u_\infty(y)|^2}{|x-y|^{n+2s}}dx\,dy \\
&\quad -\frac{1}{2_s^*}\int_\Omega |u_j(x)-u_\infty(x)|^{2_s^*}dx \\
&= \left(\frac{1}{2}-\frac{1}{2_s^*}\right)\int_{\mathbb{R}^n\times\mathbb{R}^n}\frac{|u_j(x)-u_\infty(x)-u_j(y)+u_\infty(y)|^2}{|x-y|^{n+2s}}dx\,dy+o(1) \\
&= \frac{s}{n}\int_{\mathbb{R}^n\times\mathbb{R}^n}\frac{|u_j(x)-u_\infty(x)-u_j(y)+u_\infty(y)|^2}{|x-y|^{n+2s}}dx\,dy+o(1),
\end{aligned}$$

so, recalling Step 15.4,

$$\begin{aligned}
\mathscr{J}_{s,\lambda}(u_\infty)+\frac{s}{n}\int_{\mathbb{R}^n\times\mathbb{R}^n}\frac{|u_j(x)-u_\infty(x)-u_j(y)+u_\infty(y)|^2}{|x-y|^{n+2s}}dx\,dy &= \mathscr{J}_{s,\lambda}(u_j)+o(1) \\
&= c+o(1),
\end{aligned} \tag{15.10}$$

as $j\to+\infty$, thanks to (15.3). Since the sequence $\{\|u_j\|_{\mathbb{H}_0^s(\Omega)}\}_{j\in\mathbb{N}}$ is bounded in $\mathbb{R}$ by Step 15.1, we can assume that up to a subsequence, if necessary,

$$\|u_j-u_\infty\|^2_{\mathbb{H}_0^s(\Omega)}\to L \tag{15.11}$$

so, again as a consequence of Step 15.5,

$$\int_\Omega |u_j(x)-u_\infty(x)|^{2_s^*}\,dx\to L$$

as $j\to+\infty$. Of course, $L\in[0,+\infty)$ and, by definition of S_s and $S_s(\,\cdot\,)$ (see (1.92) and (1.93)), we have that

$$S_s\le S_s(u_j-u_\infty)\to\frac{L}{L^{2/2_s^*}}$$

as $j\to+\infty$. As a consequence,

$$L\ge L^{2/2_s^*}S_s.$$

Thus,

$$\text{either}\quad L=0\quad\text{or}\quad L\ge S_s^{n/(2s)}.$$

The case $L\ge S_s^{n/(2s)}$ cannot occur. Otherwise, by (15.10), (15.11), and Step 15.3, we would get

$$c=\mathscr{J}_{s,\lambda}(u_\infty)+\frac{s}{n}L\ge\frac{s}{n}L\ge\frac{s}{n}S_s^{n/(2s)},$$

which contradicts (15.2). Thus, $L = 0$ and so, by (15.11), we obtain

$$\|u_j - u_\infty\|_{\mathbb{H}_0^s(\Omega)} \to 0$$

as $j \to +\infty$. Hence, the sequence $\{u_j\}_{j\in\mathbb{N}}$ has a (strongly) convergent subsequence, and this concludes the proof of Proposition 15.2. □

15.3 The geometry of the functional $\mathscr{J}_{s,\lambda}$

In this section, we show that the functional $\mathscr{J}_{s,\lambda}$ has the geometric structure required by [174, theorem 5.3]. In a nonlocal setting for nonlinearities with subcritical growth, a proof of this is given in Chapter 6. Essentially, here we adapt the proof given in Chapter 6 to the critical case. With respect to the classical case of the Laplacian, some technical differences arise.

Proposition 15.3 Let $\lambda \in [\lambda_{k,s}, \lambda_{k+1,s})$ and $\mathbb{P}_{k+1,s}$ be as in (15.1), for some $k \in \mathbb{N}$. Then there exist $\rho > 0$ and $\beta > 0$ such that, for any $u \in \mathbb{P}_{k+1,s}$ with $\|u\|_{\mathbb{H}_0^s(\Omega)} = \rho$, it results that $\mathscr{J}_{s,\lambda}(u) \geq \beta$.

Proof Let u be a function in $\mathbb{P}_{k+1,s}$. Thanks to the choice of λ and to the variational characterization of $\lambda_{k+1,s}$ (see Proposition 3.2(d)), we get

$$\begin{aligned}\mathscr{J}_{s,\lambda}(u) &= \frac{1}{2}\int_{\mathbb{R}^n\times\mathbb{R}^n}\frac{|u(x)-u(y)|^2}{|x-y|^{n+2s}}\,dx\,dy - \frac{\lambda}{2}\int_\Omega |u(x)|^2\,dx - \frac{1}{2_s^*}\int_\Omega |u(x)|^{2_s^*}\,dx\\ &\geq \frac{1}{2}\left(1-\frac{\lambda}{\lambda_{k+1,s}}\right)\int_{\mathbb{R}^n\times\mathbb{R}^n}\frac{|u(x)-u(y)|^2}{|x-y|^{n+2s}}\,dx\,dy - \frac{1}{2_s^*}\int_\Omega |u(x)|^{2_s^*}\,dx\\ &\geq \frac{1}{2}\left(1-\frac{\lambda}{\lambda_{k+1,s}}\right)\|u\|^2_{\mathbb{H}_0^s(\Omega)} - \frac{1}{2_s^* S_s^{2_s^*/2}}\|u\|^{2_s^*}_{\mathbb{H}_0^s(\Omega)},\end{aligned}$$

thanks also to Lemma 1.32. Hence, for a suitable positive constant κ, we have

$$\mathscr{J}_{s,\lambda}(u) \geq \frac{1}{2}\left(1-\frac{\lambda}{\lambda_{k+1,s}}\right)\|u\|^2_{\mathbb{H}_0^s(\Omega)}\left(1-\kappa\|u\|^{2_s^*-2}_{\mathbb{H}_0^s(\Omega)}\right).$$

Now let $u \in \mathbb{P}_{k+1,s}$ be such that $\|u\|_{\mathbb{H}_0^s(\Omega)} = \rho > 0$. Since $2_s^* > 2$, choosing ρ sufficiently small (i.e., ρ such that $1-\kappa\rho^{2_s^*-2} > 0$), we get

$$\inf_{\substack{u\in\mathbb{P}_{k+1,s}\\ \|u\|_{\mathbb{H}_0^s(\Omega)}=\rho}} \mathscr{J}_{s,\lambda}(u) \geq \frac{1}{2}\left(1-\frac{\lambda}{\lambda_{k+1,s}}\right)\rho^2(1-\kappa\rho^{2_s^*-2}) =: \beta > 0,$$

and this ends the proof. □

Proposition 15.4 Let $\lambda \in [\lambda_{k,s}, \lambda_{k+1,s})$, for some $k \in \mathbb{N}$. Then $\mathscr{J}_{s,\lambda}(u) \leq 0$, for any $u \in \operatorname{span}\{e_{1,s}, \ldots, e_{k,s}\}$.

Proof Let $u \in \text{span}\{e_{1,s},\dots,e_{k,s}\}$. Then

$$u(x)=\sum_{i=1}^{k} u_i e_{i,s}(x),$$

with $u_i \in \mathbb{R}$, $i=1,\dots,k$. Since $\{e_{1,s},\dots,e_{k,s},\dots\}$ is an orthonormal basis of $L^2(\Omega)$ and an orthogonal basis of $\mathbb{H}_0^s(\Omega)$, by Proposition 3.2(f), we get

$$\int_{\Omega} |u(x)|^2\,dx=\sum_{i=1}^{k} u_i^2 \tag{15.12}$$

and

$$\int_{\mathbb{R}^n\times\mathbb{R}^n} \frac{|u(x)-u(y)|^2}{|x-y|^{n+2s}}\,dx\,dy=\sum_{i=1}^{k} u_i^2\|e_{i,s}\|^2_{\mathbb{H}_0^s(\Omega)}. \tag{15.13}$$

Then, by (15.12) and (15.13) and using Proposition 3.2(b) and (e) (here we also use the fact that $e_{i,s} \in \mathbb{H}_0^s(\Omega)$, for $i=1,\dots,k$; see Proposition 3.2(f)), we get

$$\begin{aligned}\mathscr{J}_{s,\lambda}(u)&=\frac{1}{2}\sum_{i=1}^{k} u_i^2\left(\|e_i\|^2_{\mathbb{H}_0^s(\Omega)}-\lambda\right)-\frac{1}{2_s^*}\|u\|^{2_s^*}_{L^{2_s^*}(\Omega)}\\&\le\frac{1}{2}\sum_{i=1}^{k} u_i^2(\lambda_i-\lambda)\le 0\end{aligned}$$

because $\lambda_{i,s}\le\lambda_{k,s}\le\lambda$, for any $i=1,\dots,k$. Hence, the assertion follows. □

Proposition 15.5 Let $\lambda \ge 0$, and let $\mathbb{F}$ be a finite-dimensional subspace of $\mathbb{H}_0^s(\Omega)$. Then there exists $R>\rho$ such that $\mathscr{J}_{s,\lambda}(u)\le 0$, for any $u\in\mathbb{F}$ with $\|u\|_{\mathbb{H}_0^s(\Omega)}\ge R$, where ρ is given in Proposition 15.3.

Proof Let $u\in\mathbb{F}$. Then the nonnegativity of λ gives

$$\begin{aligned}\mathscr{J}_{s,\lambda}(u)&=\frac{1}{2}\|u\|^2_{\mathbb{H}_0^s(\Omega)}-\frac{\lambda}{2}\|u\|^2_{L^2(\Omega)}-\frac{1}{2_s^*}\|u\|^{2_s^*}_{L^{2_s^*}(\Omega)}\\&\le\frac{1}{2}\|u\|^2_{\mathbb{H}_0^s(\Omega)}-\frac{1}{2_s^*}\|u\|^{2_s^*}_{L^{2_s^*}(\Omega)}\\&\le\frac{1}{2}\|u\|^2_{\mathbb{H}_0^s(\Omega)}-\frac{\kappa}{2_s^*}\|u\|^{2_s^*}_{\mathbb{H}_0^s(\Omega)},\end{aligned}$$

for some positive constant κ, thanks to the fact that in any finite-dimensional space, all the norms are equivalent. Hence, if $\|u\|_{\mathbb{H}_0^s(\Omega)}\to+\infty$, then

$$\mathscr{J}_{s,\lambda}(u)\to-\infty$$

because $2_s^*>2$ by assumption, so the assertion follows. □

15.4 A nondegeneracy estimate

As in Chapter 14, given that the functional $\mathscr{J}_{s,\lambda}$ satisfies the Palais–Smale condition until a suitable threshold related to the fractional critical Sobolev constant S_s (see Proposition 15.2), we need to estimate the linking critical level of the functional itself. To do this, we need to work again with the function u_ε, defined in (14.19), and to prove the following result:

Proposition 15.6 Let $s \in (0,1)$, $n > 2s$, and u_ε be as in (14.19). Then the following estimates hold true

$$\int_{\mathbb{R}^n} |u_\varepsilon(x)|^{2_s^*-1}\,dx = O(\varepsilon^{(n-2s)/2}),$$

and

$$\int_{\mathbb{R}^n} |u_\varepsilon(x)|\,dx = O(\varepsilon^{(n-2s)/2})$$

as $\varepsilon \to 0$, for some positive constant C_s depending on s.

Proof Let us start by proving the $L^{2_s^*-1}$-estimate. By (14.16) and (14.19), we have that

$$\begin{aligned}
\int_{\mathbb{R}^n} |u_\varepsilon(x)|^{2_s^*-1}\,dx &\le \int_{B_{2\delta}} |U_\varepsilon(x)|^{2_s^*-1}\,dx \\
&= \kappa^*\left(\varepsilon^{-(n-2s)/2}\right)^{2_s^*-1} \int_{B_{2\delta}} \left[\left(\mu^2+|x/\varepsilon|^2\right)^{-(n-2s)/2}\right]^{2_s^*-1} dx \\
&= \kappa^*\varepsilon^{(n-2s)/2} \int_0^{2\delta/\varepsilon} \frac{r^{n-1}}{\left(\mu^2+r^2\right)^{(n+2s)/2}}\,dr \\
&\cong \tilde{\kappa}\varepsilon^{(n-2s)/2}\left(1+\int_1^{2\delta/\varepsilon} r^{-2s-1}\,dr\right) \\
&= O(\varepsilon^{(n-2s)/2})
\end{aligned}$$

as $\varepsilon \to 0$. In a similar way, we get

$$\begin{aligned}
\int_{\mathbb{R}^n} |u_\varepsilon(x)|\,dx &\le \int_{B_{2\delta}} |U_\varepsilon(x)|\,dx \\
&= \kappa^*\varepsilon^{-(n-2s)/2} \int_{B_{2\delta}} \left(\mu^2+|x/\varepsilon|^2\right)^{-(n-2s)/2} dx \\
&= \kappa^*\varepsilon^{(n+2s)/2} \int_0^{2\delta/\varepsilon} \frac{r^{n-1}}{\left(\mu^2+r^2\right)^{(n-2s)/2}}\,dr \\
&\cong \tilde{\kappa}\varepsilon^{(n+2s)/2}\left(1+\int_1^{2\delta/\varepsilon} r^{2s-1}\,dr\right) \\
&= O(\varepsilon^{(n-2s)/2})
\end{aligned}$$

as $\varepsilon \to 0$. In both cases, κ^* and $\tilde{\kappa}$ are positive constants. Hence, the proposition is proved. □

Now we can prove the following result:

Proposition 15.7 Let $s \in (0,1)$ and $n > 2s$. Let $\lambda \in [\lambda_{k,s}, \lambda_{k+1,s})$, for some $k \in \mathbb{N}$, and let $\mathbb{F}_\varepsilon$ be the linear space defined as

$$\mathbb{F} = \mathbb{F}_\varepsilon := \operatorname{span}\{e_{1,s}, \dots, e_{k,s}, u_\varepsilon\}, \tag{15.14}$$

where u_ε is given in (14.19). Finally, let

$$M_{\varepsilon,\lambda} := \max_{\substack{u \in \mathbb{F}_\varepsilon \\ \|u\|_{L^{2^*_s}(\Omega)}=1}} S_{s,\lambda}(u), \tag{15.15}$$

where the function $S_{s,\lambda}(\,\cdot\,)$ is defined in (14.6). Then

(a) $M_{\varepsilon,\lambda}$ is achieved in $u_M \in \mathbb{F}_\varepsilon$, and u_M can be written as follows[1]:

$$u_M = v + t u_\varepsilon, \quad \text{with } v \in \operatorname{span}\{e_{1,s}, \dots, e_{k,s}\} \text{ and } t \geq 0; \tag{15.16}$$

(b) The following estimate holds true:

$$M_{\varepsilon,\lambda} \leq \begin{cases} (\lambda_{k,s} - \lambda)\|v\|^2_{L^2(\Omega)} & \text{if } t = 0 \\ (\lambda_{k,s} - \lambda)\|v\|^2_{L^2(\Omega)} + S_{s,\lambda}(u_\varepsilon)\big(1 + O(\varepsilon^{(n-2s)/2})\|v\|_{L^2(\Omega)}\big) & \text{if } t > 0 \\ \qquad + O(\varepsilon^{(n-2s)/2}\|v\|_{L^2(\Omega)} & \end{cases}$$

as $\varepsilon \to 0$, where v and t are given in (15.16).

Proof First, note that $u_\varepsilon \in \mathbb{H}^s_0(\Omega)$ and $e_{i,s} \in \mathbb{H}^s_0(\Omega)$, for any $i = 1, \dots, k$, thanks to Proposition 3.2(b) and (e) and to Lemma 1.24. As a consequence of this, $\mathbb{F}_\varepsilon \subset \mathbb{H}^s_0(\Omega)$, and, by definition, of $S_{s,\lambda}$

$$M_{\varepsilon,\lambda} := \max_{\substack{u \in \mathbb{F}_\varepsilon \\ \|u\|_{L^{2^*_s}(\Omega)}=1}} \left(\int_{\mathbb{R}^n \times \mathbb{R}^n} \frac{|u(x) - u(y)|^2}{|x-y|^{n+2s}}\, dx\, dy - \lambda \int_\Omega |u(x)|^2\, dx \right).$$

Let us start by proving part (a). The maximum in formula (15.15) is achieved thanks to the Weierstrass theorem (applied here in a finite-dimensional space). Thus, let $u_M \in \mathbb{F}_\varepsilon$ be the function such that

$$M_{\varepsilon,\lambda} = \int_{\mathbb{R}^n \times \mathbb{R}^n} \frac{|u_M(x) - u_M(y)|^2}{|x-y|^{n+2s}}\, dx\, dy - \lambda \int_\Omega |u_M(x)|^2\, dx$$

and

$$\|u_M\|_{L^{2^*_s}(\Omega)} = 1.$$

[1] Beware that u_M, v, and t depend on ε. For simplicity, we omit this dependence in the notation.

Note that $u_M \not\equiv 0$. Moreover, since $u_M \in \mathbb{F}_\varepsilon$, by definition of $\mathbb{F}_\varepsilon$, we have that

$$u_M = v + t u_\varepsilon,$$

for some $v \in \operatorname{span}\{e_{1,s},\dots,e_{k,s}\}$ and $t \in \mathbb{R}$. We can suppose that $t \geq 0$ (otherwise, if $t \leq 0$, we can replace u_M with $-u_M$). This concludes the proof of assertion (a).

Now let us show (b). First, note that, if $t = 0$ in formula (15.16), then $u_M = v \in \operatorname{span}\{e_{1,s},\dots,e_{k,s}\}$, and

$$\begin{aligned} M_{\varepsilon,\lambda} &= \int_{\mathbb{R}^n\times\mathbb{R}^n} \frac{|u_M(x)-u_M(y)|^2}{|x-y|^{n+2s}}\,dx\,dy - \lambda\int_\Omega |u_M(x)|^2\,dx \\ &= \int_{\mathbb{R}^n\times\mathbb{R}^n} \frac{|v(x)-v(y)|^2}{|x-y|^{n+2s}}\,dx\,dy - \lambda\int_\Omega |v(x)|^2\,dx \\ &\leq (\lambda_{k,s}-\lambda)\|v\|^2_{L^2(\Omega)}, \end{aligned}$$

by Proposition 3.2. Thus, assertion (b) holds true when $t = 0$.

Now let us consider the case when $t > 0$ in (15.16). By the Hölder inequality and the properties of u_M, we can bound u_M as follows:

$$\|u_M\|^2_{L^2(\Omega)} \leq |\Omega|^{n/(2s)}\|u_M\|^2_{L^{2^*_s}(\Omega)} = |\Omega|^{n/(2s)}. \tag{15.17}$$

Also, since $u_M \in \mathbb{F}_\varepsilon$, by definition of $\mathbb{F}_\varepsilon$, we have that

$$u_M = \tilde{v} + t z_\varepsilon, \tag{15.18}$$

where

$$\tilde{v} = v + t\sum_{i=1}^k \left(\int_\Omega u_\varepsilon(x) e_{i,s}(x)\,dx\right) e_{i,s} \in \operatorname{span}\{e_{1,s},\dots,e_{k,s}\}$$

and

$$z_\varepsilon = u_\varepsilon - \sum_{i=1}^k \left(\int_\Omega u_\varepsilon(x) e_{i,s}(x)\,dx\right) e_{i,s}, \tag{15.19}$$

so $\tilde{v}$ and z_ε are orthogonal in $L^2(\Omega)$ (see Proposition 3.2(f)). As a consequence of this, we get

$$\|u_M\|^2_{L^2(\Omega)} = \|\tilde{v}\|^2_{L^2(\Omega)} + t^2\|z_\varepsilon\|^2_{L^2(\Omega)} \geq \|\tilde{v}\|^2_{L^2(\Omega)}. \tag{15.20}$$

By (15.17) and (15.20), we have obtained that $\|u_M\|_{L^2(\Omega)}$ and $\|\tilde{v}\|_{L^2(\Omega)}$ are all bounded uniformly in ε by a suitable $\tilde{c} > 0$.

Now we claim that, up to renaming such a constant,

$$t \leq \tilde{c}. \tag{15.21}$$

For this, note that, by Proposition 3.4, we have that $e_{i,s} \in L^\infty(\Omega)$, for any $i \in \mathbb{N}$, so $\tilde{v} \in \operatorname{span}\{e_{1,s},\dots,e_{k,s}\}$ also does. Hence, $\tilde{v} \in L^{2^*_s}(\Omega)$ because Ω is bounded. Moreover, by the equivalence of the norms in a finite-dimensional space, we also have

$$\|\tilde{v}\|_{L^{2^*_s}(\Omega)} \leq \tilde{c}. \tag{15.22}$$

By Propositions 3.4 and 15.6, we have

$$\left| \int_\Omega u_\varepsilon(x) e_{i,s}(x)\, dx \right| \le \|u_\varepsilon\|_{L^1(\Omega)} \|e_{i,s}\|_{L^\infty(\Omega)} = O(\varepsilon^{(n-2s)/2})$$

as $\varepsilon \to 0$. As a consequence, using (15.19) and again Proposition 15.6,

$$\begin{aligned}
\|z_\varepsilon\|_{L^{2_s^*}(\Omega)} &\ge \|u_\varepsilon\|_{L^{2_s^*}(\Omega)} - \sum_{i=1}^{k} \left| \int_\Omega u_\varepsilon(x) e_{i,s}(x)\, dx \right| \|e_{i,s}\|_{L^{2_s^*}(\Omega)} \\
&= S_s^{(n-2s)/(4s)} + O(\varepsilon^{(n-2s)/2}) \\
&\ge \frac{S_s^{(n-2s)/(4s)}}{2},
\end{aligned}$$

for ε sufficiently small. Thus, by (15.18), the fact that $t \ge 0$, and (15.22), we have

$$\frac{S_s^{(n-2s)/(4s)}\, t}{2} \le t \|z_\varepsilon\|_{L^{2_s^*}(\Omega)} \le \|u_M\|_{L^{2_s^*}(\Omega)} + \|\tilde{v}\|_{L^{2_s^*}(\Omega)} \le 1 + \tilde{c},$$

thanks to the properties of u_M. Therefore, the claim in (15.21) follows.

Finally, the convexity[2], the monotonicity properties of the integrals, (15.21), and the fact that in $\operatorname{span}\{e_{1,s}, \dots, e_{k,s}\}$ all the norms are equivalent yield

$$\begin{aligned}
1 = \|u_M\|_{L^{2_s^*}(\Omega)}^{2_s^*} &= \int_\Omega |u_M(x)|^{2_s^*}\, dx \\
&\ge \int_\Omega |tu_\varepsilon(x)|^{2_s^*}\, dx + 2_s^* \int_\Omega (tu_\varepsilon(x))^{2_s^*-1} v(x)\, dx \\
&\ge \|tu_\varepsilon\|_{L^{2_s^*}(\Omega)}^{2_s^*} - 2_s^* \tilde{c}^{2_s^*-1} \|u_\varepsilon\|_{L^{2_s^*-1}(\Omega)}^{2_s^*-1} \|v\|_{L^\infty(\Omega)} \\
&\ge \|tu_\varepsilon\|_{L^{2_s^*}(\Omega)}^{2_s^*} - \hat{c}\, \|u_\varepsilon\|_{L^{2_s^*-1}(\Omega)}^{2_s^*-1} \|v\|_{L^2(\Omega)},
\end{aligned}$$

for some positive constant $\hat{c}$, so

$$\|tu_\varepsilon\|_{L^{2_s^*}(\Omega)}^{2_s^*} \le 1 + \hat{c}\, \|u_\varepsilon\|_{L^{2_s^*-1}(\Omega)}^{2_s^*-1} \|v\|_{L^2(\Omega)}. \tag{15.23}$$

[2] If f is a differentiable convex function, then $f(y) \ge f(x) + f'(x)(y-x)$. Here we take $f(s) = s^{2_s^*}$, $x = tu_\varepsilon$, and $y = u_M = v + tu_\varepsilon$.

Now we are ready to prove the inequality in part (b). By (15.16), we get

$$
\begin{aligned}
M_{\varepsilon,\lambda} &= \int_{\mathbb{R}^n\times\mathbb{R}^n} \frac{|u_M(x)-u_M(y)|^2}{|x-y|^{n+2s}}\,dx\,dy - \lambda\int_\Omega |u_M(x)|^2\,dx \\
&= \int_{\mathbb{R}^n\times\mathbb{R}^n} \frac{|v(x)+tu_\varepsilon(x)-v(y)-tu_\varepsilon(y)|^2}{|x-y|^{n+2s}}\,dx\,dy \\
&\quad -\lambda\int_\Omega |v(x)+tu_\varepsilon(x)|^2\,dx \\
&= \int_{\mathbb{R}^n\times\mathbb{R}^n} \frac{|v(x)-v(y)|^2}{|x-y|^{n+2s}}\,dx\,dy + t^2\int_{\mathbb{R}^n\times\mathbb{R}^n} \frac{|u_\varepsilon(x)-u_\varepsilon(y)|^2}{|x-y|^{n+2s}}\,dx\,dy \\
&\quad + 2t\int_{\mathbb{R}^n\times\mathbb{R}^n} \frac{(u_\varepsilon(x)-u_\varepsilon(y))(v(x)-v(y))}{|x-y|^{n+2s}}\,dxdy \qquad (15.24) \\
&\quad -\lambda\int_\Omega |v(x)|^2\,dx - \lambda t^2\int_\Omega |u_\varepsilon(x)|^2\,dx - 2\lambda t\int_\Omega u_\varepsilon(x)v(x)\,dx \\
&\le (\lambda_{k,s}-\lambda)\|v\|^2_{L^2(\Omega)} + S_{s,\lambda}(u_\varepsilon)\|tu_\varepsilon\|^2_{L^{2^*_s}(\Omega)} \\
&\quad + 2t\int_{\mathbb{R}^n\times\mathbb{R}^n} \frac{(u_\varepsilon(x)-u_\varepsilon(y))(v(x)-v(y))}{|x-y|^{n+2s}} \\
&\quad - 2\lambda t\int_\Omega u_\varepsilon(x)v(x)\,dx,
\end{aligned}
$$

thanks to Proposition 3.2 and to the definition of $S_{s,\lambda}(\cdot)$. Now we write

$$v = \sum_{i=1}^k v_i e_{i,s},$$

for some $v_i \in \mathbb{R}$, so

$$\|v\|^2_{L^2(\Omega)} = \sum_{i=1}^k v_i^2.$$

Also,

$$\langle u_\varepsilon, v\rangle_{\mathbb{H}^s_0(\Omega)} = \sum_{i=1}^k v_i\langle u_\varepsilon, e_{i,s}\rangle_{\mathbb{H}^s_0(\Omega)} = \sum_{i=1}^k \lambda_{i,s} v_i \int_\Omega u_\varepsilon(x) e_{i,s}(x)\,dx.$$

Thus, by the Hölder inequality and the equivalence of the norms in a finite-dimensional space,

$$\begin{aligned}
|\langle u_\varepsilon, v\rangle_{\mathbb{H}^s_0(\Omega)}| &\le \sum_{i=1}^{k} \lambda_{i,s}|v_i|\, \|u_\varepsilon\|_{L^1(\Omega)} \|e_{i,s}\|_{L^\infty(\Omega)} \\
&\le \tilde{\kappa}\lambda_{k+1,s}\|u_\varepsilon\|_{L^1(\Omega)}\|v\|_{L^\infty(\Omega)} \\
&\le \bar{\kappa}\|u_\varepsilon\|_{L^1(\Omega)}\|v\|_{L^2(\Omega)},
\end{aligned}$$

for suitable $\tilde{\kappa}$ and $\bar{\kappa} > 0$ ($\bar{\kappa}$ possibly depends on k). More explicitly,

$$\left| \int_{\mathbb{R}^n\times\mathbb{R}^n} \frac{(u_\varepsilon(x)-u_\varepsilon(y))(v(x)-v(y))}{|x-y|^{n+2s}}\, dxdy \right| \le \bar{\kappa}\|v\|_{L^2(\Omega)}\|u_\varepsilon\|_{L^1(\Omega)}. \tag{15.25}$$

Gathering the results in (15.24) and (15.25) and using the Hölder inequality again, we get

$$\begin{aligned}
M_{\varepsilon,\lambda} &\le (\lambda_{k,s}-\lambda)\|v\|^2_{L^2(\Omega)} + S_{s,\lambda}(u_\varepsilon)\|tu_\varepsilon\|^2_{L^{2^*_s}(\Omega)} + 2t\bar{\kappa}\|v\|_{L^2(\Omega)}\|u_\varepsilon\|_{L^1(\Omega)} \\
&\quad + 2\lambda t\|u_\varepsilon\|_{L^1(\Omega)}\|v\|_{L^\infty(\Omega)} \\
&\le (\lambda_{k,s}-\lambda)\|v\|^2_{L^2(\Omega)} + S_{s,\lambda}(u_\varepsilon)\|tu_\varepsilon\|^2_{L^{2^*_s}(\Omega)} + \kappa\|u_\varepsilon\|_{L^1(\Omega)}\|v\|_{L^2(\Omega)},
\end{aligned}$$

for some positive κ, thanks to (15.21) and the equivalence of the norms in the finite-dimensional space $\mathrm{span}\{e_{1,s},\dots,e_{k,s}\}$. This and (15.23) yield

$$\begin{aligned}
M_{\varepsilon,\lambda} &\le (\lambda_{k,s}-\lambda)\|v\|^2_{L^2(\Omega)} + S_{s,\lambda}(u_\varepsilon)\big(1+\hat{c}\,\|u_\varepsilon\|^{2^*_s-1}_{L^{2^*_s-1}(\Omega)}\|v\|_{L^2(\Omega)}\big)^{2/2^*_s} \\
&\quad + \kappa\|u_\varepsilon\|_{L^1(\Omega)}\|v\|_{L^2(\Omega)},
\end{aligned}$$

so, by Proposition 15.6,

$$\begin{aligned}
M_{\varepsilon,\lambda} &\le (\lambda_{k,s}-\lambda)\|v\|^2_{L^2(\Omega)} + S_{s,\lambda}(u_\varepsilon)\big(1+O(\varepsilon^{(n-2s)/2})\|v\|_{L^2(\Omega)}\big)^{2/2^*_s} \\
&\quad + O(\varepsilon^{(n-2s)/2})\|v\|_{L^2(\Omega)} \\
&\le (\lambda_{k,s}-\lambda)\|v\|^2_{L^2(\Omega)} + S_{s,\lambda}(u_\varepsilon)\big(1+O(\varepsilon^{(n-2s)/2})\|v\|_{L^2(\Omega)}\big) \\
&\quad + O(\varepsilon^{(n-2s)/2})\|v\|_{L^2(\Omega)},
\end{aligned}$$

because $2 < 2^*_s$ (and $s > 0$). Then assertion (b) is proved. This ends the proof of Proposition 15.7. □

15.5 Proof of Theorem 15.1

Here we will apply the linking theorem to the functional $\mathcal{J}_{s,\lambda}$. By Proposition 15.2, we know that $\mathcal{J}_{s,\lambda}$ satisfies the Palais–Smale condition under a suitable threshold, while Propositions 15.3–15.5 ensure the geometric structure required by the linking theorem. Hence, it remains to estimate the linking critical level of $\mathcal{J}_{s,\lambda}$ in order to show that it is below the constant $(s/n)S_s^{n/(2s)}$.

15.5.1 A comment on the estimate of the critical level

The linking critical level of $\mathcal{J}_{s,\lambda}$ is given by

$$c_{\mathrm{L},\lambda} := \inf_{h\in\Gamma}\max_{u\in Q} \mathcal{J}_{s,\lambda}(h(u)), \tag{15.26}$$

where

$$\begin{aligned}&\Gamma := \{h \in C(\overline{Q};\mathbb{H}^s_0(\Omega)) : h = id \text{ on } \partial Q\}, \quad \text{and}\\ &Q := \big(\overline{B}_R \cap \operatorname{span}\{e_{1,s},\dots,e_{k,s}\}\big) \oplus \{r\tilde{z}_\varepsilon : r \in (0,R)\},\end{aligned} \tag{15.27}$$

R is given in Proposition 15.5, and $\tilde{z}_\varepsilon := z_\varepsilon/\|z_\varepsilon\|_{\mathbb{H}^s_0(\Omega)}$,

$$z_\varepsilon := u_\varepsilon - \sum_{i=1}^{k}\Big(\int_\Omega u_\varepsilon(x)e_{i,s}(x)\,dx\Big)e_{i,s},$$

and u_ε is as in (14.19). Note that $z_\varepsilon \not\equiv 0$, by the definition of u_ε.

In order to apply the linking theorem, we need to show that, for any $\lambda \geq \lambda_{1,s}$, the critical level $c_{\mathrm{L},\lambda}$ satisfies the following inequality:

$$c_{\mathrm{L},\lambda} < \frac{s}{n} S_s^{n/(2s)}, \tag{15.28}$$

where S_s is given in formula (1.92). Note that, by the definition of $c_{\mathrm{L},\lambda}$, for any $h \in \Gamma$,

$$c_{\mathrm{L},\lambda} \leq \max_{u\in Q} \mathcal{J}_{s,\lambda}(h(u)), \tag{15.29}$$

so, in particular, taking $h = id$ and using the fact that $Q \subset \mathbb{F}_\varepsilon$, we have

$$c_{\mathrm{L},\lambda} \leq \max_{u\in Q} \mathcal{J}_{s,\lambda}(u) \leq \max_{u\in\mathbb{F}_\varepsilon} \mathcal{J}_{s,\lambda}(u), \tag{15.30}$$

where $\mathbb{F}_\varepsilon$ is given in (15.14).

Because $\mathbb{F}_\varepsilon$ is a linear space,

$$\max_{u\in\mathbb{F}_\varepsilon} \mathcal{J}_{s,\lambda}(u) = \max_{\substack{u\in\mathbb{F}_\varepsilon\\ \zeta\neq 0}} \mathcal{J}_{s,\lambda}\left(|\zeta|\cdot\frac{u}{|\zeta|}\right) = \max_{\substack{u\in\mathbb{F}_\varepsilon\\ \zeta>0}} \mathcal{J}_{s,\lambda}(\zeta u) \leq \max_{\substack{u\in\mathbb{F}_\varepsilon\\ \zeta\geq 0}} \mathcal{J}_{s,\lambda}(\zeta u). \tag{15.31}$$

Hence, putting together (15.30) and (15.31), we have that

$$c_{\mathrm{L},\lambda} \leq \max_{\substack{u\in\mathbb{F}_\varepsilon\\ \zeta\geq 0}} \mathcal{J}_{s,\lambda}(\zeta u). \tag{15.32}$$

However, by Proposition 14.4, we have that, for any $u \in \mathbb{H}^s_0(\Omega)\setminus\{0\}$,

$$\max_{\zeta\geq 0} \mathcal{J}_{s,\lambda}(\zeta u) = \frac{s}{n} S_{s,\lambda}^{n/(2s)}(u) = \frac{s}{n}\left(\frac{\|u\|_{\mathbb{H}^s_0(\Omega)} - \lambda\|u\|^2_{L^2(\Omega)}}{\|u\|^{2^*_s}_{L^{2^*_s}(\Omega)}}\right)^{n/(2s)}, \tag{15.33}$$

also thanks to (14.6). Thus, by (15.32) and (15.33), we get that

$$c_{\mathrm{L},\lambda} \leq \frac{s}{n}\max_{u\in\mathbb{F}_\varepsilon} S_{s,\lambda}^{n/(2s)}(u).$$

Hence, taking into account the definition of $S_{s,\lambda}(\cdot)$ and its scale invariance, to prove (15.28), it is enough to show that

$$M_{\varepsilon,\lambda} < S_s,$$

for some $\varepsilon > 0$, where $M_{\varepsilon,\lambda}$ is defined in (15.15).

15.5.2 End of the proof of Theorem 15.1

When $\lambda \in (0, \lambda_{1,s})$ Theorem 15.1 comes from Theorem 14.1. Let us consider the case where $\lambda > \lambda_{1,s}$ and assume that $\lambda \in (\lambda_{k,s}\, \lambda_{k+1,s})$, for some $k \in \mathbb{N}$ (note that in the setting of Theorem 15.1, the parameter λ cannot be an eigenvalue of the fractional Laplace operator $(-\Delta)^s$). As we said in Subsection 15.5.1, it remains to prove that for suitable values of $\varepsilon > 0$ the following relation holds true:

$$M_{\varepsilon,\lambda} := \max_{\substack{u \in \mathbb{F}_\varepsilon \\ \|u\|_{L^{2^*_s}(\Omega)} = 1}} S_{s,\lambda}(u) < S_s. \tag{15.34}$$

By Proposition 15.7(a),

$$M_{\varepsilon,\lambda} = \int_{\mathbb{R}^n \times \mathbb{R}^n} \frac{|u_M(x) - u_M(y)|^2}{|x-y|^{n+2s}}\, dx\, dy - \lambda \int_\Omega |u_M(x)|^2\, dx,$$

with

$$u_M = v + t u_\varepsilon,$$

for some $v \in \operatorname{span}\{e_{1,s}, \dots, e_{k,s}\}$ and $t \geq 0$. If $t = 0$, by Proposition 15.7(b) and the choice of $\lambda > \lambda_{k,s}$, we get that

$$M_{\varepsilon,\lambda} \leq (\lambda_{k,s} - \lambda)\|v\|^2_{L^2(\Omega)} < 0 < S_s,$$

so (15.34) is proved.

Now suppose that $t > 0$. Again by Proposition 15.7(b), we deduce that, for ε sufficiently small,

$$\begin{aligned} M_{\varepsilon,\lambda} &\leq (\lambda_{k,s} - \lambda)\|v\|^2_{L^2(\Omega)} + S_{s,\lambda}(u_\varepsilon)\big(1 + O(\varepsilon^{(n-2s)/2})\|v\|_{L^2(\Omega)}\big) \\ &\quad + O(\varepsilon^{(n-2s)/2})\|v\|_{L^2(\Omega)} \\ &= O(\varepsilon^{n-2s}) + S_{s,\lambda}(u_\varepsilon)\big(1 + O(\varepsilon^{(n-2s)/2})\|v\|_{L^2(\Omega)}\big), \end{aligned} \tag{15.35}$$

because the parabola $(\lambda_{k,s} - \lambda)\|v\|^2_{L^2(\Omega)} + O(\varepsilon^{(n-2s)/2})\|v\|_{L^2(\Omega)}$ always stays below its vertex; that is,

$$(\lambda_{k,s} - \lambda)\|v\|^2_{L^2(\Omega)} + O(\varepsilon^{(n-2s)/2})\|v\|_{L^2(\Omega)} \leq \frac{1}{4(\lambda - \lambda_{k,s})} O(\varepsilon^{n-2s}) = O(\varepsilon^{n-2s}).$$

Note that here it is crucial that $\lambda \neq \lambda_{k,s}$.

We have to distinguish the cases $n > 4s$ and $n = 4s$. Let us start with the first one; that is, $n > 4s$. Thus, by (1.80) and Propositions 14.11 and 14.12,

$$\begin{aligned} M_{\varepsilon,\lambda} &\le O(\varepsilon^{n-2s}) + S_{s,\lambda}(u_\varepsilon)\big(1 + O(\varepsilon^{(n-2s)/2})\|v\|_{L^2(\Omega)}\big) \\ &\le O(\varepsilon^{n-2s}) + \big(S_s + O(\varepsilon^{n-2s}) - \lambda C_s \varepsilon^{2s}\big)\big(1 + O(\varepsilon^{(n-2s)/2})\|v\|_{L^2(\Omega)}\big) \\ &= S_s + O(\varepsilon^{n-2s}) - \lambda C_s \varepsilon^{2s} \\ &= S_s + \varepsilon^{2s}\big(O(\varepsilon^{n-4s}) - \lambda C_s\big) \\ &< S_s \end{aligned}$$

as $\varepsilon \to 0$. Here we also use the fact that λ is positive (i.e., $\lambda > \lambda_{k,s} > 0$). This ends the proof of (15.34) when $n > 4s$.

Now let $n = 4s$. Arguing as earlier, by (1.80) and Propositions 14.11 and 14.12, we get

$$\begin{aligned} M_{\varepsilon,\lambda} &\le O(\varepsilon^{2s}) + S_{s,\lambda}(u_\varepsilon)\big(1 + O(\varepsilon^{s})\|v\|_{L^2(\Omega)}\big) \\ &\le S_s + O(\varepsilon^{2s}) - \lambda C_s \varepsilon^{2s}|\log\varepsilon| \\ &= S_s + \varepsilon^{2s}|\log\varepsilon|\big(|\log\varepsilon|^{-1} - \lambda C_s\big) \\ &< S_s \end{aligned}$$

as $\varepsilon \to 0$. This concludes the proof of (15.34) when $n = 4s$.

Hence, condition (15.34) is satisfied when $n \ge 4s$, for any $\lambda > 0$ different from the eigenvalues of $(-\Delta)^s$. Then, by the linking theorem, we get the existence of a nontrivial solution $u \in \mathbb{H}_0^s(\Omega)$ of problem (14.2) for any $\lambda > 0$ different from the eigenvalues of $(-\Delta)^s$.

16

The Brezis–Nirenberg result in low dimension

In Chapters 14 and 15 we studied the nonlocal critical problem (14.2) in dimension $n \geq 4s$, and we showed that nontrivial solutions exist, provided that $\lambda > 0$ is different from the eigenvalues of the fractional Laplacian with homogeneous Dirichlet datum. We would remark that the dimensional restriction $n \geq 4s$ is consistent with the classical case when $s = 1$ (see [46]).

In the case of the Laplacian, the case $n = 3$ was treated in detail in [90]. The purpose of this chapter is to extend this analysis to the case of the fractional Laplacian under the dimensional restriction $n \in (2s, 4s)$. Notice once more that this condition reduces to the classical $n = 3$ when $s = 1$, but for different values of $s \in (0, 1)$ it may cover different dimensions (e.g., it boils down to $n = 1$ when $s \in (1/4, 1/2)$, but it becomes $n = 2, 3$ for $s = 7/8$).

We thought that it would be interesting to check what happens in lower dimension, that is, when $n \in (2s, 4s)$. The aim of this chapter is to cover this case so as to treat the critical setting completely. The results discussed here were obtained in [201].

16.1 Existence of a nontrivial solution

The result we obtain in this chapter is the following:

Theorem 16.1 *Let $s \in (0, 1)$, $2s < n < 4s$, and Ω be an open, bounded subset of $\mathbb{R}^n$ with continuous boundary. Then there exists $\lambda_s^\star > 0$ such that, for any $\lambda > \lambda_s^\star$ different from the eigenvalues of $(-\Delta)^s$, problem* (14.2) *admits a nontrivial weak solution $u \in \mathbb{H}_0^s(\Omega)$.*

We will prove Theorem 16.1 by again applying the mountain pass and linking theorems to the functional $\mathcal{J}_{s,\lambda}$ (see (14.4)) in a combined way. As we know by Chapters 14 and 15, the main difficulty to be overcome is the lack of compactness. To solve this problem, we first will show that the Palais–Smale condition for $\mathcal{J}_{s,\lambda}$ holds true in a suitable range related to the best fractional Sobolev constant S_s (see

Subsection 1.5.2 for more details), and then we will prove that the critical mountain pass and linking levels are below this threshold.

It is interesting to note that when $s = 1$, Theorem 16.1 reads as the results obtained for the Laplacian in [46, corollary 1.1] for a bounded domain of $\mathbb{R}^3$ and in [74, theorem 1] (see also [46, theorem 1.2]) in the case of the ball in $\mathbb{R}^3$. Hence, Theorem 16.1 may be seen as the natural extension of some classical results well know for the Laplacian (see [46, 60, 74, 114] and also [212, 220]) to the nonlocal fractional framework.

16.2 Proof of Theorem 16.1

Here we prove the main theorem of this chapter using classical critical points results. Roughly speaking, arguing as in Propositions 14.2 and 14.3 and Section 15.3, we can say that the functional $\mathcal{J}_{s,\lambda}$ has the geometry required by the mountain pass and linking theorems, respectively, when $\lambda < \lambda_{1,s}$ and $\lambda \geq \lambda_{1,s}$ (here, as usual, $\lambda_{k,s}$ and $e_{k,s}$ denote the eigenvalues and the eigenfunctions of $(-\Delta)^s$ with homogeneous Dirichlet boundary data).

Precisely, we have the following results, which are valid for any $n > 2s$:

Proposition 16.2 Let $\lambda \in (0, \lambda_{1,s})$. Then there exist $\rho > 0$, $\beta > 0$ and $e \in \mathbb{H}_0^s(\Omega)$ such that

(a) For any $u \in \mathbb{H}_0^s(\Omega)$ with $\|u\|_{\mathbb{H}_0^s(\Omega)} = \rho$, it results that $\mathcal{J}_{s,\lambda}(u) \geq \beta$.
(b) $e \geq 0$ a.e. in $\mathbb{R}^n$, $\|e\|_{\mathbb{H}_0^s(\Omega)} > \rho$, and $\mathcal{J}_{s,\lambda}(e) < 0$.

In particular, we can construct e as follows:

$$e = e_\varepsilon := \zeta_\varepsilon u_\varepsilon, \tag{16.1}$$

with ζ_ε large enough and u_ε as in (14.19) for ε sufficiently small.

Proposition 16.3 Let $\lambda \in [\lambda_{k,s}, \lambda_{k+1,s})$, for some $k \in \mathbb{N}$. Then

(a) There exist $\rho > 0$ and $\beta > 0$ such that, for any $u \in \mathbb{P}_{k+1,s}$ with $\|u\|_{\mathbb{H}_0^s(\Omega)} = \rho$, it results that $\mathcal{J}_{s,\lambda}(u) \geq \beta$, where $\mathbb{P}_{k+1,s}$ is given in (5.31).
(b) $\mathcal{J}_{s,\lambda}(u) \leq 0$, for any $u \in \operatorname{span}\{e_{1,s}, \dots, e_{k,s}\}$.
(c) For any finite-dimensional subspace $\mathbb{F}$ of $\mathbb{H}_0^s(\Omega)$, there exists $R > \rho$ such that

$$\sup_{\substack{u \in \mathbb{F} \\ \|u\|_{\mathbb{H}_0^s(\Omega)} \geq R}} \mathcal{J}_{s,\lambda}(u) \leq 0.$$

In particular, we can construct $\mathbb{F}$ as follows:

$$\mathbb{F} = \mathbb{F}_\varepsilon, \tag{16.2}$$

with $\mathbb{F}_\varepsilon$ given in (15.14) for ε small enough.

Moreover, by Proposition 15.2, the functional $\mathcal{J}_{s,\lambda}$ verifies the Palais–Smale condition only in an energy range bounded above by the constant $(s/n)\,S_s^{n/(2s)}$. Hence, as in the preceding chapters, to apply the classical critical points theorems cited earlier to $\mathcal{J}_{s,\lambda}$, it is enough to show that the mountain pass and linking critical levels of $\mathcal{J}_{s,\lambda}$ stay below this threshold.

For this, the following result will be crucial:

Proposition 16.4 Let $s \in (0,1)$, $2s < n < 4s$, and let u_ε be as in (14.19). Then there exists $\lambda_s^\star > 0$ such that, for any $\lambda > \lambda_s^\star$,

$$S_{s,\lambda}(u_\varepsilon) < S_s,$$

provided that $\varepsilon > 0$ is sufficiently small. Here $S_{s,\lambda}(\,\cdot\,)$ is the functional defined in (14.6).

Proof We observe that

$$\frac{n}{2s}\left(1-\frac{2}{2_s^*}\right)=1.$$

Hence, by (14.6) and Propositions 14.11 and 14.12, we get

$$\begin{aligned} S_{s,\lambda}(u_\varepsilon) &\leq \frac{S_s^{n/(2s)}+O(\varepsilon^{n-2s})-\lambda C_s\varepsilon^{n-2s}+O(\varepsilon^{2s})}{\left(S_s^{n/(2s)}+O(\varepsilon^n)\right)^{2/2_s^*}} \\ &\leq S_s+\varepsilon^{n-2s}\left(O(1)-\lambda\widetilde{C}_s\right)+O(\varepsilon^{2s}) < S_s, \end{aligned}$$

if λ is large enough, say, $\lambda > \lambda_s^\star > 0$, provided that $\varepsilon > 0$ is sufficiently small. Thus, the assertion is proved. □

Now we can estimate the critical level of the functional $\mathcal{J}_{s,\lambda}$. This is what we will do in the next subsection.

16.2.1 Estimates of the critical level of $\mathcal{J}_{s,\lambda}$

Let $\lambda_s^\star > 0$ be as in Proposition 16.4. Of course, only the two cases $\lambda_s^\star < \lambda_{1,s}$ and $\lambda_s^\star \geq \lambda_{1,s}$ can occur. Let us consider them separately.

Case 1: $\lambda_s^\star < \lambda_{1,s}$**.** This case will be dealt with by using either the mountain pass theorem when $\lambda \in (\lambda_s^\star, \lambda_{1,s})$ or the linking theorem when $\lambda \geq \lambda_{1,s}$. First of all, let $\lambda \in (\lambda_s^\star, \lambda_{1,s})$. Hence, by the mountain pass theorem, the critical level of $\mathcal{J}_{s,\lambda}$ is given by

$$c_{\mathrm{MP},\lambda} := \inf_{P\in\mathcal{P}}\max_{v\in P([0,1])}\mathcal{J}_{s,\lambda}(v),$$

where

$$\mathcal{P} := \left\{P \in C([0,1];\mathbb{H}_0^s(\Omega)) : P(0)=0,\ P(1)=e\right\},$$

with e as in (16.1). We claim that, for any $\lambda \in (\lambda_s^\star, \lambda_{1,s})$, it holds true that

$$\beta \leq c_{\mathrm{MP},\lambda} < \frac{s}{n}\,S_s^{n/(2s)}, \tag{16.3}$$

where β is given in Proposition 16.2 and S_s is defined in formula (1.92).

The fact that

$$c_{\mathrm{MP},\lambda} \geq \beta \tag{16.4}$$

follows exactly as in the proof of Claim 14.1. Let us show that

$$c_{\mathrm{MP},\lambda} < \frac{s}{n} S_s^{n/(2s)}.$$

For this purpose, recalling (16.1), we have that the map $[0,1] \ni t \mapsto t\zeta_\varepsilon u_\varepsilon$ belongs to $\mathscr{P}$. By Proposition 14.4, we have that

$$\inf_{P\in\mathscr{P}} \max_{v\in P([0,1])} \mathscr{J}_{s,\lambda}(v) \leq \max_{\zeta\geq 0} \mathscr{J}_{s,\lambda}(\zeta u_\varepsilon) = \frac{s}{n} S_{s,\lambda}^{n/(2s)}(u_\varepsilon),$$

while, by Proposition 16.4, we get that, for any $\lambda > \lambda_s^\star$,

$$\frac{s}{n} S_{s,\lambda}^{n/(2s)}(u_\varepsilon) < \frac{s}{n} S_s^{n/(2s)},$$

for ε sufficiently small. Hence, putting together these two inequalities, we conclude that

$$c_{\mathrm{MP},\lambda} < \frac{s}{n} S_s^{n/(2s)}, \tag{16.5}$$

provided that $\lambda \in (\lambda_s^\star, \lambda_{1,s})$. Thus, (16.3) follows from (16.4) and (16.5).

Now let $\lambda \geq \lambda_{1,s}$. We can suppose that $\lambda \in [\lambda_{k,s}, \lambda_{k+1,s})$, for some $k \in \mathbb{N}$. In fact, we suppose that $\lambda \in (\lambda_{k,s}, \lambda_{k+1,s})$ because in the setting of Theorem 16.1 the parameter λ cannot be an eigenvalue of the fractional Laplace operator $(-\Delta)^s$.

Here we have to estimate the linking critical level of $\mathscr{J}_{s,\lambda}$ given by

$$c_{\mathrm{L},\lambda} := \inf_{h\in\Gamma} \max_{u\in Q} \mathscr{J}_{s,\lambda}(h(u)),$$

where Γ and Q are as in (15.27), R is given in Proposition 16.3, and $\tilde{z}_\varepsilon$ is as in Subsection 15.5.1.

What we need in this framework is that, for any $\lambda \geq \lambda_{1,s}$,

$$c_{\mathrm{L},\lambda} < \frac{s}{n} S_s^{n/(2s)}, \tag{16.6}$$

where S_s is given in formula (1.92). By Subsection 15.5.1, it is easy to see that to prove (16.6), it is enough to show that

$$M_{\varepsilon,\lambda} < S_s, \tag{16.7}$$

for some $\varepsilon > 0$, where $M_{\varepsilon,\lambda}$ is defined in (15.15).

For proving (16.7), we proceed as in Subsection 15.5.2 (where we considered the case $n \geq 4s$), taking into account the estimates given in Proposition 14.12 when $n \in (2s, 4s)$. For the sake of clarity and for the reader's convenience, we prefer to repeat all the calculations.

By Proposition 15.7(a),

$$M_{\varepsilon,\lambda} = \|u_M\|^2_{\mathbb{H}^s_0(\Omega)} - \lambda \|u_M\|^2_{L^2(\Omega)},$$

with

$$u_M = v + tu_\varepsilon, \quad v \in \operatorname{span}\{e_{1,s}, \dots, e_{k,s}\}, \quad t \geq 0, \quad \|u_M\|_{L^{2^*_s}(\Omega)} = 1.$$

If $t = 0$, by Proposition 15.7(b) and the choice of $\lambda > \lambda_{k,s}$, we get that

$$M_{\varepsilon,\lambda} \leq (\lambda_{k,s} - \lambda)\|v\|^2_{L^2(\Omega)} < 0 < S_s,$$

so (16.7) is proved.

Now suppose that $t > 0$. By Propositions 15.7(b) and 14.12, with $2s < n < 4s$, we deduce that

$$\begin{aligned}
M_{\varepsilon,\lambda} &\leq (\lambda_{k,s} - \lambda)\|v\|^2_{L^2(\Omega)} + S_{s,\lambda}(u_\varepsilon)\big(1 + O(\varepsilon^{(n-2s)/2})\|v\|_{L^2(\Omega)}\big) \\
&\quad + O(\varepsilon^{(n-2s)/2})\|v\|_{L^2(\Omega)} \\
&\leq (\lambda_{k,s} - \lambda)\|v\|^2_{L^2(\Omega)} + \big(S_s + O(\varepsilon^{n-2s}) - \lambda C_s \varepsilon^{n-2s} + O(\varepsilon^{2s})\big)\cdot \\
&\quad \big(1 + O(\varepsilon^{(n-2s)/2})\|v\|_{L^2(\Omega)}\big) + O(\varepsilon^{(n-2s)/2})\|v\|_{L^2(\Omega)} \\
&\leq (\lambda_{k,s} - \lambda)\|v\|^2_{L^2(\Omega)} + S_s + O(\varepsilon^{n-2s}) - \lambda C_s \varepsilon^{n-2s} + O(\varepsilon^{2s}) \\
&\quad + O(\varepsilon^{(n-2s)/2})\|v\|_{L^2(\Omega)} \\
&\leq S_s + O(\varepsilon^{n-2s}) - \lambda C_s \varepsilon^{n-2s} + O(\varepsilon^{2s})
\end{aligned}$$

because the parabola $(\lambda_{k,s} - \lambda)\|v\|^2_{L^2(\Omega)} + O(\varepsilon^{(n-2s)/2})\|v\|_{L^2(\Omega)}$ always stays below its vertex (see Subsection 15.5.2). Thus,

$$\begin{aligned}
M_{\varepsilon,\lambda} &\leq S_s + O(\varepsilon^{n-2s}) - \lambda C_s \varepsilon^{n-2s} + O(\varepsilon^{2s}) \\
&= S_s + \varepsilon^{n-2s}\big(O(1) - \lambda C_s + O(\varepsilon^{4s-n})\big) < S_s
\end{aligned}$$

as $\varepsilon \to 0$, thanks to the fact that $\lambda > \lambda^\star_s$. This ends the proof of (16.7).

Ultimately, we have shown that the critical level of $\mathcal{J}_{s,\lambda}$ stays below $(s/n)\, S_s^{n/(2s)}$, provided that $\lambda > \lambda^\star_s$ is not an eigenvalue of $(-\Delta)^s$.

Case 2: $\lambda^\star_s \geq \lambda_{1,s}$. In this setting, the only critical level of $\mathcal{J}_{s,\lambda}$ to be estimated is the one given by the linking theorem. To do this, we can argue as in the last part of Case 1 (where we estimate the linking critical level). In this way, we get that, for any $\lambda > \lambda^\star_s$ different from an eigenvalue of $(-\Delta)^s$,

$$c_{\mathrm{L},\lambda} < \frac{s}{n} S_s^{n/(2s)}.$$

Now we can conclude the proof of Theorem 16.1.

16.2.2 End of the proof of Theorem 16.1

In Subsection 16.2.1, we showed that there exists $\lambda_s^\star > 0$ such that, for any $\lambda > \lambda_s^\star$, with λ not an eigenvalue of $(-\Delta)^s$, the critical level of $\mathcal{J}_{s,\lambda}$ stays below $(s/n)\, S_s^{n/(2s)}$, and therefore, the Palais–Smale condition is satisfied (see, in particular, (16.3), (16.6), and Proposition 15.2).

Let us consider two cases: either $0 < \lambda_s^\star < \lambda_{1,s}$ or $\lambda_s^\star \geq \lambda_{1,s}$. Considering these two situations separately, we have

- $0 < \lambda_s^\star < \lambda_{1,s}$. In this case, for any $\lambda \in (\lambda_s^\star, \lambda_{1,s})$, problem (14.3) admits a mountain pass solution $u \in \mathbb{H}_0^s(\Omega)$ at level $\mathcal{J}_{s,\lambda}(u) = c_{\mathrm{MP},\lambda}$. Moreover, u is nontrivial because $\mathcal{J}_{s,\lambda}(u) = c_{\mathrm{MP},\lambda} \geq \beta > 0 = \mathcal{J}_{s,\lambda}(0)$, while, for any $\lambda > \lambda_{1,s} > \lambda_s^\star$, with λ different from an eigenvalue of $(-\Delta)^s$, problem (14.3) admits a nontrivial linking solution $u \in \mathbb{H}_0^s(\Omega)$ at level $\mathcal{J}_{s,\lambda}(u) = c_{\mathrm{L},\lambda} \geq \beta$.
- $\lambda_s^\star \geq \lambda_{1,s}$. In this case, problem (14.3) admits a nontrivial linking solution u at level $\mathcal{J}_{s,\lambda}(u) = c_{\mathrm{L},\lambda} \geq \beta$, provided that $\lambda > \lambda_s^\star$ is not an eigenvalue of the fractional Laplacian.

Ultimately, for any value of $\lambda > \lambda_s^\star$ that is not an eigenvalue of $(-\Delta)^s$, we get a nontrivial weak solution for problem (14.2). This concludes the proof of Theorem 16.1.

17

The critical equation in the resonant case

In previous chapters we proved that the famous result by Brezis and Nirenberg (see [46, 60, 114, 221]) for the Laplace equation continues to hold in the nonlocal setting of (14.2), provided that λ is not an eigenvalue of $(-\Delta)^s$. With respect to the classical Brezis–Nirenberg result, the resonant case in dimensions different from $4s$ remains open, that is, the case where $n \neq 4s$ and λ is an eigenvalue of the operator $(-\Delta)^s$ with homogeneous Dirichlet boundary data.

The aim of this chapter is to consider (14.2) in this setting because we thought that it would be interesting to check what happens in this case to verify whether the classical result known for the Laplacian can be extended to the nonlocal fractional framework. In this way, the study of the critical fractional Laplace problem (14.2) is completed. This chapter is based on the paper [197].

17.1 Main results

The main result of this chapter is the following:

Theorem 17.1 *Let $s \in (0,1)$, $n > 2s$, and Ω be an open, bounded subset of $\mathbb{R}^n$ with continuous boundary. Moreover, let λ be an eigenvalue of $(-\Delta)^s$ with homogeneous Dirichlet boundary data. Then problem* (14.2) *admits a nontrivial weak solution $u \in \mathbb{H}^s_0(\Omega)$, provided that either*

- *$n > 4s$ or*
- *$2s < n < 4s$ and λ is sufficiently large.*

As a consequence of Theorem 17.1 and Theorems 14.1, 15.1 and 16.1, we get the following existence result, which extends completely to the nonlocal fractional framework the well-known Brezis–Nirenberg-type results given in [46, 60, 114, 221] for the Laplace equation:

Theorem 17.2 *Let $s \in (0,1)$, $n > 2s$, and Ω be an open, bounded subset of $\mathbb{R}^n$ with continuous boundary. Then problem* (14.2) *admits a nontrivial weak solution $u \in \mathbb{H}^s_0(\Omega)$, provided that either*

- *$n > 4s$ and $\lambda > 0$,*
- *$n = 4s$ and $\lambda > 0$ is different from the eigenvalues of $(-\Delta)^s$, or*
- *$n < 4s$ and $\lambda > 0$ is sufficiently large.*

Roughly speaking, Theorem 17.2 says that what happens in the nonlocal framework is exactly what we know in the classing setting (see [46, 60, 114, 221] and also [120, 212, 220] and references therein). Of course, the extension from the local setting to the nonlocal one is not straightforward, as we have already seen in previous chapters and we will see in this one.

We would like to note that, as it happens in the Laplacian case when $n = 4$, in the nonlocal framework there is also a dimension ($n = 4s$) where resonance creates problem. Also, when $s = 1$ (which corresponds to the Laplace case), these two dimensions are the same. In the classical setting of the Laplacian, this fact was not underlined in the original paper of Capozzi, Fortunato, and Palmieri (see [60]), but it was noticed by Zhang in [221]. For an explanation of this strange phenomenon, see also [14, 112].

To prove Theorem 17.1, we will argue as in previous chapters; that is, we will look for critical points of the Euler–Lagrange functional $\mathcal{J}_{s,\lambda}$ associated with problem (14.2).

In this chapter, we are interested in problem (14.2) when the parameter λ is an eigenvalue of $(-\Delta)^s$, say, when $\lambda = \lambda_{k,s}$, for some $k \in \mathbb{N}$. Without loss of generality, we can assume that $\lambda_{k,s}$ has multiplicity $h \in \mathbb{N}$ and that

$$\lambda_{k-1,s} < \lambda_{k,s} = \lambda_{k+1,s} = \cdots = \lambda_{k+h-1,s} < \lambda_{k+h,s}. \tag{17.1}$$

To prove Theorem 17.1, we will consider the functional $\mathcal{J}_{s,\lambda_{k,s}}$, that is, $\mathcal{J}_{s,\lambda}$, with $\lambda = \lambda_{k,s}$.

By Proposition 15.2 (which holds true for any $\lambda > 0$) and the fact that $\lambda_{k,s} \geq \lambda_{1,s} > 0$ (see Proposition 3.1(a)), we know that $\mathcal{J}_{s,\lambda_{k,s}}$ verifies the Palais–Smale condition up to a suitable threshold; that is, the following result holds true:

Proposition 17.3 Let $c \in \mathbb{R}$ be such that

$$c < \frac{s}{n} S_s^{n/(2s)},$$

where S_s is the constant defined in (1.92). Then the functional $\mathcal{J}_{s,\lambda_{k,s}}$ satisfies the Palais–Smale condition at level c.

As in previous chapters, because this condition does not hold globally, to apply the linking theorem to $\mathcal{J}_{s,\lambda_{k,s}}$, we need to estimate the critical level of this functional and show that it stays below the threshold where the Palais–Smale condition is satisfied.

17.2 Estimate of the minimax critical level

This section is devoted to estimation of the minimax critical level of the functional $\mathcal{J}_{s,\lambda_{k,s}}$. Here, in some sense, we argue as in Subsection 14.3.1, Section 15.4, and Subsection 16.2.1 even if, with respect to these sections, some differences arise due to the fact that here we consider the resonant case, that is, the case where λ is an eigenvalue of $(-\Delta)^s$. In fact, here the estimates are more delicate: on this we will be more precise in what follows.

Note that Proposition 16.3 ensures that $\mathcal{J}_{s,\lambda_{k,s}}$ has the linking geometric features (note that this result holds true without any restriction on the dimension n; that is, it is valid for any $n > 2s$). The linking critical level $c_{\mathrm{L},\lambda_{k,s}}$ of $\mathcal{J}_{s,\lambda_{k,s}}$ is given by (15.26), with $\lambda = \lambda_{k,s}$. We want to show the following result:

Proposition 17.4 Let S_s be as in (1.92). Then

$$c_{\mathrm{L},\lambda_{k,s}} < \frac{s}{n} S_s^{n/(2s)},$$

provided that either

- $n > 4s$ or
- $n < 4s$ and $\lambda_{k,s}$ is large enough.

Proof To estimate $c_{\mathrm{L},\lambda_{k,s}}$, in Subsection 15.5.1 (here we consider this result in the case where $\lambda = \lambda_{k,s}$) we showed that it is enough to prove that

$$M_{\varepsilon,\lambda_{k,s}} := \max_{\substack{u\in\mathbb{F}_\varepsilon \\ \|u\|_{L^{2^*_s}(\Omega)}=1}} \left(\|u\|^2_{\mathbb{H}^s_0(\Omega)} - \lambda_{k,s}\|u\|^2_{L^2(\Omega)}\right) < S_s, \tag{17.2}$$

where $\mathbb{F}_\varepsilon$ is the space defined in (15.14). Let us prove inequality (17.2) : this estimate is more delicate than in previous chapters owing to the fact that here we have to mange the resonant case. For this, we prefer to give all the details of the proof. To this end, let us recall that, by Proposition 15.7(a), $M_{\varepsilon,\lambda_{k,s}}$ is achieved in some $u_M \in \mathbb{F}_\varepsilon$, which can be written as

$$u_M = \tilde{v} + t z_\varepsilon, \tag{17.3}$$

where $\tilde{v} \in \mathrm{span}\{e_{1,s},\dots,e_{k,s}\}$, $t \geq 0$, and z_ε is as in Subsection 15.5.1 and such that

$$\|u_M\|_{L^{2^*_s}(\Omega)} = 1. \tag{17.4}$$

From now on, we proceed by steps.

Claim 17.1 *There exists a positive constant $\bar{c}$ such that $t = t_\varepsilon \leq \bar{c}$, for $\varepsilon > 0$ small enough.*

Proof The proof is exactly the one given in Proposition 15.7 (see formula (15.21)). □

Claim 17.2 *The function $u_M \in \mathbb{F}_\varepsilon$ can be written as*

$$u_M = v + P_k\tilde{v} + t\tilde{u}_\varepsilon,$$

where, for $t \geq 0$,

$$v = \sum_{i=1}^{k-1}\left(\int_\Omega (\tilde{v}(x) - tu_\varepsilon(x))e_{i,s}(x)\,dx\right)e_{i,s} \in \operatorname{span}\{e_{1,s},\dots,e_{k-1,s}\}, \tag{17.5}$$

$$\tilde{u}_\varepsilon = u_\varepsilon - P_k u_\varepsilon, \tag{17.6}$$

and the map $\mathbb{H}_0^s(\Omega) \ni w \mapsto P_k w$ denotes the projection of w on the direction $e_{k,s}$; that is,

$$P_k w = \left(\int_\Omega w(x)e_{k,s}(x)\,dx\right)e_{k,s}.$$

Moreover, there exists a positive constant $\bar{\kappa}$, independent of ε, such that

$$\left|\int_\Omega \tilde{u}_\varepsilon(x)v(x)\,dx\right| = \left|\int_\Omega u_\varepsilon(x)v(x)\,dx\right| \leq \bar{\kappa}\|v\|_{L^2(\Omega)}\|u_\varepsilon\|_{L^1(\Omega)} \tag{17.7}$$

and

$$\left|\int_{\mathbb{R}^n\times\mathbb{R}^n} \frac{(\tilde{u}_\varepsilon(x) - \tilde{u}_\varepsilon(y))(v(x) - v(y))}{|x-y|^{n+2s}}\,dx\,dy\right| \leq \bar{\kappa}\|v\|_{L^2(\Omega)}\|u_\varepsilon\|_{L^1(\Omega)}, \tag{17.8}$$

for any $\varepsilon > 0$.

Proof By (17.3) and the definition of z_ε (given in Subsection 15.5.1), it is easily seen that

$$\begin{aligned} u_M &= \sum_{i=1}^{k}\left(\int_\Omega \tilde{v}(x)e_{i,s}(x)\,dx\right)e_{i,s} + t\Big(u_\varepsilon - \sum_{i=1}^{k}\left(\int_\Omega u_\varepsilon(x)e_{i,s}(x)\,dx\right)e_{i,s}\Big) \\ &= \sum_{i=1}^{k-1}\left(\int_\Omega (\tilde{v}(x) - tu_\varepsilon(x))e_{i,s}(x)\,dx\right)e_{i,s} + P_k\tilde{v} + t(u_\varepsilon - P_k u_\varepsilon) \\ &= v + P_k\tilde{v} + t\tilde{u}_\varepsilon, \end{aligned}$$

with v and $\tilde{u}_\varepsilon$ as in (17.5) and (17.6), respectively. Let us start by showing that (17.7) holds true. For this, note that v and $P_k u_\varepsilon$ are orthogonal in $L^2(\Omega)$, so

$$\int_\Omega \tilde{u}_\varepsilon(x)v(x)\,dx = \int_\Omega \big(u_\varepsilon(x) - P_k u_\varepsilon\big)v(x)\,dx = \int_\Omega u_\varepsilon(x)v(x)\,dx,$$

while the Hölder inequality and the equivalence of the norm in a finite-dimensional space give

$$\begin{aligned}\left|\int_\Omega \tilde{u}_\varepsilon(x)v(x)\,dx\right| &= \left|\int_\Omega u_\varepsilon(x)v(x)\,dx\right|\\ &\le \|u_\varepsilon\|_{L^1(\Omega)}\|v\|_{L^\infty(\Omega)}\\ &\le \bar{\kappa}\|u_\varepsilon\|_{L^1(\Omega)}\|v\|_{L^2(\Omega)},\end{aligned}$$

for a suitable $\bar{\kappa} > 0$, independent of ε. Thus, (17.7) is proved.

Now let us show (17.8). For this purpose, we write

$$v = \sum_{i=1}^{k-1} v_i e_{i,s}, \tag{17.9}$$

for some $v_i \in \mathbb{R}$, so, again by Proposition 3.1(f),

$$\|v\|^2_{L^2(\Omega)} = \sum_{i=1}^{k-1} v_i^2.$$

By (17.9), the fact that $e_{i,s}$ is an eigenfunction of $(-\Delta)^s$ with eigenvalue $\lambda_{i,s}$, and the definition of scalar product in $\mathbb{H}^s_0(\Omega)$, we have

$$\begin{aligned}\langle \tilde{u}_\varepsilon, v\rangle_{\mathbb{H}^s_0(\Omega)} &= \sum_{i=1}^{k-1} v_i\langle \tilde{u}_\varepsilon, e_{i,s}\rangle_{\mathbb{H}^s_0(\Omega)}\\ &= \sum_{i=1}^{k-1}\lambda_{i,s}v_i\int_\Omega \tilde{u}_\varepsilon(x)e_{i,s}(x)\,dx\\ &= \sum_{i=1}^{k-1}\lambda_{i,s}v_i\int_\Omega u_\varepsilon(x)e_{i,s}(x)\,dx,\end{aligned}$$

also thanks to the definition of $\tilde{u}_\varepsilon$ and the orthogonality properties of $e_{i,s}$. Thus, by this and again by the Hölder inequality, we get

$$\begin{aligned}|\langle \tilde{u}_\varepsilon, v\rangle_{\mathbb{H}^s_0(\Omega)}| &\le \sum_{i=1}^{k-1}\lambda_{i,s}|v_i|\,\|u_\varepsilon\|_{L^1(\Omega)}\|e_{i,s}\|_{L^\infty(\Omega)}\\ &\le \bar{\kappa}\|u_\varepsilon\|_{L^1(\Omega)}\|v\|_{L^2(\Omega)};\end{aligned}$$

that is,

$$\left|\int_{\mathbb{R}^n\times\mathbb{R}^n}\frac{(\tilde{u}_\varepsilon(x)-\tilde{u}_\varepsilon(y))(v(x)-v(y))}{|x-y|^{n+2s}}\,dx\,dy\right| \le \bar{\kappa}\|v\|_{L^2(\Omega)}\|u_\varepsilon\|_{L^1(\Omega)},$$

for a suitable $\bar{\kappa} > 0$ possibly depending on k, but independent of ε. Hence, (17.8) is proved, and this ends the proof of Claim 17.2. □

Now we are ready to show the validity of (17.2), that is, that

$$M_{\varepsilon,\lambda_{k,s}} = \int_{\mathbb{R}^n\times\mathbb{R}^n} \frac{|u_M(x)-u_M(y)|^2}{|x-y|^{n+2s}}\,dx\,dy - \lambda_{k,s}\int_\Omega |u_M(x)|^2\,dx \tag{17.10}$$
$$< S_s.$$

In doing this, we have to take into account that $u_M = v + P_k\tilde{v} + t\tilde{u}_\varepsilon$ by Claim 17.2 and that, in particular, we have to estimate three different contributions coming from v, $P_k\tilde{v}$, and $\tilde{u}_\varepsilon$. With respect to similar calculations carried on in Subsections 15.5.2 and 16.2.1, here we have to pay attention to the contribution coming from v because of the resonance occurring in this case.

Let us show (17.10). By Claim 17.2, we have that

$$\begin{aligned}
M_{\varepsilon,\lambda_{k,s}} &= \int_{\mathbb{R}^n\times\mathbb{R}^n} \frac{|u_M(x)-u_M(y)|^2}{|x-y|^{n+2s}}\,dx\,dy - \lambda_{k,s}\int_\Omega |u_M(x)|^2\,dx \\
&= \int_{\mathbb{R}^n\times\mathbb{R}^n} \frac{|v(x)+P_k\tilde{v}(x)+t\tilde{u}_\varepsilon(x)-v(y)-P_k\tilde{v}(y)-t\tilde{u}_\varepsilon(y)|^2}{|x-y|^{n+2s}}\,dx\,dy \\
&\quad - \lambda_{k,s}\int_\Omega |v(x)+P_k\tilde{v}(x)+t\tilde{u}_\varepsilon(x)|^2\,dx \\
&= \|v\|^2_{\mathbb{H}^s_0(\Omega)} + \|P_k\tilde{v}\|^2_{\mathbb{H}^s_0(\Omega)} + t^2\|\tilde{u}_\varepsilon\|^2_{\mathbb{H}^s_0(\Omega)} \qquad (17.11) \\
&\quad + 2t\int_{\mathbb{R}^n\times\mathbb{R}^n} \frac{(\tilde{u}_\varepsilon(x)-\tilde{u}_\varepsilon(y))(v(x)-v(y))}{|x-y|^{n+2s}}\,dxdy \\
&\quad - \lambda_{k,s}\big(\|v\|^2_{L^2(\Omega)} + \|P_k\tilde{v}\|^2_{L^2(\Omega)} + t^2\|\tilde{u}_\varepsilon\|^2_{L^2(\Omega)}\big) - 2\lambda_{k,s}t\int_\Omega \tilde{u}_\varepsilon(x)v(x)\,dx \\
&= \|v\|^2_{\mathbb{H}^s_0(\Omega)} + t^2\|\tilde{u}_\varepsilon\|^2_{\mathbb{H}^s_0(\Omega)} + 2t\int_{\mathbb{R}^n\times\mathbb{R}^n} \frac{(\tilde{u}_\varepsilon(x)-\tilde{u}_\varepsilon(y))(v(x)-v(y))}{|x-y|^{n+2s}}\,dxdy \\
&\quad - \lambda_{k,s}\big(\|v\|^2_{L^2(\Omega)} + t^2\|\tilde{u}_\varepsilon\|^2_{L^2(\Omega)}\big) - 2\lambda_{k,s}t\int_\Omega \tilde{u}_\varepsilon(x)v(x)\,dx,
\end{aligned}$$

thanks to the orthogonality properties of v, $P_k\tilde{v}$, and $\tilde{u}_\varepsilon$ and also to the definition of $\lambda_{k,s}$. Now note that, by (17.6),

$$
\begin{aligned}
\|\tilde{u}_\varepsilon\|^2_{\mathbb{H}^s_0(\Omega)} - \lambda_{k,s}\|\tilde{u}_\varepsilon\|^2_{L^2(\Omega)} &= \|u_\varepsilon - P_k u_\varepsilon\|^2_{\mathbb{H}^s_0(\Omega)} - \lambda_{k,s}\|u_\varepsilon - P_k u_\varepsilon\|^2_{L^2(\Omega)} \\
&= \|u_\varepsilon\|^2_{\mathbb{H}^s_0(\Omega)} + \|P_k u_\varepsilon\|^2_{\mathbb{H}^s_0(\Omega)} \\
&\quad -2\int_{\mathbb{R}^n\times\mathbb{R}^n} \frac{(u_\varepsilon(x)-u_\varepsilon(y))(P_k u_\varepsilon(x)-P_k u_\varepsilon(y))}{|x-y|^{n+2s}}\,dxdy \\
&\quad -\lambda_{k,s}\big(\|u_\varepsilon\|^2_{L^2(\Omega)} + \|P_k u_\varepsilon\|^2_{L^2(\Omega)}\big) \\
&\quad +2\lambda_{k,s}\int_\Omega u_\varepsilon(x)\,P_k u_\varepsilon(x)\,dx \\
&= \|u_\varepsilon\|^2_{\mathbb{H}^s_0(\Omega)} - \lambda_{k,s}\|u_\varepsilon\|^2_{L^2(\Omega)} \\
&\quad -2\int_{\mathbb{R}^n\times\mathbb{R}^n} \frac{(u_\varepsilon(x)-u_\varepsilon(y))(P_k u_\varepsilon(x)-P_k u_\varepsilon(y))}{|x-y|^{n+2s}}\,dxdy \\
&\quad +2\lambda_{k,s}\int_\Omega u_\varepsilon(x)\,P_k u_\varepsilon(x)\,dx \\
&= \|u_\varepsilon\|^2_{\mathbb{H}^s_0(\Omega)} - \lambda_{k,s}\,\|u_\varepsilon\|^2_{L^2(\Omega)} \\
&\quad -2\big(\|P_k u_\varepsilon\|^2_{\mathbb{H}^s_0(\Omega)} - \lambda_{k,s}\|P_k u_\varepsilon\|^2_{L^2(\Omega)}\big) \\
&= \|u_\varepsilon\|^2_{\mathbb{H}^s_0(\Omega)} - \lambda_{k,s}\,\|u_\varepsilon\|^2_{L^2(\Omega)},
\end{aligned}
\tag{17.12}
$$

thanks to the definition of P_k.

Then, combining (17.7), (17.8), (17.11), Claim 17.1, and (17.12), we get

$$
\begin{aligned}
M_{\varepsilon,\lambda_{k,s}} &= \|v\|^2_{\mathbb{H}^s_0(\Omega)} + t^2\|\tilde{u}_\varepsilon\|^2_{\mathbb{H}^s_0(\Omega)} + 2t\int_{\mathbb{R}^n\times\mathbb{R}^n} \frac{(\tilde{u}_\varepsilon(x)-\tilde{u}_\varepsilon(y))(v(x)-v(y))}{|x-y|^{n+2s}}\,dxdy \\
&\quad -\lambda_{k,s}\big(\|v\|^2_{L^2(\Omega)} + t^2\|\tilde{u}_\varepsilon\|^2_{L^2(\Omega)}\big) - 2\lambda_{k,s}t\int_\Omega \tilde{u}_\varepsilon(x)v(x)\,dx \\
&= \|v\|^2_{\mathbb{H}^s_0(\Omega)} + t^2\|u_\varepsilon\|^2_{\mathbb{H}^s_0(\Omega)} \\
&\quad +2t\int_{\mathbb{R}^n\times\mathbb{R}^n} \frac{(\tilde{u}_\varepsilon(x)-\tilde{u}_\varepsilon(y))(v(x)-v(y))}{|x-y|^{n+2s}}\,dxdy \qquad (17.13) \\
&\quad -\lambda_{k,s}\big(\|v\|^2_{L^2(\Omega)} + t^2\|u_\varepsilon\|^2_{L^2(\Omega)}\big) - 2\lambda_{k,s}t\int_\Omega \tilde{u}_\varepsilon(x)v(x)\,dx \\
&\le \|v\|^2_{\mathbb{H}^s_0(\Omega)} - \lambda_{k,s}\|v\|^2_{L^2(\Omega)} + t^2\big(\|u_\varepsilon\|^2_{\mathbb{H}^s_0(\Omega)} - \lambda_{k,s}\|u_\varepsilon\|^2_{L^2(\Omega)}\big) \\
&\quad +\tilde{\kappa}\|v\|_{L^2(\Omega)}\|u_\varepsilon\|_{L^1(\Omega)},
\end{aligned}
$$

provided that $\varepsilon > 0$ is sufficiently small and for some $\tilde{\kappa} > 0$, independent of ε.

Since $v \in \operatorname{span}\{e_{1,s},\ldots,e_{k-1,s}\}$ and (17.1) holds true, by Proposition 3.2 and (17.13), we have

$$\begin{aligned} M_{\varepsilon,\lambda_{k,s}} &\le \|v\|^2_{\mathbb{H}^s_0(\Omega)} - \lambda_{k,s}\|v\|^2_{L^2(\Omega)} + t^2\big(\|u_\varepsilon\|^2_{\mathbb{H}^s_0(\Omega)} - \lambda_{k,s}\|u_\varepsilon\|^2_{L^2(\Omega)}\big) \\ &\quad + \tilde{\kappa}\|v\|_{L^2(\Omega)}\|u_\varepsilon\|_{L^1(\Omega)} \\ &\le \big(\lambda_{k-1,s} - \lambda_{k,s}\big)\|v\|^2_{L^2(\Omega)} \\ &\quad + t^2\big(\|u_\varepsilon\|^2_{\mathbb{H}^s_0(\Omega)} - \lambda_{k,s}\|u_\varepsilon\|^2_{L^2(\Omega)}\big) + \tilde{\kappa}\|v\|_{L^2(\Omega)}\|u_\varepsilon\|_{L^1(\Omega)} \\ &= \big(\lambda_{k-1,s} - \lambda_{k,s}\big)\|v\|^2_{L^2(\Omega)} + S_{s,\lambda_{k,s}}(u_\varepsilon)\|tu_\varepsilon\|^2_{L^{2^*_s}(\Omega)} + \tilde{\kappa}\|v\|_{L^2(\Omega)}\|u_\varepsilon\|_{L^1(\Omega)}, \end{aligned} \tag{17.14}$$

if ε is small enough, where $S_{s,\lambda_{k,s}}(\cdot)$ is the function defined in (14.6) with $\lambda = \lambda_{k,s}$.

Now note that, by the convexity, the monotonicity properties of the integrals, (17.4), Claim 17.1, and the fact that in $\operatorname{span}\{e_{1,s},\ldots,e_{k,s}\}$ all the norms are equivalent, we have

$$\begin{aligned} 1 = \|u_M\|^{2^*_s}_{L^{2^*_s}(\Omega)} &= \int_\Omega |u_M(x)|^{2^*_s}\,dx \\ &\ge \int_\Omega |tu_\varepsilon(x)|^{2^*_s}\,dx + 2^*_s\int_\Omega (tu_\varepsilon(x))^{2^*_s-1}\tilde{v}(x)\,dx \\ &\ge \|tu_\varepsilon\|^{2^*_s}_{L^{2^*_s}(\Omega)} - 2^*_s\bar{c}^{\,2^*_s-1}\|u_\varepsilon\|^{2^*_s-1}_{L^{2^*_s-1}(\Omega)}\|\tilde{v}\|_{L^\infty(\Omega)} \\ &\ge \|tu_\varepsilon\|^{2^*_s}_{L^{2^*_s}(\Omega)} - \hat{c}\,\|u_\varepsilon\|^{2^*_s-1}_{L^{2^*_s-1}(\Omega)}\|\tilde{v}\|_{L^2(\Omega)}, \end{aligned} \tag{17.15}$$

for some positive constant $\hat{c}$, so

$$\|tu_\varepsilon\|^{2^*_s}_{L^{2^*_s}(\Omega)} \le 1 + \hat{c}\,\|u_\varepsilon\|^{2^*_s-1}_{L^{2^*_s-1}(\Omega)}\|\tilde{v}\|_{L^2(\Omega)}, \tag{17.16}$$

for ε sufficiently small. Moreover, by Young's inequality for any $\sigma > 0$, we have

$$\tilde{\kappa}\|v\|_{L^2(\Omega)}\|u_\varepsilon\|_{L^1(\Omega)} \le \sigma\,\|v\|^2_{L^2(\Omega)} + \frac{\tilde{\kappa}^2}{4\sigma}\,\|u_\varepsilon\|^2_{L^1(\Omega)}, \tag{17.17}$$

for any $\varepsilon > 0$. Hence, (17.14), (17.15), and (17.17) give, for ε small enough,

$$\begin{aligned} M_{\varepsilon,\lambda_{k,s}} &= \big(\lambda_{k-1,s} - \lambda_{k,s}\big)\|v\|^2_{L^2(\Omega)} + S_{s,\lambda_{k,s}}(u_\varepsilon)\|tu_\varepsilon\|^2_{L^{2^*_s}(\Omega)} \\ &\quad + \tilde{\kappa}\|v\|_{L^2(\Omega)}\|u_\varepsilon\|_{L^1(\Omega)} \\ &\le \big(\lambda_{k-1,s} - \lambda_{k,s}\big)\|v\|^2_{L^2(\Omega)} + S_{s,\lambda_{k,s}}(u_\varepsilon)\big(1 + \hat{c}\,\|u_\varepsilon\|^{2^*_s-1}_{L^{2^*_s-1}(\Omega)}\|\tilde{v}\|_{L^2(\Omega)}\big) \\ &\quad + \sigma\,\|v\|^2_{L^2(\Omega)} + \frac{\tilde{\kappa}^2}{4\sigma}\,\|u_\varepsilon\|^2_{L^1(\Omega)} \\ &= \big(\lambda_{k-1,s} - \lambda_{k,s} + \sigma\big)\|v\|^2_{L^2(\Omega)} + S_{s,\lambda_{k,s}}(u_\varepsilon)\big(1 + O(\varepsilon^{(n-2s)/2})\big) + O(\varepsilon^{n-2s}), \end{aligned} \tag{17.18}$$

thanks to Proposition 15.6.

Now let us choose $\sigma > 0$ such that $\sigma < \lambda_{k,s} - \lambda_{k-1,s}$ (this choice is admissible because $\lambda_{k,s} - \lambda_{k-1,s} > 0$ by (17.1)). Then (17.18) yields

$$M_{\varepsilon,\lambda_{k,s}} \leq S_{s,\lambda_{k,s}}(u_\varepsilon)\big(1 + O(\varepsilon^{(n-2s)/2})\big) + O(\varepsilon^{n-2s}) \tag{17.19}$$

as $\varepsilon \to 0$. Let us distinguish the two different cases $n > 4s$ and $2s < n < 4s$.

Case 1: $n > 4s$. By Propositions 14.11 and 14.12 and by the definition of $S_{s,\lambda_{k,s}}(\cdot)$ (see (14.6) with $\lambda = \lambda_{k,s}$), we get

$$\begin{aligned}
S_{s,\lambda_{k,s}}(u_\varepsilon) &= \frac{\displaystyle\int_{\mathbb{R}^n\times\mathbb{R}^n} \frac{|u_\varepsilon(x)-u_\varepsilon(y)|^2}{|x-y|^{n+2s}}\,dx\,dy - \lambda_{k,s}\int_\Omega |u_\varepsilon(x)|^2\,dx}{\left(\displaystyle\int_\Omega |u_\varepsilon(x)|^{2_s^*}\,dx\right)^{2/2_s^*}} \\
&\leq \frac{S_s^{n/(2s)} + O(\varepsilon^{n-2s}) - \lambda_{k,s} C_s \varepsilon^{2s}}{\big(S_s^{n/(2s)} + O(\varepsilon^n)\big)^{2/2_s^*}} \\
&\leq S_s + O(\varepsilon^{n-2s}) - \lambda_{k,s} C_s \varepsilon^2
\end{aligned}$$

as $\varepsilon \to 0$. Thus, as a consequence of this and of (17.19), we deduce that

$$\begin{aligned}
M_{\varepsilon,\lambda_{k,s}} &\leq \big(S_s + O(\varepsilon^{n-2s}) - \lambda_{k,s}C_s\varepsilon^2\big)\big(1 + O(\varepsilon^{(n-2s)/2})\big) + O(\varepsilon^{n-2s}) \\
&= S_s + O(\varepsilon^{n-2s}) - \lambda_{k,s}C_s\varepsilon^2 \\
&= S_s + \varepsilon^{2s}\big(O(\varepsilon^{n-4s}) - \lambda_{k,s}C_s\big) \\
&< S_s,
\end{aligned}$$

provided that $\varepsilon > 0$ is sufficiently small. Hence, (17.10) holds true. This concludes the proof of Proposition 17.4 in the case where $n > 4s$.

Case 2: $2s < n < 4s$. Again by Propositions 14.11 and 14.12 and by the definition of $S_{s,\lambda_{k,s}}(\cdot)$, we have

$$\begin{aligned}
S_{s,\lambda_{k,s}}(u_\varepsilon) &\leq \frac{S_s^{n/(2s)} + O(\varepsilon^{n-2s}) - \lambda_{k,s}C_s\varepsilon^{n-2s} + O(\varepsilon^{2s})}{\big(S_s^{n/(2s)} + O(\varepsilon^n)\big)^{2/2_s^*}} \\
&\leq S_s + O(\varepsilon^{n-2s}) - \lambda_{k,s}\widetilde{C}_s\varepsilon^{n-2s} + O(\varepsilon^{2s})
\end{aligned}$$

as $\varepsilon \to 0$. Thus, by this and (17.19), we deduce that

$$\begin{aligned}
M_{\varepsilon,\lambda_{k,s}} &\leq \big(S_s + O(\varepsilon^{n-2s}) - \lambda_{k,s}\widetilde{C}_s\varepsilon^{n-2s} + O(\varepsilon^{2s})\big)\big(1 + O(\varepsilon^{(n-2s)/2})\big) + O(\varepsilon^{n-2s}) \\
&= S_s + O(\varepsilon^{n-2s}) - \lambda_{k,s}\widetilde{C}_s\varepsilon^{n-2s} + O(\varepsilon^{2s}) \\
&= S_s + \varepsilon^{n-2s}\big(O(1) - \lambda_{k,s}\widetilde{C}_s\big) + O(\varepsilon^{2s}) \\
&< S_s,
\end{aligned}$$

if $\lambda_{k,s}$ is large enough, say, $\lambda_{k,s} > \lambda_s > 0$, and provided that $\varepsilon > 0$ is sufficiently small. Thus, (17.10) is verified when $n < 4s$. Ultimately, the proof of Proposition 17.4 is concluded. □

To end this chapter, we have the following:

Proof of Theorem 17.1 It easily follows as an application of the linking theorem, thanks to Propositions 16.3, 17.3, and 17.4. □

18

The Brezis–Nirenberg result for a general nonlocal equation

The aim of this chapter is to study a general critical nonlocal equation and to give a Brezis–Nirenberg result for it. The main results proved in this chpater are obtained by adapting the variational and topological techniques used in the classical critical case (see [9, 12, 46, 60, 114, 212, 220]) to the nonlocal fractional setting. The main difficulties are related to the encoding of the Dirichlet boundary datum in the variational formulation and to the lack of compactness. To overcome the first difficulty, we will work in the functional space $X_0^s(\Omega)$, while for the second difficulty we will apply suitable variants of some classical critical points theorems that take into account that the Palais–Smale condition holds true in a suitable energy range related to the best fractional critical Sobolev constant S_K. Of course, different from the classical case of the Laplacian, the fact that the leading operator is nonlocal provides technical and conceptual complications, and some estimates become more delicate and involved. The results of this chapter have been obtained in the papers [196, 204, 205].

18.1 Assumptions and main results

In this chapter, we study the following general integrodifferential equation:

$$\begin{cases} \mathscr{L}_K u + \lambda u + |u|^{2_s^*-2}u + f(x,u) = 0 & \text{in } \Omega \\ u = 0 & \text{in } \mathbb{R}^n \setminus \Omega, \end{cases} \tag{18.1}$$

where $s \in (0,1)$ is fixed, $\Omega \subset \mathbb{R}^n$, $n > 2s$, is open, bounded, and with continuous boundary, $\lambda > 0$, and $\mathscr{L}_K$ is the nonlocal operator defined in (1.54), while the nonlinearity represents a lower-order perturbation of the critical power.

Precisely, the nonlinear term $f:\Omega\times\mathbb{R}\to\mathbb{R}$ is a Carathéodory function verifying (6.5) and the following conditions:

$$\sup\{|f(x,t)| : a.e.\ x\in\Omega,\ |t|\le M\} < +\infty \quad \textit{for any } M>0; \tag{18.2}$$

$$\lim_{|t|\to+\infty}\frac{f(x,t)}{|t|^{2_s^*-1}} = 0, \quad \textit{uniformly in } x\in\Omega. \tag{18.3}$$

As a model for f, we can take the nonlinearity $f(x,t)=a(x)|t|^q$, with $a\in L^\infty(\Omega)$ and $q\in(1,2_s^*-1)$. In particular, we are interested in the weak formulation of (18.1) given by the following problem:

$$\begin{cases} \displaystyle\int_{\mathbb{R}^n\times\mathbb{R}^n}(u(x)-u(y))(\varphi(x)-\varphi(y))K(x-y)\,dx\,dy - \lambda\int_\Omega u(x)\varphi(x)\,dx \\[2ex] \displaystyle\quad = \int_\Omega |u(x)|^{2_s^*-2}u(x)\varphi(x)\,dx + \int_\Omega f(x,u(x))\varphi(x)\,dx \quad \forall\,\varphi\in X_0^s(\Omega) \\[2ex] u\in X_0^s(\Omega). \end{cases} \tag{18.4}$$

Assumption (6.5) implies that $f(x,0)=0$, so the function $u\equiv 0$ is a (trivial) solution of (18.1). The aim of this chapter will be, then, to find nontrivial solutions for (18.1). For this, we will adapt the variational techniques of [46] to the nonlocal setting. More precisely, we will study the critical points of the functional $\mathcal{J}_{K,\lambda,f}: X_0^s(\Omega)\to\mathbb{R}$ defined as follows

$$\begin{aligned} \mathcal{J}_{K,\lambda,f}(u) = {} & \frac{1}{2}\int_{\mathbb{R}^n\times\mathbb{R}^n}|u(x)-u(y)|^2K(x-y)\,dx\,dy - \frac{\lambda}{2}\int_\Omega |u(x)|^2\,dx \\ & - \frac{1}{2_s^*}\int_\Omega |u(x)|^{2_s^*}\,dx - \int_\Omega F(x,u(x))\,dx, \end{aligned} \tag{18.5}$$

where F is given by

$$F(x,t) := \int_0^t f(x,\tau)\,d\tau. \tag{18.6}$$

As in the classical case of the Laplacian, in the nonlocal setting, the issue that needs to be overcome is that the functional $\mathcal{J}_{K,\lambda,f}$ does not satisfy the Palais–Smale condition. This lack of compactness is due to the fact that the embedding $X_0^s(\Omega)\hookrightarrow L^{2_s^*}(\mathbb{R}^n)$ (or, in the case of the fractional Laplacian, $H^s(\mathbb{R}^n)\hookrightarrow L^{2_s^*}(\mathbb{R}^n)$) is not compact. For this, the functional $\mathcal{J}_{K,\lambda,f}$ does not verify the Palais–Smale condition globally but – as we will see in what follows – only in a suitable range related to the best fractional critical Sobolev constant in the embedding $X_0^s(\Omega)\hookrightarrow L^{2_s^*}(\mathbb{R}^n)$.

The first result of this chapter is the following one:

Theorem 18.1 *Let $s\in(0,1)$, $n>2s$, and Ω be an open, bounded subset of $\mathbb{R}^n$ with continuous boundary, and let $\lambda\in(0,\lambda_1)$, where λ_1 is the first eigenvalue of* *the nonlocal operator* $-\mathscr{L}_K$ *with homogeneous Dirichlet boundary conditions. Let* $K : \mathbb{R}^n \setminus \{0\} \to (0, +\infty)$ *be a function satisfying conditions* (1.55) *and* (1.56), *and let* $f : \Omega \times \mathbb{R} \to \mathbb{R}$ *be a Carathéodory function verifying* (6.5) *and* (18.2) *and* (18.3). *Finally, assume that*

$$\begin{gathered}\text{there exists } u_0 \in X_0^s(\Omega) \setminus \{0\}, \text{ with } u_0 \geq 0 \text{ a.e. in } \mathbb{R}^n, \text{ such that} \\ \sup_{\zeta \geq 0} \mathscr{J}_{K,\lambda,f}(\zeta u_0) < \frac{s}{n} S_K^{n/(2s)},\end{gathered} \tag{18.7}$$

where S_K *is defined as in* (1.90). *Then problem* (18.1) *admits a nontrivial weak solution* $u \in X_0^s(\Omega)$.

Condition (18.7), which looks artificial at first glance, boils down to the a natural condition when $f \equiv 0$. In this case, problem (18.1) becomes

$$\begin{cases} \mathscr{L}_K u + \lambda u + |u|^{2_s^*-2} u = 0 & \text{in } \Omega \\ u = 0 & \text{in } \mathbb{R}^n \setminus \Omega, \end{cases} \tag{18.8}$$

whose weak formulation is given by the following critical equation:

$$\begin{cases} \displaystyle\int_{\mathbb{R}^n \times \mathbb{R}^n} (u(x) - u(y))(\varphi(x) - \varphi(y)) K(x - y)\, dx\, dy - \lambda \int_\Omega u(x)\varphi(x)\, dx \\ \displaystyle\quad = \int_\Omega |u(x)|^{2_s^*-2} u(x)\varphi(x)\, dx \quad \forall\, \varphi \in X_0^s(\Omega) \\ u \in X_0^s(\Omega). \end{cases} \tag{18.9}$$

In order to state an existence result for (18.8), we need the following constant:

$$S_{K,\lambda} := \inf_{v \in X_0^s(\Omega) \setminus \{0\}} S_{K,\lambda}(v), \tag{18.10}$$

where, for any $v \in X_0^s(\Omega) \setminus \{0\}$,

$$\begin{aligned} S_{K,\lambda}(v) &:= \frac{\displaystyle\int_{\mathbb{R}^n \times \mathbb{R}^n} |v(x) - v(y)|^2 K(x - y)\, dx\, dy - \lambda \int_\Omega |v(x)|^2\, dx}{\displaystyle\left(\int_\Omega |v(x)|^{2_s^*}\, dx\right)^{2/2_s^*}} \\ &= \frac{\displaystyle\int_{\mathbb{R}^n \times \mathbb{R}^n} |v(x) - v(y)|^2 K(x - y)\, dx\, dy - \lambda \int_{\mathbb{R}^n} |v(x)|^2\, dx}{\displaystyle\left(\int_{\mathbb{R}^n} |v(x)|^{2_s^*}\, dx\right)^{2/2_s^*}}. \end{aligned} \tag{18.11}$$

Of course, $S_{K,\lambda}(v) < S_K(v)$, for any $v \in X_0^s(\Omega) \setminus \{0\}$ and any $\lambda > 0$, so $S_{K,\lambda} \leq S_K$. The topic of research of this chapter actually will focus on the case where the *strict* inequality occurs (see (18.12)).

Now we can state the following result:

Theorem 18.2 *Let $s \in (0,1)$, $n > 2s$, and Ω be an open, bounded subset of $\mathbb{R}^n$ with continuous boundary. Let $K : \mathbb{R}^n \setminus \{0\} \to (0,+\infty)$ be a function satisfying conditions* (1.55) *and* (1.56), *and let $\lambda \in (0,\lambda_1)$, where λ_1 denotes the first eigenvalue of the nonlocal operator $-\mathcal{L}_K$ with homogeneous Dirichlet boundary conditions. Assume that*

$$S_{K,\lambda} < S_K. \tag{18.12}$$

Then problem (18.8) *admits a nontrivial weak solution $u \in X_0^s(\Omega)$.*

In the nonlocal framework, the simplest example we can deal with is given by problem (14.2), treated in the preceding chapters.

18.2 Some preliminary results

In this section, we prove some preliminary results that will be useful in what follows.

18.2.1 Estimates on the nonlinearity

We use the behavior of f at zero and at infinity to deduce some estimates on the nonlinear term and its primitive (with respect to its second variable) F defined in (6.8). This part is quite standard and does not take into account the nonlocal nature of the problem: the reader familiar with these topics may skip it and go directly to Section 18.3.

Lemma 18.3 *Assume that $f : \Omega \times \mathbb{R} \to \mathbb{R}$ is a Carathéodory function satisfying conditions* (6.5) *and* (18.2) *and* (18.3). *Then, for any $\varepsilon > 0$, there exists $M = M(\varepsilon) > 0$ such that a.e. $x \in \Omega$ and, for any $t \in \mathbb{R}$,*

$$|f(x,t)| \le 2_s^* \varepsilon |t|^{2_s^*-1} + M(\varepsilon), \tag{18.13}$$

so, as a consequence,

$$|F(x,t)| \le \varepsilon\, |t|^{2_s^*} + M(\varepsilon)|t|, \tag{18.14}$$

where F is defined as in (6.8).

Proof By assumption (18.3), for any $\varepsilon > 0$, there exists $\sigma = \sigma(\varepsilon) > 0$ such that, for any $t \in \mathbb{R}$ with $|t| > \sigma$ and a.e. $x \in \Omega$, we get

$$|f(x,t)| \le 2_s^* \varepsilon |t|^{2_s^*-1}. \tag{18.15}$$

Moreover, by (18.2), there exists $M(\varepsilon) = M(\sigma(\varepsilon)) > 0$ such that, a.e. $x \in \Omega$ and for any $t \in \mathbb{R}$ with $|t| \le \sigma$, we have

$$|f(x,t)| \le M(\varepsilon). \tag{18.16}$$

Combining (18.15) and (18.16), (18.13) easily follows. Using the definition of F, we also get (18.14). □

Lemma 18.4 *Assume that $f : \Omega \times \mathbb{R} \to \mathbb{R}$ is a Carathéodory function satisfying conditions* (6.5), (18.2), *and* (18.3). *Then, for any $\varepsilon > 0$, there exists $\delta = \delta(\varepsilon)$ such that, a.e. $x \in \Omega$ and for any $t \in \mathbb{R}$,*

$$|f(x,t)| \leq 2\varepsilon|t| + 2_s^*\delta(\varepsilon)|t|^{2_s^*-1}, \tag{18.17}$$

so, as a consequence,

$$|F(x,t)| \leq \varepsilon\,|t|^2 + \delta(\varepsilon)\,|t|^{2_s^*}, \tag{18.18}$$

where F is defined as in (6.8).

Proof By assumption (6.5), for any $\varepsilon > 0$, there exists $\sigma = \sigma(\varepsilon) > 0$ such that, for any $t \in \mathbb{R}$ with $|t| < \sigma$ and a.e. $x \in \Omega$, we get

$$|f(x,t)| \leq 2\varepsilon|t|. \tag{18.19}$$

Moreover, by (18.13) with $\varepsilon = 1$, we get

$$|f(x,t)| \leq 2_s^*|t|^{2_s^*-1} + M(1) = |t|^{2_s^*-1}\Big(2_s^* + \frac{M(1)}{|t|^{2_s^*-1}}\Big),$$

so there exists $\delta(\varepsilon) = \delta(\sigma(\varepsilon)) > 0$ such that, a.e. $x \in \Omega$ and for any $t \in \mathbb{R}$ with $|t| \geq \sigma$, we have

$$|f(x,t)| \leq 2_s^*\delta(\varepsilon)|t|^{2_s^*-1}. \tag{18.20}$$

Combining (18.19) and (18.20), (18.17) easily follows. Using the definition of F, we also deduce (18.18). □

18.2.2 Variational formulation of the problem

In this subsection, we study problem (18.4), which is the Euler–Lagrange equation of the functional $\mathcal{J}_{K,\lambda,f}$ defined, in (18.5). Notice that this functional is well defined, thanks to assumptions (18.2) and (18.3), to the fact that Ω is bounded, and to Lemma 1.31(b). Moreover, $\mathcal{J}_{K,\lambda,f}$ is Fréchet differentiable in $u \in X_0^s(\Omega)$, and, for any $\varphi \in X_0^s(\Omega)$,

$$\begin{aligned}\langle \mathcal{J}'_{K,\lambda,f}(u),\varphi\rangle = &\int_{\mathbb{R}^n\times\mathbb{R}^n}(u(x)-u(y))(\varphi(x)-\varphi(y))K(x-y)\,dx\,dy\\ &-\lambda\int_\Omega u(x)\varphi(x)\,dx - \int_\Omega |u(x)|^{2_s^*-1}(x)\varphi(x)\,dx\\ &-\int_\Omega f(x,u(x))\varphi(x)\,dx.\end{aligned}$$

Thus, critical points of $\mathcal{J}_{K,\lambda,f}$ are solutions to problem (18.4). To find these critical points, we will apply a variant of the mountain pass theorem without the Palais–Smale condition, as given in [11] (see also [46, theorem 2.2]) and as in the model case treated in Chapter 14.

Indeed, here we cannot apply the classical mountain pass theorem because, owing to the lack of compactness in the embedding $X_0^s(\Omega) \hookrightarrow L^{2_s^*}(\mathbb{R}^n)$ (see Lemma 1.31(b)), the functional $\mathcal{J}_{K,\lambda,f}$ does not verify the Palais–Smale condition globally but only in an energy range determined by the best fractional critical Sobolev constant S_K given in formula (1.90) (see also Lemma 1.31(b)).

18.3 The critical case with a lower-order perturbation

This section is devoted to problem (18.4) and to the proof of Theorem 18.1. First of all, we prove that the functional $\mathcal{J}_{K,\lambda,f}$ has the geometric features required by [46, theorem 2.2]. In doing this, we argue essentially as in Propositions 14.2 and 14.3. Here, in addition, we have to take into account the presence of the perturbation f.

Proposition 18.5 Let $\lambda \in (0, \lambda_1)$, and let f be a Carathéodory function satisfying conditions (6.5) and (18.2) and (18.3). Then there exist $\rho > 0$ and $\beta > 0$ such that, for any $u \in X_0^s(\Omega)$ with $\|u\|_{X_0^s(\Omega)} = \rho$, it results that $\mathcal{J}_{K,\lambda,f}(u) \geq \beta$.

Proof Let u be a function in $X_0^s(\Omega)$. By (18.18), we get that, for any $\varepsilon > 0$,

$$
\begin{aligned}
\mathcal{J}_{K,\lambda,f}(u) &\geq \frac{1}{2}\int_{\mathbb{R}^n\times\mathbb{R}^n} |u(x)-u(y)|^2 K(x-y)\,dx\,dy - \frac{\lambda}{2}\int_\Omega |u(x)|^2\,dx \\
&\quad - \frac{1}{2_s^*}\int_\Omega |u(x)|^{2_s^*}\,dx - \varepsilon\int_\Omega |u(x)|^2 dx - \delta(\varepsilon)\int_\Omega |u(x)|^{2_s^*}\,dx \\
&\geq \frac{1}{2}\left(1-\frac{\lambda}{\lambda_1}\right)\int_{\mathbb{R}^n\times\mathbb{R}^n} |u(x)-u(y)|^2 K(x-y)\,dx\,dy \\
&\quad - \varepsilon\|u\|^2_{L^2(\Omega)} - \left(\frac{1}{2_s^*}+\delta(\varepsilon)\right)\|u\|^{2_s^*}_{L^{2_s^*}(\Omega)} \\
&\geq \frac{1}{2}\left(1-\frac{\lambda}{\lambda_1}\right)\int_{\mathbb{R}^n\times\mathbb{R}^n} |u(x)-u(y)|^2 K(x-y)\,dx\,dy \\
&\quad - \varepsilon|\Omega|^{(2_s^*-2)/2_s^*}\|u\|^2_{L^{2_s^*}(\Omega)} - \left(\frac{1}{2_s^*}+\delta(\varepsilon)\right)\|u\|^{2_s^*}_{L^{2_s^*}(\Omega)},
\end{aligned}
\tag{18.21}
$$

thanks to Lemma 6.5 (here we need $0 < \lambda < \lambda_1$) and to the fact that $L^{2_s^*}(\Omega) \hookrightarrow L^2(\Omega)$ continuously (being Ω bounded and $2 < 2_s^*$).

Using (1.56) and Lemma 1.28(a) and (b), we deduce from (18.21) that, for any $\varepsilon > 0$,

$$
\begin{aligned}
\mathcal{J}_{K,\lambda,f}(u) &\geq \frac{1}{2}\left(1-\frac{\lambda}{\lambda_1}\right)\int_{\mathbb{R}^n\times\mathbb{R}^n}|u(x)-u(y)|^2K(x-y)\,dx\,dy \\
&\quad -\varepsilon c|\Omega|^{(2_s^*-2)/2_s^*}\int_{\mathbb{R}^n\times\mathbb{R}^n}\frac{|u(x)-u(y)|^2}{|x-y|^{n+2s}}\,dx\,dy \\
&\quad -\left(\frac{1}{2_s^*}+\delta(\varepsilon)\right)c^{2_s^*/2}\left(\int_{\mathbb{R}^n\times\mathbb{R}^n}\frac{|u(x)-u(y)|^2}{|x-y|^{n+2s}}\,dx\,dy\right)^{2_s^*/2} \\
&\geq \left[\frac{1}{2}\left(1-\frac{\lambda}{\lambda_1}\right)-\frac{\varepsilon c|\Omega|^{(2_s^*-2)/2_s^*}}{\theta}\right]\|u\|^2_{X_0^s(\Omega)} \qquad (18.22)\\
&\quad -\left(\frac{1}{2_s^*}+\delta(\varepsilon)\right)\left(\frac{c_L K}{\theta}\right)^{2_s^*/2}\|u\|^{2_s^*}_{X_0^s(\Omega)} \\
&= \left[\frac{1}{2}\left(1-\frac{\lambda}{\lambda_1}\right)-\frac{\varepsilon c|\Omega|^{(2_s^*-2)/2_s^*}}{\theta}\right]\|u\|^2_{X_0^s(\Omega)} \\
&\quad -\left(\frac{1}{2_s^*}+\delta(\varepsilon)\right)\left(\frac{c_L K}{\theta}\right)^{2_s^*/2}\|u\|^{2_s^*}_{X_0^s(\Omega)}.
\end{aligned}
$$

Choosing $\varepsilon > 0$ such that $2\varepsilon c|\Omega|^{(2_s^*-2)/2_s^*} < \theta(1-\lambda/\lambda_1)$, by (18.22), it easily follows that

$$\mathcal{J}_{K,\lambda,f}(u) \geq \alpha\|u\|^2_{X_0^s(\Omega)}\left(1-\kappa\|u\|^{2_s^*-2}_{X_0^s(\Omega)}\right),$$

for suitable positive constants α and κ.

Now let $u \in X_0^s(\Omega)$ be such that $\|u\|_{X_0^s(\Omega)} = \rho > 0$. Because $2_s^* > 2$, we can choose ρ sufficiently small (i.e., ρ such that $1-\kappa\rho^{2_s^*-2} > 0$) that

$$\inf_{\substack{u\in X_0^s(\Omega)\\ \|u\|_{X_0^s(\Omega)}=\rho}} \mathcal{J}_{K,\lambda,f}(u) \geq \alpha\rho^2(1-\kappa\rho^{2_s^*-2}) =: \beta > 0.$$

Hence, the proposition is proved. □

Proposition 18.6 Let $\lambda \in (0,\lambda_1)$, and let f be a Carathéodory function satisfying conditions (6.5), (18.2) and (18.3). Then there exists $e \in X_0^s(\Omega)$ such that $e \geq 0$ a.e. in $\mathbb{R}^n$, $\|e\|_{X_0^s(\Omega)} > \rho$, and

$$\mathcal{J}_{K,\lambda,f}(e) < 0,$$

where ρ is given in Proposition 18.5. In particular, if we assume condition (18.7), we can construct e as

$$e = \zeta_0 u_0, \tag{18.23}$$

with u_0 as in (18.7) and $\zeta_0 > 0$ large enough.

Proof We fix $u \in X_0^s(\Omega)$ such that $\|u\|_{X_0^s(\Omega)} = 1$ and $u \geq 0$ a.e. in $\mathbb{R}^n$; we remark that this choice is possible thanks to (1.27) (alternatively, one can replace any $u \in X_0^s(\Omega)$ with its positive part, which belongs to $X_0^s(\Omega)$ too, thanks to Lemma 1.22).

Now let $\zeta > 0$. By Lemma 6.5 (here we again use the fact that $\lambda < \lambda_1$) and Lemma 18.3 (see formula (18.14) used here with $\varepsilon = 1/(2 \cdot 2_s^*)$), we have

$$\begin{aligned}\mathcal{J}_{K,\lambda,f}(\zeta u) &= \frac{\zeta^2}{2}\int_{\mathbb{R}^n\times\mathbb{R}^n} |u(x)-u(y)|^2 K(x-y)\,dx\,dy - \frac{\lambda}{2}\zeta^2\int_\Omega |u(x)|^2\,dx \\ &\quad - \frac{\zeta^{2_s^*}}{2_s^*}\int_\Omega |u(x)|^{2_s^*}\,dx - \int_\Omega F(x,\zeta u(x))\,dx \\ &\leq \frac{\zeta^2}{2} - \left(\frac{1}{2_s^*} - \frac{1}{2\cdot 2_s^*}\right)\zeta^{2_s^*}\int_\Omega |u(x)|^{2_s^*}\,dx + M(1/(2\cdot 2_s^*))\int_\Omega |\zeta u(x)|\,dx \\ &= \frac{\zeta^2}{2} - \frac{\zeta^{2_s^*}}{2\cdot 2_s^*}\int_\Omega |u(x)|^{2_s^*}\,dx + \zeta M(1/(2\cdot 2_s^*))\int_\Omega |u(x)|\,dx.\end{aligned}$$

Because $2_s^* > 2 > 1$, passing to the limit as $\zeta \to +\infty$, we get that $\mathcal{J}_{K,\lambda,f}(\zeta u) \to -\infty$, so the assertion follows taking $e = \zeta u$, with ζ sufficiently large.

In particular, under condition (18.7), we can take $u = u_0/\|u_0\|_{X_0^s(\Omega)} \in X_0^s(\Omega)$ (note that $u_0 \not\equiv 0$ by assumption) so that we can choose $e = \zeta_0 u_0$ with ζ_0 large enough. □

18.3.1 End of the proof of Theorem 18.1

Propositions 18.5 and 18.6 give that the geometry of the variant of the mountain pass theorem stated in [46, theorem 2.2] is fulfilled by $\mathcal{J}_{K,\lambda,f}$. Moreover, since $F(0) = 0$, we easily get that $\mathcal{J}_{K,\lambda,f}(0) = 0 < \beta$, with β given in Proposition 18.5. Now set

$$c := \inf_{P\in\mathcal{P}} \sup_{v\in P([0,1])} \mathcal{J}_{K,\lambda,f}(v), \tag{18.24}$$

where

$$\mathcal{P} := \left\{P \in C([0,1]; X_0^s(\Omega)) : P(0) = 0,\ P(1) = e\right\},$$

with $e = \zeta_0 u_0$ given in Proposition 18.6. To prove Theorem 18.1, we proceed by steps.

Claim 18.1 *The constant c given in* (18.24) *is such that*

$$\beta \leq c < \frac{s}{n} S_K^{n/(2s)},$$

where β is given in Proposition 18.5, and S_K is defined in formula (1.90).

Proof The fact that $c \geq \beta$, where β follows exactly as in the proof of Claim 14.1 (we have just to replace $\mathcal{J}_{s,\lambda}$ with $\mathcal{J}_{K,\lambda,f}$). Now let us show the other inequality.

Since $e = \zeta_0 u_0$, thanks to (18.23), the map $[0,1] \ni t \mapsto t\,\zeta_0 u_0$ belongs to $\mathscr{P}$. Hence, by assumption (18.7), we conclude that

$$\inf_{P\in\mathscr{P}} \sup_{v\in P([0,1])} \mathscr{J}_{K,\lambda,f}(v) \le \sup_{\zeta\ge 0} \mathscr{J}_{K,\lambda,f}(\zeta u_0) < \frac{s}{n} S_K^{n/(2s)},$$

so $c < (s/n)\, S_K^{n/(2s)}$. Thus, the claim is proved. □

By [46, theorem 2.2], there exists a sequence $\{u_j\}_{j\in\mathbb{N}}$ in $X_0^s(\Omega)$ such that

$$\mathscr{J}_{K,\lambda,f}(u_j) \to c \tag{18.25}$$

and

$$\sup\left\{|\langle \mathscr{J}'_{K,\lambda,f}(u_j),\varphi\rangle| \,:\, \varphi \in X_0^s(\Omega), \|\varphi\|_{X_0^s(\Omega)} = 1\right\} \to 0 \tag{18.26}$$

as $j \to +\infty$. The following facts hold true:

Claim 18.2 *The sequence $\{u_j\}_{j\in\mathbb{N}}$ is bounded in $X_0^s(\Omega)$.*

Proof For any $j \in \mathbb{N}$, by (18.25) and (18.26), it easily follows that there exists $\kappa > 0$ such that

$$|\mathscr{J}_{K,\lambda,f}(u_j)| \le \kappa \tag{18.27}$$

and

$$\left|\left\langle \mathscr{J}'_{K,\lambda,f}(u_j), \frac{u_j}{\|u_j\|_{X_0^s(\Omega)}}\right\rangle\right| \le \kappa. \tag{18.28}$$

As a consequence of (18.27) and (18.28), we have

$$\mathscr{J}_{K,\lambda,f}(u_j) - \frac{1}{2}\langle \mathscr{J}'_{K,\lambda,f}(u_j), u_j\rangle \le \kappa\left(1 + \|u_j\|_{X_0^s(\Omega)}\right). \tag{18.29}$$

By the Hölder inequality, Lemma 1.28(a), and (1.56),

$$\begin{aligned} \int_\Omega |u_j(x)|\,dx &\le |\Omega|^{(2_s^*-1)/2_s^*}\left(\int_\Omega |u_j(x)|^{2_s^*}\,dx\right)^{1/2_s^*} \\ &\le \widetilde{\kappa}|\Omega|^{(2_s^*-1)/2_s^*}\|u_j\|_{X_0^s(\Omega)}, \end{aligned} \tag{18.30}$$

for some positive constant $\widetilde{\kappa}$. Accordingly, by Lemma 18.3 and (18.30), we obtain

$$\begin{aligned} &\mathscr{J}_{K,\lambda,f}(u_j) - \frac{1}{2}\langle \mathscr{J}'_{K,\lambda,f}(u_j), u_j\rangle \\ &= \left(\frac{1}{2} - \frac{1}{2_s^*}\right)\|u_j\|^{2_s^*}_{L^{2_s^*}(\Omega)} - \int_\Omega F(x,u_j(x))\,dx \\ &\quad + \frac{1}{2}\int_\Omega f(x,u_j(x))u_j(x)\,dx \\ &\ge \frac{s}{n}\|u_j\|^{2_s^*}_{L^{2_s^*}(\Omega)} - \varepsilon\left(1 + \frac{2_s^*}{2}\right)\|u_j\|^{2_s^*}_{L^{2_s^*}(\Omega)} - \frac{3M(\varepsilon)}{2}\int_\Omega |u_j(x)|\,dx \\ &\ge \frac{s}{n}\|u_j\|^{2_s^*}_{L^{2_s^*}(\Omega)} - \varepsilon\left(1 + \frac{2_s^*}{2}\right)\|u_j\|^{2_s^*}_{L^{2_s^*}(\Omega)} - C_\varepsilon\|u_j\|_{X_0^s(\Omega)}, \end{aligned} \tag{18.31}$$

for a suitable $C_\varepsilon > 0$, possibly depending on $|\Omega|$. Choosing $\varepsilon > 0$ small enough (i.e., ε such that $(s/n) > \varepsilon\big(1+(2_s^*/2)\big)$), (18.29) and (18.31) imply that, for any $j \in \mathbb{N}$,

$$\|u_j\|_{L^{2_s^*}(\Omega)}^{2_s^*} \le \kappa_* \Big(1 + \|u_j\|_{X_0^s(\Omega)}\Big), \tag{18.32}$$

for a suitable positive constant κ_*.

Finally, by Lemmas 6.5 (which holds true because $0 < \lambda < \lambda_1$) and 18.3 (used here with $\varepsilon = 1$), it follows that

$$\begin{aligned}
\mathcal{J}_{K,\lambda,f}(u_j) &\ge \frac{1}{2}\left(1-\frac{\lambda}{\lambda_1}\right)\|u_j\|_{X_0^s(\Omega)}^2 - \frac{1}{2_s^*}\|u_j\|_{L^{2_s^*}(\Omega)}^{2_s^*} - \int_\Omega F(x,u_j(x))\,dx \\
&\ge \frac{1}{2}\left(1-\frac{\lambda}{\lambda_1}\right)\|u_j\|_{X_0^s(\Omega)}^2 - \left(\frac{1}{2_s^*}+1\right)\|u_j\|_{L^{2_s^*}(\Omega)}^{2_s^*} \\
&\quad - M(1)\int_\Omega |u_j(x)|\,dx.
\end{aligned} \tag{18.33}$$

Combining (18.27), (18.32), and (18.33), we conclude that

$$\begin{aligned}
\kappa \ge \mathcal{J}_{K,\lambda,f}(u_j) &\ge \frac{1}{2}\left(1-\frac{\lambda}{\lambda_1}\right)\|u_j\|_{X_0^s(\Omega)}^2 - \left(\frac{1}{2_s^*}+1\right)\|u_j\|_{L^{2_s^*}(\Omega)}^{2_s^*} \\
&\quad - M(1)\int_\Omega |u_j(x)|\,dx \\
&\ge \frac{1}{2}\left(1-\frac{\lambda}{\lambda_1}\right)\|u_j\|_{X_0^s(\Omega)}^2 - \kappa_*\left(\frac{1}{2_s^*}+1\right)\Big(1+\|u_j\|_{X_0^s(\Omega)}\Big) \\
&\quad - M(1)\int_\Omega |u_j(x)|\,dx,
\end{aligned}$$

so, recalling (18.30), we see that, for any $j \in \mathbb{N}$,

$$\|u_j\|_{X_0^s(\Omega)}^2 \le \kappa_{**}\Big(1+\|u_j\|_{X_0^s(\Omega)}\Big),$$

for a suitable positive constant κ_{**}. Hence, the proof of the claim is complete. □

Claim 18.3 *Problem* (18.4) *admits a solution* $u_\infty \in X_0^s(\Omega)$.

Proof Since $\{u_j\}_{j\in\mathbb{N}}$ is bounded in $X_0^s(\Omega)$ and $X_0^s(\Omega)$ is a reflexive space (being a Hilbert space, by Lemma 1.28(c)), up to a subsequence, still denoted by u_j, there exists $u_\infty \in X_0^s(\Omega)$ such that $u_j \to u_\infty$ weakly in $X_0^s(\Omega)$; that is, for any $\varphi \in X_0^s(\Omega)$,

$$\begin{aligned}
&\int_{\mathbb{R}^n\times\mathbb{R}^n} (u_j(x)-u_j(y))(\varphi(x)-\varphi(y))K(x-y)\,dx\,dy \\
&\quad \to \int_{\mathbb{R}^n\times\mathbb{R}^n} (u_\infty(x)-u_\infty(y))(\varphi(x)-\varphi(y))K(x-y)\,dx\,dy
\end{aligned} \tag{18.34}$$

as $j \to +\infty$. Moreover, since $\{u_j\}_{j\in\mathbb{N}}$ is bounded in $X_0^s(\Omega)$, we deduce from (18.32) that

$$\{u_j\}_{j\in\mathbb{N}} \quad \text{is bounded in} \quad L^{2_s^*}(\Omega). \tag{18.35}$$

Consequently, by Lemma 1.31(b) and the fact that $L^{2_s^*}(\mathbb{R}^n)$ is a reflexive space, we have that, up to a subsequence,

$$u_j \to u_\infty \quad \text{weakly in} \quad L^{2_s^*}(\mathbb{R}^n) \tag{18.36}$$

as $j \to +\infty$, while, by Lemma 1.31(a), up to a subsequence,

$$u_j \to u_\infty \quad \text{in } L^\nu(\mathbb{R}^n) \tag{18.37}$$

$$u_j \to u_\infty \quad \text{a.e. in } \mathbb{R}^n \tag{18.38}$$

as $j \to +\infty$, for any $\nu \in [1, 2_s^*)$ (see, e.g., [43, theorem IV.9]).

By (18.36) and the fact that $\{|u_j|^{2_s^*-2}u_j\}_{j\in\mathbb{N}}$ is bounded in $L^{2_s^*/(2_s^*-1)}(\Omega)$, we see that

$$|u_j|^{2_s^*-2}u_j \to |u_\infty|^{2_s^*-2}u_\infty \quad \text{weakly in} \quad L^{2_s^*/(2_s^*-1)}(\Omega) \tag{18.39}$$

as $j \to +\infty$. Furthermore, by (18.35), Lemma 18.3 (see formula (18.13)), and the fact that Ω is bounded, we get that

$$f(\cdot, u_j(\cdot)) \quad \text{is bounded in} \quad L^{2_s^*/(2_s^*-1)}(\Omega). \tag{18.40}$$

Moreover, by (18.38) and the fact that the map $t \mapsto f(\cdot, t)$ is continuous in $t \in \mathbb{R}$, we get

$$f(\cdot, u_j(\cdot)) \to f(\cdot, u_\infty(\cdot)) \quad \text{a.e. in } \Omega \tag{18.41}$$

as $j \to +\infty$. Then (18.40) and (18.41) yield that

$$f(\cdot, u_j(\cdot)) \to f(\cdot, u_\infty(\cdot)) \quad \text{weakly in} \quad L^{2_s^*/(2_s^*-1)}(\Omega)$$

as $j \to +\infty$. As a consequence of this, because the dual of $L^{2_s^*/(2_s^*-1)}(\Omega)$ is $L^{2_s^*}(\Omega)$, it is easily seen that

$$\int_\Omega f(x, u_j(x))\varphi(x)\,dx \to \int_\Omega f(x, u_\infty(x))\varphi(x)\,dx \quad \text{for any } \varphi \in L^{2_s^*}(\Omega),$$

so, in particular,

$$\int_\Omega f(x, u_j(x))\varphi(x)\,dx \to \int_\Omega f(x, u_\infty(x))\varphi(x)\,dx \quad \text{for any } \varphi \in X_0^s(\Omega) \tag{18.42}$$

as $j \to +\infty$ (here we use Lemma 1.31(b)).

Since (18.26) holds true, for any $\varphi \in X_0^s(\Omega)$,

$$0 \leftarrow \langle \mathscr{J}'_{K,\lambda,f}(u_j), \varphi \rangle,$$

so, passing to the limit in this expression as $j \to +\infty$ and taking into account (18.34), (18.37), (18.39), and (18.42), we get

$$\begin{aligned}\int_{\mathbb{R}^n \times \mathbb{R}^n} (u_\infty(x) - u_\infty(y))(\varphi(x) - \varphi(y))K(x-y)\,dx\,dy - \lambda \int_\Omega u_\infty(x)\varphi(x)\,dx \\ - \int_\Omega |u_\infty(x)|^{2_s^*-2} u_\infty(x)\varphi(x)\,dx - \int_\Omega f(x, u_\infty(x))\varphi(x)\,dx = 0\end{aligned}$$

for any $\varphi \in X_0^s(\Omega)$; that is, u_∞ is a solution of problem (18.2), and the claim follows. □

Claim 18.4 *The solution u_∞ is nontrivial; that is, $u_\infty \not\equiv 0$ in Ω.*

Proof Suppose, by contradiction, that $u_\infty \equiv 0$ in Ω (so $u_\infty \equiv 0$ in $\mathbb{R}^n$ because $u_\infty \in X_0^s(\Omega)$). Using Lemma 18.3 and (18.35), we see that, for any $\varepsilon > 0$ and $j \in \mathbb{N}$,

$$\begin{aligned}\left| \int_\Omega f(x, u_j(x)) u_j(x)\,dx \right| &\le 2_s^* \varepsilon \int_\Omega |u_j(x)|^{2_s^*}\,dx + M(\varepsilon) \int_\Omega |u_j(x)|\,dx \\ &\le 2_s^* \varepsilon C + M(\varepsilon) \|u_j\|_{L^1(\Omega)}\end{aligned} \tag{18.43}$$

and

$$\begin{aligned}\left| \int_\Omega F(x, u_j(x))\,dx \right| &\le \varepsilon \int_\Omega |u_j(x)|^{2_s^*}\,dx + M(\varepsilon) \int_\Omega |u_j(x)|\,dx \\ &\le \varepsilon C + M(\varepsilon) \|u_j\|_{L^1(\Omega)}.\end{aligned} \tag{18.44}$$

Keeping ε fixed and taking the limit in j, we obtain

$$\limsup_{j \to +\infty} \left| \int_\Omega f(x, u_j(x)) u_j(x)\,dx \right| \le 2_s^* \varepsilon C$$

and

$$\limsup_{j \to +\infty} \left| \int_\Omega F(x, u_j(x))\,dx \right| \le \varepsilon C,$$

due to (18.37). Hence, by sending $\varepsilon \to 0^+$, we obtain

$$\int_\Omega f(x, u_j(x)) u_j(x)\,dx \to 0 \tag{18.45}$$

and

$$\int_\Omega F(x, u_j(x))\,dx \to 0 \tag{18.46}$$

as $j \to +\infty$. Moreover, because $\{u_j\}_{j\in\mathbb{N}}$ bounded in $X_0^s(\Omega)$ by Claim 18.2, from (18.26), it follows that

$$0 \leftarrow \langle \mathscr{J}'_{K,\lambda}(u_j), u_j\rangle = \int_{\mathbb{R}^n\times\mathbb{R}^n} |u_j(x)-u_j(y)|^2 K(x-y)\,dx\,dy - \lambda\int_\Omega |u_j(x)|^2\,dx$$
$$- \int_\Omega |u_j(x)|^{2_s^*}\,dx - \int_\Omega f(x,u_j(x))u_j(x)\,dx,$$

which, thanks to (18.37) and (18.45), gives

$$\int_{\mathbb{R}^n\times\mathbb{R}^n} |u_j(x)-u_j(y)|^2 K(x-y)\,dx\,dy - \int_\Omega |u_j(x)|^{2_s^*}\,dx \to 0 \tag{18.47}$$

as $j \to +\infty$.

Now, by Claim 18.2, the sequence $\{\|u_j\|_{X_0^s(\Omega)}\}_{j\in\mathbb{N}}$ is bounded in $\mathbb{R}$. Hence, up to a subsequence, if necessary, we can assume that

$$\|u_j\|^2_{X_0^s(\Omega)} = \int_{\mathbb{R}^n\times\mathbb{R}^n} |u_j(x)-u_j(y)|^2 K(x-y)\,dx\,dy \to L, \tag{18.48}$$

so, as a consequence of (18.47),

$$\int_\Omega |u_j(x)|^{2_s^*}\,dx \to L \tag{18.49}$$

as $j \to +\infty$. Of course, $L \in [0,+\infty)$. Furthermore, by (18.25), we have that, as $j \to +\infty$,

$$\frac{1}{2}\int_{\mathbb{R}^n\times\mathbb{R}^n} |u_j(x)-u_j(y)|^2 K(x-y)\,dx\,dy - \frac{\lambda}{2}\int_\Omega |u_j(x)|^2\,dx$$
$$- \frac{1}{2_s^*}\int_\Omega |u_j(x)|^{2_s^*}\,dx - \int_\Omega F(x,u_j(x))\,dx \to c,$$

so, using (18.37) with $u_\infty \equiv 0$ in $\mathbb{R}^n$, (18.46), (18.48), and (18.49), it follows that

$$c = \left(\frac{1}{2} - \frac{1}{2_s^*}\right)L = \frac{s}{n}L. \tag{18.50}$$

Since $c \geq \beta > 0$ by Claim 18.1, it is easily seen that $L > 0$. Moreover, by Lemma 1.31(b) and (1.90),

$$\int_{\mathbb{R}^n\times\mathbb{R}^n} |u_j(x)-u_j(y)|^2 K(x-y)\,dx\,dy \geq S_K \|u_j\|^2_{L^{2_s^*}(\Omega)},$$

so, passing to the limit as $j \to +\infty$ and taking into account (18.48) and (18.49), we get

$$L \geq S_K L^{2/2_s^*},$$

which, combined with (18.50), gives

$$c \geq \frac{s}{n} S_K^{2_s^*/(2_s^*-2)} = \frac{s}{n} S_K^{n/(2s)}.$$

This contradicts Claim 18.1. Hence, $u_\infty \not\equiv 0$ in Ω, and this ends the proof of Claim 18.4. Thus, the proof of Theorem 18.1 is complete. □

18.4 The critical case $\mathscr{L}_K u + \lambda u + |u|^{2_s^*-2}u = 0$

In this section, we consider problem (18.8), and we prove Theorem 18.2. In order to do this, it is enough to show that when $f \equiv 0$, conditions (18.7) and (18.12) are equivalent. For this, we need the following proposition, which is the generalization of Proposition 14.4:

Proposition 18.7 For any $u_0 \in X_0^s(\Omega) \setminus \{0\}$,

$$\sup_{\zeta\geq 0}\left(\frac{\zeta^2}{2}\int_{\mathbb{R}^n\times\mathbb{R}^n}|u_0(x)-u_0(y)|^2K(x-y)\,dx\,dy-\frac{\zeta^2\lambda}{2}\int_\Omega |u_0(x)|^2\,dx\right.$$
$$\left.-\frac{\zeta^{2_s^*}}{2_s^*}\int_\Omega |u_0(x)|^{2_s^*}\,dx\right)=\frac{s}{n}\,S_{K,\lambda}^{n/(2s)}(u_0), \tag{18.51}$$

where the function $X_0^s(\Omega)\setminus\{0\} \ni v \mapsto S_{K,\lambda}(v)$ is defined in (18.11).

Proof Let $g : [0,+\infty) \to \mathbb{R}$ be the following function:

$$g(\zeta)=\frac{\zeta^2}{2}\int_{\mathbb{R}^n\times\mathbb{R}^n}|u_0(x)-u_0(y)|^2K(x-y)\,dx\,dy-\frac{\zeta^2\lambda}{2}\int_\Omega |u_0(x)|^2\,dx$$
$$-\frac{\zeta^{2_s^*}}{2_s^*}\int_\Omega |u_0(x)|^{2_s^*}\,dx.$$

Note that g is differentiable in $(0,+\infty)$, and

$$g'(\zeta)=\zeta\int_{\mathbb{R}^n\times\mathbb{R}^n}|u_0(x)-u_0(y)|^2K(x-y)\,dx\,dy-\zeta\lambda\int_\Omega |u_0(x)|^2\,dx$$
$$-\zeta^{2_s^*-1}\int_\Omega |u_0(x)|^{2_s^*}\,dx,$$

so $g'(\zeta)\geq 0$ if and only if

$$\zeta\leq\bar{\zeta}=\left(\frac{\displaystyle\int_{\mathbb{R}^n\times\mathbb{R}^n}|u_0(x)-u_0(y)|^2K(x-y)\,dx\,dy-\lambda\int_\Omega |u_0(x)|^2\,dx}{\displaystyle\int_\Omega |u_0(x)|^{2_s^*}\,dx}\right)^{1/(2_s^*-2)}.$$

Therefore, $\bar{\zeta}$ is a maximum for g and

$$\sup_{\zeta\geq 0} g(\zeta)=\max_{\zeta\geq 0} g(\zeta)=g(\bar{\zeta})=\frac{s}{n}\,S_{K,\lambda}^{n/(2s)}(u_0).$$

This concludes the proof. □

18.4.1 End of the proof of Theorem 18.2

By Proposition 18.7, we deduce that when $f \equiv 0$ in $\Omega \times \mathbb{R}$, condition (18.7) reads as follows

$$\begin{gathered} \text{there exists } u_0 \in X_0^s(\Omega) \setminus \{0\}, \text{ with } u_0 \geq 0 \text{ a.e. in } \mathbb{R}^n, \\ \text{such that } S_{K,\lambda}(u_0) < S_K. \end{gathered} \tag{18.52}$$

It is easily seen that conditions (18.52) and (18.12) are equivalent. Indeed, suppose that (18.52) holds true. Then

$$S_{K,\lambda} := \inf_{v \in X_0^s(\Omega) \setminus \{0\}} S_{K,\lambda}(v) \leq S_{K,\lambda}(u_0) < S_K, \tag{18.53}$$

while, if (18.12) is valid, then there exists $\widetilde{v} \in X_0^s(\Omega) \setminus \{0\}$ such that

$$S_{K,\lambda}(\widetilde{v}) < S_K. \tag{18.54}$$

Thus, condition (18.52) follows by taking $u_0 = |\widetilde{v}| \in X_0^s(\Omega)$; we remark that this choice is possible because, for any $v \in X_0^s(\Omega)$, its positive part $v^+ = \max\{v, 0\}$ and its negative part $v^- = \max\{-v, 0\}$ belongs to $X_0^s(\Omega)$ too, by Lemma 1.22 (hence, $|v| = v^+ + v^- \in X_0^s(\Omega)$, because $X_0^s(\Omega)$ a linear space). Indeed, by triangle inequality, a.e. $x, y \in \mathbb{R}^n$,

$$\big||\widetilde{v}(x)| - |\widetilde{v}(y)|\big| \leq |\widetilde{v}(x) - \widetilde{v}(y)|.$$

This and (18.54) give

$$S_{K,\lambda}(u_0) = S_{K,\lambda}(|\widetilde{v}|) \leq S_{K,\lambda}(\widetilde{v}) < S_K.$$

Hence, conditions (18.52) and (18.12) are equivalent, so Theorem 18.2 comes from Theorem 18.1.

18.5 The general case $\lambda > 0$

In earlier sections of this chapter, we studied problem 18.1 in the case where $\lambda \in (0, \lambda_1)$. Here we consider $\lambda \geq \lambda_1$; in this setting, we assume that, in addition to (6.5) and (18.2) and (18.3), the nonlinearity f also verifies the following conditions:

$$F(x,t) \geq 0 \quad \text{a.e.} \quad x \in \Omega, \quad t \in \mathbb{R}; \tag{18.55}$$

$$F(x,t) \leq \frac{1}{2} f(x,t)t + \frac{s}{n}|t|^{2_s^*} \quad \text{a.e.} \quad x \in \Omega,\ t \in \mathbb{R}. \tag{18.56}$$

Note that the model $f(x,t) = a(x)|t|^{q-2}t$, with $a \in L^\infty(\Omega)$, $a(x) \geq 0$ a.e. $x \in \Omega$, and $q \in (2, 2_s^*)$, also satisfies assumptions (18.55) and (18.56).

Assumption (18.55) is natural, and it is also assumed in the classical case of the Laplacian, where the parameter λ is greater or equal to the first eigenvalue of $-\Delta$ with homogeneous Dirichlet boundary data (see, e.g., [114]).

Condition (18.56) will be crucial in the proof of the Palais–Smale condition (see Step 18.3 in the proof of Proposition 18.9). This is only the generalization of the assumption required in the classical case of the Laplacian (see, e.g., [46, formula (2.34)] and [114, section 3]).

Our aim is to show that in this framework, problem (18.1) has nontrivial solutions for any $\lambda > 0$. To this end, we will again use some variants of the classical mountain pass and linking theorems.

The main result of this section is the following:

Theorem 18.8 *Let $s \in (0,1)$, $n > 2s$, and Ω be an open, bounded subset of $\mathbb{R}^n$ with continuous boundary. Let $K : \mathbb{R}^n \setminus \{0\} \to (0,+\infty)$ be a function satisfying conditions* (1.55) *and* (1.56), *and let $f : \Omega \times \mathbb{R} \to \mathbb{R}$ be a Carathéodory function verifying* (6.5), (18.2), (18.3), (18.55) *and* (18.56). *Finally, assume that*

$$\begin{gathered}\text{there exists } u_0 \in X_0^s(\Omega) \setminus \{0\}, \text{ with } u_0 \geq 0 \text{ a.e. in } \mathbb{R}^n, \text{ such that}\\ S_{K,\lambda}(u) < S_K \quad \text{for any } u \in \mathscr{U},\\ \text{where } \mathscr{U} := \begin{cases} \operatorname{span}\{u_0\} & \text{if } \lambda \in (0, \lambda_1)\\ \operatorname{span}\{e_1, \dots, e_k, u_0\} & \text{if } \lambda \in [\lambda_k, \lambda_{k+1}),\ k \in \mathbb{N}. \end{cases}\end{gathered} \tag{18.57}$$

Then, for any $\lambda > 0$, problem (18.1) *admits a nontrivial weak solution $u \in X_0^s(\Omega)$.*

18.5.1 Some comments on the main theorems

The existence theorems proved in this chapter may be seen as the natural extension to the nonlocal framework of well-know results obtained in [12, 46, 60, 114] (see also [9, 212, 220]), where the classical critical Laplace equation was considered. The aim of this section is to complete the study of problem (18.1), started in Sections 18.3 and 18.4, where only the case where the positive parameter λ is less than the first eigenvalue of $-\mathscr{L}_K$ was considered. Note that when $\lambda \in (0, \lambda_1)$, the assertion of Theorem 18.8 holds true under weaker assumptions on the data of the problem. To be precise, Theorem 18.8 is still valid if we do not assume condition (18.55) and if we replace (18.57) with the following:

$$\begin{gathered}\text{there exists } u_0 \in X_0^s(\Omega) \setminus \{0\}, \text{ with } u_0 \geq 0 \text{ a.e. in } \mathbb{R}^n, \text{ such that}\\ \sup_{\zeta \geq 0} \mathscr{J}_{K,\lambda,f}(\zeta u_0) < \frac{s}{n} S_K^{n/(2s)},\end{gathered} \tag{18.58}$$

which is weaker than (18.57), and it is the assumption considered in Theorem 18.1. In this sense, Theorem 18.8 may be seen as an extension of Theorem 18.1.

In the context of nonlocal problems, we also would like to recall the recent paper [22], where the authors studied the effect of a lower-order perturbation in a critical elliptic problem driven by a nonlocal operator defined starting from the power of the eigenvalues of $-\Delta$. In this framework, we also refer to [214], where a nonlocal version of the Brezis–Nirenberg result was given in the case of a regional square root of the Laplacian (which is an operator related to but different from the one considered here).

18.5.2 The Palais–Smale condition for the functional $\mathcal{J}_{K,\lambda,f}$

We start by proving that the functional $\mathcal{J}_{K,\lambda,f}$ satisfies the Palais–Smale condition at any level c, whenever c stays below a suitable threshold depending on S_K.

Proposition 18.9 Let $\lambda \in [\lambda_k, \lambda_{k+1})$, for some $k \in \mathbb{N}$, and let f be a Carathéodory function satisfying conditions (6.5), (18.2), (18.3), (18.55), and (18.56). Let $c \in \mathbb{R}$ be such that

$$c < \frac{s}{n} S_K^{n/(2s)}. \tag{18.59}$$

Then the functional $\mathcal{J}_{K,\lambda,f}$ satisfies the Palais–Smale condition at level c.

Proof For this proof, we adapt the arguments used in the classical elliptic setting to the nonlocal fractional framework. When $f \equiv 0$, the Palais–Smale condition for $\mathcal{J}_{K,\lambda,f}$ was proved in [196, proposition 4.1], while in the case when $\lambda < \lambda_1$, a proof can be found in [204, subsection 3.1.1]. Here some nontrivial differences arise with respect to these cases. Now we proceed by steps.

Let $\{u_j\}_{j\in\mathbb{N}}$ be a Palais–Smale sequence for the functional $\mathcal{J}_{K,\lambda,f}$, that is, a sequence satisfying (18.25) and (18.26). As in Subsection 18.3.1, we have that

Step 18.1 *The sequence $\{u_j\}_{j\in\mathbb{N}}$ is bounded in $X_0^s(\Omega)$.*

Step 18.2 *Problem* (18.4) *admits*[1] *a solution $u_\infty \in X_0^s(\Omega)$.*

Note that for this part, we do not need conditions (18.55) and (18.56). Furthermore, the next steps are valid. Here the argue as in the proof of Proposition 15.2, taking into account the presence of the term f.

Step 18.3 *The following equality holds true:*

$$\mathcal{J}_{K,\lambda,f}(u_\infty) = \frac{s}{n}\int_\Omega |u_\infty(x)|^{2_s^*}dx + \frac{1}{2}\int_\Omega f(x,u_\infty(x))u_\infty(x)dx$$
$$- \int_\Omega F(x,u_\infty(x))dx \geq 0.$$

Proof By Step 18.2, taking $\varphi = u_\infty$ as a test function in (18.4), we get

$$\int_{\mathbb{R}^n\times\mathbb{R}^n} |u_\infty(x)-u_\infty(y)|^2 K(x-y)dx\,dy - \lambda\int_\Omega |u_\infty(x)|^2 dx$$
$$= \int_\Omega |u_\infty(x)|^{2_s^*}dx + \int_\Omega f(x,u_\infty(x))u_\infty(x)dx,$$

[1] As a matter of fact, the solution u_∞ constructed in Step 18.2 will turn out to be the limit of $\{u_j\}_{j\in\mathbb{N}}$ in the topology of $X_0^s(\Omega)$, but this will be achieved only at the end of the proof (see (18.81)). Notice that, at this stage, the existence of the solution given by Step 18.2 does not end the proof of Theorem 18.8 because (18.25) and (18.26) are in use.

so

$$\begin{aligned}\mathscr{J}_{K,\lambda,f}(u_\infty) &= \left(\frac{1}{2}-\frac{1}{2_s^*}\right)\int_\Omega |u_\infty(x)|^{2_s^*}dx + \frac{1}{2}\int_\Omega f(x,u_\infty(x))u_\infty(x)dx \\ &\quad - \int_\Omega F(x,u_\infty(x))dx \\ &= \frac{s}{n}\int_\Omega |u_\infty(x)|^{2_s^*}dx + \frac{1}{2}\int_\Omega f(x,u_\infty(x))u_\infty(x)dx \\ &\quad - \int_\Omega F(x,u_\infty(x))dx \geq 0.\end{aligned}$$

The last inequality follows from assumption (18.56). Hence, Step 18.3 is proved. □

Step 18.4 *The following equality holds true*:

$$\mathscr{J}_{K,\lambda,f}(u_j) = \mathscr{J}_{K,\lambda,f}(u_\infty) + \frac{1}{2}\|u_j - u_\infty\|^2_{X_0^2(\Omega)} - \frac{1}{2_s^*}\|u_j - u_\infty\|^{2_s^*}_{L^{2_s^*}(\Omega)} + o(1)$$

as $j \to +\infty$.

Proof First of all, we observe that, by Step 18.1 and Lemma 1.31, the sequence $\{u_j\}_{j\in\mathbb{N}}$ is bounded in $X_0^s(\Omega)$ and in $L^{2_s^*}(\mathbb{R}^n)$. Since $L^{2_s^*}(\mathbb{R}^n)$ is a reflexive space, we have that, up to a subsequence,

$$u_j \to u_\infty \quad \text{weakly in } L^{2_s^*}(\mathbb{R}^n) \tag{18.60}$$

as $j \to +\infty$, while, by Lemma 1.30, up to a subsequence,

$$u_j \to u_\infty \quad \text{in } L^\nu(\mathbb{R}^n) \tag{18.61}$$

$$u_j \to u_\infty \quad \text{a.e. in } \mathbb{R}^n \tag{18.62}$$

as $j \to +\infty$, for any $\nu \in [1, 2_s^*)$ (see, e.g., [43, theorem IV.9]).

Moreover, since (18.38) holds true, by the Brezis–Lieb lemma (see [45, theorem 1]), we get

$$\begin{aligned}&\int_{\mathbb{R}^n\times\mathbb{R}^n} |u_j(x)-u_j(y)|^2 K(x-y)\,dx\,dy \\ &= \int_{\mathbb{R}^n\times\mathbb{R}^n} |u_j(x)-u_\infty(x)-u_j(y)+u_\infty(y)|^2 K(x-y)\,dx\,dy \\ &\quad + \int_{\mathbb{R}^n\times\mathbb{R}^n} |u_\infty(x)-u_\infty(y)|^2 K(x-y)\,dx\,dy + o(1)\end{aligned} \tag{18.63}$$

and

$$\int_\Omega |u_j(x)|^{2_s^*}dx = \int_\Omega |u_j(x)-u_\infty(x)|^{2_s^*}dx + \int_\Omega |u_\infty(x)|^{2_s^*}dx + o(1) \tag{18.64}$$

as $j \to +\infty$.

Finally, we claim that

$$\int_\Omega F(x,u_j(x))dx \to \int_\Omega F(x,u_\infty(x))dx \tag{18.65}$$

as $j \to +\infty$. For this purpose, note that since $\{u_j\}_{j\in\mathbb{N}}$ is bounded in $L^{2_s^*}(\Omega)$, there exists $\kappa > 0$ such that

$$\|u_j\|_{L^{2_s^*}(\Omega)}^{2_s^*} \le \kappa \quad \text{for any } j \in \mathbb{N}. \tag{18.66}$$

Now let us fix $\delta > 0$. By Lemma 18.3 (used here with $\varepsilon = \delta/(2\kappa)$), we have that

$$|F(x,u_j(x))| \le \frac{\delta}{2\kappa}|u_j(x)|^{2_s^*} + M(\delta/(2\kappa))|u_j(x)| \quad \text{a.e. in } \Omega. \tag{18.67}$$

Then, thanks to the Sobolev embeddings of $X_0^s(\Omega)$, we have that $F(\cdot,u_j(\cdot)) \in L^1(\Omega)$, and putting

$$\eta(\delta) := \left(\frac{\delta}{2M(\delta/(2\kappa))\kappa^{1/2_s^*}}\right)^{2_s^*/(2_s^*-1)},$$

for any measurable subset Ω' of Ω such that

$$|\Omega'| \le \eta(\delta), \tag{18.68}$$

we get

$$\begin{aligned}\int_{\Omega'} |F(x,u_j(x))|\,dx &\le \frac{\delta}{2\kappa}\int_{\Omega'} |u_j(x)|^{2_s^*}\,dx + M(\delta/(2\kappa))\int_{\Omega'} |u_j(x)|\,dx \\ &\le \frac{\delta}{2\kappa}\|u_j\|_{L^{2_s^*}(\Omega)}^{2_s^*} + M(\delta/(2\kappa))\|u_j\|_{L^1(\Omega')} \\ &\le \frac{\delta}{2} + M(\delta/(2\kappa))|\Omega'|^{(2_s^*-1)/2_s^*}\|u_j\|_{L^{2_s^*}(\Omega)} \\ &\le \frac{\delta}{2} + M(\delta/(2\kappa))\eta(\delta)^{(2_s^*-1)/2_s^*}\kappa^{1/2_s^*} \\ &= \delta,\end{aligned} \tag{18.69}$$

thanks to (18.66)–(18.68) and the Hölder inequality. Hence, $F(\cdot,u_j(\cdot))$ is uniformly integrable[2] in Ω. Moreover, by (18.38) and the fact that the map $t \mapsto F(\cdot,t)$ is continuous in $t \in \mathbb{R}$, we get

$$F(\cdot,u_j(\cdot)) \to F(\cdot,u_\infty(\cdot)) \quad \text{a.e. in } \Omega \tag{18.70}$$

as $j \to +\infty$. Thus, by (18.69) and (18.74) and the Vitali convergence theorem, (18.65) holds true.

[2] Or, according to the different terminologies, absolutely continuous in Ω, uniformly with respect to $j \in \mathbb{N}$.

Therefore, also using the definition of $\mathcal{J}_{K,\lambda,f}$, by (18.37) and (18.63)–(18.65), we deduce that

$$\begin{aligned}\mathcal{J}_{K,\lambda,f}(u_j) = &\frac{1}{2}\int_{\mathbb{R}^n\times\mathbb{R}^n} |u_j(x)-u_\infty(x)-u_j(y)+u_\infty(y)|^2K(x-y)\,dx\,dy\\ &+\frac{1}{2}\int_{\mathbb{R}^n\times\mathbb{R}^n}|u_\infty(x)-u_\infty(y)|^2K(x-y)\,dx\,dy-\frac{\lambda}{2}\int_\Omega|u_\infty(x)|^2dx\\ &-\frac{1}{2_s^*}\int_\Omega|u_j(x)-u_\infty(x)|^{2_s^*}\,dx-\frac{1}{2_s^*}\int_\Omega|u_\infty(x)|^{2_s^*}\,dx\\ &-\int_\Omega F(x,u_\infty(x))dx+o(1)\\ =\ &\mathcal{J}_{K,\lambda,f}(u_\infty)+\frac{1}{2}\int_{\mathbb{R}^n\times\mathbb{R}^n}|u_j(x)\\ &-u_\infty(x)-u_j(y)+u_\infty(y)|^2K(x-y)\,dx\,dy\\ &-\frac{1}{2_s^*}\int_\Omega|u_j(x)-u_\infty(x)|^{2_s^*}\,dx+o(1)\end{aligned}$$

as $j\to+\infty$, so the assertion of Step 18.4 follows. □

Step 18.5 *The following equality holds true:*

$$\|u_j-u_\infty\|^2_{X_0^2(\Omega)}=\|u_j-u_\infty\|^{2_s^*}_{L^{2_s^*}(\Omega)}+o(1)$$

as $j\to+\infty$.

Proof First, we note that, as a consequence of (18.60) and (18.64), it holds true that

$$\begin{aligned}&\int_\Omega\Big(|u_j(x)|^{2_s^*-2}u_j(x)-|u_\infty(x)|^{2_s^*-2}u_\infty(x)\Big)(u_j(x)-u_\infty(x))\,dx\\ &=\int_\Omega|u_j(x)|^{2_s^*}\,dx-\int_\Omega|u_\infty(x)|^{2_s^*-2}u_\infty(x)u_j(x)\,dx\\ &\quad-\int_\Omega|u_j(x)|^{2_s^*-2}u_j(x)u_\infty(x)\,dx+\int_\Omega|u_\infty(x)|^{2_s^*}\,dx\\ &=\int_\Omega|u_j(x)|^{2_s^*}\,dx-\int_\Omega|u_\infty(x)|^{2_s^*}\,dx+o(1)\\ &=\int_\Omega|u_j(x)-u_\infty(x)|^{2_s^*}\,dx+o(1)\end{aligned}\tag{18.71}$$

as $j\to+\infty$. Furthermore, arguing exactly as in the proof of (18.65), we deduce that

$$\int_\Omega f(x,u_j(x))u_j(x)\,dx\to\int_\Omega f(x,u_\infty(x))u_\infty(x)\,dx\tag{18.72}$$

as $j \to +\infty$. Also, by (18.38) and the fact that the map $t \mapsto f(\cdot,t)$ is continuous in $t \in \mathbb{R}$, it is easily seen that

$$f(\cdot,u_j(\cdot)) \to f(\cdot,u_\infty(\cdot)) \quad \text{a.e. in } \Omega \tag{18.73}$$

as $j \to +\infty$. Now, by Lemma 18.3, the fact that Ω is bounded, and (18.66), it follows that $f(\cdot,u_j(\cdot))$ is bounded in $L^{2_s^*/(2_s^*-1)}(\Omega)$, so, also thanks to (18.73), we get

$$f(\cdot,u_j(\cdot)) \to f(\cdot,u_\infty(\cdot)) \quad \text{weakly in } L^{2_s^*/(2_s^*-1)}(\Omega), \tag{18.74}$$

which yields

$$\int_\Omega f(x,u_j(x))u_\infty(x)\,dx \to \int_\Omega f(x,u_\infty(x))u_\infty(x)\,dx \tag{18.75}$$

as $j \to +\infty$, because $u_\infty \in L^{2_s^*}(\Omega)$.

Again by [204, lemma 5], we get that $f(\cdot,u_\infty(\cdot))$ belongs to $L^{2_s^*/(2_s^*-1)}(\Omega)$, so, by (18.60), we deduce that

$$\int_\Omega f(x,u_\infty(x))u_j(x)\,dx \to \int_\Omega f(x,u_\infty(x))u_\infty(x)\,dx \tag{18.76}$$

as $j \to +\infty$. Now, by (18.26) and Steps 18.1 and 18.2, it is easily seen that

$$o(1) = \langle \mathcal{J}'_{K,\lambda}(u_j), u_j - u_\infty \rangle = \langle \mathcal{J}'_{K,\lambda}(u_j) - \mathcal{J}'_{K,\lambda}(u_\infty), u_j - u_\infty \rangle \tag{18.77}$$

as $j \to +\infty$. Moreover,

$$\begin{aligned}
&\langle \mathcal{J}'_{K,\lambda}(u_j) - \mathcal{J}'_{K,\lambda}(u_\infty), u_j - u_\infty \rangle \\
&\quad = \int_{\mathbb{R}^n \times \mathbb{R}^n} |u_j(x) - u_\infty(x) - u_j(y) + u_\infty(y)|^2 K(x-y)\,dx\,dy \\
&\qquad - \lambda \int_\Omega |u_j(x) - u_\infty(x)|^2\,dx \\
&\qquad - \int_\Omega \left(|u_j(x)|^{2_s^*-2}u_j(x) - |u_\infty(x)|^{2_s^*-2}u_\infty(x)\right)(u_j(x) - u_\infty(x))\,dx \\
&\qquad - \int_\Omega \left(f(x,u_j(x)) - f(x,u_\infty(x))\right)(u_j(x) - u_\infty(x))\,dx \\
&\quad = \int_{\mathbb{R}^n \times \mathbb{R}^n} |u_j(x) - u_\infty(x) - u_j(y) + u_\infty(y)|^2 K(x-y)\,dx\,dy \\
&\qquad - \int_\Omega |u_j(x) - u_\infty(x)|^{2_s^*}\,dx + o(1)
\end{aligned} \tag{18.78}$$

as $j \to +\infty$, thanks to (18.37), (18.71), (18.72), (18.75), and (18.76). Putting together (18.77) and (18.78), we get the assertion of Step 18.5. □

Now we can conclude the proof of Proposition 18.9. By Step 18.5, it is easy to see that

$$\begin{aligned}
&\frac{1}{2}\int_{\mathbb{R}^n\times\mathbb{R}^n}|u_j(x)-u_\infty(x)-u_j(y)+u_\infty(y)|^2K(x-y)\,dx\,dy\\
&\qquad -\frac{1}{2^*_s}\int_\Omega|u_j(x)-u_\infty(x)|^{2^*_s}dx\\
&\qquad =\left(\frac{1}{2}-\frac{1}{2^*_s}\right)\int_{\mathbb{R}^n\times\mathbb{R}^n}|u_j(x)-u_\infty(x)-u_j(y)+u_\infty(y)|^2K(x-y)\,dx\,dy+o(1)\\
&\qquad =\frac{s}{n}\int_{\mathbb{R}^n\times\mathbb{R}^n}|u_j(x)-u_\infty(x)-u_j(y)+u_\infty(y)|^2K(x-y)\,dx\,dy+o(1),
\end{aligned}$$

so, as a consequence of this, Step 18.4, and (18.25), we obtain

$$\begin{aligned}
\mathcal{J}_{K,\lambda,f}(u_\infty)&+\frac{s}{n}\int_{\mathbb{R}^n\times\mathbb{R}^n}|u_j(x)-u_\infty(x)-u_j(y)+u_\infty(y)|^2K(x-y)\,dx\,dy\\
&=\mathcal{J}_{K,\lambda,f}(u_j)+o(1)\\
&=c+o(1),
\end{aligned}\tag{18.79}$$

as $j\to+\infty$. Note that, by Step 18.1, the sequence $\{\|u_j\|_{X^s_0(\Omega)}\}_{j\in\mathbb{N}}$ is bounded in $\mathbb{R}$. Hence, up to a subsequence, if necessary, we can assume that

$$\|u_j-u_\infty\|^2_{X^s_0(\Omega)}\to L,\tag{18.80}$$

so, again as a consequence of Step 18.5,

$$\int_\Omega|u_j(x)-u_\infty(x)|^{2^*_s}\,dx\to L$$

as $j\to+\infty$. Of course, $L\in[0,+\infty)$. Moreover, by definition of S_K, we have that

$$L\geq L^{2/2^*_s}S_K,$$

so

$$L=0\quad\text{or}\quad L\geq S_K^{n/(2s)}.$$

The case $L\geq S_K^{n/(2s)}$ cannot occur. Otherwise, by (18.79), (18.80), and Step 18.3, we would get

$$c=\mathcal{J}_{K,\lambda,f}(u_\infty)+\frac{s}{n}L\geq\frac{s}{n}L\geq\frac{s}{n}S_K^{n/(2s)},$$

which contradicts (18.59). Thus, $L = 0$, so (18.80) yields

$$\|u_j - u_\infty\|_{X_0^s(\Omega)} \to 0 \tag{18.81}$$

as $j \to +\infty$. This ends the proof of Proposition 18.9. □

18.5.3 The geometry of the functional $\mathcal{J}_{K,\lambda,f}$

This subsection is devoted to study of the geometric structure of the functional $\mathcal{J}_{K,\lambda,f}$. Precisely, we show that it fits with the geometric assumptions of the linking theorem given in [174, theorem 5.3]. Remember that we consider only the case where $\lambda \geq \lambda_1$ because the other one was treated in Section 18.3.

Here we argue essentially as in the classical case of the Laplacian (see [9, 212, 220] and references therein), but some nontrivial technical differences arise. In the pure nonlocal critical case, that is, when $f \equiv 0$, proof of the geometric features of $\mathcal{J}_{K,\lambda,f}$ was done in [196, propositions 5.1–5.3] (see also Section 15.3 for the case $\mathcal{L}_K = -(-\Delta)^s$). Here we generalize this result to the perturbed setting. The argument follows those used in Section 18.3.

Proposition 18.10 Let $\lambda \in [\lambda_k, \lambda_{k+1})$, for some $k \in \mathbb{N}$, and let f be a Carathéodory function satisfying conditions (6.5), (18.2), and (18.3). Then, there exist $\rho > 0$ and $\beta > 0$ such that, for any $u \in \mathbb{P}_{k+1}$ with $\|u\|_{X_0^s(\Omega)} = \rho$, it results that $\mathcal{J}_{K,\lambda,f}(u) \geq \beta$.

Proof Let u be a function in $\mathbb{P}_{k+1}$. By Lemma 18.4, we get that, for any $\varepsilon > 0$,

$$\begin{aligned}
\mathcal{J}_{K,\lambda,f}(u) &\geq \frac{1}{2}\int_{\mathbb{R}^n \times \mathbb{R}^n} |u(x)-u(y)|^2 K(x-y)\,dx\,dy - \frac{\lambda}{2}\int_\Omega |u(x)|^2\,dx \\
&\quad - \frac{1}{2_s^*}\int_\Omega |u(x)|^{2_s^*}\,dx - \varepsilon \int_\Omega |u(x)|^2 dx \\
&\quad - \delta(\varepsilon)\int_\Omega |u(x)|^{2_s^*}\,dx \\
&\geq \frac{1}{2}\left(1 - \frac{\lambda}{\lambda_{k+1}}\right)\int_{\mathbb{R}^n \times \mathbb{R}^n} |u(x)-u(y)|^2 K(x-y)\,dx\,dy \qquad (18.82) \\
&\quad - \varepsilon \|u\|^2_{L^2(\Omega)} - \left(\frac{1}{2_s^*} + \delta(\varepsilon)\right)\|u\|^{2_s^*}_{L^{2_s^*}(\Omega)} \\
&\geq \frac{1}{2}\left(1 - \frac{\lambda}{\lambda_{k+1}}\right)\int_{\mathbb{R}^n \times \mathbb{R}^n} |u(x)-u(y)|^2 K(x-y)\,dx\,dy \\
&\quad - \varepsilon |\Omega|^{(2_s^*-2)/2_s^*}\|u\|^2_{L^{2_s^*}(\Omega)} - \left(\frac{1}{2_s^*} + \delta(\varepsilon)\right)\|u\|^{2_s^*}_{L^{2_s^*}(\Omega)},
\end{aligned}$$

thanks to Proposition 3.1 and to the fact that $L^{2_s^*}(\Omega) \hookrightarrow L^2(\Omega)$ continuously (because Ω is bounded and $2 < 2_s^*$, so we can use the Hölder inequality).

By (1.56) and Lemma 1.28, from (18.82), we deduce that, for any $\varepsilon > 0$,

$$\begin{aligned}\mathscr{J}_{K,\lambda,f}(u) \geq & \frac{1}{2}\left(1-\frac{\lambda}{\lambda_{k+1}}\right)\int_{\mathbb{R}^n\times\mathbb{R}^n}|u(x)-u(y)|^2K(x-y)\,dx\,dy \\ & -\varepsilon c|\Omega|^{(2_s^*-2)/2_s^*}\int_{\mathbb{R}^n\times\mathbb{R}^n}\frac{|u(x)-u(y)|^2}{|x-y|^{n+2s}}\,dx\,dy \\ & -\left(\frac{1}{2_s^*}+\delta(\varepsilon)\right)c^{2_s^*/2}\left(\int_{\mathbb{R}^n\times\mathbb{R}^n}\frac{|u(x)-u(y)|^2}{|x-y|^{n+2s}}\,dx\,dy\right)^{2_s^*/2} \qquad (18.83)\\ \geq & \left[\frac{1}{2}\left(1-\frac{\lambda}{\lambda_{k+1}}\right)-\frac{\varepsilon c|\Omega|^{(2_s^*-2)/2_s^*}}{\theta}\right]\|u\|^2_{X_0^s(\Omega)} \\ & -\left(\frac{1}{2_s^*}+\delta(\varepsilon)\right)\left(\frac{c_L K}{\theta}\right)^{2_s^*/2}\|u\|^{2_s^*}_{X_0^s(\Omega)}.\end{aligned}$$

We recall that θ was introduced in (1.56). Now we can choose $\varepsilon > 0$ such that $2\varepsilon c|\Omega|^{(2_s^*-2)/2_s^*} < \theta(1-\lambda/\lambda_{k+1})$, so, by (18.83), it is easily seen that

$$\mathscr{J}_{K,\lambda,f}(u) \geq \alpha\|u\|^2_{X_0^s(\Omega)}\left(1-\kappa\|u\|^{2_s^*-2}_{X_0^s(\Omega)}\right),$$

for suitable positive constants α and κ.

Finally, let $u \in \mathbb{P}_{k+1}$ be such that $\|u\|_{X_0^s(\Omega)} = \rho > 0$. Because $2_s^* > 2$, choosing ρ sufficiently small (i.e., ρ such that $1-\kappa\rho^{2_s^*-2} > 0$), we get

$$\inf_{\substack{u\in\mathbb{P}_{k+1}\\ \|u\|_{X_0^s(\Omega)}=\rho}} \mathscr{J}_{K,\lambda,f}(u) \geq \alpha\rho^2(1-\kappa\rho^{2_s^*-2}) =: \beta > 0,$$

so the proof is complete. □

Proposition 18.11 Let $\lambda \in [\lambda_k, \lambda_{k+1})$, for some $k \in \mathbb{N}$, and let f be a Carathéodory function satisfying conditions (6.5), (18.2), and (18.3) and (18.55). Then $\mathscr{J}_{K,\lambda,f}(u) \leq 0$, for any $u \in \operatorname{span}\{e_1,\ldots,e_k\}$.

Proof Let $u \in \operatorname{span}\{e_1,\ldots,e_k\}$. Then

$$u(x) = \sum_{i=1}^{k} u_i e_i(x),$$

with $u_i \in \mathbb{R}$, $i = 1,\ldots,k$. Since $\{e_1,\ldots,e_k,\ldots\}$ is an orthonormal basis of $L^2(\Omega)$ and an orthogonal basis of $X_0^s(\Omega)$, by Proposition 3.1(f), we get

$$\int_\Omega |u(x)|^2\,dx = \sum_{i=1}^{k} u_i^2 \qquad (18.84)$$

and

$$\int_{\mathbb{R}^n\times\mathbb{R}^n}|u(x)-u(y)|^2K(x-y)\,dx\,dy=\sum_{i=1}^{k}u_i^2\|e_i\|^2_{X_0^s(\Omega)}. \tag{18.85}$$

Then, by (18.84) and (18.85) and using Proposition 3.1(b) and (e), the fact that $e_i\in X_0^s(\Omega)$, for any $i\in\mathbb{N}$, and (18.55), we get

$$\begin{aligned}\mathcal{J}_{K,\lambda,f}(u)&=\frac{1}{2}\sum_{i=1}^{k}u_i^2\left(\|e_i\|^2_{X_0^s(\Omega)}-\lambda\right)-\frac{1}{2_s^*}\int_\Omega|u(x)|^{2_s^*}dx-\int_\Omega F(x,u(x))\,dx\\&\le\frac{1}{2}\sum_{i=1}^{k}u_i^2\left(\|e_i\|^2_{X_0^s(\Omega)}-\lambda\right)\\&=\frac{1}{2}\sum_{i=1}^{k}u_i^2(\lambda_i-\lambda)\le 0\end{aligned}$$

because $\lambda_i\le\lambda_k\le\lambda$, for any $i=1,\ldots,k$. This concludes the proof. □

Proposition 18.12 Let $\lambda\ge 0$, let f be a Carathéodory function satisfying conditions (6.5), (18.2), (18.3), and (18.55), and let $\mathbb{F}$ be a finite-dimensional subspace of $X_0^s(\Omega)$. Then there exist $R>\rho$ such that $\mathcal{J}_{K,\lambda,f}(u)\le 0$, for any $u\in\mathbb{F}$ with $\|u\|_{X_0^s(\Omega)}\ge R$, where ρ is given in Proposition 18.10.

Proof Let $u\in\mathbb{F}$. Since $\lambda\ge 0$ and (18.55) holds true, we obtain that

$$\begin{aligned}\mathcal{J}_{K,\lambda,f}(u)&\le\frac{1}{2}\|u\|^2_{X_0^s(\Omega)}-\frac{1}{2_s^*}\|u\|^{2_s^*}_{L^{2_s^*}(\Omega)}\\&\le\frac{1}{2}\|u\|^2_{X_0^s(\Omega)}-\frac{\kappa}{2_s^*}\|u\|^{2_s^*}_{X_0^s(\Omega)},\end{aligned}$$

for some positive constant κ, thanks to the fact that in any finite-dimensional space all the norms are equivalent. Hence, if $\|u\|_{X_0^s(\Omega)}\to+\infty$, then

$$\mathcal{J}_{K,\lambda,f}(u)\to-\infty$$

because $2_s^*>2$ by assumption. This ends the proof. □

By Propositions 18.10–18.12, we deduce that the functional $\mathcal{J}_{K,\lambda,f}$ has the geometric features required by the linking theorem [174, theorem 5.3]. Now we have to choose carefully the finite-dimensional space $\mathbb{F}$ in which we apply such theorem in order to have the suitable estimate on the critical level (i.e., an estimate that fits with the range where the Palais–Smale condition holds true).

To this end, let u_0 be the function given in assumption (18.57) and $u_0^{(k+1)}$ its projection on $\mathbb{P}_{k+1}$, and $\tilde{u}_0^{(k+1)}$ the normalization of $u_0^{(k+1)}$ in $X_0^s(\Omega)$; that is,

$$u_0^{(k+1)}=u_0-\sum_{i=1}^{k}\left(\int_\Omega u_0(x)e_i(x)\,dx\right)e_i,\quad \tilde{u}_0^{(k+1)}=u_0^{(k+1)}/\|u_0^{(k+1)}\|_{X_0^s(\Omega)}. \tag{18.86}$$

Here we take

$$\mathbb{F} = V \oplus \operatorname{span}\left\{\tilde{u}_0^{(k+1)}\right\},$$

where $V = \operatorname{span}\{e_1, \ldots, e_k\}$. Note that V is finite dimensional and that $V \oplus W = X_0^s(\Omega)$, with $W = \mathbb{P}_{k+1}$, thanks to Proposition 3.1(f). Moreover, by Propositions 18.10–18.12, we get that there exist $\rho > 0$, $\beta > 0$ and $R > \rho$ such that

$$\mathscr{J}_{K,\lambda}(u) \geq \beta > 0 \quad \text{for any } u \in W \text{ with } \|u\|_{X_0^s(\Omega)} = \rho > 0,$$

$$\mathscr{J}_{K,\lambda}(u) \leq 0 \quad \text{for any } u \in V,$$

and

$$\mathscr{J}_{K,\lambda}(u) \leq 0 \quad \text{for any } u \in \mathbb{F} \text{ with } \|u\|_{X_0^s(\Omega)} \geq R.$$

Hence, the geometric structure of $\mathscr{J}_{K,\lambda}$ fits with the one required by the linking theorem (see assumptions (I_1') and (I_5) of [174, theorem 5.3] as well as [174, remark 5.5(iii)]).

The linking critical level c_L of $\mathscr{J}_{K,\lambda,f}$ is given by

$$c_L := \inf_{h\in\Gamma} \max_{u\in Q} \mathscr{J}_{K,\lambda,f}(h(u)),$$

where

$$\Gamma := \{h \in C(\overline{Q}; X_0^s(\Omega)) : h = id \text{ on } \partial Q\}$$

and

$$Q := \left(\overline{B}_R \cap V\right) \oplus \left\{r\,\tilde{u}_0^{(k+1)} : r \in (0,R)\right\}.$$

In the next subsection, we will estimate the critical level c_L.

18.5.4 Estimates of the critical level of $\mathscr{J}_{K,\lambda,f}$

Given to the critical nature of our problem, we have to estimate the linking critical level c_L to show that it stays below the threshold under which the Palais–Smale condition is satisfied by $\mathscr{J}_{K,\lambda,f}$. For this, first note that, by definition of c_L, for any $h \in \Gamma$,

$$c_L \leq \max_{u\in Q} \mathscr{J}_{K,\lambda,f}(h(u)),$$

from which, taking $h = id$ in $\overline{Q}$, we deduce that

$$c_L \leq \max_{u\in Q} \mathscr{J}_{K,\lambda,f}(u). \tag{18.87}$$

Furthermore, recall that $\mathbb{F}$ is a linear space, so, as a consequence,

$$\begin{aligned}\max_{u\in\mathbb{F}} \mathscr{J}_{K,\lambda,f}(u) &= \max_{\substack{u\in\mathbb{F}\\ \zeta\neq 0}} \mathscr{J}_{K,\lambda,f}\left(|\zeta|\cdot\frac{u}{|\zeta|}\right)\\ &= \max_{\substack{u\in\mathbb{F}\\ \zeta>0}} \mathscr{J}_{K,\lambda,f}(\zeta u)\\ &\leq \max_{\substack{u\in\mathbb{F}\\ \zeta\geq 0}} \mathscr{J}_{K,\lambda,f}(\zeta u).\end{aligned} \tag{18.88}$$

Hence, by (18.87), (18.88), and the fact that $Q \subset \mathbb{F}$, we obtain

$$c_L \leq \max_{u\in Q} \mathscr{J}_{K,\lambda,f}(u) \leq \max_{u\in\mathbb{F}} \mathscr{J}_{K,\lambda,f}(u) \leq \max_{\substack{u\in\mathbb{F}\\ \zeta\geq 0}} \mathscr{J}_{K,\lambda,f}(\zeta u). \tag{18.89}$$

Thanks to (18.55) and Proposition 18.7, we also have

$$\begin{aligned}\max_{\zeta\geq 0} \mathscr{J}_{K,\lambda,f}(\zeta u) \leq \max_{\zeta\geq 0}\Bigg(&\frac{\zeta^2}{2}\int_{\mathbb{R}^n\times\mathbb{R}^n} |u(x)-u(y)|^2 K(x-y)\,dx\,dy\\ &-\frac{\zeta^2\lambda}{2}\int_\Omega |u(x)|^2\,dx - \frac{\zeta^{2^*_s}}{2^*_s}\int_\Omega |u(x)|^{2^*_s}\,dx\Bigg)\\ =\frac{s}{n} S_{K,\lambda}^{n/(2s)}(u),&\end{aligned} \tag{18.90}$$

for any $u \in X_0^s(\Omega)\setminus\{0\}$.

Finally, taking into account that $\mathbb{F} = V \oplus \operatorname{span}\{\tilde{u}_0^{(k+1)}\} = \operatorname{span}\{e_1,\ldots,e_k,u_0\} = \mathscr{U}$ (here we use the construction of $\tilde{u}_0^{(k+1)}$ and $\mathscr{U}$ defined in (18.57)), by (18.57), (18.89), and (18.90), we deduce that

$$c_L \leq \max_{\substack{u\in\mathbb{F}\\ \zeta\geq 0}} \mathscr{J}_{K,\lambda,f}(\zeta u) \leq \frac{s}{n}\max_{u\in\mathbb{F}} S_{K,\lambda}^{n/(2s)}(u) < \frac{s}{n} S_K^{n/(2s)}. \tag{18.91}$$

Ultimately, we have shown that the linking critical level of $\mathscr{J}_{K,\lambda,f}$ stays below the threshold where the Palais–Smale condition is verified.

Proof of Theorem 18.8 The conclusion of Theorem 18.8 follows from the linking theorem [174, theorem 5.3], thanks to Propositions 18.9–18.12 and subsection 18.5.4. □

18.6 The model case

When dealing with problem (18.1) with $\mathscr{L}_K = -(-\Delta)^s$, we get the following problem:

$$\begin{cases} (-\Delta)^s u - \lambda u = |u|^{2^*_s-2}u + f(x,u) & \text{in } \Omega\\ u = 0 & \text{in } \mathbb{R}^n\setminus\Omega. \end{cases} \tag{18.92}$$

The weak formulation of (18.92) is given by

$$\begin{cases} \displaystyle\int_{\mathbb{R}^n\times\mathbb{R}^n} \frac{(u(x)-u(y))(\varphi(x)-\varphi(y))}{|x-y|^{n+2s}}\,dx\,dy - \lambda\int_\Omega u(x)\varphi(x)\,dx \\[2ex] \displaystyle\quad = \int_\Omega |u(x)|^{2_s^*-2}u(x)\varphi(x)dx + \int_\Omega f(x,u(x))\varphi(x)\,dx \quad \forall\,\varphi\in\mathbb{H}_0^s(\Omega) \\[2ex] u\in\mathbb{H}_0^s(\Omega), \end{cases} \tag{18.93}$$

which represents the Euler–Lagrange equation of the functional $\mathcal{J}_{s,\lambda,f}:\mathbb{H}_0^s(\Omega)\to\mathbb{R}$ defined as

$$\begin{aligned} \mathcal{J}_{s,\lambda,f}(u) = &\frac{1}{2}\int_{\mathbb{R}^n\times\mathbb{R}^n} \frac{|u(x)-u(y)|^2}{|x-y|^{n+2s}}\,dx\,dy \\ &-\frac{\lambda}{2}\int_\Omega |u(x)|^2\,dx - \frac{1}{2_s^*}\int_\Omega |u(x)|^{2_s^*}\,dx - \int_\Omega F(x,u(x))\,dx, \end{aligned}$$

with F given in (6.8).

In this framework, Theorem 18.8 reads as follows:

Theorem 18.13 *Let $s\in(0,1)$, $n>2s$, and Ω be an open, bounded subset of $\mathbb{R}^n$ with continuous boundary. Let $f:\Omega\times\mathbb{R}\to\mathbb{R}$ be a Carathéodory function verifying* (6.5), (18.2), (18.3), *and* (18.55) *and* (18.56), *and assume that*

$$\begin{aligned} &\textit{there exists } u_0\in\mathbb{H}_0^s(\Omega)\setminus\{0\}, \textit{ with } u_0\geq 0 \textit{ a.e. in } \Omega, \textit{ such that} \\ &\qquad S_{s,\lambda}(u) < S_s \textit{ for any } u\in\mathcal{U}, \\ &\textit{where } \mathcal{U} := \begin{cases} \operatorname{span}\{u_0\} & \textit{if } \lambda\in(0,\lambda_1) \\ \operatorname{span}\{e_1,\dots,e_k,u_0\} & \textit{if } \lambda\in[\lambda_k,\lambda_{k+1}),\ k\in\mathbb{N}. \end{cases} \end{aligned} \tag{18.94}$$

Then, for any $\lambda>0$, problem (18.92) *admits a nontrivial weak solution $u\in\mathbb{H}_0^s(\Omega)$.*

Note that Theorem 18.13 generalizes Theorems 15.1, 16.1, and 17.1 to the perturbed case.

19

Existence of multiple solutions

In this book, starting from Chapter 14, we have considered the nonlocal counterpart of the famous Brezis–Nirenberg result in [46], and we have given some existence theorems. The first multiplicity result for problem (14.1) was proved by Cerami, Fortunato, and Struwe in [62], where it was shown that in a suitable left neighborhood of any eigenvalue of $-\Delta$ (with Dirichlet boundary data), the number of solutions is at least twice the multiplicity of the eigenvalue. The authors also gave an estimate of the length of this neighborhood, which depends on the best critical Sobolev constant, on the Lebesgue measure of the set where the problem is set, and on the space dimension.

Later, in [63] the authors proved that in dimension $n \geq 6$ and for $\lambda > 0$ less than the first eigenvalue of $-\Delta$ (with homogeneous Dirichlet boundary conditions), problem (14.1) has at least two nontrivial solutions, one of which is a changing-sign solution (for other results on changing-sign solutions, see, e.g., [70, 191, 213]). More recently, in [88] the authors improved the result in [63], while in [87] Devillanova and Solimini proved the existence of infinitely many solutions for (14.1), provided that the dimension $n \geq 7$ and the parameter λ is positive. Finally, in [70] the authors showed that for $n \geq 4$, problem (14.1) has at least $(n+1)/2$ pairs of nontrivial solutions, provided that $\lambda > 0$ is not an eigenvalue of $-\Delta$, and $(n+1-m)/2$ pairs of nontrivial solutions if λ is an eigenvalue of $-\Delta$ with multiplicity $m < n+2$. When $m \geq n+2$, [70] gave no information about the multiplicity of solutions for (14.1) when λ is an eigenvalue of $-\Delta$. A partial answer to this question was given in [67], where the authors showed that when $n \geq 5$ and $\lambda \geq \lambda_1$, then problem (14.1) has at least $(n+1)/2$ pairs of nontrivial solutions.

A natural question is whether all these results can be extended to the fractional nonlocal counterpart of (14.1), that is, to the problem (14.2) or its generalization.

The aim of this chapter is to focus attention on the multiplicity of solutions for (18.8). In particular, the starting point is the paper [62], and the goal is to extend the results obtained there to the nonlocal fractional setting. This chapter is based on the paper [100] (see also the recent paper [164]).

19.1 A multiplicity result

The main feature of this chapter concerns the existence of multiple solutions for problems (18.8). Precisely, with the notation introduced in Chapter 3, the main result of this chapter reads as follows:

Theorem 19.1 *Let $s \in (0,1)$, $n > 2s$, and Ω be an open, bounded subset of $\mathbb{R}^n$ with continuous boundary. Let $K : \mathbb{R}^n \setminus \{0\} \to (0,+\infty)$ be a function satisfying assumptions* (1.55) *and* (1.56). *Let $\lambda \in \mathbb{R}$, let $\lambda^\star$ be the eigenvalue of problem* (3.1) *given by*

$$\lambda^\star := \min\{\lambda_k : \lambda < \lambda_k\}, \tag{19.1}$$

and let $m \in \mathbb{N}$ be its multiplicity. Assume that

$$\lambda \in (\lambda^\star - S_K|\Omega|^{-\frac{2s}{n}}, \lambda^\star), \tag{19.2}$$

where S_K is the best fractional critical Sobolev constant defined in (1.90). *Then problem* (18.8) *has at least m pairs of nontrivial weak solutions $\{-u_i(\lambda), u_i(\lambda)\}$ such that*

$$\|u_i(\lambda)\|_{X_0^s(\Omega)} \to 0 \quad \text{as } \lambda \to \lambda^\star,$$

for any $i = 1, \ldots, m$.

This theorem represents the fractional nonlocal counterpart of the famous multiplicity result gotten by Cerami, Fortunato, and Struwe in [62], using essentially an abstract critical point theorem due to Bartolo, Benci, and Fortunato in [26], whose main tool is a pseudoindex theory introduced in [33] for studying indefinite functionals. Theorem 19.1 proves that, in a suitable left neighborhood of any eigenvalue of the integrodifferential operator $-\mathscr{L}_K$ (with homogeneous Dirichlet boundary condition), the number of nontrivial solutions for problem (18.8) is at least twice the multiplicity of the eigenvalue. Hence, we show that there is a bifurcation from any eigenvalue of the operator $-\mathscr{L}_K$. In addition, we give an estimate of the length of this left neighborhood, in which the existence of multiple solutions occurs. This estimate depends on the best fractional critical Sobolev constant S_K, on the Lebesgue measure of Ω, and on n and s, as stated in (19.2). We would point out that this condition is crucial to show that the energy functional associated with (18.8) satisfies all the geometric assumptions required by the abstract critical point theorem we will use.

To prove Theorem 19.1, we use the following result due to Bartolo, Benci, and Fortunato (see [62, theorem 2.5] and [26, theorem 2.4]), which, as usual for abstract critical points theorems, gives the existence of critical points for a functional (sufficiently smooth), provided that it satisfies suitable geometric and compactness conditions:

Theorem 19.2 ([26, 62]) *Let H be a real Hilbert space with norm $\|\cdot\|$, and suppose that $\mathscr{I} \in C^1(H,\mathbb{R})$ is a functional on H satisfying the following conditions:*

(I_1) *$\mathscr{I}(u) = \mathscr{I}(-u)$ and $\mathscr{I}(0) = 0$.*

(I_2) *There exists a constant $\beta > 0$ such that the Palais–Smale condition for $\mathscr{I}$ holds in $(0, \beta)$.*

(I_3) *There exist two closed subspaces $V, W \subset H$ and positive constants ρ, δ, β', with $\delta < \beta' < \beta$, such that*

(a) *$\mathscr{I}(u) \leq \beta'$, for any $u \in W$.*

(b) *$\mathscr{I}(u) \geq \delta$, for any $u \in V$ with $\|u\| = \rho$.*

(c) *Codim $V < +\infty$ and dim $W \geq$ codim V.*

Then there exist at least dim W − codim V pairs of critical points of $\mathscr{I}$, with critical values belonging to the interval $[\delta, \beta']$.

The abstract result gotten in [26, theorem 2.4] is a generalization of [11, theorem 2.13] obtained using a pseudoindex theory introduced in [33] for exploiting the existence of multiple critical points of functionals that are bounded neither above nor below on a Hilbert space.

19.2 Proof of Theorem 19.1

The idea consists of applying Theorem 19.2 to the functional $\mathcal{J}_{K,\lambda}$, defined as

$$\mathcal{J}_{K,\lambda} := \mathcal{J}_{K,\lambda,0};$$

that is, $\mathcal{J}_{K,\lambda,f}$ in (13.1.2) with $f \equiv 0$. Let $\lambda^\star$ be as in (19.1). Then

$$\lambda^\star = \lambda_k \quad \text{for some } k \in \mathbb{N}.$$

Since $\lambda^\star$ has multiplicity $m \in \mathbb{N}$, by assumption, we have that

$$\begin{array}{lll} \lambda^\star & = \lambda_1 < \lambda_2 & \text{if } k = 1 \\ \lambda_{k-1} < \lambda^\star & = \lambda_k = \cdots = \lambda_{k+m-1} < \lambda_{k+m} & \text{if } k \geq 2. \end{array} \tag{19.3}$$

Also, before going on with the proof of Theorem 19.1, we would note that under condition (19.2), the parameter λ is such that

$$\lambda > 0. \tag{19.4}$$

Indeed, by definition of $\lambda^\star$ and taking into account (3.7), it is easily seen that

$$\lambda^\star \geq \lambda_1. \tag{19.5}$$

In addition, the variational characterization of the first eigenvalue λ_1 (see Proposition 3.1), the definition of S_K (see (1.90)), and the fact that

$$\int_\Omega |u(x)|^2 \, dx \leq |\Omega|^{2s/n} \left(\int_\Omega |u(x)|^{2^*_s} \, dx \right)^{2/2^*_s}$$

give that

$$\lambda_1 \geq S_K |\Omega|^{-2s/n},$$

which, combined with (19.5), yields

$$\lambda^\star \geq S_K|\Omega|^{-2s/n}.$$

Hence, as a consequence of this and of (19.2), we get (19.4).

Now, with the notation of the abstract result stated in Theorem 19.2, we set

$$W = \operatorname{span}\{e_1,\dots,e_{k+m-1}\}$$

and

$$V = \begin{cases} X_0^s(\Omega) & \text{if } k=1 \\ \left\{u \in X_0^s(\Omega) : \langle u, e_j\rangle_{X_0^s(\Omega)} = 0 \ \forall j = 1,\dots,k-1\right\} & \text{if } k \geq 2. \end{cases}$$

Note that both W and V are closed subsets of $X_0^s(\Omega)$ and

$$\dim W = k+m-1 \quad \operatorname{codim} V = k-1, \tag{19.6}$$

so $(I_3)(c)$ of Theorem 19.2 is satisfied.

Now we study the geometric structure of the functional $\mathcal{J}_{K,\lambda}$.

Claim 19.1 *The functional $\mathcal{J}_{K,\lambda}$ has the geometric features required by Theorem 19.2.*

Proof Let us show that the functional $\mathcal{J}_{K,\lambda}$ verifies assumptions (I_3)(a) and (b) of Theorem 19.2 (here condition (19.2) will be crucial). For this, let $u \in W$. Then

$$u(x) = \sum_{i=1}^{k+m-1} u_i e_i(x),$$

with $u_i \in \mathbb{R}$, $i = 1,\dots,k+m-1$.

Since $\{e_1,\dots,e_k,\dots\}$ is an orthonormal basis of $L^2(\Omega)$ and an orthogonal basis of $X_0^s(\Omega)$ (see Proposition 3.1), taking into account (19.3), we get

$$\|u\|^2_{X_0^s(\Omega)} = \sum_{i=1}^{k+m-1} u_i^2\|e_i\|^2_{X_0^s(\Omega)} = \sum_{i=1}^{k+m-1} \lambda_i u_i^2 \leq \lambda_k \sum_{i=1}^{k+m-1} u_i^2 = \lambda_k\|u\|^2_{L^2(\Omega)} = \lambda^\star\|u\|^2_{L^2(\Omega)},$$

so, by this and the Hölder inequality, we have

$$\begin{aligned} \mathcal{J}_{K,\lambda}(u) &= \frac{1}{2}\int_{\mathbb{R}^n\times\mathbb{R}^n} |u(x)-u(y)|^2 K(x-y)\,dx\,dy \\ &\quad - \frac{\lambda}{2}\int_\Omega |u(x)|^2 dx - \frac{1}{2_s^*}\int_\Omega |u(x)|^{2_s^*} dx \\ &\leq \frac{1}{2}(\lambda^\star - \lambda)\int_\Omega |u(x)|^2 dx - \frac{1}{2_s^*}\int_\Omega |u(x)|^{2_s^*} dx \\ &\leq \frac{1}{2}(\lambda^\star-\lambda)|\Omega|^{\frac{2s}{n}}\left(\int_\Omega |u(x)|^{2_s^*} dx\right)^{\frac{2}{2_s^*}} - \frac{1}{2_s^*}\int_\Omega |u(x)|^{2_s^*} dx. \end{aligned} \tag{19.7}$$

Now, for $t \geq 0$, let

$$g(t) = \frac{1}{2}(\lambda^\star - \lambda)|\Omega|^{\frac{2s}{n}} t^2 - \frac{1}{2_s^*} t^{2_s^*}.$$

Note that the function g is differentiable in $(0, +\infty)$ and

$$g'(t) = (\lambda^\star - \lambda)|\Omega|^{\frac{2s}{n}} t - t^{2_s^*-1},$$

so $g'(t) \geq 0$ if and only if

$$t \leq \bar{t} = \left[(\lambda^\star - \lambda)|\Omega|^{\frac{2s}{n}}\right]^{1/(2_s^*-2)}.$$

As a consequence, $\bar{t}$ is a maximum point for g, so, for any $t \geq 0$,

$$g(t) \leq \max_{t \geq 0} g(t) = g(\bar{t}) = \frac{s}{n}(\lambda^\star - \lambda)^{\frac{n}{2s}} |\Omega|. \tag{19.8}$$

By (19.7) and (19.8), we get

$$\sup_{u \in W} \mathscr{J}_{K,\lambda}(u) \leq \max_{t \geq 0} g(t) = \frac{s}{n}(\lambda^\star - \lambda)^{\frac{n}{2s}} |\Omega|. \tag{19.9}$$

We observe that

$$\frac{s}{n}(\lambda^\star - \lambda)^{\frac{n}{2s}} |\Omega| > 0$$

because $\lambda < \lambda^\star$ by (19.2).

Finally, let $u \in V$. Then

$$\|u\|^2_{X_0^s(\Omega)} \geq \lambda^\star \|u\|^2_{L^2(\Omega)}. \tag{19.10}$$

Indeed, if $u \equiv 0$, then the assertion is trivial, while if $u \in V \setminus \{0\}$, it follows from the variational characterization of $\lambda^\star = \lambda_k$ given in Proposition 3.1.

Thus, by (1.90) and (19.10) and taking into account that $\lambda > 0$ (see (19.4)), it follows that

$$\begin{aligned} \mathscr{J}_{K,\lambda}(u) &\geq \frac{1}{2}\left(1 - \frac{\lambda}{\lambda^\star}\right) \|u\|^2_{X_0^s(\Omega)} - \frac{1}{2_s^* S_K^{1/2}} \|u\|^{2_s^*}_{X_0^s(\Omega)} \\ &= \|u\|^2_{X_0^s(\Omega)} \left(\frac{1}{2}\left(1 - \frac{\lambda}{\lambda^\star}\right) - \frac{1}{2_s^* S_K^{1/2}} \|u\|^{2_s^*-2}_{X_0^s(\Omega)}\right). \end{aligned} \tag{19.11}$$

Now let $u \in V$ be such that $\|u\|_{X_0^s(\Omega)} = \rho > 0$. Since $2_s^* > 2$, we can choose ρ sufficiently small, say, $\rho \leq \bar{\rho}$, with $\bar{\rho} > 0$, that

$$\frac{1}{2}\left(1 - \frac{\lambda}{\lambda^\star}\right) - \frac{1}{2_s^* S_K^{1/2}} \rho^{2_s^*-2} > 0 \tag{19.12}$$

and

$$\rho^2\left(\frac{1}{2}\left(1-\frac{\lambda}{\lambda^\star}\right)-\frac{1}{2_s^* S_K^{1/2}}\rho^{2_s^*-2}\right)<\frac{\rho^2}{2}\left(1-\frac{\lambda}{\lambda^\star}\right)<\frac{s}{n}(\lambda^\star-\lambda)^{\frac{n}{2s}}|\Omega|. \quad (19.13)$$

□

Now we can conclude the proof of Theorem 19.1. At this purpose, note that $\mathcal{J}_{K,\lambda}$ is even and $\mathcal{J}_{K,\lambda}(0)=0$, so condition (I_1) of Theorem 19.2 is verified by $\mathcal{J}_{K,\lambda}$. Moreover, by Proposition 18.9, condition (I_2) of Theorem 19.2 is also satisfied, with

$$\beta=\frac{s}{n}S_K^{n/(2s)}>0.$$

Finally, by Claim 19.1 (see (19.9)–(19.13)), we get that $\mathcal{J}_{K,\lambda}$ verifies (I_3), with

$$\rho=\bar{\rho},$$

$$\beta'=\frac{s}{n}(\lambda^\star-\lambda)^{\frac{n}{2s}}|\Omega|,$$

and

$$\delta=\bar{\rho}^2\left(\frac{1}{2}\left(1-\frac{\lambda}{\lambda^\star}\right)-\frac{1}{2_s^* S_K^{1/2}}\bar{\rho}^{2_s^*-2}\right).$$

Note that

$$0<\delta<\beta'<\beta,$$

thanks to (19.12), (19.13), and assumption (19.2).

All in all, the functional $\mathcal{J}_{K,\lambda}$ satisfies both the compactness assumption and the geometric features required by the abstract critical points theorem stated in Theorem 19.2. As a consequence, $\mathcal{J}_{K,\lambda}$ has m pairs $\{-u_i(\lambda), u_i(\lambda)\}$ of critical points whose critical value $\mathcal{J}_{K,\lambda}(\pm u_i(\lambda))$ is such that

$$\begin{aligned}0&<\bar{\rho}^2\left(\frac{1}{2}\left(1-\frac{\lambda}{\lambda^\star}\right)-\frac{1}{2_s^* S_K^{1/2}}\bar{\rho}^{2_s^*-2}_{X_0^s(\Omega)}\right)\\&\le\mathcal{J}_{K,\lambda}(\pm u_i(\lambda))\\&\le\frac{s}{n}(\lambda^\star-\lambda)^{\frac{n}{2s}}|\Omega|,\end{aligned} \quad (19.14)$$

for any $i=1,\ldots,m$.

Since $\mathcal{J}_{K,\lambda}(0)=0$ and (19.14) holds true, it is easy to see that these critical points are all different from the trivial function. Hence, problem (18.8) admits m pairs of nontrivial weak solutions.

Now fix $i \in \{1,\dots,m\}$. By (19.14), we obtain

$$\begin{aligned}
(\lambda^\star - \lambda)^{n/2s}|\Omega|\frac{s}{n} &\geq \mathscr{J}_{K,\lambda}(u_i(\lambda)) \\
&= \mathscr{J}_{K,\lambda}(u_i(\lambda)) - \frac{1}{2}\langle \mathscr{J}'_\lambda(u_i(\lambda)), u_i(\lambda)\rangle \\
&= \left(\frac{1}{2} - \frac{1}{2_s^*}\right)\|u_i(\lambda)\|_{L^{2_s^*}(\Omega)}^{2_s^*} \\
&= \frac{s}{n}\|u_i(\lambda)\|_{L^{2_s^*}(\Omega)}^{2_s^*},
\end{aligned} \tag{19.15}$$

so, passing to the limit as $\lambda \to \lambda^\star$ in (19.15), it follows that

$$\|u_i(\lambda)\|_{L^{2_s^*}(\Omega)}^{2_s^*} \to 0 \quad \text{as} \quad \lambda \to \lambda^\star. \tag{19.16}$$

Then, by (19.16), since $L^{2_s^*}(\Omega) \hookrightarrow L^2(\Omega)$ continuously (being Ω bounded), we also get

$$\|u_i(\lambda)\|_{L^2(\Omega)}^2 \to 0 \quad \text{as } \lambda \to \lambda^\star. \tag{19.17}$$

So, arguing as earlier, we have

$$\begin{aligned}
(\lambda^\star - \lambda)^{n/2s}|\Omega|\frac{s}{n} &\geq \mathscr{J}_{K,\lambda}(u_i(\lambda)) \\
&= \frac{1}{2}\|u_i(\lambda)\|_{X_0^s(\Omega)}^2 - \frac{\lambda}{2}\|u_i(\lambda)\|_{L^2(\Omega)}^2 - \frac{1}{2_s^*}\|u_i(\lambda)\|_{L^{2_s^*}(\Omega)}^{2_s^*},
\end{aligned}$$

which, combined with (19.16) and (19.17), gives

$$\|u_i(\lambda)\|_{X_0^s(\Omega)} \to 0 \quad \text{as } \lambda \to \lambda^\star.$$

This concludes the proof of Theorem 19.1.

20

Nonlocal critical equations with concave-convex nonlinearities

In this chapter, we focus our attention on the following critical nonlocal fractional problem:

$$\begin{cases} (-\Delta)^s u = \lambda u^q + u^{2_s^*-1} & \text{in } \Omega \\ u > 0 & \text{in } \Omega \\ u = 0 & \text{in } \mathbb{R}^n \setminus \Omega, \end{cases} \tag{20.1}$$

where $s \in (0, 1)$ is fixed, and $(-\Delta)^s$ is the fractional Laplace operator defined, up to normalization factors, as in (1.20), while $\Omega \subset \mathbb{R}^n$, $n > 2s$, is open, bounded and with continuous boundary and $\lambda > 0$. The main results of this chapter show the existence and multiplicity of solutions to problem (20.1) for different values of λ. The dependency on this parameter changes according to whether we consider the concave power case ($0 < q < 1$) or the convex power case ($1 < q < 2_s^* - 1$). These two cases will be treated separately.

One also can define a fractional power of the Laplacian using spectral decomposition, as discussed in Chapter 5. The same problem is considered here, but this spectral fractional Laplacian was treated in [22]. Some related problems involving this operator have been studied in [42, 51, 73, 214]. As in [22], the purpose of this chapter is to study the existence of weak solutions for (20.1).

Problems similar to (20.1) have been also studied in the local setting with different elliptic operators. As far as we know, the first example in this direction was given in [111] for the p-Laplacian operator, Δ_p (see Chapter 1). Other results, this time for the Laplacian (or essentially the classical Laplacian) operator, can be found in [2, 8, 22, 39]. More generally, the case of fully nonlinear operators was studied in [66].

It is worth noting here that problem (20.1), with $\lambda = 0$, has no solution whenever Ω is a star-shaped domain. This was proved in [103, 185] using a Pohozaev identity for the operator $(-\Delta)^s$. This fact motivates the perturbation term λu^q, $\lambda > 0$. The results of this chapter were obtained in [23].

20.1 Main results

We now summarize the main results of this chapter. First, in Section 20.2, we will look at problem (20.1) in the concave case $q < 1$, and we will prove the following result:

Theorem 20.1 *Let $s \in (0,1)$, $n > 2s$, and $q \in (0,1)$. Then there exists $0 < \Lambda < +\infty$ such that*

(*a*) *Problem* (20.1) *has no solution for $\lambda > \Lambda$.*
(*b*) *Problem* (20.1) *has a minimal solution for any $0 < \lambda < \Lambda$; moreover, the family of minimal solutions is increasing with respect to λ.*
(*c*) *Problem* (20.1) *has at least one weak solution for $\lambda = \Lambda$.*
(*d*) *Problem* (20.1) *has at least two weak solutions for $0 < \lambda < \Lambda$.*

The convex case is treated in Section 20.3. The corresponding existence result for problem (20.1) is given by the following:

Theorem 20.2 *Let $s \in (0,1)$, $n > 2s$, and $q \in (0, 2_s^* - 1)$. Then problem* (20.1) *admits at least one weak solution provided that either*

- $n > \dfrac{2s(q+3)}{q+1}$ *and $\lambda > 0$, or*
- $n \leq \dfrac{2s(q+3)}{q+1}$ *and λ is sufficiently large.*

Theorem 20.1 corresponds to the nonlocal version of the main result of [8], while Theorem 20.2 may be seen as the nonlocal counterpart of the results obtained for the standard Laplace operator in [46, subsections 2.3–2.5], (see also [111, theorems 3.2 and 3.3] for the case of the p-Laplacian operator). Note, in particular, that when $s = 1$, one has $2s(q+3)/(q+1) = 2(q+3)/(q+1) < 4$ owing to the choice of $q > 1$.

20.1.1 Variational formulation of the problem

To present the weak formulation of (20.1), and taking into account that we are looking for positive solutions, we consider the following Dirichlet problem:

$$\begin{cases} (-\Delta)^s u = \lambda (u_+)^q + (u_+)^{2_s^*-1} & \text{in } \Omega \\ u = 0 & \text{in } \mathbb{R}^n \setminus \Omega, \end{cases} \tag{20.2}$$

where $u_+ := \max\{u, 0\}$ denotes the positive part of u. The crucial observation here is that, by the maximum principle [206, proposition 2.2.8], if u is a solution of (20.2), then u is strictly positive in Ω, and therefore, it is also a solution of (20.1).

To find solutions of (20.2), we use a variational approach. More precisely, we associate with problem (20.2) the energy functional $\mathcal{J}_{s,\lambda,q} : \mathbb{H}_0^s(\Omega) \to \mathbb{R}$, defined as follows:

$$\mathcal{J}_{s,\lambda,q}(u) = \frac{1}{2} \int_{\mathbb{R}^n \times \mathbb{R}^n} \frac{|u(x) - u(y)|^2}{|x - y|^{n+2s}} \, dx \, dy \\ - \frac{\lambda}{q+1} \int_{\Omega} (u_+)(x)^{q+1} \, dx - \frac{1}{2_s^*} \int_{\Omega} (u_+)(x)^{2_s^*} \, dx.$$

Note that $\mathcal{J}_{s,\lambda}$ is C^1 and that its critical points correspond to solutions of (20.2).

In both cases, $q < 1$ and $q > 1$, we use the mountain pass theorem by Ambrosetti and Rabinowitz (see [11]). To do this, we will show that $\mathcal{J}_{s,\lambda,q}$ satisfies a compactness property and has suitable geometric features. The fact that the functional has the suitable geometry is easy to check. Hence, the difficulty in applying the mountain pass theorem lies in proving a local Palais–Smale condition at level $c \in \mathbb{R}$. Moreover, since the Palais–Smale condition does not hold globally, we have to prove that the mountain pass critical level of $\mathcal{J}_{s,\lambda,q}$ lies below the threshold of application of this compactness condition.

In the concave setting, $q < 1$, the idea is to prove the existence of at least two positive solutions for an admissible small range of λ. For this, we will use a contradiction argument inspired by [8]. The proof is divided into several steps: we will first show that we have a solution that is a local minimum for the functional $\mathcal{J}_{s,\lambda,q}$. In the next step, to find a second solution, we will suppose that this local minimum is the only critical point of the functional, and then we will prove a local Palais–Smale condition under a critical level related to the best fractional critical Sobolev constant given in (1.92). Also, we will find a path under this critical level localizing the Sobolev minimizers at the possible concentration on Dirac deltas. These deltas will be obtained by the concentration-compactness result in [162, theorem 1.5] inspired in the classical result by P. L. Lions in [135, 136]. Applying the mountain pass theorem given in [11] and its refined version given in [116], we will reach a contradiction.

In the convex case, $q > 1$, we will also apply the mountain pass theorem to obtain the existence of at least one solution for (20.2) for suitable values of λ depending on the dimension n. As before, we will prove a local Palais–Smale condition in an appropriate range related with the constant S_s defined on (1.92). The strategy to obtain a solution follows the ideas given in [46] (see also [212, 220]) adapted to the nonlocal functional framework.

20.2 The critical and concave case $0 < q < 1$

This section is devoted to the study of problem (20.1) in the case of the exponent $0 < q < 1$. We point out that the result of Theorem 20.1 in the subcritical case could be obtained by the arguments given in this chapter. However, in this subcritical case, the Palais–Smale condition is easier to prove – it is indeed satisfied for any energy level – and the separation of solutions, presented in Lemma 20.4, are not needed. This approach was carried out in [25], where the authors obtained the equivalent to Theorem 20.1 for a related problem using a technique developed in [5].

We begin with the following result that uses, in its proof, a standard comparison method, as well as some ideas given in [8, lemmas 3.1 and 3.4].

Lemma 20.3 *Let* $0 < q < 1$*, and let* Λ *be defined by*

$$\Lambda := \sup\left\{\lambda > 0 \; : \; \textit{problem } (20.1) \textit{ has a weak solution}\right\}. \tag{20.3}$$

Then, $0 < \Lambda < \infty$*, and the critical concave problem* (20.1) *has at least one weak solution for every* $0 < \lambda \le \Lambda$*. Moreover, for* $0 < \lambda < \Lambda$*, we get a family of minimal solutions increasing with respect to* λ*.*

By Lemma 20.3, we easily deduce statements (a)–(c) of Theorem 20.1. Hence, in what follows, we will focus on proving statement (d) of that theorem, that is, on the existence of a second solution for (20.1).

As we said at the beginning of this chapter, to prove the existence of the second solution, we will first show that the minimal solution $u_\lambda > 0$ given by Lemma 20.3 is a local minimum for the functional $\mathcal{J}_{s,\lambda,q}$. For this, following the ideas given in [22], we establish a separation lemma in the topology of the class

$$\mathscr{C}_s(\Omega) := \left\{ w \in C^0(\overline{\Omega}) : \|w\|_{\mathscr{C}_s(\Omega)} := \left\| \frac{w}{\delta^s} \right\|_{L^\infty(\Omega)} < \infty \right\}, \tag{20.4}$$

where $\delta(x) := \text{dist}(x, \partial\Omega)$. Precisely, we have the following result:

Lemma 20.4 *Assume that* $0 < \check{\lambda}_1 < \check{\lambda}_0 < \check{\lambda}_2 < \Lambda$*. Let* $u_{\check{\lambda}_1}$*,* $u_{\check{\lambda}_0}$*, and* $u_{\check{\lambda}_2}$ *be the corresponding minimal solutions to* (20.1)*, for* $\lambda = \check{\lambda}_1, \check{\lambda}_0$*, and* $\check{\lambda}_2$*, respectively. If*

$$Z := \{u \in \mathscr{C}_s(\Omega) | \, u_{\check{\lambda}_1} \le u \le u_{\check{\lambda}_2}\},$$

then there exists $\varepsilon > 0$ *such that*

$$\{u_{\check{\lambda}_0}\} + \varepsilon B_1 \subset Z,$$

with

$$B_1 := \left\{ w \in C^0(\overline{\Omega}) : \left\| \frac{w}{\delta^s} \right\|_{L^\infty(\Omega)} < 1 \right\}.$$

Proof Let u be an arbitrary solution of (20.1) for $0 < \lambda < \Lambda$. Then, by Hopf's lemma (see [53, proposition 2.7] and [184, lemma 3.2]), there exists a positive constant c such that

$$u(x) \ge c\delta(x)^s, \quad x \in \Omega. \tag{20.5}$$

However, by [184, proposition 1.1], we get that there exists a positive constant C such that

$$u(x) \le C\delta(x)^s \quad x \in \Omega. \tag{20.6}$$

Thus, by (20.5) and (20.6), the assertion follows. □

Using this preceding result, we now obtain a local minimum of the functional $\mathcal{J}_{s,\lambda,q}$ in the $\mathscr{C}_s(\Omega)$-topology (this is the first step to get a local minimum in $\mathbb{H}^s_0(\Omega)$):

Lemma 20.5 *For all $\lambda \in (0,\Lambda)$, the minimal solution u_λ is a local minimum of the functional $\mathcal{J}_{s,\lambda,q}$ in the $\mathscr{C}_s$-topology.*

Proof The proof follows in a similar way as in [8] (see also [22, lemma 3.3]). In our case, we have to consider the nonlocal operator $(-\Delta)^s$ instead of $-\Delta$ and the space $\mathscr{C}_s(\Omega)$ instead of $C^1_0(\Omega)$. We omit the details. □

Now, to prove that we already have a minimum in the space $\mathbb{H}^s_0(\Omega)$, we will show that the result obtained by Brezis and Nirenberg in [46] is also valid in our context.

Proposition 20.6 Let $z_0 \in \mathbb{H}^s_0(\Omega)$ be a local minimum of $\mathcal{J}_{s,\lambda,q}$ in $\mathscr{C}_s(\Omega)$; that is, there exists $r_1 > 0$ such that

$$\mathcal{J}_{s,\lambda,q}(z_0) \le \mathcal{J}_{s,\lambda,q}(z_0+z) \quad \forall z \in \mathscr{C}_s(\Omega) \text{ with } \|z\|_{\mathscr{C}_s(\Omega)} \le r_1. \tag{20.7}$$

Then z_0 is also a local minimum of $\mathcal{J}_{s,\lambda,q}$ in $\mathbb{H}^s_0(\Omega)$; that is, there exists $r_2 > 0$ such that

$$\mathcal{J}_{s,\lambda,q}(z_0) \le \mathcal{J}_{s,\lambda,q}(z_0+z) \quad \forall z \in \mathbb{H}^s_0(\Omega) \text{ with } \|z\|_{\mathbb{H}^s_0(\Omega)} \le r_2.$$

Proof We follow the ideas given in [22, theorem 5.1]. Let z_0 be as in (20.7), and set, for $\varepsilon > 0$,

$$B_\varepsilon(z_0) := \left\{ z \in \mathbb{H}^s_0(\Omega) : \|z - z_0\|_{\mathbb{H}^s_0(\Omega)} \le \varepsilon \right\}.$$

Now we argue by contradiction, and we suppose that, for every $\varepsilon > 0$, we have

$$\min_{v \in B_\varepsilon(z_0)} \mathcal{J}_{s,\lambda}(v) < \mathcal{J}_{s,\lambda,q}(z_0). \tag{20.8}$$

We pick $v_\varepsilon \in B_\varepsilon(z_0)$ such that $\min_{v \in B_\varepsilon(z_0)} \mathcal{J}_{s,\lambda}(v) = \mathcal{J}_{s,\lambda,q}(v_\varepsilon)$. The existence of v_ε comes from a standard argument of weak lower semicontinuity. We want to prove that

$$v_\varepsilon \to z_0 \quad \text{in} \quad \mathscr{C}_s(\Omega) \quad \text{as } \varepsilon \searrow 0 \tag{20.9}$$

because this would imply that there is $z \in \mathscr{C}_s(\Omega)$, arbitrarily close to z_0 in the metric of $\mathscr{C}_s(\Omega)$ (in fact, $z = v_\varepsilon$, for some ε), such that

$$\mathcal{J}_{s,\lambda,q}(z) < \mathcal{J}_{s,\lambda,q}(z_0).$$

This contradicts our hypothesis (20.7).

Let $0 < \varepsilon \ll 1$. Note that the Euler–Lagrange equation satisfied by v_ε involves a Lagrange multiplier ξ_ε such that

$$\langle \mathcal{J}'_{s,\lambda}(v_\varepsilon), \varphi \rangle = \xi_\varepsilon \langle v_\varepsilon, \varphi \rangle_{\mathbb{H}^s_0(\Omega)} \quad \forall \varphi \in \mathbb{H}^s_0(\Omega). \tag{20.10}$$

As a consequence, since v_ε is a minimum of $\mathscr{J}_{s,\lambda}$ in $B_\varepsilon(z_0)$, we have

$$\xi_\varepsilon = \frac{\langle \mathscr{J}'_{s,\lambda}(v_\varepsilon), v_\varepsilon\rangle}{\|v_\varepsilon\|^2_{\mathbb{H}^s_0(\Omega)}} \le 0. \tag{20.11}$$

By (20.10), we easily get that v_ε satisfies

$$\begin{cases} (-\Delta)^s v_\varepsilon = \dfrac{f_\lambda(v_\varepsilon)}{1-\xi_\varepsilon} =: f^\varepsilon_\lambda(v_\varepsilon) & \text{in } \Omega \\ v_\varepsilon = 0 & \text{in } \mathbb{R}^n \setminus \Omega, \end{cases}$$

where $f_\lambda(t) := \lambda(t_+)^q + (t_+)^{2^*_s-1}$.

Since $v_\varepsilon > 0$ and $\|v_\varepsilon\|_{\mathbb{H}^s_0(\Omega)} \le C$, by Proposition 4.11, there exists a constant $C_1 > 0$ independent of ε such that

$$\|v_\varepsilon\|_{L^\infty(\Omega)} \le C_1.$$

Moreover, by (20.11), it follows that $\|f^\varepsilon_\lambda(v_\varepsilon)\|_{L^\infty(\Omega)} \le C$. Therefore, by [184, proposition 1.1] (see also Proposition 4.5), we get that $\|v_\varepsilon\|_{C^s(\overline{\Omega})} \le C_2$, for some C_2 independent of ε. Here C^s.denotes the space of Hölder continuous functions with exponent s.

Thus, by the Ascoli–Arzelà theorem, there exists a subsequence, still denoted by v_ε, such that $v_\varepsilon \to z_0$ uniformly as $\varepsilon \searrow 0$. Moreover, by [184, theorem 1.2], we obtain that, for a suitable positive constant C,

$$\left\| \frac{v_\varepsilon - z_0}{\delta^s} \right\|_{L^\infty(\Omega)} \le C \sup_\Omega |f^\varepsilon_\lambda(v_\varepsilon) - f_\lambda(z_0)|.$$

Since the latter tends to zero as $\varepsilon \searrow 0$, (20.9) is proved. □

Lemma 20.5 and Proposition 20.6 provide us with the existence of a positive local minimum in $\mathbb{H}^s_0(\Omega)$ of $\mathscr{J}_{s,\lambda}$ that will be denoted by u_0. We now make a translation as in [8] to simplify the calculations. For $0 < \lambda < \Lambda$, we consider the functions

$$g_\lambda(x,t) = \begin{cases} \lambda(u_0+t)^q - \lambda u_0^q + (u_0+t)^{2^*_s-1} - u_0^{2^*_s-1} & \text{if } t \ge 0 \\ 0, & \text{if } t < 0, \end{cases} \tag{20.12}$$

and

$$G_\lambda(x,\xi) = G_\lambda(\xi) = \int_0^\xi g_\lambda(x,t)\,dt. \tag{20.13}$$

The associated energy functional $\widetilde{\mathscr{J}}_{s,\lambda,q} : \mathbb{H}^s_0(\Omega) \to \mathbb{R}$ is given by

$$\widetilde{\mathscr{J}}_{s,\lambda,q}(u) = \frac{1}{2}\|u\|^2_{\mathbb{H}^s_0(\Omega)} - \int_\Omega G_\lambda(x,u(x))\,dx. \tag{20.14}$$

Note that because $u \in \mathbb{H}^s_0(\Omega)$, the functional $\widetilde{\mathscr{J}}_{s,\lambda,q}$ is well defined.

Finally, we define the translate problem as follows:

$$\begin{cases} (-\Delta)^s u = g_\lambda(x,u) & \text{in } \Omega \\ u = 0 & \text{in } \mathbb{R}^n \setminus \Omega. \end{cases} \tag{20.15}$$

We know that if $\widetilde{u} \not\equiv 0$ is a critical point of $\widetilde{\mathcal{J}}_{s,\lambda,q}$, then it is a solution of (20.15), and by the maximum principle (see [206, proposition 2.2.8]), this implies that $\widetilde{u} > 0$. Therefore, $u = u_0 + \widetilde{u} > 0$ will be a second solution of (20.2) and, consequently, a second solution of (20.1). Hence, to prove statement (d) of Theorem 20.1, it is enough to study the existence of a nontrivial critical point for $\widetilde{\mathcal{J}}_{s,\lambda,q}$.

First, we have the following result:

Lemma 20.7 *The function $u \equiv 0$ is a local minimum of $\widetilde{\mathcal{J}}_{s,\lambda,q}$ in $X_0^s(\Omega)$.*

Proof The proof follows along the lines of [8, lemma 4.2] (see also [22, lemma 3.4]), so we omit the details. □

20.2.1 The Palais–Smale condition for $\widetilde{\mathcal{J}}_{s,\lambda,q}$

In this subsection, assuming that we have a unique critical point, we prove that the functional $\widetilde{\mathcal{J}}_{s,\lambda,q}$ satisfies a local Palais–Smale condition. The main tool for proving this fact is an extension of the concentration-compactness principle by Lions (see [135, 136]) for nonlocal fractional operators given in [162, theorem 1.5].

We also need some technical results related to the behavior of the fractional Laplacian of a product. We start with the following lemma:

Lemma 20.8 *Let ϕ be a regular function that satisfies*

$$|\phi(x)| \leq \frac{\tilde{C}}{1+|x|^{n+s}} \quad x \in \mathbb{R}^n \tag{20.16}$$

and

$$|\nabla\phi(x)| \leq \frac{\tilde{C}}{1+|x|^{n+s+1}} \quad x \in \mathbb{R}^n, \tag{20.17}$$

for some $\tilde{C} > 0$. Let $B : \mathbb{H}_0^{s/2}(\Omega) \times \mathbb{H}_0^{s/2}(\Omega) \to \mathbb{R}$ be the bilinear form defined by

$$B(f,g)(x) := 2\int_{\mathbb{R}^n} \frac{(f(x)-f(y))(g(x)-g(y))}{|x-y|^{n+s}}\,dy. \tag{20.18}$$

Then, for every $s \in (0,1)$, there exist positive constants C_1 and C_2 such that, for $x \in \mathbb{R}^n$, one has

$$|(-\Delta)^{s/2}\phi(x)| \leq \frac{C_1}{1+|x|^{n+s}}$$

and

$$|B(\phi,\phi)(x)| \leq \frac{C_2}{1+|x|^{n+s}}.$$

Proof Let

$$I(x) := \int_{\mathbb{R}^n} \frac{|\phi(x)-\phi(y)|}{|x-y|^{n+s}}\,dy.$$

For any $x \in \mathbb{R}^n$, it is clear that

$$|(-\Delta)^{s/2}\phi(x)| \leq 2I(x).$$

Also, since $|\phi(x)| \le \tilde{C}$, we have

$$|B(\phi,\phi)(x)| \le 2\tilde{C}I(x).$$

Hence, it suffices to prove that

$$I(x) \le \frac{C}{1+|x|^{n+s}} \quad \forall x \in \mathbb{R}^n, \tag{20.19}$$

for a suitable positive constant C.

Since ϕ is a regular function, for $|x| < 1$, we obtain that

$$\begin{aligned} I(x) &\le \|\nabla\phi\|_{L^\infty(\mathbb{R}^n)} \int_{|y|<2} \frac{dy}{|x-y|^{n+s-1}} + C\int_{|y|\ge 2} \frac{dy}{|y|^{n+s}} \\ &\le C \\ &\le \frac{C}{1+|x|^{n+s}}, \end{aligned} \tag{20.20}$$

so (20.19) is proved. Now let $|x| \ge 1$. Then

$$I(x) := I_{A_1}(x) + I_{A_2}(x) + I_{A_3}(x), \tag{20.21}$$

where

$$I_{A_i}(x) := \int_{A_i} \frac{|\phi(x)-\phi(y)|}{|x-y|^{n+s}}\,dy, \quad i = 1,2,3,$$

with

$$A_1 := \left\{y : |x-y| \le \frac{|x|}{2}\right\}, \quad A_2 := \left\{y : |x-y| > \frac{|x|}{2}, |y| \le 2|x|\right\},$$

and

$$A_3 := \left\{y : |x-y| > \frac{|x|}{2}, |y| > 2|x|\right\}.$$

Therefore, since for $|x| \ge 1$ and $y \in A_1$, $|\phi(x)-\phi(y)| \le |\nabla\phi(\xi)||x-y|$, with $|x|/2 \le |\xi| \le 3/2|x|$, by (20.17), we obtain that

$$I_{A_1}(x) \le \frac{C}{|x|^{n+s+1}} \int_{A_1} \frac{dy}{|x-y|^{n+s-1}} \le C|x|^{-(n+2s)}. \tag{20.22}$$

Using that, for any x, $y \in \mathbb{R}^n$, the inequality

$$|\phi(x)| + |\phi(y)| \le \frac{C}{1+\min\{|x|^{n+s}, |y|^{n+s}\}}$$

holds true, we get

$$I_{A_2}(x) \le \frac{C}{|x|^{n+s}} \int_{A_2} \frac{dy}{(1+|y|^{n+s})} \le C|x|^{-(n+s)} \tag{20.23}$$

and

$$I_{A_3}(x) \le \frac{C}{|x|^{n+s}} \int_{A_3} \frac{dy}{|y|^{n+s}} \le C|x|^{-(n+2s)}. \tag{20.24}$$

Note that the last estimate follows from the fact that $(x,y) \in A_3$ implies $|x-y| \ge |y|/2$. Then, by (20.21)–(20.24), we get that

$$I(x) \le C|x|^{-(n+s)} \le \frac{C}{1+|x|^{n+s}} \quad |x| \ge 1. \tag{20.25}$$

Hence, by (20.20) and (20.25), we conclude (20.19). □

To establish the next auxiliary result, we consider a radial, nonincreasing cutoff function $\phi \in C_0^\infty(\mathbb{R}^n)$ and

$$\phi_\varepsilon(x) := \phi(x/\varepsilon) \quad x \in \mathbb{R}^n. \tag{20.26}$$

Now we get the following result:

Lemma 20.9 *Let $\{z_j\}_{j\in\mathbb{N}}$ be a uniformly bounded sequence in $\mathbb{H}_0^s(\Omega)$, and let ϕ_ε be the function defined in* (20.26). *Then*

$$\lim_{\varepsilon\to 0}\lim_{j\to+\infty}\left|\int_{\mathbb{R}^n} z_j(x)(-\Delta)^{s/2}\phi_\varepsilon(x)(-\Delta)^{s/2}z_j(x)\,dx\right| = 0. \tag{20.27}$$

Proof First, note that as a consequence of the fact that $\{z_j\}_{j\in\mathbb{N}}$ is uniformly bounded in the reflexive space $\mathbb{H}_0^s(\Omega)$, say, by M, we get that there exists $z \in \mathbb{H}_0^s(\Omega)$ such that, up to a subsequence,

$$\begin{aligned} z_j &\rightharpoonup z \quad \text{weakly in } \mathbb{H}_0^s(\Omega), \\ z_j &\to z \quad \text{strongly in } L^r(\Omega), \quad 1 \le r < 2_s^*, \\ z_j &\to z \quad \text{a.e. in } \Omega \end{aligned} \tag{20.28}$$

as $j \to +\infty$. Also, it is clear that

$$|(-\Delta)^{s/2}\phi_\varepsilon(x)| = \varepsilon^{-s}\left|((-\Delta)^{s/2}\phi)\left(\frac{x}{\varepsilon}\right)\right| \le C\varepsilon^{-s}. \tag{20.29}$$

Therefore, defining

$$I_1 := \left|\int_{\mathbb{R}^n} z_j(x)(-\Delta)^{s/2}\phi_\varepsilon(x)(-\Delta)^{s/2}z_j(x)\,dx\right|,$$

from (20.29) and the fact that $\|z_j\|_{\mathbb{H}_0^s(\Omega)} < M$, we get

$$\begin{aligned} I_1 &\le \|(-\Delta)^{s/2}z_j\|_{L^2(\mathbb{R}^n)}\|z_j(-\Delta)^{s/2}\phi_\varepsilon\|_{L^2(\Omega)} \\ &\le M\|(z_j-z)(-\Delta)^{s/2}\phi_\varepsilon\|_{L^2(\Omega)} + M\|z(-\Delta)^{s/2}\phi_\varepsilon\|_{L^2(\Omega)} \\ &\le C\varepsilon^{-s}\|z_j-z\|_{L^2(\Omega)} + M\|z(-\Delta)^{s/2}\phi_\varepsilon\|_{L^2(\Omega)}. \end{aligned} \tag{20.30}$$

Since $\|z\|_{\mathbb{H}_0^s(\Omega)} \le M$, then $\|z\|_{L^{2_s^*}(\Omega)} \le C$; that is, $z^2 \in L^{n/(n-2s)}(\Omega)$. Hence, for every $\rho > 0$, there exits $\eta \in C_0^\infty(\Omega)$ such that

$$\|z^2 - \eta\|_{L^{\frac{n}{n-2s}}(\Omega)} \le \rho. \tag{20.31}$$

Then, by (20.29), (20.31), and the Hölder inequality, with $p = n/n - 2s$, we obtain that

$$\begin{aligned}
\|z(-\Delta)^{s/2}\phi_\varepsilon\|^2_{L^2(\Omega)} &\le \int_{\mathbb{R}^n} |z^2(x) - \eta(x)||(-\Delta)^{s/2}\phi_\varepsilon(x)|^2\,dx \\
&\quad + \int_{\mathbb{R}^n} |\eta(x)||(-\Delta)^{s/2}\phi_\varepsilon(x)|^2\,dx \\
&\le \|z^2 - \eta\|_{L^{n/(n-2s)}(\Omega)}\|(-\Delta)^{s/2}\phi_\varepsilon\|^2_{L^{(n/s)}(\mathbb{R}^n)} \\
&\quad + \|\eta\|_{L^\infty(\Omega)}\|(-\Delta)^{s/2}\phi_\varepsilon\|^2_{L^2(\mathbb{R}^n)} \\
&\le \rho\varepsilon^{-2s}\left(\int_{\mathbb{R}^n} \left|((-\Delta)^{s/2}\phi)\left(\frac{x}{\varepsilon}\right)\right|^{n/s} dx\right)^{2s/n} \\
&\quad + C\varepsilon^{-2s}\int_{\mathbb{R}^n} \left|((-\Delta)^{s/2}\phi)\left(\frac{x}{\varepsilon}\right)\right|^2 dx \\
&\le \rho\left(\int_{\mathbb{R}^n} |(-\Delta)^{s/2}\phi(z)|^{n/s}\,dz\right)^{2s/n} \\
&\quad + C\varepsilon^{n-2s}\int_{\mathbb{R}^n} |(-\Delta)^{s/2}\phi(z)|^2\,dz \\
&\le C\rho + C\varepsilon^{n-2s}.
\end{aligned} \tag{20.32}$$

Hence, using (20.28), from (20.30), (20.32), and the fact that $n > 2s$, it follows that

$$\lim_{\varepsilon\to 0}\lim_{j\to+\infty} I_1 \le \lim_{\varepsilon\to 0} C\left(\rho + \varepsilon^{n-2s}\right)^{\frac{1}{2}} = C\rho^{\frac{1}{2}}.$$

Since $\rho > 0$ is fixed but arbitrarily small, we conclude the proof. □

Also, we have the following:

Lemma 20.10 *Let $\{z_j\}_{j\in\mathbb{N}}$ be a uniformly bounded sequence in $\mathbb{H}_0^s(\Omega)$, and let ϕ_ε be the function defined in* (20.26). *Then*

$$\lim_{\varepsilon\to 0}\lim_{j\to+\infty}\left|\int_{\mathbb{R}^n} (-\Delta)^{s/2}z_j(x)B(z_j,\phi_\varepsilon)(x)\,dx\right| = 0, \tag{20.33}$$

where B is defined in (20.18).

Proof Let

$$I_2 := \left| \int_{\mathbb{R}^n} (-\Delta)^{s/2} z_j(x) B(z_j, \phi_\varepsilon)(x)\, dx \right|.$$

Since $\|z_j\|_{\mathbb{H}_0^s(\Omega)} \leq M$, then

$$\begin{aligned} I_2 &\leq M \|B(z_j, \phi_\varepsilon)\|_{L^2(\mathbb{R}^n)} \\ &\leq M \|B(z_j - z, \phi_\varepsilon)\|_{L^2(\mathbb{R}^n)} + M \|B(z, \phi_\varepsilon)\|_{L^2(\mathbb{R}^n)}, \end{aligned} \tag{20.34}$$

where z is, as in Lemma 20.9, the weak limit of the sequence $\{z_j\}_{j\in\mathbb{N}}$ in $\mathbb{H}_0^s(\Omega)$. We estimate each of the terms in inequality (20.33).

Let

$$\psi(x) := \frac{1}{1+|x|^{n+s}} \quad \text{and} \quad \psi_\varepsilon(x) := \psi\left(\frac{x}{\varepsilon}\right), \quad x \in \mathbb{R}^n. \tag{20.35}$$

By Lemma 20.8 applied to ϕ, we note that

$$B(\phi_\varepsilon, \phi_\varepsilon)(x) = \varepsilon^{-s} B(\phi, \phi)\left(\frac{x}{\varepsilon}\right) \leq C\varepsilon^{-s} \psi\left(\frac{x}{\varepsilon}\right) = C \frac{\varepsilon^{-s}}{1+\left|\frac{x}{\varepsilon}\right|^{n+s}} \leq C\varepsilon^{-s}. \tag{20.36}$$

Therefore, by the Cauchy–Schwarz inequality and (20.36), it follows that

$$\begin{aligned} \|B(z_j - z, \phi_\varepsilon)\|^2_{L^2(\mathbb{R}^n)} &\leq \int_{\mathbb{R}^n} B(z_j - z, z_j - z)(x) B(\phi_\varepsilon, \phi_\varepsilon)(x)\, dx \\ &\leq C\varepsilon^{-s} \int_{\mathbb{R}^n} B(z_j - z, z_j - z)(x)\, dx \\ &= C\varepsilon^{-s} \|z_j - z\|^2_{X_0^{\frac{s}{2}}(\Omega)} \\ &= C\varepsilon^{-s} \int_{\mathbb{R}^n} (z_j - z)(x)(-\Delta)^{s/2}(z_j - z)(x)\, dx \\ &\leq C\varepsilon^{-s} \|z_j - z\|_{L^2(\Omega)} \|(-\Delta)^{s/2}(z_j - z)\|_{L^2(\mathbb{R}^n)} \\ &\leq C\varepsilon^{-s} \|z_j - z\|_{L^2(\Omega)}. \end{aligned} \tag{20.37}$$

However, for a suitable function f, we have that

$$\begin{aligned} \int_{\mathbb{R}^n} z^2(x)(-\Delta)^{s/2} f(x)\, dx &= \int_{\mathbb{R}^n} f(x)(-\Delta)^{s/2} z^2(x)\, dx \\ &= \int_{\mathbb{R}^n} f(x) \left(2z(x)(-\Delta)^{s/2} z(x) - B(z,z)(x)\right) dx. \end{aligned} \tag{20.38}$$

Then, arguing as in (20.37) and applying (20.38), with $f := \psi_\varepsilon(x)$, from (20.36), we get that

$$\begin{aligned}
&\|B(z,\phi_\varepsilon)\|^2_{L^2(\mathbb{R}^n)} \\
&\quad \le \int_{\mathbb{R}^n} B(z,z)(x) B(\phi_\varepsilon,\phi_\varepsilon)(x)\,dx \\
&\quad \le C\varepsilon^{-s} \int_{\mathbb{R}^n} B(z,z)(x)\psi_\varepsilon(x)\,dx \qquad (20.39)\\
&\quad \le C\varepsilon^{-s} \int_{\mathbb{R}^n} \left(-z(x)^2(-\Delta)^{s/2}\psi_\varepsilon(x) + 2z(x)\psi_\varepsilon(x)(-\Delta)^{s/2}z(x)\right) dx \\
&\quad := I_{2,1} + I_{2,2}.
\end{aligned}$$

We estimate now $I_{2,1}$ and $I_{2,2}$ separately. Let $\rho > 0$. By Lemma 20.8 applied to ψ and (20.29), it follows that

$$\begin{aligned}
|I_{2,1}| &\le C\varepsilon^{-2s} \int_{\mathbb{R}^n} z(x)^2 \left| \left((-\Delta)^{s/2}\psi\right)\left(\frac{x}{\varepsilon}\right)\right| dx \\
&\le C\varepsilon^{-2s} \int_{\mathbb{R}^n} z(x)^2 \psi\left(\frac{x}{\varepsilon}\right) dx \qquad (20.40)\\
&\le C\varepsilon^{-2s} \int_{\mathbb{R}^n} (z^2-\eta)(x)\psi_\varepsilon(x)\,dx + \varepsilon^{-2s}\int_{\mathbb{R}^n} \eta(x)\psi_\varepsilon(x)\,dx,
\end{aligned}$$

where $\eta \in C_0^\infty(\Omega)$ is the function that satisfies (20.31). Then, from (20.40), we obtain

$$\begin{aligned}
|I_{2,1}| &\le C\rho\varepsilon^{-2s}\|\psi_\varepsilon\|_{L^{\frac{n}{2s}}(\mathbb{R}^n)} + C\varepsilon^{-2s}\|\eta\|_{L^\infty(\mathbb{R}^n)}\|\psi_\varepsilon\|_{L^1(\mathbb{R}^n)} \\
&\le C\rho\|\psi\|_{L^{\frac{n}{2s}}(\mathbb{R}^n)} + C\varepsilon^{n-2s}\|\eta\|_{L^\infty(\mathbb{R}^n)}\|\psi\|_{L^1(\mathbb{R}^n)}.
\end{aligned} \qquad (20.41)$$

However,

$$|I_{2,2}| \le C\varepsilon^{-s}\|(-\Delta)^{s/2}z\|_{L^2(\mathbb{R}^n)}\|z\psi_\varepsilon\|_{L^2(\Omega)} \le C\varepsilon^{-s}\|z\psi_\varepsilon\|_{L^2(\Omega)}. \qquad (20.42)$$

Therefore, by (20.31), we get

$$\begin{aligned}
|I_{2,2}|^2 &\le C\varepsilon^{-2s}\left(\int_\Omega |(z^2-\eta)(x)|\,|\psi_\varepsilon(x)|^2\,dx + \int_{\mathbb{R}^n} \eta\,|\psi_\varepsilon(x)|^2\,dx\right) \\
&\le C\varepsilon^{-2s}\left(\rho\|\psi_\varepsilon\|^2_{L^{\frac{n}{s}}(\mathbb{R}^n)} + \|\eta\|_{L^\infty(\mathbb{R}^n)}\|\psi_\varepsilon\|^2_{L^2(\mathbb{R}^n)}\right) \qquad (20.43)\\
&\le C\rho\|\psi\|^2_{L^{\frac{n}{s}}(\mathbb{R}^n)} + C\varepsilon^{n-2s}\|\eta\|_{L^\infty(\mathbb{R}^n)}\|\psi\|^2_{L^2(\mathbb{R}^n)}.
\end{aligned}$$

Then, by (20.41) and (20.43), it follows from (20.39) that

$$\|B(z,\phi_\varepsilon)\|^2_{L^2(\mathbb{R}^n)} \leq C\left(\rho+\rho^{\frac{1}{2}}\right)+C\left(\varepsilon^{n-2s}+\varepsilon^{\frac{n-2s}{2}}\right). \tag{20.44}$$

Hence, from (20.28), (20.37), and (20.44), since $n > 2s$, we obtain

$$\lim_{\varepsilon\to 0}\lim_{j\to+\infty}\left(\|B(z_j-z,\phi_\varepsilon)\|^2_{L^2(\mathbb{R}^n)}+\|B(z,\phi_\varepsilon)\|^2_{L^2(\mathbb{R}^n)}\right)$$

$$\leq \lim_{\varepsilon\to 0} C\left(\rho^{1/2}+\varepsilon^{(n-2s)/2}\right)=C\rho^{1/2}.$$

Thus, since ρ is an arbitrary positive value,

$$\lim_{\varepsilon\to 0}\lim_{j\to+\infty}\left(\|B(z_j-z,\phi_\varepsilon)\|^2_{L^2(\mathbb{R}^n)}+\|B(z,\phi_\varepsilon)\|^2_{L^2(\mathbb{R}^n)}\right)=0. \tag{20.45}$$

Finally, by (20.34) and (20.45), we conclude that

$$\lim_{\varepsilon\to 0}\lim_{j\to+\infty}|I_2|=0,$$

and this ends the proof. □

Now we can prove the main result of this subsection.

Lemma 20.11 *If $u\equiv 0$ is the only critical point of $\widetilde{\mathcal{J}}_{s,\lambda,q}$ in $\mathbb{H}^s_0(\Omega)$, then $\widetilde{\mathcal{J}}_{s,\lambda,q}$ satisfies the Palais–Smale condition at level c, provided that $c < c^*$, where c^* is defined as*

$$c^* := \frac{s}{n}S_s^{n/2s}. \tag{20.46}$$

Here S_s denotes the Sobolev constant defined in (1.92).

Proof Let $\{u_j\}_{j\in\mathbb{N}}$ be a Palais–Smale sequence for $\widetilde{\mathcal{J}}_{s,\lambda,q}$ verifying

$$\widetilde{\mathcal{J}}_{s,\lambda,q}(u_j)\to c<c^* \quad \text{and} \quad \widetilde{\mathcal{J}}'_{s,\lambda,q}(u_j)\to 0 \tag{20.47}$$

as $j\to+\infty$. Then, since there exists $M>0$ such that $\|u_j\|_{\mathbb{H}^s_0(\Omega)}\leq M$, for any $j\in\mathbb{N}$ and, by hypothesis, $u\equiv 0$ is the unique critical point of $\widetilde{\mathcal{J}}_{s,\lambda,q}$, it follows that

$$\begin{aligned} &u_j \rightharpoonup 0 \quad \text{weakly in } \mathbb{H}^s_0(\Omega),\\ &u_j\to 0 \quad \text{strongly in } L^r(\Omega),\ 1\leq r<2^*_s,\\ &u_j\to 0 \quad \text{a.e. in } \Omega \end{aligned} \tag{20.48}$$

as $j\to+\infty$.

Also, since u_0 is a critical point of $\mathcal{J}_{s,\lambda,q}$, we have that

$$\begin{aligned}\mathcal{J}_{s,\lambda,q}(z_j) = {} & \widetilde{\mathcal{J}}_{s,\lambda,q}(u_j) + \mathcal{J}_{s,\lambda,q}(u_0) + \frac{\lambda}{q+1}\int_\Omega (u_0 + (u_j)_+)(x)^{q+1}\,dx \\ & + \lambda \int_\Omega u_0(x)^q(u_j - (u_j)_+)(x)\,dx \\ & - \frac{\lambda}{q+1}\int_\Omega (u_0 + u_j)_+(x)^{q+1}\,dx \\ & + \frac{1}{2_s^*}\int_\Omega (u_0 + (u_j)_+)(x)^{2_s^*}\,dx \\ & + \int_\Omega \left(u_0(x)^{2_s^*-1}(u_j - (u_j)_+)(x) - \frac{1}{2_s^*}(u_0 + u_j)_+(x)^{2_s^*} \right) dx \\ \leq {} & \widetilde{\mathcal{J}}_{s,\lambda,q}(u_j) + \mathcal{J}_{s,\lambda,q}(u_0),\end{aligned} \tag{20.49}$$

where

$$z_j = u_j + u_0. \tag{20.50}$$

Moreover, for every $\varphi \in \mathbb{H}_0^s(\Omega)$,

$$\begin{aligned}\langle \mathcal{J}'_{s,\lambda,q}(z_j), \varphi\rangle = {} & \langle \widetilde{\mathcal{J}}'_{s,\lambda,q}(u_j), \varphi\rangle \\ & + \int_\Omega \left(\lambda (u_0 + (u_j)_+)(x)^q + (u_0 + (u_j)_+)(x)^{2_s^*-1}\right)\varphi(x)\,dx \\ & - \int_\Omega \left(\lambda (u_0 + u_j)_+(x)^q + (u_0 + u_j)_+(x)^{2_s^*-1}\right)\varphi(x)\,dx.\end{aligned} \tag{20.51}$$

Then, by (20.47), (20.48), and (20.51), we obtain that

$$\mathcal{J}'_{s,\lambda}(z_j) \to 0 \tag{20.52}$$

as $j \to +\infty$.

From (20.49) and (20.52), we get that the sequence $\{z_j\}_{j\in\mathbb{N}}$ is uniformly bounded in $\mathbb{H}_0^s(\Omega)$. As a consequence of this, and taking into account that $u \equiv 0$ is the unique critical point of $\mathcal{J}_{s,\lambda,q}$, up to a subsequence, we get that

$$\begin{aligned} z_j &\rightharpoonup u_0 \quad \text{weakly in } \mathbb{H}_0^s(\Omega), \\ z_j &\to u_0 \quad \text{strongly in } L^r(\Omega),\ 1 \leq r < 2_s^*, \\ z_j &\to u_0 \quad \text{a.e. in } \Omega \end{aligned} \tag{20.53}$$

as $j \to +\infty$.

Following [142], it can be proved that $\mathbb{H}_0^s(\Omega)$ also could be defined as the closure of $C_0^\infty(\Omega)$ with respect to the $\mathbb{H}_0^s(\Omega)$-norm (see also Chapter 2). Hence, applying [162, theorem 1.5], we have that there exist an index set $I \subseteq \mathbb{N}$, a sequence of points $\{x_k\}_{k\in I}\subset\Omega$, and two sequences of nonnegative real numbers $\{\mu_k\}_{k\in I}$ $\{\nu_k\}_{k\in I}$ such that

$$|(-\Delta)^{s/2}(z_j)_+|^2 \to \mu \ge |(-\Delta)^{s/2}u_0|^2 + \sum_{k\in I}\mu_k\delta_{x_k}. \tag{20.54}$$

Moreover,

$$|(z_j)_+|^{2_s^*} \to \nu = |u_0|^{2_s^*} + \sum_{k\in I}\nu_k\delta_{x_k}, \tag{20.55}$$

in the sense of measures, with

$$\nu_k \le S_s^{-2_s^*/2}\mu_k^{2_s^*/2} \quad \text{for every } k\in I. \tag{20.56}$$

Here δ_{x_k} denotes the Dirac delta at x_k, while S_s is the constant given in (1.92).

We fix $k_0\in I$, and we consider $\phi\in C_0^\infty(\mathbb{R}^n)$ a nonincreasing cutoff function satisfying

$$\phi = 1 \text{ in } B_1(x_{k_0}) \quad \text{and} \quad \phi = 0 \text{ in } B_2(x_{k_0})^c.$$

Now set

$$\phi_\varepsilon(x) = \phi(x/\varepsilon) \quad x\in\mathbb{R}^n. \tag{20.57}$$

Taking the derivative of the identity given in (4.35) (see also Lemma 1.26), for any $u,\varphi\in\mathbb{H}_0^s(\Omega)$, we obtain that

$$\int_{\mathbb{R}^n\times\mathbb{R}^n}\frac{(u(x)-u(y))(\varphi(x)-\varphi(y))}{|x-y|^{n+2s}}\,dx\,dy = \int_{\mathbb{R}^n}\varphi(x)(-\Delta)^s u(x)\,dx. \tag{20.58}$$

Then, using $\phi_\varepsilon(z_j)_+$ as a test function in (20.52), by (20.58) and the fact that

$$\int_{\mathbb{R}^n}(\phi_\varepsilon(z_j)_+)(x)(-\Delta)^s z_j(x)\,dx \ge \int_{\mathbb{R}^n}(\phi_\varepsilon(z_j)_+)(x)(-\Delta)^s(z_j)_+(x)\,dx,$$

we have that

$$\begin{aligned} 0\ge \lim_{j\to+\infty}\Bigg(&\int_{\mathbb{R}^n}(\phi_\varepsilon(z_j)_+)(x)(-\Delta)^s(z_j)_+(x)\,dx \\ &-\left(\lambda\int_{B_{2\varepsilon}(x_{k_0})}(z_j)_+(x)^{q+1}\phi_\varepsilon(x)\,dx + \int_{B_{2\varepsilon}(x_{k_0})}(z_j)_+(x)^{2_s^*}\phi_\varepsilon(x)\,dx\right)\Bigg). \end{aligned}$$

Hence,

$$\begin{aligned} \lim_{j\to+\infty}\Bigg(&\int_{\mathbb{R}^n}(z_j)_+(x)(-\Delta)^{s/2}(z_j)_+(x)(-\Delta)^{s/2}\phi_\varepsilon(x)\,dx \\ &-2\int_{\mathbb{R}^n}(-\Delta)^{s/2}(z_j)_+(x)\int_{\mathbb{R}^n}\frac{(\phi_\varepsilon(x)-\phi_\varepsilon(y))((z_j)_+(x)-(z_j)_+(y))}{|x-y|^{n+s}}\,dx\,dy\Bigg) \\ \le \lim_{j\to+\infty}\Bigg(&\lambda\int_{B_{2\varepsilon}(x_{k_0})}(z_j)_+(x)^{q+1}\phi_\varepsilon(x)\,dx + \int_{B_{2\varepsilon}(x_{k_0})}(z_j)_+(x)^{2_s^*}\phi_\varepsilon(x)\,dx \\ &-\int_{B_{2\varepsilon}(x_{k_0})}((-\Delta)^{s/2}(z_j)_+)(x)^2\phi_\varepsilon(x)\,dx\Bigg). \end{aligned}$$

Therefore, by (20.53), (20.54), and (20.55), we get

$$\lim_{\varepsilon\to 0}\lim_{j\to+\infty}\left(\int_{\mathbb{R}^n}(z_j)_+(x)(-\Delta)^{s/2}(z_j)_+(x)(-\Delta)^{s/2}\phi_\varepsilon(x)\,dx\right.$$
$$\left.-2\int_{\mathbb{R}^n}(-\Delta)^{s/2}(z_j)_+(x)\int_{\mathbb{R}^n}\frac{(\phi_\varepsilon(x)-\phi_\varepsilon(y))((z_j)_+(x)-(z_j)_+(y))}{|x-y|^{n+s}}\,dx\,dy\right)$$
$$\le\lim_{\varepsilon\to 0}\left(\lambda\int_{B_{2\varepsilon}(x_{k_0})}u_0(x)^{q+1}\phi_\varepsilon(x)\,dx+\int_{B_{2\varepsilon}(x_{k_0})}\phi_\varepsilon(x)\,d\nu-\int_{B_{2\varepsilon}(x_{k_0})}\phi_\varepsilon(x)\,d\mu\right).\tag{20.59}$$

Since ϕ is a regular function with compact support, it is clear that it satisfies the hypothesis of Lemma 20.8. Therefore, by Lemmas 20.9 and 20.10 applied to the sequence $\{(z_j)_+\}_{j\in\mathbb{N}}$, it follows that the left-hand side of (20.59) goes to zero; that is, we obtain that

$$\int_{B_{2\varepsilon}(x_{k_0})}\phi_\varepsilon(x)\,d\nu+\lambda\int_{B_{2\varepsilon}(x_{k_0})}u_0(x)^{q+1}\phi_\varepsilon(x)\,dx-\int_{B_{2\varepsilon}(x_{k_0})}\phi_\varepsilon(x)\,d\mu\to\nu_{k_0}-\mu_{k_0}$$

as $\varepsilon\to 0$, and $\nu_{k_0}-\mu_{k_0}\geq 0$. Thus, from (20.56), we have that either $\nu_{k_0}=0$ or

$$\nu_{k_0}\geq S_s^{n/2s}.\tag{20.60}$$

Suppose now that $\nu_{k_0}\neq 0$. By (20.49), (20.52), and (20.60), we get that

$$\begin{aligned}c+\mathcal{J}_{s,\lambda,q}(u_0)&\geq\lim_{j\to+\infty}\left(\mathcal{J}_{s,\lambda,q}(z_j)-\frac{1}{2}\langle\mathcal{J}'_{s,\lambda,q}(z_j),z_j\rangle\right)\\&\geq\lambda\left(\frac{1}{2}-\frac{1}{q+1}\right)\int_\Omega u_0(x)^{q+1}\,dx+\frac{s}{n}\int_\Omega u_0(x)^{2_s^*}\,dx+\frac{s}{n}\nu_{k_0}\\&\geq\mathcal{J}_{s,\lambda,q}(u_0)+\frac{s}{n}S_s^{\frac{n}{2s}}\\&=\mathcal{J}_{s,\lambda,q}(u_0)+c^*.\end{aligned}$$

This is a contradiction with (20.47).

Since k_0 was arbitrary, we deduce that $\nu_k=0$, for all $k\in I$. As a consequence, we get that $(u_j)_+\to 0$ in $L^{2_s^*}(\Omega)$. Note that since u_j is equal to zero outside Ω, indeed, we have that $(u_j)_+\to 0$ in $L^{2_s^*}(\mathbb{R}^n)$. This implies convergence of $\lambda((u_j)_+)^q+((u_j)_+)^{2_s^*-1}$ in $L^{2n/n+2s}(\mathbb{R}^n)$. Finally, using the continuity of the inverse operator $(-\Delta)^{-s}$, we obtain strong convergence of $\{u_j\}_{j\in\mathbb{N}}$ in $\mathbb{H}_0^s(\Omega)$. The proof is complete. □

20.2.2 Proof of statement (d) of Theorem 20.1

In Lemma 20.11, we proved that if $u\equiv 0$ is the only critical point of the functional $\widetilde{\mathcal{J}}_{s,\lambda,q}$, then $\widetilde{\mathcal{J}}_{s,\lambda,q}$ verifies the Palais–Smale condition at any level $c<c^*$,

where c^* is the critical level defined in (20.46). Now we want to show that we can obtain a local Palais–Smale sequence for $\widetilde{\mathscr{J}}_{s,\lambda,q}$ under the critical level c^*. For this, assume, without loss of generality, that $0 \in \Omega$. By [82] (see also [42, 132] and Chapter 14), the infimum in (1.92) is attained at the function

$$u_\varepsilon(x) := \frac{\varepsilon^{(n-2s)/2}}{(|x|^2+\varepsilon^2)^{(n-2s)/2}} \quad (\varepsilon > 0); \tag{20.61}$$

that is,

$$\|(-\Delta)^{s/2} u_\varepsilon\|^2_{L^2(\mathbb{R}^n)} = \int_{\mathbb{R}^n\times\mathbb{R}^n} \frac{|u_\varepsilon(x)-u_\varepsilon(y)|^2}{|x-y|^{n+2s}}\, dx\, dy = S_s \|u_\varepsilon\|^2_{L^{2^*_s}(\mathbb{R}^n)}. \tag{20.62}$$

Also let us introduce a cutoff function $\phi_0 \in C^\infty(\mathbb{R})$, nonincreasing and satisfying

$$\phi_0(t) := \begin{cases} 1 & \text{if } 0 \le t \le \frac{1}{2} \\ 0 & \text{if } t \ge 1. \end{cases}$$

For a fixed $r > 0$ small enough such that $\overline{B}_r \subset \Omega$, set $\phi(x) = \phi_r(x) = \phi_0(|x|/r)$, and consider the family of nonnegative truncated functions

$$\eta_\varepsilon := \frac{\phi u_\varepsilon}{\|\phi u_\varepsilon\|_{L^{2^*_s}(\Omega)}} \in \mathbb{H}^s_0(\Omega). \tag{20.63}$$

Then we have the following result:

Lemma 20.12 *There exists $\varepsilon > 0$ small enough such that*

$$\sup_{t\ge 0} \widetilde{\mathscr{J}}_{s,\lambda,q}(t\eta_\varepsilon) < c^*. \tag{20.64}$$

Proof We follow the proof of [8, lemma 4.4] (see also [22, lemma 3.9]). Assume that $n \ge 4s$. Since

$$(a+b)^p \ge a^p + b^p + \mu a^{p-1} b \quad \text{for some } \mu > 0 \text{ and every } a, b \ge 0,\ p > 1, \tag{20.65}$$

then the function G_λ defined in (20.13) satisfies

$$G_\lambda(u) \ge \frac{1}{2^*_s}(u_+)^{2^*_s} + \frac{\mu}{2}(u_+)^2 u_0^{2^*_s-2}. \tag{20.66}$$

Therefore,

$$\widetilde{\mathscr{J}}_{s,\lambda,q}(t\eta_\varepsilon) \le \frac{t^2}{2}\|\eta_\varepsilon\|^2_{\mathbb{H}^s_0(\Omega)} - \frac{t^{2^*_s}}{2^*_s} - \frac{t^2}{2}\mu \int_\Omega u_0(x)^{2^*_s-2}\eta_\varepsilon(x)^2 dx.$$

Since $u_0 \ge a_0 > 0$ in $\operatorname{supp}(\eta_\varepsilon)$, we get, for any $t \ge 0$ and $\varepsilon > 0$ small enough,

$$\widetilde{\mathscr{J}}_{s,\lambda,q}(t\eta_\varepsilon) \le \frac{t^2}{2}\|\eta_\varepsilon\|^2_{\mathbb{H}^s_0(\Omega)} - \frac{t^{2^*_s}}{2^*_s} - \frac{t^2}{2}\tilde{\mu}\|\eta_\varepsilon\|^2_{L^2(\Omega)}. \tag{20.67}$$

Moreover, since $\|u_\varepsilon\|_{L^{2^*_s}(\mathbb{R}^n)}$ is independent of ε, by Proposition 14.11, we have

$$\|\eta_\varepsilon\|^2_{\mathbb{H}^s_0(\Omega)} = \frac{\|\phi u_\varepsilon\|^2_{\mathbb{H}^s_0(\Omega)}}{\|\phi u_\varepsilon\|^2_{L^{2^*_s}(\Omega)}}$$

$$\leq \frac{\displaystyle\int_{\mathbb{R}^n\times\mathbb{R}^n} \frac{|u_\varepsilon(x)-u_\varepsilon(y)|^2}{|x-y|^{n+2s}}\,dx\,dy}{\|\phi u_\varepsilon\|^2_{L^{2^*_s}(\Omega)}} + O(\varepsilon^{n-2s}) \tag{20.68}$$

$$= S_s + O(\varepsilon^{n-2s}).$$

Furthermore, by [22, lemma 3.8] (see also Proposition 14.12), it follows that

$$\|\eta_\varepsilon\|^2_{L^2(\Omega)} \geq \begin{cases} C\varepsilon^{2s} & \text{if } n > 4s \\ C\varepsilon^{2s}\log(1/\varepsilon) & \text{if } n = 4s. \end{cases} \tag{20.69}$$

Therefore, from (20.67), (20.68), and (20.69), we get

$$\widetilde{\mathcal{J}}_{s,\lambda,q}(t\eta_\varepsilon) \leq \frac{t^2}{2}(S_s + C\varepsilon^{n-2s}) - \frac{t^{2^*_s}}{2^*_s} - \frac{t^2}{2}\widetilde{C}\varepsilon^{2s} := g(t), \tag{20.70}$$

with $\widetilde{C} > 0$. Since $\lim_{t\to+\infty} g(t) = -\infty$, $\sup_{t\geq 0} g(t)$ is attained at some $t_{\varepsilon,\lambda} := t_\varepsilon \geq 0$. If $t_\varepsilon = 0$, then

$$\sup_{t\geq 0} \widetilde{\mathcal{J}}_{s,\lambda,q}(t\eta_\varepsilon) \leq \sup_{t\geq 0} g(t) = g(0) = 0,$$

for any $0 < \lambda < \Lambda$, and (20.64) is trivially verified. Now we suppose that $t_\varepsilon > 0$. Differentiating the preceding function $g(t)$, we obtain that

$$0 = g'(t_\varepsilon) = t_\varepsilon(S_s + C\varepsilon^{n-2s}) - t_\varepsilon^{2^*_s-1} - t_\varepsilon\widetilde{C}\varepsilon^{2s}, \tag{20.71}$$

which implies

$$t_\varepsilon \leq (S_s + C\varepsilon^{n-2s})^{(1/2^*_s-2)}. \tag{20.72}$$

Also, we have, for $\varepsilon > 0$ small enough,

$$t_\varepsilon \geq c > 0. \tag{20.73}$$

Indeed, from (20.71), we get

$$t_\varepsilon^{2^*_s-2} = S_s + C\varepsilon^{n-2s} - \widetilde{C}\varepsilon^{2s} \geq c > 0,$$

provided that ε is small enough. Moreover, the function

$$t \mapsto \frac{t^2}{2}(S_s + C\varepsilon^{n-2s}) - \frac{t^{2^*_s}}{2^*_s}$$

is increasing on $\left[0,(S_s+C\varepsilon^{n-2s})^{(1/2_s^*-2)}\right]$. Hence, by (20.72) and (20.73), we obtain

$$\sup_{t\geq 0} g(t)=g(t_\varepsilon)\leq \frac{s}{n}(S_s+C\varepsilon^{n-2s})^{\frac{n}{2s}}-\overline{C}\varepsilon^{2s},$$

for some $\overline{C}>0$. Therefore, by (20.70), for $n>4s$, we get that

$$\sup_{t\geq 0} \widetilde{\mathcal{J}}_{s,\lambda,q}(t\eta_\varepsilon)\leq g(t_\varepsilon)\leq \frac{s}{n}S_s^{n/2s}+C\varepsilon^{n-2s}-\overline{C}\varepsilon^{2s}<\frac{s}{n}S_s^{n/2s}=c^*. \tag{20.74}$$

If $n=4s$, the same conclusion follows.

The last case, $2s<n<4s$, follows by using estimate (20.65), which gives

$$G_\lambda(u)\geq \frac{1}{2_s^*}(u_+)^{2_s^*}+\frac{\mu}{2_s^*-1}u_0(u_+)^{2_s^*-1}. \tag{20.75}$$

Then (20.75) jointly with the inequality [22, (3.28)], instead of (20.69), and arguing in a similar way as earlier, finishes the proof. □

To complete the existence of the second solution, that is, statement (d) in Theorem 20.1, in view of the previous results, we look for a path with energy below the critical level c^*. Let us fix $\lambda\in(0,\Lambda)$. We consider $M_\varepsilon>0$ large enough that $\widetilde{\mathcal{J}}_{s,\lambda,q}(M_\varepsilon\eta_\varepsilon)<\widetilde{\mathcal{J}}_{s,\lambda,q}(0)$. Note that such M_ε exists because $\lim_{t\to+\infty}\widetilde{\mathcal{J}}_{s,\lambda,q}(t\eta_\varepsilon)=-\infty$. Furthermore, by Lemma 20.7, there exists $\alpha>0$ such that if $\|u\|_{\mathbb{H}_0^s(\Omega)}=\alpha$, then $\widetilde{\mathcal{J}}_{s,\lambda,q}(u)\geq\widetilde{\mathcal{J}}_{s,\lambda,q}(0)$. We define

$$\Gamma_\varepsilon:=\left\{\gamma\in C([0,1];\mathbb{H}_0^s(\Omega)):\ \gamma(0)=0,\ \gamma(1)=M_\varepsilon\eta_\varepsilon\right\}$$

and the minimax value

$$c_\varepsilon=\inf_{\gamma\in\Gamma_\varepsilon}\sup_{0\leq t\leq 1}\widetilde{\mathcal{J}}_{s,\lambda,q}(\gamma(t)). \tag{20.76}$$

By the preceding arguments, $c_\varepsilon\geq\widetilde{\mathcal{J}}_{s,\lambda,q}(0)$. Also, by Lemma 20.12, for $\varepsilon\ll 1$, we obtain that

$$c_\varepsilon\leq\sup_{0\leq t\leq 1}\widetilde{\mathcal{J}}_{s,\lambda,q}(tM_\varepsilon\eta_\varepsilon)=\sup_{t\geq 0}\widetilde{\mathcal{J}}_{s,\lambda,q}(t\eta_\varepsilon)<c^*.$$

Therefore, by Lemma 20.11 and the mountain pass theorem in [11], if $c_\varepsilon>\widetilde{\mathcal{J}}_{s,\lambda,q}(0)$, or the corresponding refinement given in [116] if the minimax level $c_\varepsilon=\widetilde{\mathcal{J}}_{s,\lambda,q}(0)$, we obtain the existence of a nontrivial solution of (20.15), provided that $u\equiv 0$ is its unique solution. Of course, this is a contradiction.

Thus, $\widetilde{\mathcal{J}}_{s,\lambda,q}$ admits a critical point $\tilde{u}$ different from the trivial function. As a consequence, $u=u_0+\tilde{u}$ is a solution, different from u_0, of problem (20.1). This concludes the proof of Theorem 20.1.

20.3 The critical and convex case $q > 1$

In this section, we discuss problem (20.1) in the convex setting $q > 1$. Here we will argue essentially as in previous chapters (see Chapter 14 and the following), where we studied the linear case $q = 1$ again using variational techniques. With respect to the case $q = 1$, there are some extra difficulties to prove the Palais–Smale condition and to obtain the estimates of the mountain pass critical value, while it is easy to check the good geometry of the functional. Let us start with this geometric feature.

Proposition 20.13 Assume that $\lambda > 0$ and $1 < q < 2_s^* - 1$. Then there exist $\alpha > 0$ and $\beta > 0$ such that

(a) For any $u \in X_0^s(\Omega)$ with $\|u\|_{X_0^s(\Omega)} = \alpha$, one has that $\mathcal{J}_{s,\lambda,q}(u) \geq \beta$.
(b) There exists a positive function $e \in X_0^s(\Omega)$ such that $\|e\|_{X_0^s(\Omega)} > \alpha$ and $\mathcal{J}_{s,\lambda,q}(e) < \beta$.

Proof By the Sobolev embedding theorem, since $q + 1 < 2_s^*$, it can be easily seen that

$$\mathcal{J}_{s,\lambda,q}(u) \geq g(||u||_{X_0^s(\Omega)}),$$

where $g(t) = C_1 t^2 - \lambda C_2 t^{q+1} - C_3 t^{2_s^*}$, for some positive constants C_1, C_2, and C_3. Therefore, there exists $\alpha > 0$ such that $\beta := g(\alpha) > 0$. Then $\mathcal{J}_{s,\lambda,q}(u) \geq \beta$, for $u \in X_0^s(\Omega)$, with $||u||_{X_0^s(\Omega)} = \alpha$. Hence, assertion (a) is proved.

For part (b), fix a positive function $u_0 \in X_0^s(\Omega)$ such that $\|u_0\|_{X_0^s(\Omega)} = 1$, and consider $t > 0$. Since $2_s^* > 2$, it follows that

$$\lim_{t\to\infty} \mathcal{J}_{s,\lambda,q}(tu_0) = -\infty.$$

Then there exists t_0 large enough such that, for $e := t_0 u_0$, we get that $\|e\|_{X_0^s(\Omega)} > \alpha$ and $\mathcal{J}_{s,\lambda,q}(e) < \beta$. Thus assertion (b) is proved. □

By a similar argument, it follows that

$$\lim_{t\to 0^+} \mathcal{J}_{s,\lambda,q}(tu_0) = 0. \tag{20.77}$$

20.3.1 The Palais–Smale condition for $\mathcal{J}_{s,\lambda,q}$

In this subsection, we will show that the functional $\mathcal{J}_{s,\lambda,q}$ satisfies the Palais–Smale condition in a suitable energy range involving the best fractional critical Sobolev constant S_s given in (1.92).

Proposition 20.14 Let $\lambda > 0$ and $1 < q < 2_s^* - 1$. Let $c \in \mathbb{R}$ be such that $c < c^*$, where c^* is given in (20.46). Then the functional $\mathcal{J}_{s,\lambda,q}$ satisfies the Palais–Smale condition at level c.

Proof This proof is quite standard, and we will use the same arguments for it that we considered, for instance, in Section 15.2. We prefer to repeat it for the sake of clarity and for reader's convenience.

Let $\{u_j\}_{j\in\mathbb{N}}$ be a Palais–Smale sequence at level c for $\mathscr{I}_{s,\lambda,q}$ in $\mathbb{H}_0^s(\Omega)$; that is,

$$\mathscr{I}_{s,\lambda,q}(u_j) \to c \tag{20.78}$$

and

$$\mathscr{I}'_{s,\lambda,q}(u_j) \to 0. \tag{20.79}$$

Arguing as in the proof of Proposition 15.2, we have the following claims:

Claim 20.1 *The sequence $\{u_j\}_{j\in\mathbb{N}}$ is bounded in $\mathbb{H}_0^s(\Omega)$.*

Claim 20.2 *Problem* (20.2) *admits a weak solution $u_\infty \in \mathbb{H}_0^s(\Omega)$.*

Claim 20.3 *The following equality holds:*

$$\begin{aligned}\mathscr{I}_{s,\lambda,q}(u_j) &= \mathscr{I}_{s,\lambda,q}(u_\infty) \\ &\quad + \frac{1}{2}\|u_j - u_\infty\|^2_{\mathbb{H}_0^s(\Omega)} - \frac{1}{2_s^*}\int_\Omega |(u_j)_+(x) - (u_\infty)_+(x)|^{2_s^*}\,dx + o(1)\end{aligned}$$

as $j \to +\infty$.

Moreover, we have

Claim 20.4 *The following estimate holds*:

$$\begin{aligned}\|u_j - u_\infty\|^2_{\mathbb{H}_0^s(\Omega)} &= \int_\Omega |(u_j)_+(x) - (u_\infty)_+(x)|^{2_s^*}\,dx + o(1) \\ &\leq \int_\Omega |(u_j)(x) - (u_\infty)(x)|^{2_s^*}\,dx + o(1).\end{aligned}$$

Proof First, note that, by Claim 20.1, we have that, up to a subsequence,

$$u_j \to u_\infty \quad \text{weakly in } L^{2_s^*}(\mathbb{R}^n), \tag{20.80}$$

while, by Lemma 1.31(a),

$$u_j \to u_\infty \quad \text{in } L^\nu(\mathbb{R}^n),\ \nu \in [1, 2_s^*) \tag{20.81}$$

as $j \to +\infty$. Moreover, by the Brezis–Lieb lemma, we have that

$$\begin{aligned}\int_\Omega |(u_j)_+(x)|^{2_s^*}\,dx &= \int_\Omega |(u_j)_+(x) - (u_\infty)_+(x)|^{2_s^*}\,dx \\ &\quad + \int_\Omega |(u_\infty)_+(x)|^{2_s^*}\,dx + o(1)\end{aligned} \tag{20.82}$$

as $j \to +\infty$. As a consequence of (20.80) and (20.82), we get

$$\begin{aligned}
&\int_\Omega \left((u_j)_+(x)^{2_s^*-1} - (u_\infty)_+(x)^{2_s^*-1}\right)\left(u_j(x) - u_\infty(x)\right) dx \\
&= \int_\Omega (u_j)_+(x)^{2_s^*}\, dx - \int_\Omega (u_\infty)_+(x)^{2_s^*-1} u_j(x)\, dx \\
&- \int_\Omega (u_j)_+(x)^{2_s^*-1} u_\infty(x)\, dx + \int_\Omega (u_\infty)_+(x)^{2_s^*}\, dx \qquad (20.83)\\
&= \int_\Omega (u_j)_+(x)^{2_s^*}\, dx - \int_\Omega (u_\infty)_+(x)^{2_s^*}\, dx + o(1) \\
&= \int_\Omega |(u_j)_+(x) - (u_\infty)_+(x)|^{2_s^*}\, dx + o(1)
\end{aligned}$$

as $j \to +\infty$. Furthermore, Claim 20.1 and (20.81) give

$$\begin{aligned}
&\int_\Omega \left((u_j)_+(x)^q - (u_\infty)_+(x)^q\right)\left(u_j(x) - u_\infty(x)\right) dx \\
&= \int_\Omega (u_j)_+(x)^{q+1}\, dx - \int_\Omega (u_\infty)_+(x)^q u_j(x)\, dx \\
&- \int_\Omega (u_j)_+(x)^q u_\infty(x)\, dx + \int_\Omega (u_\infty)_+(x)^{q+1}\, dx \qquad (20.84)\\
&= o(1).
\end{aligned}$$

as $j \to +\infty$. Then, by (20.79), Claim 20.2, (20.83), and (20.84), we conclude that

$$\begin{aligned}
o(1) &= \langle \mathscr{J}'_{s,\lambda}(u_j), u_j - u_\infty \rangle \\
&= \langle \mathscr{J}'_{s,\lambda}(u_j) - \mathscr{J}'_{s,\lambda}(u_\infty), u_j - u_\infty \rangle \\
&= \|u_j - u_\infty\|^2_{\mathbb{H}_0^s(\Omega)} - \lambda \int_\Omega \left((u_j)_+(x)^q - (u_\infty)_+(x)^q\right)(u_j(x) - u_\infty(x))\, dx \\
&\quad - \int_\Omega \left((u_j)_+(x)^{2_s^*-1} - (u_\infty)_+(x)^{2_s^*-1}\right)(u_j(x) - u_\infty(x))\, dx \\
&= \|u_j - u_\infty\|^2_{\mathbb{H}_0^s(\Omega)} - \int_\Omega |(u_j)_+(x) - (u_\infty)_+(x)|^{2_s^*}\, dx + o(1)
\end{aligned}$$

as $j \to +\infty$. Hence, the proof of Claim 20.4 is complete. □

Now we can finish the proof of Proposition 20.14. By Claim 20.4, we know that

$$\begin{aligned}
&\frac{1}{2}\|u_j - u_\infty\|^2_{\mathbb{H}_0^s(\Omega)} - \frac{1}{2_s^*}\int_\Omega |(u_j)_+(x) - (u_\infty)_+(x)|^{2_s^*}\, dx \\
&= \frac{s}{n}\|u_j - u_\infty\|^2_{\mathbb{H}_0^s(\Omega)} + o(1). \qquad (20.85)
\end{aligned}$$

Then, by (20.78), Claim 20.3, and (20.85), we obtain

$$\mathscr{I}_{s,\lambda,q}(u_\infty)+\frac{s}{n}\|u_j-u_\infty\|^2_{\mathbb{H}^s_0(\Omega)}=\mathscr{I}_{s,\lambda,q}(u_j)+o(1)=c+o(1). \tag{20.86}$$

However, by (20.1), up to a subsequence, we can assume that

$$\|u_j-u_\infty\|^2_{\mathbb{H}^s_0(\Omega)}\to L\geq 0, \tag{20.87}$$

and then, as a consequence of Claim 20.4,

$$\int_\Omega |u_j(x)-u_\infty(x)|^{2^*_s}\,dx\to\tilde{L}\geq L.$$

By the definition of S_s given in (1.92), we have

$$L\geq S_s\tilde{L}^{2/2^*_s}\geq S_sL^{2/2^*_s},$$

so

$$L=0 \quad \text{or} \quad L\geq S_s^{n/2s}.$$

We now prove that the case $L\geq S_s^{n/2s}$ cannot occur. Indeed, taking $\varphi=u_\infty\in\mathbb{H}^s_0(\Omega)$ as a test function in Claim 20.2, we have that

$$\|u_\infty\|^2_{\mathbb{H}^s_0(\Omega)}=\lambda\int_\Omega (u_\infty)_+(x)^{q+1}\,dx+\int_\Omega (u_\infty)_+(x)^{2^*_s}\,dx.$$

Hence,

$$\mathscr{I}_{s,\lambda,q}(u_\infty)=\lambda\left(\frac{1}{2}-\frac{1}{q+1}\right)\|(u_\infty)_+\|^{q+1}_{L^{q+1}(\Omega)}+\frac{s}{n}\|(u_\infty)_+\|^{2^*_s}_{L^{2^*_s}(\Omega)}\geq 0, \tag{20.88}$$

thanks to the positivity of λ and the fact that $q>1$. Therefore, if $L\geq S_s^{n/2s}$, by (20.86), (20.87), and (20.88), we would get

$$c=\mathscr{I}_{s,\lambda,q}(u_\infty)+\frac{s}{n}L\geq\frac{s}{n}L\geq\frac{s}{n}S_s^{n/2s},$$

which contradicts the choice of c. Thus, $L=0$, so, by (20.87), we obtain that

$$\|u_j-u_\infty\|_{\mathbb{H}^s_0(\Omega)}\to 0.$$

The proof is complete. □

Remark 20.15 Note that the proof of Proposition 20.14 also could be obtained by the concentration-compactness theory of Subsection 20.2.1. This simply means that the arguments performed in the last part of the proof of Lemma 20.11 can be adapted to the convex setting.

20.3.2 Proof of Theorem 20.2

By Proposition 20.13 and (20.77), we get that $\mathcal{J}_{s,\lambda}$ satisfies the geometric features required by the mountain pass theorem (see [11]). Moreover, by Proposition 20.14, the functional $\mathcal{J}_{s,\lambda,q}$ verifies the Palais–Smale condition at any level c under a suitable threshold.

Now, as in the concave case, we find a path with energy below the critical level c^*. Precisely, the following result holds true:

Lemma 20.16 *Let $\lambda > 0$, c^* be as in (20.46), and let η_ε be the nonnegative function defined in (20.63). Then there exists $\varepsilon > 0$ small enough such that*

$$\sup_{t\geq 0} \mathcal{J}_{s,\lambda,q}(t\eta_\varepsilon) < c^*,$$

provided that either

- $n > \dfrac{2s(q+3)}{q+1}$ *and $\lambda > 0$, or*
- $n \leq \dfrac{2s(q+3)}{q+1}$ *and $\lambda > \lambda_s^\star$, for a suitable $\lambda_s^\star > 0$.*

Proof Let us start by considering the case where $n > 2s(q+3)/(q+1)$. First, note that since $q > 1$, we get that $n > 2s\,(1+1/q)$. Therefore, denoting

$$N := -(n-(n-2s)(q+1)) > 0,$$

for some positive constants c and $\tilde{C}$, it follows that

$$\begin{aligned}
\int_{\mathbb{R}^n} \eta_\varepsilon(x)^{q+1}\,dx &= C\int_{|x|<r} u_\varepsilon(x)^{q+1}\,dx \\
&= C\varepsilon^{((n-2s)/2)(q+1)} \int_{|x|<r} \frac{dx}{(|x|^2+\varepsilon^2)^{(n-2s)(q+1)/2}} \\
&= C\varepsilon^{-((n-2s)/2)(q+1)} \int_0^r \frac{\rho^{n-1}}{\left(1+\left(\frac{\rho}{\varepsilon}\right)^2\right)^{(n-2s)(q+1)/2}}\,d\rho \\
&= C\varepsilon^{n-(n-2s/2)(q+1)} \int_0^{r/\varepsilon} \frac{t^{n-1}}{(1+t^2)^{(n-2s)(q+1)/2}}\,dt \\
&\geq C\varepsilon^{n-\left(\frac{n-2s}{2}\right)(q+1)} \int_1^{r/\varepsilon} t^{n-1-(n-2s)(q+1)}\,dt \\
&= \frac{C\varepsilon^{n-(n-2s/2)(q+1)}}{N}\left(1-\left(\frac{\varepsilon}{r}\right)^N\right) \\
&\geq \tilde{C}\varepsilon^{n-((n-2s)/2)(q+1)}.
\end{aligned} \tag{20.89}$$

Then, by (20.68) and (20.89), for any $t \geq 0$ and $\varepsilon > 0$ small enough, we obtain

$$\begin{aligned}\mathscr{J}_{s,\lambda}(t\eta_\varepsilon) &= \frac{t^2}{2}\|\eta_\varepsilon\|^2_{\mathbb{H}^s_0(\Omega)} - \frac{t^{2^*_s}}{2^*_s} - \lambda\frac{t^{q+1}}{q+1}\int_\Omega \eta_\varepsilon(x)^{q+1}\,dx \\ &\leq \frac{t^2}{2}(S_s + C\varepsilon^{n-2s}) - \frac{t^{2^*_s}}{2^*_s} - \tilde{C}\lambda\frac{t^{q+1}}{q+1}\varepsilon^{n-\left(\frac{n-2s}{2}\right)(q+1)} =: g(t).\end{aligned} \tag{20.90}$$

It is clear that

$$\lim_{t\to\infty} g(t) = -\infty;$$

therefore, $\sup_{t\geq 0} g(t)$ is attained at some $t_{\varepsilon,\lambda} := t_\varepsilon \geq 0$.

As we commented in the proof of Lemma 20.12, we could suppose that $t_\varepsilon > 0$. Differentiating $g(t)$ and setting it to zero, we obtain that

$$0 = g'(t_\varepsilon) = t_\varepsilon(S_s + C\varepsilon^{n-2s}) - t_\varepsilon^{2^*_s-1} - \tilde{C}\lambda t_\varepsilon^q \varepsilon^{n-((n-2s)/2)(q+1)}. \tag{20.91}$$

Hence,

$$t_\varepsilon < (S_s + C\varepsilon^{n-2s})^{1/(2^*_s-2)}.$$

Moreover, we have that, for $\varepsilon > 0$ small enough,

$$t_\varepsilon \geq c > 0. \tag{20.92}$$

Indeed, from (20.91), it follows that

$$t_\varepsilon^{2^*_s-2} + \tilde{C}\lambda t_\varepsilon^{q-1}\varepsilon^{n-((n-2s)/2)(q+1)} = S_s + C\varepsilon^{n-2s} \geq c > 0, \quad \text{for } \varepsilon > 0 \text{ small enough.}$$

Also, since the function

$$t \mapsto \frac{t^2}{2}(S_s + C\varepsilon^{n-2s}) - \frac{t^{2^*_s}}{2^*_s}$$

is increasing on $\left[0, (S_s + C\varepsilon^{n-2s})^{1/(2^*_s-2)}\right]$, by (20.90) and (20.92), we obtain

$$\begin{aligned}\sup_{t\geq 0} g(t) = g(t_\varepsilon) &\leq \frac{s}{n}(S_s + C\varepsilon^{n-2s})^{\frac{n}{2s}} - \overline{C}\varepsilon^{n-(n-2s/2)(q+1)} \\ &\leq \frac{s}{n}S_s^{\frac{n}{2s}} + C\varepsilon^{n-2s} - \overline{C}\varepsilon^{n-((n-2s)/2)(q+1)},\end{aligned} \tag{20.93}$$

for some $\overline{C} > 0$. Finally, thanks to the hypothesis on n, we conclude from (20.93) that

$$\sup_{t\geq 0} \mathscr{J}_{s,\lambda,q}(t\eta_\varepsilon) \leq g(t_\varepsilon) < \frac{s}{n}S_s^{n/2s}.$$

Consider now the case $n \leq 2s(q+3)/(q+1)$. Arguing exactly as in the preceding case, we get that

$$(S_s + C\varepsilon^{n-2s}) = t_{\varepsilon,\lambda}^{2^*_s-2} + \tilde{C}\lambda t_{\varepsilon,\lambda}^{q-1}\varepsilon^{n-((n-2s)/2)(q+1)}, \tag{20.94}$$

with $t_{\varepsilon,\lambda} > 0$ the point where the $\sup_{t\geq 0} g(t)$ is attained.

We claim that

$$t_{\varepsilon,\lambda} \to 0 \quad \text{when } \lambda \to +\infty. \tag{20.95}$$

To see this, assume that $\limsup_{\lambda\to+\infty} t_{\varepsilon,\lambda} = \ell > 0$. Then, passing to the limit when $\lambda \to +\infty$ in (20.94), we would get $(S_s + C\varepsilon^{n-2s}) = +\infty$, which is a contradiction, and (20.95) follows.

If we take now β the positive number given in Proposition 20.13, by (20.95), we obtain that

$$\begin{aligned}
0 &\leq \sup_{t\geq 0} \mathcal{J}_{s,\lambda,q}(t\eta_\varepsilon) \\
&\leq g(t_{\varepsilon,\lambda}) \\
&= \frac{t_{\varepsilon,\lambda}^2}{2}(S_s + C\varepsilon^{n-2s}) - \frac{t_{\varepsilon,\lambda}^{2_s^*}}{2_s^*} - \widetilde{C}\lambda\frac{t_{\varepsilon,\lambda}^{q+1}}{q+1}\varepsilon^{n-(n-2s/2)(q+1)} \\
&\leq \frac{t_{\varepsilon,\lambda}^2}{2}(S_s + C\varepsilon^{n-2s}) - \frac{t_{\varepsilon,\lambda}^{2_s^*}}{2_s^*} \to 0,
\end{aligned}$$

when $\lambda \to +\infty$. Then

$$\lim_{\lambda\to+\infty} \sup_{t\geq 0} \mathcal{J}_{s,\lambda,q}(t\eta_\varepsilon) = 0,$$

which easily yields the desired conclusion for the case $n \leq 2s(q+3)/(q+1)$. □

Now we can conclude the proof of Theorem 20.2. To do so, we define

$$\Gamma_\varepsilon := \{\gamma \in C([0,1]; \mathbb{H}_0^s(\Omega)) : \gamma(0) = 0, \gamma(1) = M_\varepsilon\eta_\varepsilon\},$$

for some $M_\varepsilon > 0$ big enough such that $\mathcal{J}_{s,\lambda}(M_\varepsilon\eta_\varepsilon) < 0$. Observe that, for every $\gamma \in \Gamma_\varepsilon$, the function $t \to \|\gamma(t)\|_{\mathbb{H}_0^s(\Omega)}$ is continuous in $[0,1]$. Therefore, for the α given in Proposition 20.13, since $\|\gamma(0)\|_{\mathbb{H}_0^s(\Omega)} = 0 < \alpha$ and $\|\gamma(1)\|_{\mathbb{H}_0^s(\Omega)} = \|M_\varepsilon\eta_\varepsilon\|_{\mathbb{H}_0^s(\Omega)} > \alpha$ for M_ε sufficiently large, there exists $t_0 \in (0,1)$ such that $\|\gamma(t_0)\|_{\mathbb{H}_0^s(\Omega)} = \alpha$. As a consequence,

$$\sup_{0\leq t\leq 1} \mathcal{J}_{s,\lambda,q}(\gamma(t)) \geq \mathcal{J}_{s,\lambda,q}(\gamma(t_0)) \geq \inf_{\|v\|_{\mathbb{H}_0^s(\Omega)}=\alpha} \mathcal{J}_{s,\lambda,q}(v) \geq \beta > 0,$$

where β is the positive value given in Proposition 20.13. Hence,

$$c_\varepsilon := \inf_{\gamma\in\Gamma_\varepsilon} \sup_{0\leq t\leq 1} \mathcal{J}_{s,\lambda,q}(\gamma(t)) > 0.$$

Then, by Lemma 20.16, Proposition 20.14, and the mountain pass theorem given in [11], we conclude that the functional $\mathcal{J}_{s,\lambda,q}$ admits a critical point $u \in \mathbb{H}_0^s(\Omega)$, provided that $n > 2s(q+3)/(q+1)$ and $\lambda > 0$ or $n \leq 2s(q+3)/q+1$ and $\lambda > \lambda_s^\star$, for a suitable $\lambda_s^\star > 0$. Moreover, since $\mathscr{J}_{s,\lambda,q}(u) = c_\varepsilon \geq \beta > 0$ and $\mathscr{J}_{s,\lambda,q}(0) = 0$, the function u is not the trivial one. This concludes the proof of Theorem 20.2.

Remark 20.17 Some of the results obtained in Sections 20.2 and 20.3 are true for integrodifferential operators more general than the fractional Laplacian, such as, for instance, $\mathscr{L}_K$ given in (1.54).

Bibliography

[1] N. Abatangelo and E. Valdinoci. A notion of nonlocal curvature. *Numer. Funct. Anal. Optim.* 35: 793–815, 2014.

[2] B. Abdellaoui, E. Colorado, and I. Peral. Effect of the boundary conditions in the behavior of the optimal constant of some Caffarelli–Kohn–Nirenberg inequalities. Application to some doubly critical nonlinear elliptic problems. *Adv. Diff. Equations* 11: 667–720, 2006.

[3] R. A. Adams. *Sobolev Spaces*. Academic Press, New York, 1975.

[4] S. Agmon. Lectures on elliptic boundary value problems. *Mathematical Studies*, Vol. 2. Van Nostrand, Princeton, NJ, 1965.

[5] S. Alama. Semilinear elliptic equations with sublinear indefinite nonlinearities. *Adv. Diff. Equations* 4: 813–42, 1999.

[6] F. J. Almgren and E. H. Lieb. Symmetric decreasing rearrangement is sometimes continuous. *J. Am. Math. Soc.* 2: 683–773, 1989.

[7] H. Amann and E. Zehnder. Nontrivial solutions for a class of nonresonance problems and applications to nonlinear differential equations. *Ann. Scuola Norm. Sup. Pisa* 7: 539–603, 1980.

[8] A. Ambrosetti, H. Brézis, and G. Cerami. Combined effects of concave and convex nonlinearities in some elliptic problems. *J. Funct. Anal.* 122: 519–543, 1994.

[9] A. Ambrosetti and A. Malchiodi. *Nonlinear Analysis and Semilinear Elliptic Problems*. Cambridge Studies Advanced Mathematics 104. Cambridge University Press, 2007.

[10] A. Ambrosetti and G. Prodi, *A Primer of Nonlinear Analysis*. Cambridge Studies in Advanced Mathematics 34. Cambridge University Press, 1993.

[11] A. Ambrosetti and P. H. Rabinowitz. Dual variational methods in critical point theory and applications. *J. Funct. Anal.* 14: 349–81, 1973.

[12] A. Ambrosetti and M. Struwe. A note on the problem $-\Delta u = \lambda u + u|u|^{2^*-2}$. *Manusc. Math.* 54: 373–79, 1986.

[13] D. Applebaum. *Lévy Processes and Stochastic Calculus*, 2nd ed. Cambridge Studies in Advanced Mathematics 116. Cambridge University Press, 2009.

[14] G. Arioli, F. Gazzola, H.-Ch. Grunau, and E. Sassone. The second bifurcation branch for radial solutions of the Brezis-Nirenberg problem in dimension four. *NoDEA Nonlinear Differential Equations Appl.* 15: 69–90, 2008.

[15] N. Aronszajn. Boundary values of functions with finite Dirichlet integral. Technical Report 14, University of Kansas, 1955, pp. 77–94.

[16] G. Autuori, A. Fiscella, and P. Pucci. Stationary Kirchhoff problems involving a fractional operator and a critical nonlinearity. *Nonlinear Anal.* (in press).

[17] G. Autuori and P. Pucci. Elliptic problems involving the fractional Laplacian in $\mathbb{R}^N$. *J. Differential Equations* 255: 2340–62, 2013.

[18] M. Badiale and E. Serra. *Semilinear Elliptic Equations for Beginners*. Springer, Berlin, 2011.

[19] A. Bahri and H. Berestycki. A perturbation method in critical point theory and applications. *Trans. Am. Math. Soc.* 267: 1–32, 1981.

[20] A. Bahri and P. L. Lions. Morse index of some min-max critical points. I. Application to multiplicity results. *Commun. Pure Appl. Math.* 41: 1027–37, 1988.

[21] G. Barles, E. Chasseigne, and C. Imbert. On the Dirichlet problem for second-order elliptic integrodifferential equations. *Indiana Univ. Math. J.* 57: 213–46, 2008.

[22] B. Barrios, E. Colorado, A. De Pablo, and U. Sánchez. On some critical problems for the fractional Laplacian operator. *J. Differential Equations* 252: 6133–62, 2012.

[23] B. Barrios, E. Colorado, R. Servadei, and F. Soria. A critical fractional equation with concave-convex power nonlinearities. *Ann. Inst. H. Poincaré Anal. Non Linéaire* 31: 875–900, 2015.

[24] B. Barrios Barrera, A. Figalli, and E. Valdinoci. Bootstrap regularity for integrodifferential operators and its application to nonlocal minimal surfaces. *Ann. Sc. Norm. Super. Pisa Cl. Sci.* 5, doi: 10.2422/2036-2145.201202_007.

[25] B. Barrios, M. Medina, and I. Peral. Some remarks on the solvability of non-local elliptic problems with the Hardy potential. *Commun. Contemp. Math.* 16(4): 1350046, 2014.

[26] P. Bartolo, V. Benci, and D. Fortunato. Abstract critical point theorems and applications to some nonlinear problems with "strong" resonance at infinity. *Nonlinear Anal.* 7: 981–1012, 1983.

[27] R. Bartolo, A. M. Candela, and A. Salvatore. Perturbed asymptotically linear problems. *Ann. Mat. Pura Appl.* 193: 89–101, 2014.

[28] R. Bartolo and G. Molica Bisci. A pseudo-index approach to fractional equations. *Expo. Math.* 33: 502–516, 2015.

[29] T. Bartsch, Z. Liu, and T. Weth. Sign changing solutions of superlinear Schrödinger equations. *Comm. Partial Differential Equations* 29: 25–42, 2005.

[30] T. Bartsch, A. Pankov, and Z.-Q. Wang. Nonlinear Schrödinger equations with steep potential well. *Commun. Contemp. Math.* 4: 981–1012, 2001.

[31] T. Bartsch and Z.-Q. Wang. Existence and multiplicity results for some superlinear elliptic problems in $\mathbb{R}^N$. *Commun. Partial Differential Equations* 20: 1725–41, 1995.

[32] W. Beckner. Sobolev inequalities, the Poisson semigroup, and analysis on the sphere on S^N. *Proc. Natl. Acad. Sci. USA* 89: 4816–19, 1992.

[33] V. Benci. On the critical point theory for indefinite functionals in the presence of symmetries. *Trans. Am. Math. Soc.* 274: 533–72, 1982.

[34] S. Bernstein. Sur une classe d'équations fonctionnelles aux dérivées partielles. In Russian with French summary. *Bull. Acad. Sci. URSS, Set. Math.* 4: 17–26, 1940.

[35] J. Bertoin. Lévy Processes (Cambridge Tracts in Mathematics 121). Cambridge University Press, 1996.

[36] Z. Binlin, G. Molica Bisci, and R. Servadei. Superlinear nonlocal fractional problems with infinitely many solutions. *Nonlinearity* 28: 2247–64, 2015.

[37] C. Bjorland, L. Caffarelli, and A. Figalli. Non-local gradient dependent operators. *Adv. Math.* 230: 1859–94, 2012.

[38] R. M. Blumenthal, R. K. Getoor, and D. B. Ray. On the distribution of first hits for the symmetric stable processes. *Trans. Am. Math. Soc.* 99: 540–54, 1961.

[39] L. Boccardo, M. Escobedo, and I. Peral. A Dirichlet problem involving critical exponents. *Nonlinear Anal.* 24: 1639–48, 1995.

[40] L. Boccardo and T. Gallouet. Problèmes unilatéraux avec données dans L^1 (Unilateral problems with L^1 data). *C. R. Acad. Sci., Paris, Sér.* I, 311: 617–19, 1990.

[41] M. Bonforte, Y. Sire, and J. L. Vázquez. Existence, uniqueness and asymptotic behaviour for fractional porous medium equations on bounded domains. Preprint available at http://arxiv.org/pdf/1404.6195v3.pdf.

[42] C. Brändle, E. Colorado, A. De Pablo, and U. Sánchez. A concave-convex elliptic problem involving the fractional Laplacian. *Proc. Roy. Soc. Edinburgh Sect. A*, 143: 39–71, 2013.

[43] H. Brézis. *Analyse fonctionelle: Théorie et applications*. Masson, Paris, 1983.

[44] H. Brézis, J. M. Coron, and L. Nirenberg. Free vibrations for a nonlinear wave equation and a theorem of P. Rabinowitz. *Comm. Pure Appl. Math.* 33: 667–84, 1980.

[45] H. Brézis and E. Lieb. A relation between pointwise convergence of functions and convergence of functionals. *Proc. Am. Math. Soc.* 88: 486–90, 1983.

[46] H. Brézis and L. Nirenberg. Positive solutions of nonlinear elliptic equations involving critical Sobolev exponents. *Comm. Pure Appl. Math.* 36: 437–77, 1983.

[47] C. Bucur and E. Valdinoci. Nonlocal diffusion and applications. Preprint available at http://arxiv.org/pdf/1504.08292v2.pdf.

[48] X. Cabré and Y. Sire. Nonlinear equations for fractional Laplacians: I. Regularity, maximum principles, and Hamiltonian estimates. *Ann. Inst. H. Poincaré Anal. Non Linéaire* 31: 23–53, 2014.

[49] X. Cabré and Y. Sire. Nonlinear equations for fractional Laplacians: II. Existence, uniqueness, and qualitative properties of solutions. *Trans. Am. Math. Soc.* 367: 911–41, 2015.

[50] X. Cabré and J. Solà-Morales. Layer solutions in a half-space for boundary reactions. *Commun. Pure Appl. Math.* 58: 1678–1732, 2005.

[51] X. Cabré and J. Tan. Positive solutions of nonlinear problems involving the square root of the Laplacian. *Adv. Math.* 224: 2052–93, 2010.

[52] L. Caffarelli. Nonlocal equations, drifts and games. In *Nonlinear Partial Differential Equations*, Abel Symposia 7: 37–52, 2012.

[53] L. Caffarelli, J. M. Roquejoffre, and Y. Sire. Variational problems with free boundaries for the fractional laplacian. *J. Eur. Math. Soc.* 12: 1151–79, 2010.

[54] L. Caffarelli, S. Salsa, and L. Silvestre. Regularity estimates for the solution and the free boundary of the obstacle problem for the fractional Laplacian. *Invent. Math.* 171: 425–61, 2008.

[55] L. Caffarelli and L. Silvestre. An extension problem related to the fractional Laplacian. *Commun. Partial Differential Equations* 32: 1245–60, 2007.

[56] L. Caffarelli and L. Silvestre. Regularity theory for fully nonlinear integrodifferential equations. *Commun. Pure Appl. Math.* 62: 597–638, 2009.

[57] L. Caffarelli and L. Silvestre. Regularity results for nonlocal equations by approximation. *Arch. Ration. Mech. Anal.* 200: 59–88, 2011.

[58] A. Capella. Solutions of a pure critical exponent problem involving the half-Laplacian in annular-shaped domains. *Commun. Pure Appl. Anal.* 10: 1645–62, 2011.

[59] A. Capella, J. Dávila, L. Dupaigne, and Y. Sire. Regularity of radial extremal solutions for some non-local semilinear equations. *Commun. Partial Differential Equations* 36: 1353–84, 2011.

[60] A. Capozzi, D. Fortunato, and G. Palmieri. An existence result for nonlinear elliptic problems involving critical Sobolev exponent. *Ann. Inst. H. Poincaré Anal. Non Linéaire* 2: 463–70, 1985.

[61] A. Castro. Metodos variacionales y analisis functional no linear. *X Colóquio Colombiano de Matemáticas* 1980.

[62] G. Cerami, D. Fortunato, and M. Struwe. Bifurcation and multiplicity results for nonlinear elliptic problems involving critical Sobolev exponents. *Ann. Inst. H. Poincaré Anal. Non Linéaire* 1: 341–50, 1984.

[63] G. Cerami, S. Solimini, and M. Struwe. Some existence results for superlinear elliptic boundary value problems involving critical exponents. *J. Funct. Anal.* 69: 289–306, 1986.

[64] S. Challal, A. Lyaghfouri, and J. F. Rodrigues. On the A-obstacle problem and the Hausdorff measure of its free boundary. *Ann. Mat. Pura Appl.* 191: 113–65, 2012.

[65] M. Chang. Ground sate solutions of asymptotically linear fractional Schrödinger equations. *J. Math. Phys.* 54, 061504, 2013.

[66] F. Charro, E. Colorado, and I. Peral. Multiplicity of solutions to uniformly elliptic fully nonlinear equations with concave-convex right-hand side. *J. Differential Equations* 246: 4221–48, 2009.

[67] Z. Chen, N. Shioji, and W. Zou. Ground state and multiple solutions for a critical exponent problem. *NoDEA Nonlinear Differential Equations Appl.* 19: 253–77, 2012.

[68] M. Cheng. Bound state for the fractional Schrödinger equation with unbounded potential. *J. Math. Phys.* 53, 043507, 2012.

[69] Ph. G. Ciarlet. *Linear and Nonlinear Functional Analysis with Applications*. Society for Industrial and Applied Mathematics, Philadelphia, 2013.

[70] M. Clapp and T. Weth Multiple solutions for the Brezis–Nirenberg problem. *Adv. Differential Equations* 10: 463–80, 2005.

[71] D. C. Clark. A variant of the Lusternik–Schnirelman theory. *Indiana Univ. Math. J.* 22: 65–74, 1972.

[72] F. Colasuonno and P. Pucci. Multiplicity of solutions for $p(x)$-polyharmonic elliptic Kirchhoff equations. *Nonlinear Anal.* 74: 5962–74, 2011.

[73] E. Colorado, A. De Pablo, and U. Sánchez. Perturbations of a critical fractional equation. *Pacific J. Math.* 271: 65–85, 2014.

[74] M. Comte. Solutions of elliptic equations with critical Sobolev exponent in dimension three. *Nonlinear Anal.* 17: 445–55, 1991.

[75] R. Cont and P. Tankov. *Financial Modelling with Jump Processes*, Chapman & Hall/CRC Financial Mathematics Series. Boca Raton, FL, 2004.

[76] F. J. S. A. Corrêa and D. G. Costa. On a bi-nonlocal $p(x)$-Kirchhoff equation via Krasnoselskii's genus. *Math. Meth. Appl. Sci.* 38: 87–93, 2014.

[77] F. J. S. A. Corrêa and G. M. Figueiredo. On a p-Kirchhoff equation via Krasnoselskii's genus. *Appl. Math. Lett.* 22: 819–22, 2009.

[78] C. Cortazar, M. Elgueta, J. Rossi, and N. Wolanski. Asymptotic behavior for nonlocal diffusion equations. *J. Math. Pure Appl.* 86: 271–91, 2006.

[79] C. Cortazar, M. Elgueta, J. Rossi, and N. Wolanski. Boundary fluxes for nonlocal diffusion. *J. Differential Equations* 234: 360–90, 2007.

[80] C. Cortazar, M. Elgueta, J. Rossi, and N. Wolanski. How to approximate the heat equation with Neumann boundary conditions by nonlocal diffusion problems. *Arch. Rat. Mech. Anal.* 187: 137–56, 2008.

[81] V. Coti Zelati and M. Nolasco. Existence of ground states for nonlinear, pseudorelativistic Schrödinger equations. *Atti Accad. Naz. Lincei, Cl. Sci. Fis., Mat. Nat., Rend. Lincei, Mat. Appl.* 22: 51–72, 2011.

[82] A. Cotsiolis and N. Tavoularis. Best constants for Sobolev inequalities for higher order fractional derivatives. *J. Math. Anal. Appl.* 295: 225–36, 2004.

[83] E. Di Nezza, G. Palatucci, and E. Valdinoci. Hitchhiker's guide to the fractional Sobolev spaces. *Bull. Sci. Math.* 136: 521–73, 2012.

[84] S. Dipierro, G. Palatucci, and E. Valdinoci. Existence and symmetry results for a Schrödinger type problem involving the fractional Laplacian. *Le Matematiche* 68: 201–16, 2013.

[85] S. Dipierro, X. Ros-Oton, and E. Valdinoci. Nonlocal problems with Neumann boundary conditions. WIAS Preprint No. 1986, (2014). Preprint available at http://arxiv.org/pdf/1407.3313v3.pdf.

[86] S. Dipierro, O. Savin, and E. Valdinoci. All functions are locally s-harmonic up to a small error. Preprint available at http://arxiv.org/pdf/1404.3652v1.pdf.

[87] G. Devillanova and S. Solimini. Concentration estimates and multiple solutions to elliptic problems at critical growth. *Adv. Differential Equations* 7: 1257–80, 2002.

[88] G. Devillanova and S. Solimini. A multiplicity result for elliptic equations at critical growth in low dimension. *Commun. Contemp. Math.* 5: 171–7, 2003.

[89] H. Dong and D. Kim. On L_p-estimates for a class of non-local elliptic equations. *J. Funct. Anal.* 262: 1166–99, 2012.

[90] O. Druet. Elliptic equations with critical Sobolev exponents in dimension 3. *Ann. Inst. H. Poincaré Anal. Non Linéaire* 19: 125–42, 2002.

[91] L. C. Evans. Partial differential equations. In *Graduate Studies in Mathematics*, Vol. 19. American Mathematical Society, Providence, RI, 1998.

[92] E. B. Fabes, C. E. Kenig, and R. P. Serapioni. The local regularity of solutions of degenerate elliptic equations. *Comm. Partial Differential Equations* 7: 77–116, 1982.

[93] M. M. Fall. Semilinear elliptic equations for the fractional Laplacian with Hardy potential. Preprint available at http://arxiv.org/pdf/1109.5530v4.pdf.

[94] M. M. Fall and V. Felli. Unique continuation property and local asympotics of solutions to fractional elliptic equations. *Comm. Partial Differential Equations* 39: 354–97, 2014.

[95] P. Felmer, A. Quaas, and J. Tan. Positive solutions of nonlinear Schrödinger equation with the fractional Laplacian. *Proc. Roy. Soc. Edinburgh Sect. A* 142: 1237–62, 2012.

[96] M. Ferrara and G. Molica Bisci. Some applications of a Pucci–Serrin result. In *Minimax Theory and Its Applications* (in press) Helderman, Germany.

[97] M. Ferrara, G. Molica Bisci, and B. Zhang. Existence of weak solutions for non-local fractional problems via Morse theory. *Discrete and Continuous Dynamical Systems Series B*, 19: 2493–99, 2014.

[98] G. M. Figueiredo, G. Molica Bisci, and R. Servadei. On a fractional Kirchhoff-type equation via Krasnoselskii's genus. *Asymptot. Anal.* 94: 347–361, 2015.

[99] A. Fiscella. Saddle point solutions for non-local elliptic operators. *Topol. Methods Nonlinear Anal.* 44: 527–38, 2014.

[100] A. Fiscella, G. Molica Bisci, and R. Servadei. Bifurcation and multiplicity results for critical nonlocal fractional Laplacian problems. *Bull. Sci. Math.* 140: 13–35, 2016.

[101] A. Fiscella and P. Pucci. On certain nonlocal Hardy-Sobolev critical elliptic Dirichlet problems. Preprint.

[102] A. Fiscella, R. Servadei, and E. Valdinoci. A resonance problem for non-local elliptic operators. *Z. Anal. Anwendungen* 32: 411–31, 2013.

[103] A. Fiscella, R. Servadei, and E. Valdinoci. Density properties for fractional Sobolev spaces. *Ann. Acad. Sci. Fenn. Math.* 40: 235–53, 2015.

[104] A. Fiscella, R. Servadei, and E. Valdinoci. Asymptotically linear problems driven by fractional Laplacian operators. *Math. Methods Appl. Sci.* 38: 3551–3563, 2015.

[105] A. Fiscella and E. Valdinoci. A critical Kirchhoff type problem involving a nonlocal operator. *Nonlinear Anal.* 94: 156–70, 2014.

[106] G. B. Folland. *Real Analysis: Modern Techniques and Their Applications*. Wiley, New York, 1984.

[107] G. Franzina and G. Palatucci. Fractional p-eigenvalues. *Riv. Math. Univ. Parma* 5: 373–86, 2014.

[108] R. L. Frank and E. Lenzmann. Uniqueness and nondegeneracy of ground states for $(-\Delta)^s Q + Q - Q^{\alpha+1} = 0$ in $\mathbb{R}$. *Ann. Math.* (in press).

[109] M. F. Furtado, L. A. Maia, and E. A. B. Silva. On a double resonant problem in $\mathbb{R}^N$. *Differential Integral Equations* 15: 1335–44, 2002.

[110] E. Gagliardo. Proprietà di alcune classi di funzioni in più variabili. *Ricerche Mat.* 7: 102–37, 1958.

[111] J. García-Azorero and I. Peral. Multiplicity of solutions for elliptic problems with critical exponent or with non-symetric term. *Trans. Am. Math. Soc.* 323: 877–95, 1991.

[112] F. Gazzola and H.-Ch. Grunau. On the role of space dimension $n = 2+2\sqrt{2}$ in the semilinear Brezis–Nirenberg eigenvalue problem. *Analysis (Munich)* 20: 395–9, 2000.

[113] F. Gazzola and V. Rădulescu. A nonsmooth critical point theory approach to some nonlinear elliptic equations in $\mathbb{R}^N$. *Differential Integral Equations* 13: 47–60, 2002.

[114] F. Gazzola and B. Ruf. Lower-order perturbations of critical growth nonlinearities in semilinear elliptic equations. *Adv. Differential Equations* 2: 555–72, 1997.

[115] R. K. Getoor. First passage times for symmetric stable processes in space. *Trans. Am. Math. Soc.* 101: 75–90, 1961.

[116] N. Ghoussoub, D. Preiss, A general mountain pass principle for locating and classifying critical points. *Ann. Inst. H. Poincaré Anal. Non Linéaire* 6: 321–30, 1989.

[117] D. Gilbarg and N. S. Trudinger. *Elliptic Partial Differential Equations of Second Order*, 2nd Ed. Springer-Verlag, Berlin, 1983.

[118] P. Grisvard. *Elliptic Problems in Nonsmooth Domains*. Pitman, Boston, 1985.

[119] L. Hörmander. *The Analysis of Linear Partial Differential Operators*, Vols. I and II. Springer-Verlag, Berlin, 1983.

[120] O. Kavian. Introduction à la théorie des points critiques et applications aux problèmes elliptiques. In *Mathématiques & Applications*. Springer-Verlag, Paris, 1993.

[121] B. Kawohl. *Rearrangements and Convexity of Level Sets in PDE*. Springer-Verlag, Berlin, 1985.

[122] G. R. Kirchhoff. *Vorlesungen über Mathematische Physik: Mechanik*. Teubner, Leipzig, 1883.

[123] M. A. Krasnoselskii. *Topological Methods in the Theory of Nonlinear Integral Equations*. Macmillan, New York, 1964.

[124] A. Kristály. A double eigenvalue problem for Schrödinger equations involving sublinear nonlinearities at infinity. *Electron. J. Differential Equations* 42: 1–11, 2007.

[125] A. Kristály, V. Rădulescu, and Cs. Varga. Variational principles in mathematical physics, geometry, and economics: Qualitative analysis of nonlinear equations and unilateral problems. In *Encyclopedia of Mathematics and Its Applications*, Vol. 136. Cambridge University Press, 2010.

[126] A. Kristály and Cs. Varga. Multiple solutions for elliptic problems with singular and sublinear potentials. *Proc. Am. Math. Soc.* 135: 2121–6, 2007.

[127] T. Kuusi, G. Mingione, and Y. Sire. Nonlocal equations with measure data. *Commun. Math. Phys.* 337: 1317–68, 2015.

[128] T. Kuusi, G. Mingione, and Y. Sire. Nonlocal self-improving properties. *Anal. PDE* 8: 57–114, 2015.

[129] N. S. Landkof. *Foundations of Modern Potential Theory*, trans. from Russian (Grundlehren der Mathematischen Wissenschaften [Fundamental Principles of Mathematical Sciences] 180). Springer-Verlag, Berlin, 1973.

[130] N. Laskin. Fractional quantum mechanics and Lévy path integrals. *Phys. Lett. A* 268: 298–305, 2000.

[131] N. Laskin. Fractional Schrödinger equation. *Phys. Rev. E* 66, 056108, 2002.

[132] E. H. Lieb and M. Loss. *Analysis*. American Mathematical Society, Providence, 1997.

[133] E. Lindgren and P. Lindqvist. Fractional eigenvalues. *Calc. Var.* 49: 795–826, 2014.

[134] J. L. Lions. On some questions in boundary value problems of mathematical physics. In *Proceedings of International Symposium on Continuum Mechanics and Partial Differential Equations*, Rio de Janeiro, 1977. *Math. Stud.* 30: 284–346, 1978.

[135] P.-L. Lions. The concentration-compactness principle in the calculus of variations: The limit case I. *Rev. Mat. Iberoam.* 1: 145–201, 1985.

[136] P.-L. Lions. The concentration-compactness principle in the calculus of variations: The limit case II. *Rev. Mat. Iberoam.* 1: 45–121, 1985.

[137] J. L. Lions and E. Magenes. Problemi ai limiti non omogenei (III). *Ann. Scuola Norm. Sup. Pisa* 15: 41–103, 1961.

[138] J. L. Lions and E. Magenes. Problémes aux limites non homogénes et applications. In *Travaux et Recherches Mathématiques*. Dunod, Paris, 1968.

[139] R. de la Llave and E. Valdinoci. A generalization of Aubry–Mather theory to partial differential equations and pseudo-differential equations. *Ann. Inst. H. Poincaré Anal. Non Linéaire* 26: 1309–44, 2009.

[140] A. Majda and E. Tabak. A two-dimensional model for quasigeostrophic flow: comparision with the two-dimensional Euler flow. Nonlinear phenomena in ocean dynamics. *Phys. D* 98(2–4): 515–22, 1996.

[141] W. McLean. *Strongly Elliptic Systems and Boundary Integral Equations*. Cambridge University Press, 2000.

[142] N. Meyers and J. Serrin. $H = W$. *Proc. Nat. Acad. Sci. USA* 51: 1055–6, 1964.

[143] L. Modica. A gradient bound and a Liouville theorem for nonlinear Poisson equations. *Commun. Pure Appl. Math.* 38: 679–84, 1985.

[144] G. Molica Bisci. Fractional equations with bounded primitive. *Appl. Math. Lett.* 27: 53–8, 2014.

[145] G. Molica Bisci. Sequence of weak solutions for fractional equations. *Math. Res. Lett.* 21: 241–53, 2014.

[146] G. Molica Bisci and B. A. Pansera. Three weak solutions for nonlocal fractional equations. *Adv. Nonlinear Stud.* 14: 619–29, 2014.

[147] G. Molica Bisci and V. Rădulescu. Multiplicity results for elliptic fractional equations with subcritical term. *NoDEA Nonlinear Differential Equations Appl.* 22: 721–739, 2015.

[148] G. Molica Bisci and V. Rădulescu. A sharp eigenvalue theorem for fractional elliptic equations. Preprint 2015.

[149] G. Molica Bisci and V. Rădulescu. Ground state solutions of scalar field fractional Schrödinger equations. *Calc. Var. Partial Differential Equations* 54: 2985–3008, 2015

[150] G. Molica Bisci and D. Repovš. Existence and localization of solutions for nonlocal fractional equations. *Asymptot. Anal.* 90: 367–78, 2014.

[151] G. Molica Bisci and D. Repovš. Fractional nonlocal problems involving nonlinearities with bounded primitive. *J. Math. Anal. Appl.* 420: 167–76, 2014.

[152] G. Molica Bisci and D. Repovš. On doubly nonlocal fractional elliptic equations. *Atti Accad. Naz. Lincei Rend. Lincei Mat. Appl.* 26: 161–76, 2015.

[153] G. Molica Bisci, D. Repovš, and R. Servadei. Nontrivial solutions of superlinear nonlocal problems. *Forum Math.*, to appear.

[154] G. Molica Bisci, D. Repovš, and L. Vilasi. Integrodifferential fractional problems with infinitely many solutions. Preprint 2015.

[155] G. Molica Bisci and R. Servadei. A bifurcation result for nonlocal fractional equations. *Anal. Appl.* 13: 371–94, 2015.

[156] G. Molica Bisci and R. Servadei. Lower semicontinuity of functionals of fractional type and applications to nonlocal equations with critical Sobolev exponent. *Adv. Differential Equations* 20: 635–60, 2015.

[157] G. Molica Bisci and R. Servadei. A Brezis–Nirenberg splitting approach for nonlocal fractional equations. *Nonlinear Anal.* 119: 341–53, 2015.

[158] G. Molica Bisci and F. Tulone. An existence result for fractional Kirchhoff-type equations. *Z. Anal. Anwendungen* (in press).

[159] G. Molica Bisci and L. Vilasi. On a fractional degenerate Kirchhoff-type problem. *Commun. Contemp. Math.* DOI: 10.1142/S0219199715500881.

[160] E. Montefusco, B. Pellacci, and G. Verzini. Fractional diffusion with Neumann boundary conditions: the logistic equation. *Disc. Cont. Dyn. Syst. Ser. B* 18: 2175–202, 2013.

[161] R. Musina and A. I. Nazarov. On fractional Laplacians. *Commun. Partial Differential Equations* 39: 1780–90, 2014.

[162] G. Palatucci and A. Pisante. Improved Sobolev embeddings, profile decomposition and concentration-compactness for fractional Sobolev spaces. *Calc. Var. Partial Differential Equations* 50: 799–829, 2014.

[163] Y. J. Park. Fractional Polya-Szegö inequality. *J. Chungcheong Math. Soc.* 24: 267–71, 2011.

[164] K. Perera, M. Squassina, and Y. Yang. Bifurcation and multiplicity results for critical fractional p-Laplacian problems. *Math. Nachr.* (in press).

[165] P. Piersanti and P. Pucci. Existence theorems for fractional p-Laplacian problems. Anal. *Appl. (Singap.)*, in press. Preprint 2015.

[166] S. Pohozaev. On a class of quasilinear hyperbolic equations. *Math. Sborniek* 96:152–66, 1975.

[167] P. Pucci and S. Saldi. Critical stationary Kirchhoff equations in $\mathbb{R}^n$ involving nonlocal operators. *Rev. Mat. Iberoam.* (in press).

[168] P. Pucci and S. Saldi. Multiple solutions for an eigenvalue problem involving non-local elliptic p-Laplacian operators. In *Geometric Methods in PDEs* (Springer INdAM Series), Vol. 13, ed. by G. Citti, M. Manfredini, D. Morbidelli, S. Polidoro, and F. Uguzzoni. Springer-Verlag, Berlin, 2015, pp. 159–76.

[169] P. Pucci and J. Serrin. A mountain pass theorem. *J. Differential Equations* 60:142–9, 1985.

[170] P. Pucci, M. Xiang, and B. Zhang. Existence and multiplicity of entire solutions for fractional p-Kirchhoff equations. *Adv. Nonlinear Anal.*, in press. Preprint 2015.

[171] P. Pucci, M. Xiang, and B. Zhang. Multiple solutions for nonhomogeneous Schrödinger-Kirchhoff type equations involving the fractional p-Laplacian in $\mathbb{R}^N$. *Calc. Var. Partial Differential Equations* (in press).

[172] P. H. Rabinowitz. Some critical point theorems and applications to semilinear elliptic partial differential equations. *Ann. Scuola Norm. Sup. Pisa Cl. Sci.* 4: 215–23, 1978.

[173] P. H. Rabinowitz. Some minimax theorems and applications to nonlinear partial differential equations. In *Nonlinear Analysis: A Collection of Papers in Honor of Erich H. Rothe*, ed. by L. Cesari et al. New York: Academic Press, 1978, pp. 161–77.

[174] P. H. Rabinowitz. Minimax methods in critical point theory with applications to differential equations (CBMS Reg. Conf. Ser. Math. 65). American Mathematical Society, Providence, RI, 1986.

[175] P. H. Rabinowitz. On a class of nonlinear Schrödinger equations. *Z. Angew. Math. Phys.* 43: 270–91, 1992.

[176] B. Ricceri. A general variational principle and some of its applications. Fixed point theory with applications in nonlinear analysis. *J. Comput. Appl. Math.* 113: 401–10, 2000.

[177] B. Ricceri. On a three critical points theorem. *Archiv der Mathematik* 75: 220–6, 2000.

[178] B. Ricceri. A three critical points theorem revisited. *Nonlinear Anal.* 70: 3084–9, 2009.

[179] B. Ricceri. A further three critical points theorem. *Nonlinear Anal.* 71: 4151–7, 2009.

[180] B. Ricceri. Nonlinear eigenvalue problems. In D. Y. Gao and D. Motreanu (eds.), *Handbook of Nonconvex Analysis and Applications*. International Press, 2010, pp. 543–95.

[181] B. Ricceri. A multiplicity result for nonlocal problems involving nonlinearities with bounded primitive. *Stud. Univ. Babeş-Bolyai, Math.* 55: 107–14, 2010.

[182] B. Ricceri. A further refinement of a three critical points theorem. *Nonlinear Anal.* 74: 7446–54, 2011.

[183] B. Ricceri. A new existence and localization theorem for Dirichlet problem. *Dynam. Systems Appl.* 22: 317–24, 2013.

[184] X. Ros-Oton and J. Serra. The Dirichlet problem for the fractional Laplacian: regularity up to the boundary. *J. Math. Pure Appl.* 101: 275–302, 2014.

[185] X. Ros-Oton and J. Serra. The Pohozaev identity for the fractional Laplacian. *Arch. Ration. Mech. Anal.* 213: 587–628, 2014.

[186] X. Ros-Oton and J. Serra. Fractional Laplacian: Pohozaev identity and nonexistence results. Preprint available at http://arxiv.org/pdf/1205.0494.pdf.

[187] S. Salsa. *Partial Differential Equations in Action: From Modelling to Theory*. Springer-Verlag Italia, Milano, 2008.

[188] S. Salsa. Optimal regularity in lower dimensional obstacle problems. In *Subelliptic PDEs and Applications to Geometry and Finance* (Lect. Notes Semin. Interdiscip. Mat. 6, Semin. Interdiscip. Mat. (S.I.M.)). Potenza, 2007, pp. 217–26.

[189] S. Salsa. The problems of the obstacle in lower dimension and for the fractional Laplacian. In *Regularity Estimates for Nonlinear Elliptic and Parabolic Problems* (Lecture Notes in Math. 2045). Springer, Heidelberg, 2012, pp. 153–244.

[190] M. Schechter. A variation of the mountain pass lemma and applications. *J. Lond. Math. Soc.* 44: 491–502, 1991.

[191] M. Schechter and W. Zou. On the Brezis–Nirenberg problem. *Arch. Ration. Mech. Anal.* 197: 337–56, 2010.

[192] S. Secchi. Ground state solutions for nonlinear fractional Schröinger equations in $\mathbb{R}^N$. *J. Math. Phys.* 54, 031501, 2013.

[193] S. Secchi. Perturbation results for some nonlinear equations involving fractional operators. *Differ. Equ. Appl.* 5: 221–36, 2013.

[194] S. Secchi. On fractional Schrödinger equations in $\mathbb{R}^N$ without the Ambrosetti–Rabinowitz condition. *Topol. Methods Nonlinear Anal.* (in press).

[195] R. Servadei. Infinitely many solutions for fractional Laplace equations with subcritical nonlinearity. *Contemp. Math.* 595: 317–40, 2013.

[196] R. Servadei. The Yamabe equation in a non-local setting. *Adv. Nonlinear Anal.* 2: 235–70, 2013.

[197] R. Servadei. A critical fractional Laplace equation in the resonant case. *Topol. Methods Nonlinear Anal.* 43: 251–67, 2014.

[198] R. Servadei and E. Valdinoci. Mountain pass solutions for non-local elliptic operators. *J. Math. Anal. Appl.* 389: 887–98, 2012.

[199] R. Servadei and E. Valdinoci. Lewy–Stampacchia type estimates for variational inequalities driven by (non)local operators. *Rev. Mat. Iberoam.* 29: 1091–126, 2013.

[200] R. Servadei and E. Valdinoci. Variational methods for non-local operators of elliptic type. *Discrete Contin. Dyn. Syst.* 33: 2105–37, 2013.

[201] R. Servadei and E. Valdinoci. A Brezis–Nirenberg result for non-local critical equations in low dimension. *Commun. Pure Appl. Anal.* 12(6): 2445–64, 2013.

[202] R. Servadei and E. Valdinoci. Weak and viscosity solutions of the fractional Laplace equation. *Publ. Mat.* 58: 133–54, 2014.

[203] R. Servadei and E. Valdinoci. On the spectrum of two different fractional operators. *Proc. Roy. Soc. Edinburgh Sect. A.* 144A: 1–25, 2014.

[204] R. Servadei and E. Valdinoci. The Brezis–Nirenberg result for the fractional Laplacian. *Trans. Am. Math. Soc.* 367: 67–102, 2015.

[205] R. Servadei and E. Valdinoci. Fractional Laplacian equations with critical Sobolev exponent. *Rev. Mat. Comput.* 28: 655–676, 2015.

[206] L. Silvestre. Regularity of the obstacle problem for a fractional power of the Laplace operator. *Commun. Pure Appl. Math.* 60: 67–112, 2007.

[207] L. N. Slobodeckij. Generalized Sobolev spaces and their applications to boundary value problems of partial differential equations. *Leningrad. Gos. Ped. Inst. Učep. Zap.* 197: 54–112, 1958.

[208] P. R. Stinga and J. L. Torrea. Extension problem and Harnack's inequality for some fractional operators. *Commun. Partial Differ. Equations* 35: 2092–122, 2010.

[209] P. Stinga and B. Volzone. Fractional semilinear Neumann problems arising from a fractional Keller-Segel model. *Calc. Var. Partial Differential Equations* 54: 1009–42, 2015.

[210] W. A. Strauss. Existence of solitary waves in higher dimensions. *Commun. Math. Phys.* 55: 149–62, 1977.

[211] M. Struwe. Infinitely many critical points for functionals which are not even and applications to superlinear boundary value problems. *Manuscripta Math.* 32: 335–64, 1980.

[212] M. Struwe. *Variational Methods: Applications to Nonlinear Partial Differential Equations and Hamiltonian Systems* (Ergebnisse der Mathematik und ihrer Grenzgebiete 3). Springer-Verlag, Berlin, 1990.

[213] A. Szulkin, T. Weth, and M. Willem. Ground state solutions for a semilinear problem with critical exponent. *Differential Integral Equations* 22: 913–26, 2009.

[214] J. Tan. The Brezis–Nirenberg type problem involving the square root of the Laplacian. *Calc. Var. Partial Differential Equations* 36: 21–41, 2011.

[215] K. Teng. Multiple solutions for a class of fractional Schrödinger equations in $\mathbb{R}^N$. *Nonlinear Anal. Real World Appl.* 21: 76–86, 2015.

[216] E. Valdinoci. From the long jump random walk to the fractional Laplacian. *Bol. Soc. Esp. Mat. Apl. S$\vec{e}$MA* 49: 33–44, 2009.

[217] J. L. Vázquez. Nonlinear diffusion with fractional laplacian operators. In *Nonlinear Partial Differential Equations*, Abel Symp. 7: 271–98, 2012.

[218] J. L. Vázquez. Recent progress in the theory of nonlinear diffusion with fractional Laplacian operators. *Discrete Contin. Dyn. Syst. Ser. S* 7(4): 857–85, 2014.

[219] L. Vlahos, H. Isliker, Y. Kominis, and K. Hizonidis. Normal and anomalous diffusion: a tutorial In *Order and Chaos*, Vol. 10, ed. by T. Bountis. Patras University Press, 2008.

[220] M. Willem. *Minimax Theorems* (Progress in Nonlinear Differential Equations and Their Applications 24). Birkhäuser, Boston, 1996.

[221] D. Zhang. On multiple solutions of $\Delta u + \lambda u + |u|^{4/(n-2)}u = 0$. *Nonlinear Anal.* 13: 353–72, 1989.

Index

www.ingramcontent.com/pod-product-compliance
Lightning Source LLC
LaVergne TN
LVHW102009210926
848483LV00010B/238
9781107111943